Advances in Artificial Intelligence: Models, Optimization, and Machine Learning

Advances in Artificial Intelligence: Models, Optimization, and Machine Learning

Editors

Florin Leon
Mircea Hulea
Marius Gavrilescu

MDPI • Basel • Beijing • Wuhan • Barcelona • Belgrade • Manchester • Tokyo • Cluj • Tianjin

Editors
Florin Leon
"Gheorghe Asachi" Technical
University of Iași
Romania

Mircea Hulea
"Gheorghe Asachi" Technical
University of Iași
Romania

Marius Gavrilescu
"Gheorghe Asachi" Technical
University of Iași
Romania

Editorial Office
MDPI
St. Alban-Anlage 66
4052 Basel, Switzerland

This is a reprint of articles from the Special Issue published online in the open access journal *Mathematics* (ISSN 2227-7390) (available at: https://www.mdpi.com/journal/mathematics/special_issues/Artificial_Intelligence_Models_Optimization_Machine_Learning).

For citation purposes, cite each article independently as indicated on the article page online and as indicated below:

LastName, A.A.; LastName, B.B.; LastName, C.C. Article Title. *Journal Name* **Year**, *Volume Number*, Page Range.

ISBN 978-3-0365-4515-8 (Hbk)
ISBN 978-3-0365-4516-5 (PDF)

© 2022 by the authors. Articles in this book are Open Access and distributed under the Creative Commons Attribution (CC BY) license, which allows users to download, copy and build upon published articles, as long as the author and publisher are properly credited, which ensures maximum dissemination and a wider impact of our publications.

The book as a whole is distributed by MDPI under the terms and conditions of the Creative Commons license CC BY-NC-ND.

Contents

About the Editors vii

Florin Leon, Mircea Hulea and Marius Gavrilescu
Preface to the Special Issue on "Advances in Artificial Intelligence: Models, Optimization, and Machine Learning"
Reprinted from: *Mathematics* **2022**, *10*, 1721, doi:10.3390/math10101721 1

Silvia Curteanu, Florin Leon, Andra-Maria Mircea-Vicoveanu and Doina Logofătu
Regression Methods Based on Nearest Neighbors with Adaptive Distance Metrics Applied to a Polymerization Process
Reprinted from: *Mathematics* **2021**, *9*, 547, doi:10.3390/math9050547 5

Florin Leon and Marius Gavrilescu
A Review of Tracking and Trajectory Prediction Methods for Autonomous Driving
Reprinted from: *Mathematics* **2021**, *9*, 660, doi:10.3390/math9060660 25

Jui-Sheng Chou, Dinh-Nhat Truong and Chih-Fong Tsai
Solving Regression Problems with Intelligent Machine Learner for Engineering Informatics
Reprinted from: *Mathematics* **2021**, *9*, 686, doi:10.3390/math9060686 63

Carlos M. Castorena, Itzel M. Abundez, Roberto Alejo, Everardo E. Granda-Gutiérrez, Eréndira Rendón and Octavio Villegas
Deep Neural Network for Gender-Based Violence Detection on Twitter Messages
Reprinted from: *Mathematics* **2021**, *9*, 807, doi:10.3390/math9080807 89

Seokho Kang
k-Nearest Neighbor Learning with Graph Neural Networks
Reprinted from: *Mathematics* **2021**, *9*, 830, doi:10.3390/math9080830 101

Ángel Luis Muñoz Castañeda, Noemí DeCastro-García and David Escudero García
RHOASo: An Early Stop Hyper-Parameter Optimization Algorithm
Reprinted from: *Mathematics* **2021**, *9*, 2334, doi:10.3390/math9182334 113

Elena Niculina Dragoi and Vlad Dafinescu
Review of Metaheuristics Inspired from the Animal Kingdom
Reprinted from: *Mathematics* **2021**, *9*, 2335, doi:10.3390/math9182335 165

Xinglong Feng, Xianwen Gao and Ling Luo
A ResNet50-Based Method for Classifying Surface Defects in Hot-Rolled Strip Steel
Reprinted from: *Mathematics* **2021**, *9*, 2359, doi:10.3390/math9192359 217

Amelia Bădică and Costin Bădică and Ion Buligiu and Liviu Ion Ciora and Doina Logofătu
Dynamic Programming Algorithms for Computing Optimal Knockout Tournaments
Reprinted from: *Mathematics* **2021**, *9*, 2480, doi:10.3390/math9192480 233

Krešimir Kušić, Edouard Ivanjko, Filip Vrbanić, Martin Gregurić and Ivana Dusparic
Spatial-Temporal Traffic Flow Control on Motorways Using Distributed Multi-Agent Reinforcement Learning
Reprinted from: *Mathematics* **2021**, *9*, 3081, doi:10.3390/math9233081 257

Florin Leon
ActressMAS, a .NET Multi-Agent Framework Inspired by the Actor Model
Reprinted from: *Mathematics* **2022**, *10*, 382, doi:10.3390/math10030382 285

Fahman Saeed, Muhammad Hussain and Hatim A. Aboalsamh
Automatic Fingerprint Classification Using Deep Learning Technology (DeepFKTNet)
Reprinted from: *Mathematics* **2022**, *10*, 1285, doi:10.3390/math10081285 . . . . **319**

Subhajit Chatterjee, Debapriya Hazra, Yung-Cheol Byun and Yong-Woon Kim
Enhancement of Image Classification Using Transfer Learning and GAN-Based Synthetic Data Augmentation
Reprinted from: *Mathematics* **2022**, *10*, 1541, doi:10.3390/math10091541 . . . . **337**

About the Editors

Florin Leon

Florin Leon, Ph.D., is currently a Full Professor at the Department of Computer Science and Engineering of the "Gheorghe Asachi" University of Iasi, Romania. He received a doctoral degree in computer science from the same university, followed by a postdoctoral fellowship completed in 2007. In 2015, he defended his habilitation thesis. He has authored and co-authored more than 180 journal articles, book chapters and conference papers, and 14 books. He has 629 citations with an h-index of 12 according to Scopus. He was a member of the guest editorial boards for three journal Special Issues, and he has participated in 29 national and international research projects, 3 of which were as the principal investigator. His scientific interests include: artificial intelligence, machine learning, multiagent systems and software design. In his research, he used various machine learning techniques for modelling, such as simple, stacked and deep neural networks, instance-based methods, and large-margin nearest neighbor regression. He also addressed optimization problems using different types of evolutionary algorithms, quantum-inspired algorithms and combinations of global and local search methods. Moreover, he studied multiagent systems with complex behaviors and performed various agent-based simulations. Prof. Leon was a member of the organizing committees or program committee chair of five conferences. He is currently a member of IEEE Systems, Man and Cybernetics Society: Computational Collective Intelligence Technical Community and the Romanian Association for Artificial Intelligence.

Mircea Hulea

Mircea Hulea, Ph.D., is currently an Associate Professor at the Department of Computer Science and Engineering of the "Gheorghe Asachi" Technical University of Iasi, Romania. He received his M.S. and Ph.D. degrees in Computer Engineering and Automatic Control from the same university, in 2004 and 2008, respectively. In this institution, he was also a Postdoctoral Researcher when he worked on the project of Biomimetic hardware and software systems and their applications, from 2010 to 2013. He is the author of over 40 technical publications, proceedings, editorials and books, with more than 20 being indexed in the Web of Science. His research interests include brain modelling, humanoid robotics and optical wireless communications. He is the coordinator of the collaborative research network on spiking neural networks at the host university, and a member in the management committee of the European project COST Action 19111 (NEWFOCUS).

Marius Gavrilescu

Marius Gavrilescu, Ph.D., is a Lecturer at the Department of Computer Science and Engineering of the "Gheorghe Asachi" Technical University of Iasi, Romania. His research activity and interests involve the following: machine learning—classification and regression models, object recognition, deep learning neural network architectures applied mainly for the processing of medical images and data originating from natural sciences; computer graphics and data visualization: visual representations and rendering of multidimensional data from medical imaging; volume graphics; GPU programming and parallel algorithms; filtering, enhancement and analysis of images from the automotive field, natural sciences, climatology and meteorology; modelling and simulation of physical phenomena: fluid dynamics, collisions, optical models based on ray tracing or path tracing, real-time graphics and physics engines.

Editorial

Preface to the Special Issue on "Advances in Artificial Intelligence: Models, Optimization, and Machine Learning"

Florin Leon *, Mircea Hulea * and Marius Gavrilescu *

Faculty of Automatic Control and Computer Engineering, "Gheorghe Asachi" Technical University of Iasi, Bd. Mangeron 27, 700050 Iasi, Romania
* Correspondence: florin.leon@academic.tuiasi.ro (F.L.); mircea.hulea@academic.tuiasi.ro (M.H.); marius.gavrilescu@academic.tuiasi.ro (M.G.)

Citation: Leon, F.; Hulea, M.; Gavrilescu, M. Preface to the Special Issue on "Advances in Artificial Intelligence: Models, Optimization, and Machine Learning". *Mathematics* 2022, *10*, 1721. https://doi.org/10.3390/math10101721

Received: 15 May 2022
Accepted: 16 May 2022
Published: 18 May 2022

Publisher's Note: MDPI stays neutral with regard to jurisdictional claims in published maps and institutional affiliations.

Copyright: © 2022 by the authors. Licensee MDPI, Basel, Switzerland. This article is an open access article distributed under the terms and conditions of the Creative Commons Attribution (CC BY) license (https://creativecommons.org/licenses/by/4.0/).

Recent advancements in artificial intelligence and machine learning have led to the development of powerful tools for use in problem solving in a wide array of scientific and technical fields. In particular, supervised models allow for the searching, optimization, and classification of data with high complexity, high dimensionality, and vast solution spaces. Problems that had proven challenging or nearly impossible are now solvable given sufficient training time and computational resources, allowing for learning, knowledge discovery, and decision making that easily outperform human abilities. In particular, machine learning models obtained using high-complexity deep neural networks have seen a huge increase in popularity due to their ability to learn functions, rules, and correlations within massive and diverse data sets. Currently, machine learning has demonstrated and is continually demonstrating its potential to solve complex and important real-world problems. Consequently, following a careful and thorough peer-review process, this Special Issue offers valuable contributions to modeling, optimization, classification, and regression for solving problems in a wide variety of technical fields, using modern artificial intelligence and machine learning means.

In the following paragraphs, we provide summaries of these contributions.

Curteanu et al. [1] perform an extensive comparative study between three regression algorithms for the prediction of monomer conversion, numerical average molecular weight, and gravimetrical average molecular weight for the free radical polymerization of methyl methacrylate achieved in a batch bulk process. The first two algorithms are based on the concept of a large margin, typical of support vector machines, but used here for regression in conjunction with an instance-based method, where the learning of problem-specific distance metrics can be achieved either with an evolutionary algorithm or with an approximate differential approach. Another original regression method is based on the idea of denoising autoencoders, i.e., prototype weights and positions are set in such a way as to minimize the error on a slightly corrupted version of the training set.

Leon and Gavrilescu [2] provide a survey of modern methods for prediction and tracking in automotive applications. The paper covers a wide range of methods applied in various scenarios, where pedestrians, vehicles, and other obstacles are found in difficult-to-handle configurations. The scientific contributions analyzed in the survey offer methods using deep neural networks, stochastic methods, motion models, as well as many hybrid approaches to solve problems within scenarios comprising multiple interacting agents with variable or multi-modal behavior, occlusion, high reaction times, and, generally speaking, any contexts encountered within autonomous driving. While the current state of the art generally favors neural-network-based approaches, many non-neural-network solutions are explored as well.

The work by Chou et al. [3] uses an intelligent machine learner to build prediction models with applications in industrial experiments such as resource planning for software projects, the comparison of processor performance, and the estimation of bicycle rentals per day and resources demand for increasing productivity and efficient customer service. The proposed approach matches or obtains better results than the existing methods reported in the literature for the same applications.

The problem of the detection of gender-based violence is tackled by Castorena et al. [4] based on language used on Twitter. The artificial intelligence systems used for this goal are deep neural networks that require minimal preprocessing of data based on feature extraction. The success rate of the proposed method in identifying gender-based violence in the Spanish language in Mexico is about 80%, which is encouraging for future improvements of the proposed method.

Kang [5] proposes an improvement of the k-nearest neighbors (kNN) algorithm using a graph neural network (GNN) to improve the learning process. The resulting contribution is called kNNGNN and consists of generating a GNN that learns kNN rules from a graph representation of the data. The author evaluates both weighted and unweighted versions of kNN using various similarity metrics and demonstrates the applicability of the proposed method for both classification and regression problems.

The work by Muñoz Castañeda et al. [6] describes a new algorithm for the hyper-parameter optimization (HPO) of machine learning algorithms based on the conditional optimization of concave asymptotic functions. It is shown that the size of the data subset does not have a great impact on its performance, and the algorithm only requires an upper bound on the number of iterations to perform.

Drăgoi and Dafinescu [7] review a large number of metaheuristic optimization algorithms inspired by animal behavior, both vertebrates and invertebrates, proposed between 2006 and 2021. The authors note that despite many critiques of the metaheuristic community, the trend of proposing algorithms based on new sources of inspiration remains stable because of the many areas of applicability and the tendency to offer the source code in order to increase the ease of use. Exotic inspiration sources and uncommon behaviors seem to have a greater probability of devising new optimization techniques.

Feng et al. [8] use a model based on the ResNet50 architecture to identify surface defects on rolled strip steel for automotive manufacturing. ResNet50 is combined with other models such as the convolutional block attention module (CBAM) and FcaNet for improved accuracy. The resulting hybrid method is tested using a data set exhibiting defect patterns such as surface scratches, cracks, tears, spots, or oxidation layers. The proposed hybrid model performs slightly better than similar approaches. However, as the authors themselves note, the method requires more computational power than competing lightweight models, considering that the rolled steel coils are evaluated using images acquired in real time at very high rates.

Bădică et al. [9] study hierarchically shaped single-elimination tournaments and propose a dynamic programming algorithm for use in computing optimal tournaments that maximize attractiveness, e.g., where the best players have the chance to meet in the later stages of the competition. The authors also develop more efficient deterministic and sub-optimal stochastic versions of the algorithm.

The goal of the paper by Kušić et al. [10] is to improve artificial intelligence techniques for traffic control by dynamically setting zones with variable speed limits. In addition, method validation is performed using four agents instead of two, as in previous research. This work is important in reducing traffic congestion by automatically adjusting the speed limits and the position of these zones.

In his work, Leon [11] describes the architecture of ActressMAS, a .NET multi-agent framework which allows the implementation of two sub-paradigms in multi-agent systems, i.e., one focused on autonomy and planning, and another focused on interactions and emergent behaviors in agent simulations. Its main advantages are conceptual simplicity and ease of use, which make it particularly suitable for teaching agent-based concepts. However, the framework proves to be sufficiently powerful to implement a large number of algorithms, protocols, and simulations characteristic of intelligent agents and multi-agent systems. The framework and the examples are open-source and publicly available.

The paper by Saeed et al. [12] presents a method to optimize the structure of convolutional neural networks (CNNs) by determining the number of filters and layers for the classification of fingerprints using multiple sensors. This research is important for improving the cost and response time of systems based on CNNs.

Chatterjee et al. [13] propose a method for the automatic identification of plastic bottles from images for recycling purposes. To this end, a model based on a generative adversarial network (GAN) augments a data set consisting of a few original images, while the actual classification is handled by an ensemble based on transfer learning from the InceptionV3 and Xception models. The proposed solution is shown to have very high accuracy, and it is worth mentioning that it seems to handle rotation and translation quite well. However, both training and evaluation are carried out on relatively simple images, each containing a single plastic bottle against a relatively homogeneous background. It would be interesting to see whether in future work the authors will improve their method to handle more diverse and realistic scenarios, such as plastic bottles found among other waste, multiple plastic bottles arranged in piles where they occlude one another, etc.

These 13 papers in this Special Issue have been selected following a process with an acceptance rate of 62%. The authors' geographical distribution is displayed in Table 1, which shows 42 authors from 12 countries.

Table 1. Geographic distribution of authors by country.

Country	Number of Authors
China	3
Croatia	4
Germany	2
India	2
Ireland	1
Mexico	6
Romania	12
Saudi Arabia	3
South Korea	3
Spain	3
Taiwan	2
Vietnam	1

The guest editors wish to thank the authors for their contributions and for their commitment to improving their work, the reviewers for investing time and effort into analyzing and providing valuable comments and corrections, and last but not least, the editorial staff for managing the review and publication process efficiently and thoroughly. We hope that the selected publications will have a lasting impact on the scientific community and that they will be motivating factors for other researchers to pursue their scientific goals.

Funding: This research received no external funding.

Conflicts of Interest: The authors declare no conflict of interest.

References

1. Curteanu, S.; Leon, F.; Mircea-Vicoveanu, A.; Logofătu, D. Regression Methods Based on Nearest Neighbors with Adaptive Distance Metrics Applied to a Polymerization Process. *Mathematics* **2021**, *9*, 547. [CrossRef]
2. Leon, F.; Gavrilescu, M. A Review of Tracking and Trajectory Prediction Methods for Autonomous Driving. *Mathematics* **2021**, *9*, 660. [CrossRef]
3. Chou, J.; Truong, D.; Tsai, C. Solving Regression Problems with Intelligent Machine Learner for Engineering Informatics. *Mathematics* **2021**, *9*, 686. [CrossRef]
4. Castorena, C.; Abundez, I.; Alejo, R.; Granda-Gutiérrez, E.; Rendón, E.; Villegas, O. Deep Neural Network for Gender-Based Violence Detection on Twitter Messages. *Mathematics* **2021**, *9*, 807. [CrossRef]
5. Kang, S. k-Nearest Neighbor Learning with Graph Neural Networks. *Mathematics* **2021**, *9*, 830. [CrossRef]
6. Muñoz Castañeda, Á.; DeCastro-García, N.; Escudero García, D. RHOASo: An Early Stop Hyper-Parameter Optimization Algorithm. *Mathematics* **2021**, *9*, 2334. [CrossRef]

7. Drăgoi, E.; Dafinescu, V. Review of Metaheuristics Inspired from the Animal Kingdom. *Mathematics* **2021**, *9*, 2335. [CrossRef]
8. Feng, X.; Gao, X.; Luo, L. A ResNet50-Based Method for Classifying Surface Defects in Hot-Rolled Strip Steel. *Mathematics* **2021**, *9*, 2359. [CrossRef]
9. Bădică, A.; Bădică, C.; Buligiu, I.; Ciora, L.; Logofătu, D. Dynamic Programming Algorithms for Computing Optimal Knockout Tournaments. *Mathematics* **2021**, *9*, 2480. [CrossRef]
10. Kušić, K.; Ivanjko, E.; Vrbanić, F.; Gregurić, M.; Dusparic, I. Spatial-Temporal Traffic Flow Control on Motorways Using Distributed Multi-Agent Reinforcement Learning. *Mathematics* **2021**, *9*, 3081. [CrossRef]
11. Leon, F. ActressMAS, a .NET Multi-Agent Framework Inspired by the Actor Model. *Mathematics* **2022**, *10*, 382. [CrossRef]
12. Saeed, F.; Hussain, M.; Aboalsamh, H. Automatic Fingerprint Classification Using Deep Learning Technology (DeepFKTNet). *Mathematics* **2022**, *10*, 1285. [CrossRef]
13. Chatterjee, S.; Hazra, D.; Byun, Y.; Kim, Y. Enhancement of Image Classification Using Transfer Learning and GAN-Based Synthetic Data Augmentation. *Mathematics* **2022**, *10*, 1541. [CrossRef]

Article

Regression Methods Based on Nearest Neighbors with Adaptive Distance Metrics Applied to a Polymerization Process

Silvia Curteanu [1], Florin Leon [2,*], Andra-Maria Mircea-Vicoveanu [1] and Doina Logofătu [3]

[1] Faculty of Chemical Engineering and Environmental Protection, "Gheorghe Asachi" Technical University of Iași, Bd. Mangeron 73, 700050 Iași, Romania; scurtean@ch.tuiasi.ro (S.C.); andramircea@yahoo.com (A.-M.M.-V.)
[2] Faculty of Automatic Control and Computer Engineering, "Gheorghe Asachi" Technical University of Iași, Bd. Mangeron 27, 700050 Iași, Romania
[3] Faculty of Computer Science and Engineering, Frankfurt University of Applied Sciences, Nibelungenplatz 1, 60318 Frankfurt am Main, Germany; logofatu@fb2.fra-uas.de
* Correspondence: florin.leon@academic.tuiasi.ro

Citation: Curteanu, S.; Leon, F.; Mircea-Vicoveanu, A.-M.; Logofătu, D. Regression Methods Based on Nearest Neighbors with Adaptive Distance Metrics Applied to a Polymerization Process. *Mathematics* 2021, 9, 547. https://doi.org/10.3390/math9050547

Academic Editor: Radi Romansky

Received: 28 January 2021
Accepted: 1 March 2021
Published: 5 March 2021

Publisher's Note: MDPI stays neutral with regard to jurisdictional claims in published maps and institutional affiliations.

Copyright: © 2021 by the authors. Licensee MDPI, Basel, Switzerland. This article is an open access article distributed under the terms and conditions of the Creative Commons Attribution (CC BY) license (https://creativecommons.org/licenses/by/4.0/).

Abstract: Empirical models based on sampled data can be useful for complex chemical engineering processes such as the free radical polymerization of methyl methacrylate achieved in a batch bulk process. In this case, the goal is to predict the monomer conversion, the numerical average molecular weight and the gravimetrical average molecular weight. This process is characterized by non-linear gel and glass effects caused by the sharp increase in the viscosity as the reaction progresses. To increase accuracy, one needs more samples in the areas with higher variation and this is achieved with adaptive sampling. An extensive comparative study is performed between three regression algorithms for this chemical process. The first two are based on the concept of a large margin, typical of support vector machines, but used for regression, in conjunction with an instance-based method. The learning of problem-specific distance metrics can be performed by means of either an evolutionary algorithm or an approximate differential approach. Having a set of prototypes with different distance metrics is especially useful when a large number of instances should be handled. Another original regression method is based on the idea of denoising autoencoders, i.e., the prototype weights and positions are set in such a way as to minimize the mean square error on a slightly corrupted version of the training set, where the instances inputs are slightly changed with a small random quantity. Several combinations of parameters and ways of splitting the data into training and testing sets are used in order to assess the performance of the algorithms in different scenarios.

Keywords: large margin nearest neighbor regression; distance metrics; prototypes; evolutionary algorithm; approximate differential optimization; multiple point hill climbing; adaptive sampling; free radical polymerization

1. Introduction

There are many situations when one needs to discover a relationship between one or more independent variables, the inputs, and one real-valued dependent variable, i.e., the output—from data samples. This type of problem is known as regression, and a large number of algorithms have been proposed by researchers. Among the most popular ones, one can mention neural networks, support vector machine regression (ε-SVR, ν-SVR), decision trees (M5P, random forest, REPTree) or methods based on instances (k-nearest neighbor) or rules (M5, decision table).

The Large Margin Nearest Neighbor for Regression (LMNNR) algorithm [1] has been used in several studies so far for a variety of applications and its performance has been compared to that of classic regression methods implemented in the popular collection of machine learning algorithms Weka [2]. Thus, in [1,3], it was used for the prediction of corrosion resistance of some alloys containing titanium and molybdenum, widely used in dental applications. The material corrosion was quantified by the polarization resistance of the TiMo alloys.

The LMNNR algorithm was also applied in a different field, that of predicting students' performance based on their active use of social media tools during the learning process [4,5]. The training data were collected over six winter semesters in consecutive years, from a total of 343 students. Almost 19,000 social media contributions were recorded and used to compute 14 numeric features for each student. Based on these, the final grade was predicted.

The results of LMNNR have been generally shown to be better than those of the other regression algorithms used for comparison. Although more variants for model training and model representation have been proposed, its main disadvantage is that its sensitivity to local optimal often requires multiple runs, and thus increases the training time.

In [6], a modified nearest-neighbor regression method (kNN) is proposed for modeling the photocatalytic degradation of the Reactive Red 184 dye for which insufficient data are available. It can handle partial information without "filling in" additional computed values (mean values) or ignoring the incomplete instances. In the case of the photocatalytic degradation process, the kNN method recorded correlations of over 0.9.

A study based on an adaptive regression model appropriate for cases with insufficient or missing data was also performed in [7]. Its aim was to investigate the electrochemical behavior of ZrTi alloys in artificial saliva. This method has only one internal parameter whose optimal value is found automatically.

The prediction of the sublimation rate of naphthalene in various working conditions was studied in [8]. Different regression methods were applied and the performance of the original Large Margin Nearest Neighbor Regression algorithm (LMNNR) proved superior to those of other classical ones.

In the present study, three regression variants are applied for the free radical polymerization of methyl methacrylate (MMA) achieved in a batch bulk process. The first two variants are based on LMNNR trained either with an evolutionary algorithm or by gradient descent, where the derivatives are approximated by means of the central difference method. This is the first time that LMNNR has been applied for this process. Its difficulty is caused by the gel effect, corresponding to an abrupt conversion and molecular mass jump which may be missed by less accurate regression techniques. The third variant is a new, original algorithm named Nearest Neighbor Regression with Adaptive Distance Metrics Trained by Multiple Point Hill Climbing on Noisy Training Set Error (RADIAN) which is not based on the concept of a large margin but is inspired by denoising autoencoders used in deep learning [9], where the input data are slightly corrupted, and the model is forced to learn the correct data from the corrupted version in order to prevent overfitting.

Regarding the results obtained for the polymerization of MMA with the abovementioned methods, it is important to point out not only that they are very good, but also that they follow previous sustained efforts, made with different other methods whose results were inferior to those reported here.

2. Dataset

In order to test the functionality of the regression algorithms mentioned in the article, a real-world problem, namely the free radical polymerization of methyl methacrylate, was chosen as a case study. Two reasons justify this choice: the complexity of the process, so the difficulties in modeling and also the fact that our group has tried different modeling methods for this system, so that the obtained results can be compared with those reported in this paper.

Polymerization reactions present some difficulties in modeling and optimization actions, because of their specific features, as well as the general characteristics of the chemical processes. Reactions are complex and their mechanism is often not fully known. Developing accurate models implies precise knowledge of the phenomenology of the process, as well as of the physical and chemical laws that govern them. A series of approximations are often needed, influencing the accuracy of the model results. In addition, the complexity of the mathematical models causes supplementary difficulties regarding the solution mode and the time required for this operation, given the requirement of using the models in online optimal control procedures. Under these conditions, empirical models that use input–output data sets can be considered a preferable alternative to mechanistic models, both in terms of working methodology and the accuracy of results.

The free radical polymerization is characterized by diffusion-controlled effects. As the viscosity of the reaction mass increases, there is a sudden increase in the conversion and molecular masses, as a result of the diffusion difficulties encountered by the increasing macroradicals. The so-called glass and gel phenomena appear as a result of decreasing the values of the propagation and termination rate constants. The result of the manifestation of controlled diffusion phenomena is the end of polymerization reaction before the complete consumption of the reactants.

From the point of view of the modeling action, the diffusion-controlled effects are more difficult to model, especially since their phenomenology is not completely elucidated. Various models have been proposed, with a pronounced empirical character, e.g., [10] or [11], but their efficiency and application are limited. In addition, they have a pronounced empirical character, including many constants that can be determined by matching the experimental data, which means their dependence on each set of reaction conditions (temperature, concentrations of reactants etc.).

Some of the reasons listed above involve the need to apply modeling methods leading to better results, a variant being represented by neural networks, applied in the form of different methodologies. The following examples belong to our working group and have the role of justifying the new methodology described and applied in this paper and highlighting the results obtained, better than in the previous approaches. Therefore, several previous results will be presented.

The first series of attempts [12,13] implied the design of neural networks of feedforward type, by the method of successive trials, to correlate the conversion and molecular masses with the reaction conditions. If satisfactory results were obtained for the conversion, for the molecular masses, especially for the average gravimetric molecular mass, the accuracy was below the minimum required. A more complex approach, which led to better results [14] was based on combining a simplified phenomenological model with neural networks, obtaining hybrid models. Several modeling modalities were considered, namely the neural networks have replaced different parts of the model—in general the parts difficult to model due to diffusion-controlled phenomena. The results obtained were much better than the models represented by single neural networks, but also not very satisfactory for gravimetrical molecular weight.

Another example is represented by the use of a hybrid stacked recurrent neural model for a batch MMA polymerization reactor [15]. Stacked recurrent neural networks are developed for modeling the gel effect, and they are associated with a simplified phenomenological model to obtain a complete model, improved in performance and robustness because of the multiple neural networks included in the model. The results are satisfactory, but there is still room for improvement.

Regarding the mentioned methods, their complexity should be noted, deriving from the need to determine optimal neural networks and their combination with other instruments.

3. Standard and Large Margin Nearest Neighbor Methods
3.1. Standard Nearest Neighbor-Based Regression

Learning methods based on instances are among the simplest machine learning algorithms but provide remarkable results for a large variety of tasks, especially when the training data are not affected by noise and when there is a proper correspondence between the dimensionality of the problem and the size of the training dataset. For regression problems, the goal is to approximate a function between a real-valued dependent variable given one or more independent variables based on a training set of examples $S = \{\mathbf{x}_1, \ldots, \mathbf{x}_m\}$.

In this case, k-Nearest Neighbor (kNN), the value of a query instance can be computed as the mean value of the function of the nearest neighbors:

$$\tilde{f}(\mathbf{x}) = \frac{1}{k} \sum_{\mathbf{x}' \in N(\mathbf{x})} f(\mathbf{x}'), \tag{1}$$

where $N(\mathbf{x}) \subseteq S$ is the set of the closest k instances (i.e., neighbors) of $\mathbf{x}$ in the dataset S.

Nevertheless, finding the most appropriate value of k may not be straightforward, so another possibility is to give weights to the training instances such that the weights of each neighbor depend on the distance to the query point:

$$\tilde{f}(\mathbf{x}) = \frac{1}{z} \sum_{\mathbf{x}' \in S} w_d(\mathbf{x}, \mathbf{x}') \cdot f(\mathbf{x}'), \tag{2}$$

where z is a normalization factor. The inverse of the square Euclidean distance is often used to determine these weights:

$$w_d(\mathbf{x}, \mathbf{x}') = \frac{1}{d(\mathbf{x}, \mathbf{x}')^2} = \frac{1}{\sum_{i=1}^{n}(x_i - x'_i)^2}. \tag{3}$$

kNN works well for many problems, especially when the number of instances is large, the dimensionality of the data is not too big and there is little noise in the data. A crucial component of such a method is the distance metric, because the new instances are evaluated based on their similarity to the training instances. The Euclidean distance is the most common, but different particularizations of the general Minkowski distance, such as the Manhattan distance, or more advanced distance metrics such as the exponentially negative distance function, can also be used [16,17].

However, the standard approach does not take into account problem-specific information, but only considers some general optimizations, such as choosing the best number of neighbors k by cross-validation, or normalizing instance values on each dimension. There is little problem-specific knowledge embedded into the method.

3.2. Large Margin Nearest Neighbor for Classification

To increase performance, researchers have tried to find various methods to tune the distance metric for the different problems being considered. Problem-specific distance metrics have been recognized as capable of significantly improving the results. This process is known as distance metric learning. In general, distance metric learning can be defined as finding a linear transformation $\mathbf{x}' = \mathbf{L}\mathbf{x}$, which transforms the distance between two vectors $\mathbf{x}_i$ and $\mathbf{x}_j$ to:

$$d_L(\mathbf{x}_i, \mathbf{x}_j) = \|\mathbf{L}(\mathbf{x}_i - \mathbf{x}_j)\|_2. \tag{4}$$

Since all operations in classification or regression based on k nearest neighbors can be performed with square distances, a transformation based on a square matrix may be easier to use: $\mathbf{M} = \mathbf{L}^T \mathbf{L}$, such that the square distance becomes:

$$d_M(\mathbf{x}_i, \mathbf{x}_j) = (\mathbf{x}_i - \mathbf{x}_j)^T \mathbf{M} (\mathbf{x}_i - \mathbf{x}_j). \tag{5}$$

In the following algorithms, we investigate the use of the concept of "large margins", well known from the support vectors machines (SVM). We adapt it for regression problems, from ideas pertaining to classification problems.

As presented above, a problem-specific distance metric means that the actual distance is multiplied by a matrix that is computed from the training set, resulting in warping the problem space such that, e.g., for a classification problem, the instances that belong to the same class are closer together than the instances that belong to different classes, and also possibly insuring an arbitrary separation margin between classes.

The idea of a large margin was transferred from the SVM domain to kNN to perform classification tasks in the Large-Margin Nearest Neighbor algorithm (LMNN) [18]. The optimization problem is solved by semi-defined programming and the method can be extended to be invariant to multivariate polynomial transformations [19].

In [18], the **M** matrix is computed such that the distance between an instance x_i and its k neighbors from the same class x_j (named a "target") is minimized. However, this is not sufficient, because reducing all distances to 0 would satisfy this condition. The second constraint is that the distance between instance x_i and its k neighbors from a different class x_l (named an "imposter") is maximized. An additional idea is that the minimum distance to an imposter should be greater than any distance to a target plus some additional value:

$$d_M(\mathbf{x}_i, \mathbf{x}_l) \geq 1 + d_M(\mathbf{x}_i, \mathbf{x}_j), \tag{6}$$

where 1 is arbitrary and has the significance of an imposed margin between the classes.

These two constraints are conflicting; therefore the authors introduce some weights, following an analogy from physics regarding forces of attraction and repulsion. Ultimately, in their examples, they consider the weights of the constraints to be equal.

3.3. Large Margin Nearest Neighbor for Regression

For a regression problem, the same concept can be applied by taking into account the actual real values of the output instead of discrete class values. This is the main idea of the Large Margin Nearest Neighbor for Regression (LMNNR) algorithm [3]. The distance metric is computed by optimizing an objective function composed of two conflicting criteria which can be given different weights in the final objective functions or can be scaled using different functions [5]. LMNNR also allows that different matrices can be used for different regions of the problem space, identified by special points known as prototypes. The locations of the prototypes can be statically initialized, e.g., by using a clustering algorithm such as k-means [3] or dynamically learned, together with the distance metrics [1].

The optimization of the objective function can be carried out either using an evolutionary algorithm [3], which can be rather slow, but has a good chance of finding a global optimum, or by using an approach based on gradient descent [1], which is much faster, but may need several different runs in order to converge to a good solution.

We assume that **M** is a diagonal matrix. This has the advantage that the m_{ii} elements can be interpreted as the weights of the inputs. Because of the definition $\mathbf{M} = \mathbf{L}^T \mathbf{L}$, **M** must be symmetrical and positive semidefinite. Using **M**, Equation (3) is still valid, but becomes:

$$w_{d_M}(\mathbf{x}, \mathbf{x}') = \frac{1}{d_M(\mathbf{x}, \mathbf{x}')} = \frac{1}{\sum_{i=1}^{n} m_{ii} \cdot (x_i - x'_i)^2}. \tag{7}$$

In this formulation, there is a single **M** for all instances. However, it is possible to use different distance metrics for different groups of instances identified by special center points which we call prototypes. Each prototype P can have its own matrix $\mathbf{M}^P$. When calculating the weight of the distance for a new point, an instance will use the weights for the closest prototype m_{ii}^P instead of m_{ii} in Equation (7). The optimization problem assumes the minimization of the following objective function:

$$F = \phi_1(F_1) + \phi_2(F_1) + \phi_3(F_3), \tag{8}$$

in which $\phi_i(\cdot)$ are custom functions, e.g., in the simplest case, $\phi_i(x) = w_i \cdot x$.

We will use the following notations when defining the components of F: $d_{ij} = d_M(\mathbf{x}_i, \mathbf{x}_j)$, $d_{ik} = d_M(\mathbf{x}_i, \mathbf{x}_k)$, $g_{ij} = |f(\mathbf{x}_i) - f(\mathbf{x}_j)|$ and $g_{ik} = |f(\mathbf{x}_i) - f(\mathbf{x}_k)|$.

Thus, the first criterion is:

$$F_1 = \sum_{i=1}^{n} \sum_{j \in N(i)} d_{ij} \cdot (1 - g_{ij}), \qquad (9)$$

where $N(i)$ is the set of the k nearest neighbors for an instance i (e.g., $k = 3$). It ensures that the closer the instances in the input space, the closer their output values should be. Conversely, distant instances should have different output values.

The second criterion is:

$$F_2 = \sum_{i=1}^{n} \sum_{j \in N(i)} \sum_{l \in N(i)} \Delta d_{ijl} \cdot \Delta y_{ijl} \cdot \left(1 + \frac{1}{\Delta d_{ijl} \cdot \Delta y_{ijl} + \varepsilon}\right), \qquad (10)$$

where $\Delta d_{ijl} = \max(d_{il} - d_{ij}, 0)$, $\Delta y_{ijl} = \max(g_{ij} - g_{il}, 0)$ and where ε is a small positive real number. This criterion imposes penalties when the "proximity order" is violated, i.e., when $g_{ij} < g_{ik}$ but $d_{ij} > d_{ik}$. It attempts to minimize the proximity order breaks [20], i.e., the cases when $g_{ij} < g_{ik}$ but $d_{ij} > d_{ik}$. Given an instance i, the optimization reduces the number of situations when another instance l is farther than another one j, but its output value is closer to that of i than the output value of j.

The third criterion can be optionally used for regularization:

$$F_3 = \sum_{j=1}^{n_p} \sum_{i=1}^{n_i} m_{ii}(j). \qquad (11)$$

From our experimental studies, we empirically found that a good objective function is:

$$F = F_1 + \sqrt{F_2}, \qquad (12)$$

because F_2 is usually larger than F_1. Additionally, it was found that regularization was not needed for the presented case studies.

In the following sections, we will give a complete description of two algorithms which implement the LMNNR idea, together with a third algorithm which relies on a different principle.

4. Description of the Algorithms

4.1. Large Margin Nearest Neighbor Regression Trained with an Evolutionary Algorithm (LMNNR-EA)

The first method solves the problem defined by Equation (7) by means of an evolutionary algorithm. The advantages of applying an evolutionary algorithm for optimization is that prototypes, with different weight values, can be used instead of a single set of weights. The prototype positions are precomputed using the k-means algorithm. Since the resulting clusters tend to be (hyper-)spherical, the method is suitable for the complex problems addressed, as the distances are computed with different variants of Euclidean metrics.

4.2. Large Margin Nearest Neighbor Regression Trained with Approximate Gradient Descent (LMNNR-AGD)

Gradient descent optimization is a more-or-less de facto standard for recent machine learning methods, especially (deep) neural networks. It is usually faster than evolutionary optimization, however it is sensitive to the initial value estimate which may cause it to converge to a local optimum.

For the problems addressed here, the standard gradient descent method cannot be applied for several reasons. First, the objective function is not continuous because of the *max* function in Equation (10). In addition, for the expressions of the components of the objective functions defined by Equations (9) and (10) the analytical form of the gradient is difficult to express. Thirdly, the positions of the prototypes need to be optimized simultaneously with the weights.

In the evolutionary approach, the space of the problem has proved to be large enough so that finding the position of the prototypes is not feasible, and that is why the compromise solution using clustering was used. With gradient-based optimization this becomes possible. Still, when the position of the prototypes changes, the neighbor instances also change. Therefore, the objective function is calculated differently, considering different or similar instances, with different weights. This is also difficult to express in the analytical formulation of the gradient.

Thus, we decided to use an approximate differential method, following the definition of the central difference for the derivative. That is, for a very small value ε:

$$f'(x) \approx \frac{f(x+\varepsilon) - f(x-\varepsilon)}{2\varepsilon}, \qquad (13)$$

in which the truncation error is $O(\varepsilon^2)$.

The value of the step size γ is very important for the convergence speed. Therefore, it can be dynamically adjusted, so that it is higher at the beginning and decreasing as the algorithm approaches the solution. It was considered that the value of the step starts from about 1 and then decreases with the number of iterations, following a quadratic reciprocal evolution, but preventing it from going below 0.1:

$$\gamma_n = \max\left(\frac{1}{a + bn + cn^2}, 0.1\right), \qquad (14)$$

where n is the number of iterations, and the values of the a, b and c parameters can be set to adjust the slope of the curve.

4.3. Nearest Neighbor Regression with Adaptive Distance Metrics Trained by Multiple Point Hill Climbing on Noisy Training Set Error (RADIAN)

The Nearest Neighbor Regression with Adaptive Distance Metrics Trained by Multiple Point Hill Climbing on Noisy Training Set Error (RADIAN) algorithm is not based on the concept of a large margin but is inspired by a technique used in deep learning, i.e., denoising autoencoders. This is a type of deep network that learns the presented data itself, but transfers it through an intermediate layer, usually a bottleneck with a number of neurons smaller than the dimensionality of the problem, so this layer will capture the essential characteristics of the data. In order to prevent the phenomenon of overfitting, the input data are slightly corrupted and the autoencoder is forced to learn the correct data from the corrupted version.

The overall concept of learning a distance metric and the distribution of the prototypes in the problem space are the same as for the previous algorithms. However, the prototype weights and positions are set so as to minimize the mean square error on a slightly corrupted version of the training set, where the instances inputs are slightly changed with a small random quantity:

$$x'_{ij} = x_{ij} + 2(r - 1/2) \cdot \varepsilon, \qquad (15)$$

where x_{ij} is the value of the data, r is a random uniform number in the [0, 1) interval and ε is a small number, e.g., $\varepsilon = 0.001$.

The minimization is performed here by multiple point hill climbing. This method combines the approach of gradient descent with the search for a solution from multiple initial points, similar in a way to the parallel search performed by evolutionary algorithms. In standard hill climbing, several neighbors of the current point are generated, e.g., by an equation similar to (15). The point with a better (lower) objective function is selected as the new current point and the procedure is repeated for a number of steps. However, this behavior is similar to gradient descent, which is prone to local optima. Therefore, the search is performed from multiple starting points, so the probability of starting in a neighborhood of the global optimum is increased.

An idea for future investigation is to initialize the positions of the prototypes using the k-means clustering algorithm, instead of random initialization, such that they fill the problem space more uniformly.

From the initial results, it seems that this simple algorithm outperforms the other two for the problems under study.

5. Modeling Methodology

The free radical polymerization of methyl methacrylate achieved in a batch bulk process has non-linear gel and glass effects, i.e., regions where the values of the output parameters have a high variation. The data necessary for the application of the regression methods were obtained on a simulator [21], and their number is very large if we consider the faithful rendering of the gel effect corresponding to an abrupt conversion and molecular mass jump. Using constant sampling may miss these regions. In order to address this problem, an adaptive sampling technique was devised [22], which can select more samples around the critical regions. This algorithm first computes the local differences between successive points in the output function. This has the same meaning as a derivative. Then, it uses a running sum to determine whether the next point will be sampled. In regions with approximately constant values, the derivative is small, therefore the space between the sampled points will be larger. In regions with high variation, the derivative will be higher, and thus more points will be sampled. A number of points between 300 and 400 were obtained for all modeled parameters, instead of thousands of data when a constant step is applied. Monomer conversion (x), numerical average molecular weight (M_n), gravimetrical average molecular weight (M_w) are determined as function of reaction conditions (initiator concentration I_0, temperature T and time t).

Figures 1 and 2 show two examples of applying the adaptive procedure for two of the considered outputs, x and M_n. The input conditions of the reaction are I_0 = 15 mol/m^3 and T = 343 K.

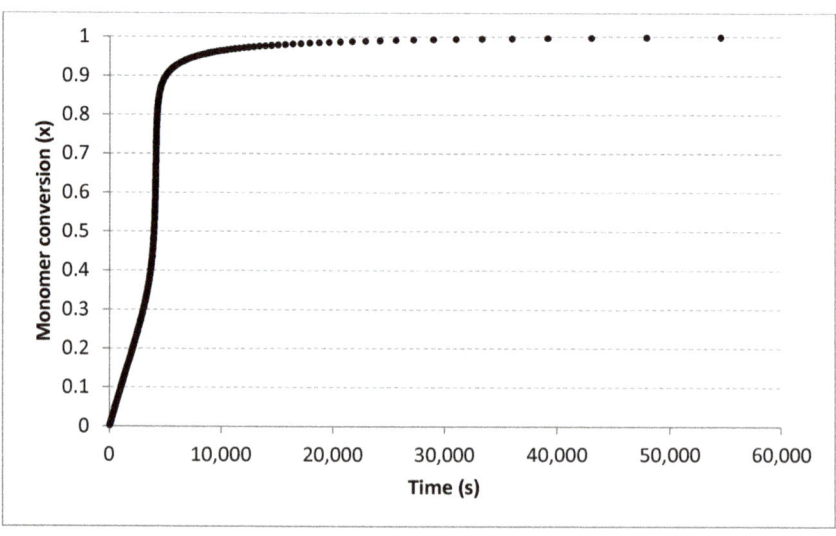

Figure 1. Adaptively sampled points (337) for the monomer conversion (x).

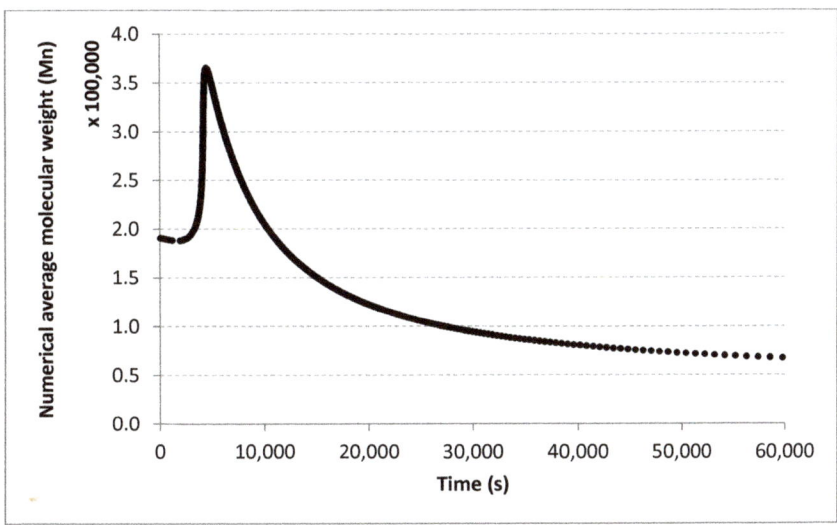

Figure 2. Adaptively sampled points (414) for the numerical average molecular weight (M_n).

6. Results and Discussion

In a previous study [22] we compared the results of several well-known algorithms from the Weka collection [2] with those of LMNNR for the three formulated problems, i.e., the conversion, average numerical and gravimetric molecular weights. The algorithms considered from Weka were: kNN, random forest, REPTree, M5 rules, additive regression, ε-SVR and ν-SVR, each with different values for their parameters. The data were split into 2/3 for training and 1/3 for testing. Table 1 presents a concise comparison between the best classical algorithm (random forest) and LMNNR (with one prototype, five optimization neighbors and five regression neighbors) in terms of the coefficient of determination (R^2) obtained for both the training and testing sets.

Table 1. Performance of the best Weka algorithm for the monomer conversion (x), numerical average molecular weight (M_n) and gravimetrical average molecular weight (M_w) vs. Large Margin Nearest Neighbor for Regression (LMNNR).

Algorithm	Training x	Testing x	Training M_n	Testing M_n	Training M_w	Testing M_w
Random forest with 100 trees	0.999800	0.999800	0.999400	0.998600	0.999800	0.999600
LMNNR	1	0.999953	1	0.999638	1	0.999816

The fact that LMNNR obtains a perfect correlation for the training set is not surprising, since it is an instance-based method. It is, however, commendable that it outperforms the best Weka algorithm on the testing set. Therefore, in the present study, we focus only on the results of LMNNR in two variants and the newly introduced algorithm RADIAN and perform a comprehensive experimental study with different settings regarding the distribution of data and the parameters of the algorithms.

In order to assess the performance of the algorithms, cross-validation was used. The most common method is to split the data into ten groups (or bins) and use nine groups for training and one group for testing, and repeat the process ten times, every time with a different test group. In this study, in order to assess different aspects of the learning process, we use three cross-validation variants:

- Standard cross-validation with 10 groups, and in each step the data are split 90% for training and 10% for testing;
- Cross-validation with three groups, and in each step the data are split 67% for training and 33% for testing: this is similar to the simpler 2/3–1/3 split (e.g., used in [22], but more relevant statistically. This is a means to roughly compare results to those obtained in the previous work. However, a direct comparison is not possible;
- A cross-validation-like procedure with 10 groups, and in each step the data are split 10% for training and 90% for testing. This scenario is used to assess the generalization capability of the models more "aggressively".

In terms of the parameters used for the algorithms, two settings were used, as displayed in Table 2:

- Setting 1: the parameters allow for greater, more general search capabilities, the values are larger, but this leads to a longer execution time;
- Setting 2: the values of the parameters are smaller. This setting allows us to see whether a shorter execution time can still provide acceptable results.

Table 2. The settings of the three regression algorithms. LMNNR-EA: Large Margin Nearest Neighbor Regression Trained with an Evolutionary Algorithm; LMNNR-AGD: Large Margin Nearest Neighbor Regression Trained with Approximate Gradient Descent; RADIAN: Nearest Neighbor Regression with Adaptive Distance Metrics Trained by Multiple Point Hill Climbing on Noisy Training Set Error.

	Algorithm 1: LMNNR-EA	Algorithm 2: LMNNR-AGD	Algorithm 3: RADIAN
Setting 1	No. groups = 10 No. prototypes = 2 No. regression neighbors = 3 No. optimization neighbors = 3 No. trials = 20 Population size = 40 Min. gene value = 0.001 Max. gene value = 10 Tournament size = 2 Crossover rate = 0.95 Mutation rate = 0.05 No. generations = 500	No. groups = 10 No. prototypes = 2 No. regression neighbors = 3 No. optimization neighbors = 3 No. trials = 20 Epsilon = 0.000001 Learning rate = 0.1 Dynamic learning rate = 0 Max. gradient descent steps = 1000	No. groups = 10 No. prototypes = 2 No. starting points = 20 No. regression neighbors = 5 No. hill-climbing steps = 30 No. hill-climbing neighbors = 20 Training set noise = 0.001 Hill-climbing noise = 0.01 Noise on output = 1
Setting 2	No. groups = 10 No. prototypes = 1 No. regression neighbors = 3 No. optimization neighbors = 3 No. trials = 10 Population size = 30 Min. gene value = 0.001 Max. gene value = 10 Tournament size = 2 Crossover rate = 0.95 Mutation rate = 0.05 No. generations = 100	No. groups = 10 No. prototypes = 1 No. regression neighbors = 3 No. optimization neighbors = 3 No. trials = 10 Epsilon = 0.000001 Learning rate = 0.1 Dynamic learning rate = 0 Max. G.D. steps = 200	No. groups = 10 No. prototypes = 1 No. starting points = 10 No. regression neighbors = 3 No. hill-climbing steps = 10 No. hill-climbing neighbors = 10 Training set noise = 0.001 Hill-climbing noise = 0.01 Noise on output = 1

Tables 3 and 4 present the experimental results obtained for some combinations of data splits and algorithm parameter configurations, for the three considered problems, i.e., x, M_n and M_w. Each algorithm was run ten times, and its best performance was evaluated using the coefficient of correlation (r) and the mean squared error (MSE).

Table 3. The results of the algorithms with setting 1. MSE: mean squared error; r: coefficient of correlation.

	x		M_n		M_w	
	r	MSE	r	MSE	r	MSE
Large training sets (90–10% data split)						
Algorithm 1	0.999425	0.000124	0.999466	5.0707403	0.999459	78.1390448
Algorithm 2	*0.999667*	*0.000074*	*0.999499*	*4.7406073*	*0.999680*	*46.6716341*
Algorithm 3	0.999527	0.000120	0.999349	6.1569147	0.999396	79.0907101
Small training sets (10–90% data split)						
Algorithm 1	*0.962001*	*0.007573*	0.987506	11.71332089	*0.960302*	*566.565343*
Algorithm 2	0.946293	0.010619	*0.988202*	*11.08952839*	0.959521	578.709620
Algorithm 3	0.91255	0.017117	0.986627	12.64085094	0.952598	574.958058

Table 4. The results of the algorithms with setting 2.

	x		M_n		M_w	
	r	MSE	r	MSE	r	MSE
Large training sets (90–10% data split)						
Algorithm 1	0.999644	0.000080	0.999469	5.0488433	0.999548	66.0130052
Algorithm 2	*0.999665*	*0.000075*	*0.999503*	*4.6982778*	*0.999667*	*48.5893648*
Algorithm 3	0.999663	0.000078	0.999345	6.1986015	0.998214	66.1359322
Small training sets (10–90% data split)						
Algorithm 1	*0.958308*	*0.008293*	0.987379	11.889027	0.961123	555.051717
Algorithm 2	0.944513	0.010950	*0.988040*	*11.244562*	*0.962062*	*541.849968*
Algorithm 3	0.920547	0.015464	0.982747	16.316099	0.957143	550.783154
Average training sets (67–33% data split)						
Algorithm 1	0.998997	0.000222	0.999171	7.843631	0.999286	*10.425242*
Algorithm 2	*0.999017*	*0.000216*	*0.999197*	*7.587175*	*0.999471*	17.444913
Algorithm 3	0.999004	0.000221	0.999191	7.634568	0.998949	15.377001

The best results in these tables are emphasized with italic font. One can see that the performance of the three algorithms is comparable, however, the third one is an order of magnitude faster. The differential approach converges faster, that is why Algorithm 2 is most of the time better than Algorithm 1.

Obviously, the best results are obtained if a comprehensive data set (data 1) and high values of the parameters specific to each algorithm (settings 1) are used. Of the three output parameters, the conversion has the best models, compared to the average molecular masses.

When only 10% of the data is used for training, the results are less accurate than those obtained for 90%, but it must be underlined that they are still quite good, with a correlation coefficient above 0.9. Finally, with the 67–33% split, the results are almost as good as those obtained with the 90–10% split. This proves that the generalization capabilities of the model are very good. It is important to mention that with the distributions 67–33% and 90–10%, good results are obtained even with setting 2 (shorter execution time).

The results obtained for different algorithms, parameter setting, data splitting or amount of data are rendered suggestively in Figures 3–14 where predicted data are compared with the experimental data.

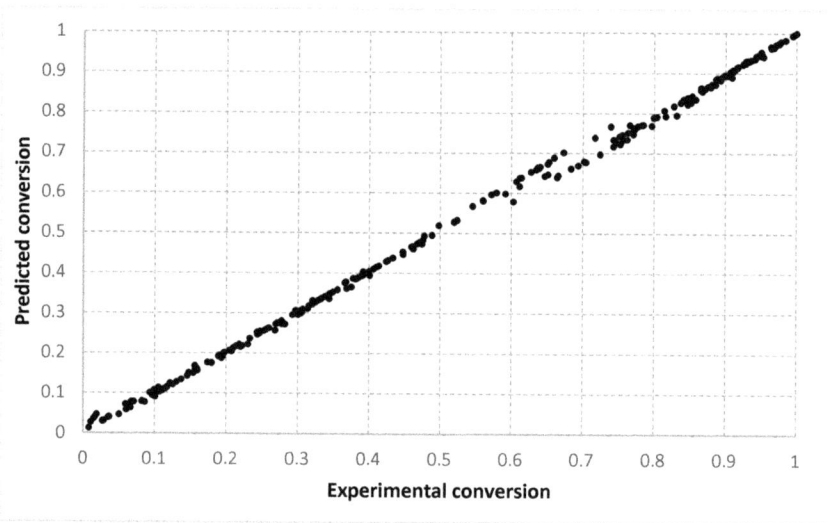

Figure 3. Predicted conversion with the LMNNR-EA algorithm versus experimental conversion, setting 1, with data split into 90% for training and 10% for testing.

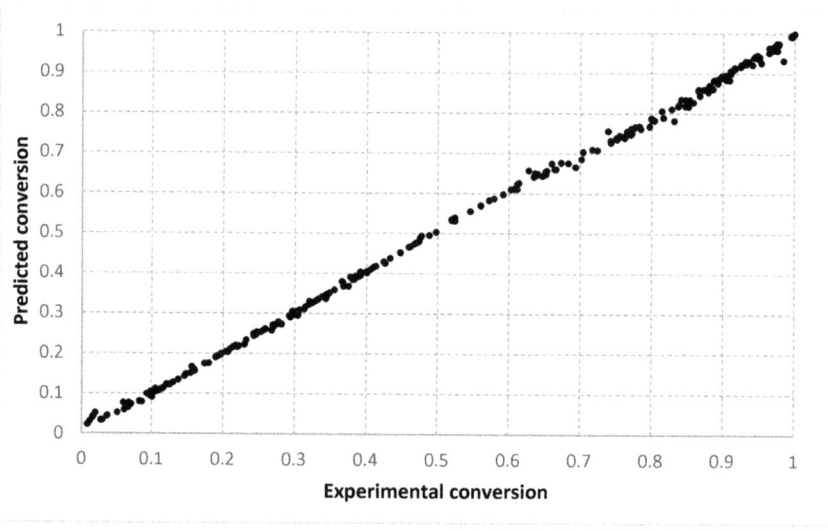

Figure 4. Predicted conversion with the RADIAN algorithm versus experimental conversion, setting 2, with data split into 90% for training and 10% for testing.

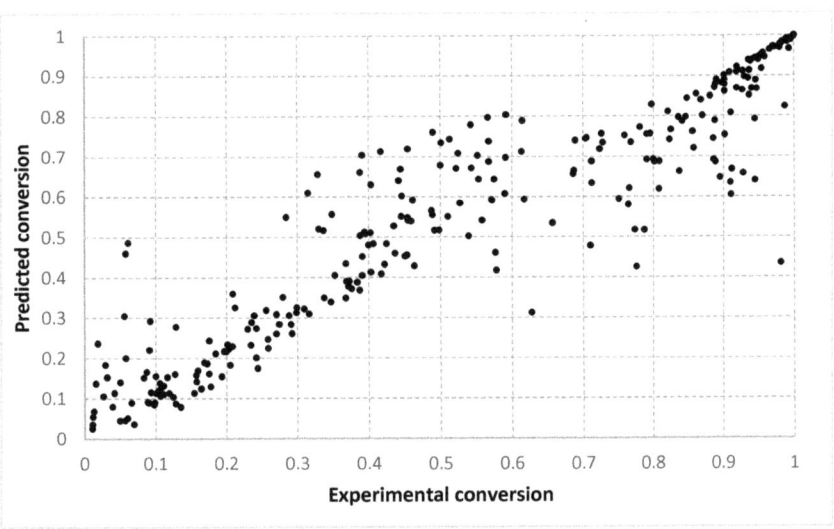

Figure 5. Predicted conversion with the LMNNR-AGD algorithm versus experimental conversion, setting 2, with data split into 10% for training and 90% for testing.

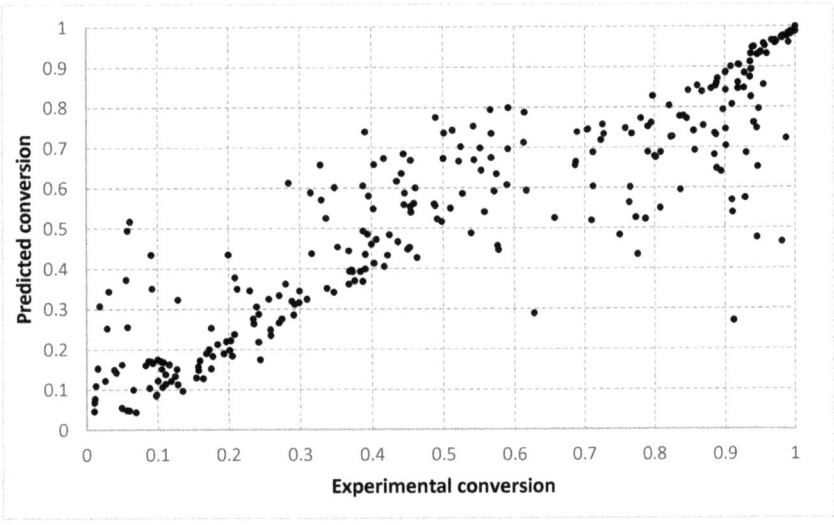

Figure 6. Predicted conversion with the RADIAN algorithm versus experimental conversion, setting 1, with data split into 10% for training and 90% for testing.

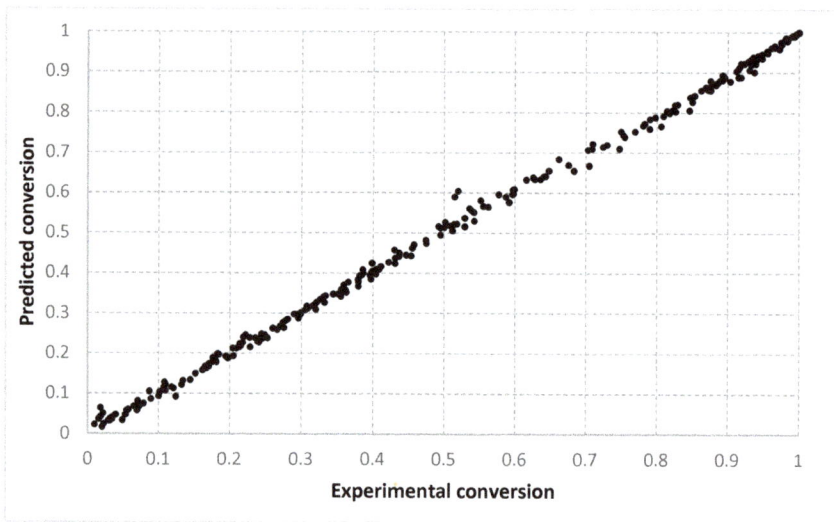

Figure 7. Predicted conversion with the LMNNR-AGD algorithm versus experimental conversion, setting 2, with data split into 67% for training and 33% for testing.

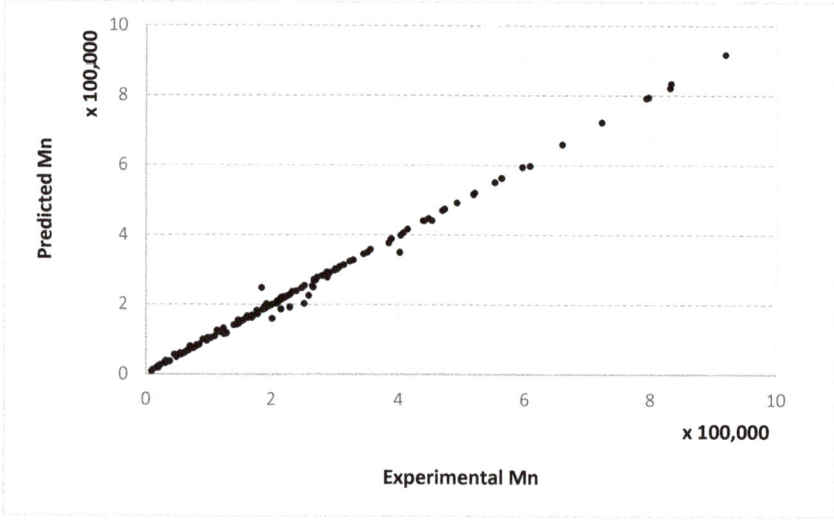

Figure 8. Predicted numerical average molecular weight with the RADIAN algorithm versus experimental data, setting 1, with data split into 90% for training and 10% for testing.

For the conversion, very good results were obtained with all three algorithms, with both types of settings. As shown in Figures 3–7, the best fit is obtained with the 90–10% split. With setting 2, slightly better solutions are found. We consider that this is because with two prototypes (i.e., setting 1), the search is being performed in a much larger space. When only one prototype is used (i.e., setting 2), the results are also more continuous. With a 10–90% split, the dispersion of the desired vs. predicted plot is greater. Visually, similar outcomes are achieved for both the LMNNR and RADIAN algorithms. The distribution of the 67–33% split, as seen in Tables 3 and 4, has an aspect more similar to the 90–10% than to the 10–90% split, with an only slightly larger dispersion. This shows that training with two thirds of the data is enough for good generalization, compared with the extreme case when training only with a tenth of the data (Figure 7 vs. Figure 5).

For molecular masses, the conclusions are similar to those obtained for the conversion, with no significant differences between the algorithms. However, one can see here that most of the data in the first half of the domain are much denser than the data from the second half. For M_n, Figures 8–10 only display the performance of the RADIAN algorithm (i.e., Algorithm 3). In this case, it is also the data split that has the strongest effect on the results.

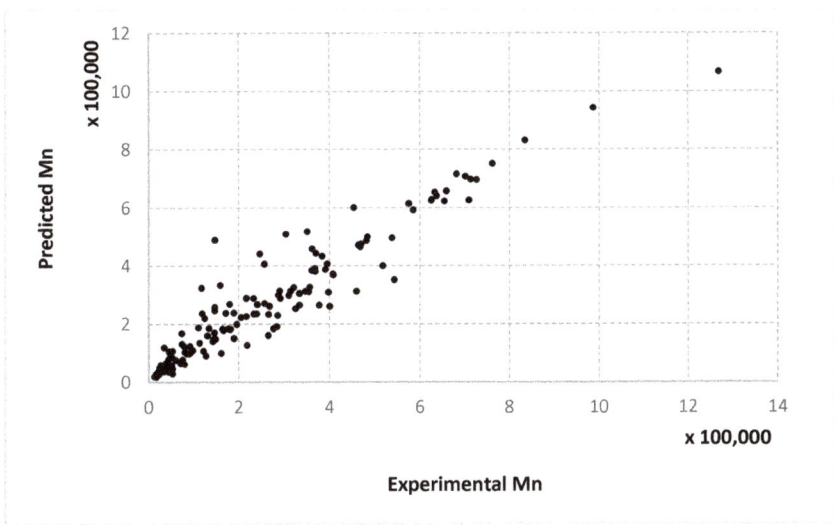

Figure 9. Predicted numerical average molecular weight with the RADIAN algorithm versus experimental data, setting 2, with data split into 10% for training and 90% for testing.

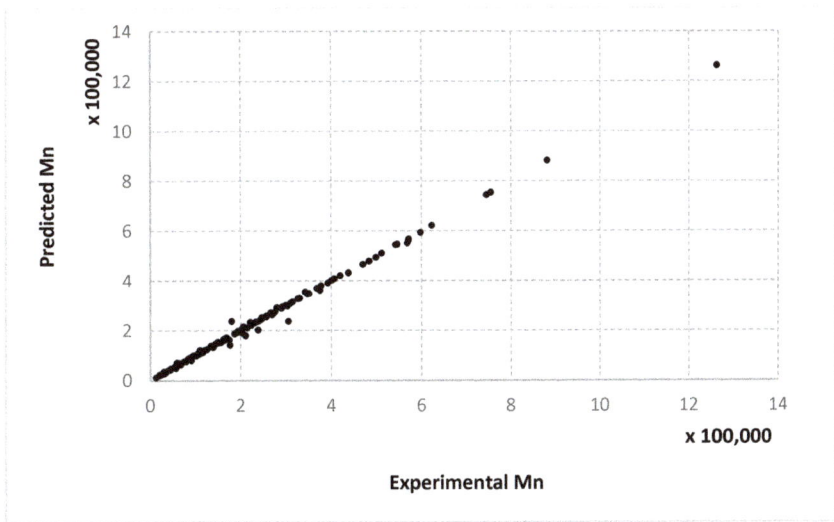

Figure 10. Predicted numerical average molecular weight with the RADIAN algorithm versus experimental data, setting 2, with data split into 67% for training and 33% for testing.

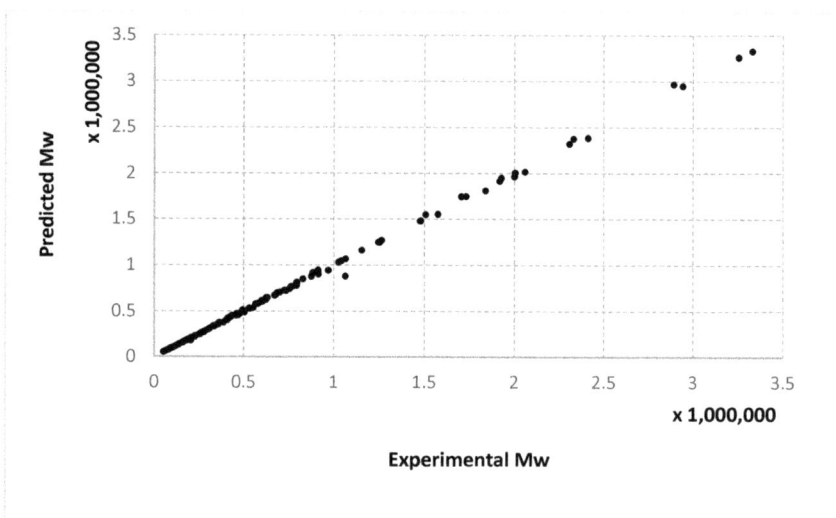

Figure 11. Predicted gravimetrical average molecular weight with the LMNNR-EA algorithm versus experimental data, setting 1, with data split into 90% for training and 10% for testing.

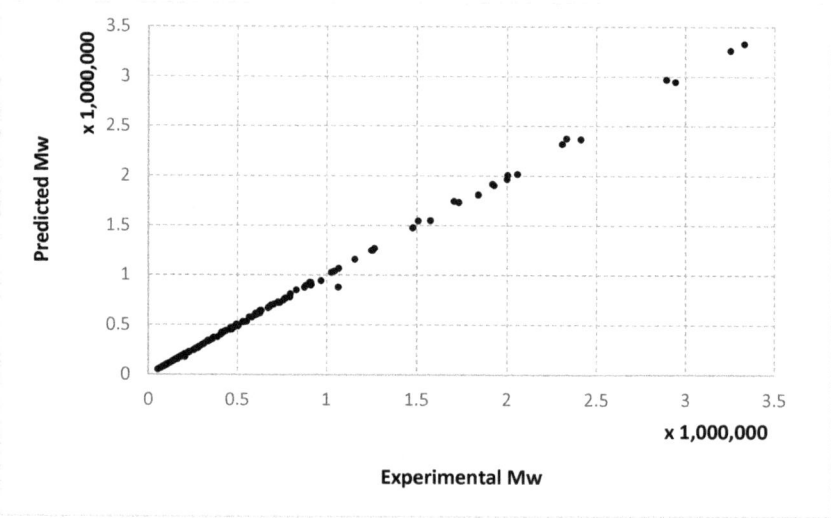

Figure 12. Predicted gravimetrical average molecular weight with the LMNNR-EA algorithm versus experimental data, setting 2, with data split into 90% for training and 10% for testing.

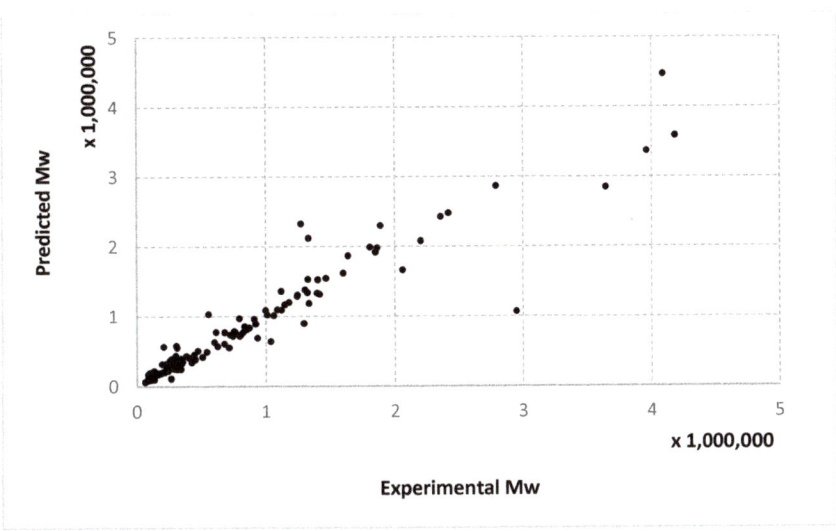

Figure 13. Predicted gravimetrical average molecular weight with the LMNNR-EA algorithm versus experimental data, setting 2, with data split into 10% for training and 90% for testing.

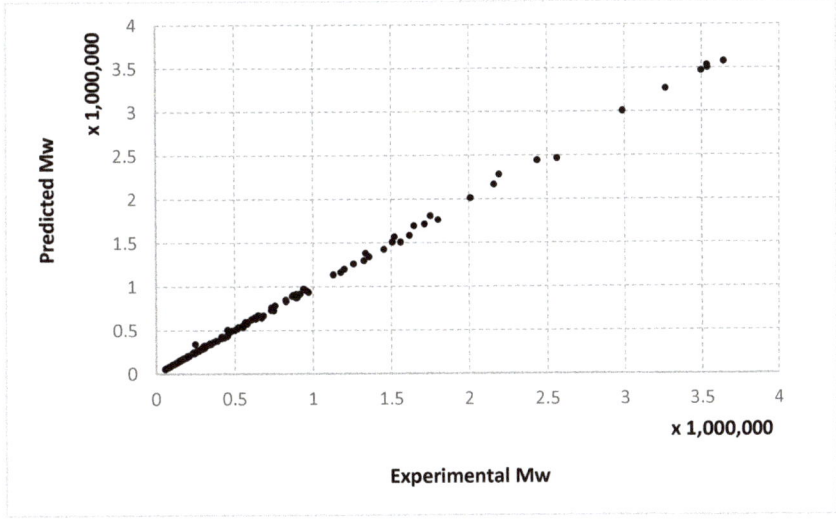

Figure 14. Predicted gravimetrical average molecular weight with the LMNNR-EA algorithm versus experimental data, setting 2, with data split into 67% for training and 33% for testing.

Very similar outcomes are encountered for the M_w output and the LMNNR-EA algorithm (i.e., Algorithm 1). However, in this case, the dispersion of the graph for the 10–90% split is less than those for x an M_n. This is in fact an indicator about the minimum amount of training information needed in order to generalize well.

We remind the reader that the data presented in Figures 3–14 refer only to the test data, i.e., the aggregated predictions of the models for the test groups.

The results obtained can also be analyzed in terms of variance of the performance metrics. For all algorithms, the variance decreases when the size of the training set increases. For example, for conversion, in the case of the 10–90% split, the standard deviation σ is about 3–5% of the mean value μ. In the case of the 67–33% split, it becomes 0.1–2% of the mean value and decreases even more for the 90–10% split. The variance can be further decreased by identifying outliers. Since the methods are heuristic, some runs simply fail to provide good solutions. By removing data outside the $\mu \pm 2\sigma$ range, the variance is reduced especially for larger training sets, e.g., the new resulting standard deviation is about 10 times smaller. Algorithm 1 has the largest variance among the three methods, while algorithms 2 and 3 have comparable variance.

In terms of execution time, the computations take longer as the size of the training set increases. For example, one fold of cross-validation for conversion takes about 6.7 s for the 10–90% split and 54 s for the 90–10% split for Algorithm 1. Algorithm 2 takes about 5 s and 32 s, respectively. Algorithm 3 requires about 0.7 s and 5.8 s, respectively. These studies were made using a computer with a 4-core 2 GHz Intel processor and 8 GB of RAM. Of course, specific times depend on the particular structure of the training data, especially the number of attributes, and the parameters of the algorithms. In particular, the complexity of Algorithm 1 mainly depends on the number of individuals in the population and the number of generations of the evolutionary algorithm. The complexity of Algorithm 2 mainly depends on the number of iterations of the approximate gradient descent procedure. The complexity of Algorithm 3 mainly depends on the number of hill-climbing steps and the number of neighbors that are generated in each step.

7. Conclusions

Instance-based classification and regression algorithms can provide very good results for complex decision boundaries, especially when the size of the training dataset is big and the number of dimensions of the problem space is not very large. The distance metric is crucial for this class of algorithms, and a significant increase in performance can be achieved by changing it depending on the specific problem under study. The optimal distance metric can be obtained by solving an optimization problem that tries to decrease the distance between the instances with similar output values and increase the distance between the instances with different output values. This process is actually equivalent to maximizing the margin between the instances with different output values. In this way, the concept of a large margin, introduced in the context of support vector machines, can also be applied to instance-based regression. The LMNNR algorithm uses this idea together with prototypes, where each prototype can have its own custom distance metric, which can be helpful when a large number of instances are available.

The corresponding optimization problem can be solved either by an evolutionary algorithm or by an approximate gradient descent method. The results are competitive compared with those obtained by classical algorithms such as support vector machines, k-nearest neighbor and random forest.

Another original regression method is designed by considering an idea from denoising autoencoders, a kind of deep neural networks. In our case, the weights that define the custom distance metric and the positions of the prototypes are computed so as to minimize the mean square error on a corrupted version of the training data created by adding a small amount of noise.

The quality of the results is also supported by the adaptive sampling technique that provides the machine learning algorithms with the most relevant data by taking into account the rate of variation of the outputs involved in the chemical process.

As future directions of research, one may investigate whether similar results can be obtained using alternative techniques, e.g., principal analysis decomposition or (deep) neural networks [9,23,24].

Concerning the case study of the free radical polymerization of MMA, the conclusions that give the necessary practical indications are the following:

- The accuracy of the modeling for the variables of interest—monomer conversion and molecular masses of the polymer—depends on the applied algorithm, the set-

tings of their parameters and the way of splitting the data. The last factor has the greater importance;
- Conversion is easier to model than molecular weights, but, with a proper combination of settings and data sharing, very good results can be obtained for all parameters of interest;
- Although good results have been identified for all three algorithms, RADIAN is preferred because it is considerably faster than the other two, which is an important factor for online optimal control procedures.

Author Contributions: Conceptualization and methodology, F.L., S.C.; software, F.L.; data curation, S.C.; investigation: S.C., A.-M.M.-V., D.L.; writing, F.L., S.C.; funding acquisition, S.C. All authors have read and agreed to the published version of the manuscript.

Funding: This work was supported by Exploratory Research Projects PN-III-P4-ID-PCE-2020-0551, financed by UEFISCDI.

Conflicts of Interest: The authors declare no conflict of interest.

References

1. Leon, F.; Curteanu, S. Large Margin Nearest Neighbour Regression Using Different Optimization Techniques. *J. Intell. Fuzzy Syst.* **2017**, *32*, 1321–1332. [CrossRef]
2. Hall, M.; Frank, E.; Holmes, G.; Pfahringer, B.; Reutemann, P.; Witten, I.H. The WEKA Data Mining Software: An Update. *Acm Sigkdd Explor.* **2009**, *11*, 10–18. [CrossRef]
3. Leon, F.; Curteanu, S. Evolutionary Algorithm for Large Margin Nearest Neighbour Regression. In Proceedings of the 7th International Conference on Computational Collective Intelligence Technologies and Applications, Madrid, Spain, 21–23 September 2015.
4. Leon, F.; Popescu, E. Using Large Margin Nearest Neighbor Regression Algorithm to Predict Student Grades Based on Social Media Traces. In *Methodologies and Intelligent Systems for Technology Enhanced Learning*; Vittorini, P., Gennari, R., Di Mascio, T., Rodríguez, S., De la Prieta, F., Ramos, C., Silveira, R.A., Eds.; MIS4TEL 2017; Book Series: Advances in Intelligent Systems and Computing; Springer: Cham, Switzerland, 2017; Volume 617.
5. Popescu, E.; Leon, F. Predicting Academic Performance Based on Learner Traces in a Social Learning Environment. *IEEE Access* **2018**, *6*, 72774–72785. [CrossRef]
6. Leon, F.; Piuleac, C.G.; Curteanu, S.; Poulios, I. Instance-based regression with missing data applied to a photocatalitic oxidation process. *Cent. Eur. J. Chem.* **2012**, *10*, 1149–1156.
7. Mareci, D.; Sutiman, D.; Chelariu, R.; Leon, F.; Curteanu, S. Evaluation of the corrosion resistance of new TiZr binary alloys by experiment and simulation based on regression model with incomplete data. *Corros. Sci.* **2013**, *73*, 106–122. [CrossRef]
8. Curteanu, S.; Leon, F.; Lupu, A.S.; Floria, S.A.; Logofatu, D. An Evaluation of Regression Algorithms Performance for the Chemical Process of Naphthalene Sublimation. In Proceedings of the 14th International Conference on Artificial Intelligence Applications and Innovations (AIAI 2018), Rhodes, Greece, 25–27 May 2018; Volume 519, pp. 219–230.
9. Goodfellow, I.; Bengio, Y.; Courville, A. *Deep Learning*; MIT Press: Cambridge, MA, USA, 2016.
10. Chiu, W.Y.; Carratt, G.M.; Soong, D.S. A computer model for the gel effect in free-radical polymerization. *Macromolecules* **1983**, *16*, 348–359. [CrossRef]
11. Curteanu, S.; Bulacovschi, V.; Lisa, C. Algorithms for using some models of gel and glass effects in free-radical polymerization of methyl methacrylate. *Polym. Plast. Technol. Eng.* **1999**, *38*, 1121–1136. [CrossRef]
12. Curteanu, S.; Leon, F.; Gâlea, D. Neural network models for free radical polymerization of methyl methacrylate. *Eurasian Chem. Technol. J.* **2003**, *5*, 225–231.
13. Curteanu, S. Direct and inverse neural network modeling in free radical polymerization. *Cent. Eur. J. Chem.* **2004**, *2*, 113–140. [CrossRef]
14. Curteanu, S.; Leon, F. Hybrid neural network models applied to a free radical polymerization process. *Polym. Plast. Technol. Eng.* **2006**, *45*, 1013–1023. [CrossRef]
15. Tian, Y.; Zhang, J.; Morris, J. Modeling and Optimal Control of a Batch Polymerization Reactor Using a Hybrid Stacked Recurrent Neural Network Model. *Ind. Eng. Chem. Res.* **2001**, *40*, 4525–4535. [CrossRef]
16. Shepard, R.N. Psychological representations of speech sounds. In *Human Communication: A Unified View*; David, E.E., Denes, P.B., Eds.; McGraw-Hill: New York, NY, USA, 1972.
17. Rumelhart, D.E.; Abrahamsen, A.A. A model for analogical reasoning. *Cognit. Psychol.* **1973**, *5*, 1–28. [CrossRef]
18. Weinberger, K.Q.; Saul, L.K. Distance metric learning for large margin nearest neighbor classification. *J. Mach. Learn. Res.* **2009**, *10*, 207–244.
19. Kumar, M.P.; Torr, P.H.S.; Zisserman, A. An Invariant Large Margin Nearest Neighbour Classifier. In Proceedings of the IEEE 11th International Conference on Computer Vision (ICCV 2007), Rio de Janeiro, Brazil, 14–21 October 2007. [CrossRef]
20. Assi, K.C.; Labelle, H.; Cheriet, F. Modified large margin nearest neighbor metric learning for regression. *IEEE Signal Process. Lett.* **2014**, *21*, 292–296. [CrossRef]

21. Curteanu, S.; Bulacovschi, V.; Constantinescu, M. Free radical polymerization of methyl methacrylate. Modelling and simulation at high conversion. *Hung. J. Ind. Chem.* **1999**, *27*, 287–292.
22. Leon, F.; Curteanu, S. Performance Comparison of Different Regression Methods for a Polymerization Process with Adaptive Sampling. In Proceedings of the 18th International Conference on Computational Intelligence and Systems Sciences, Prague, Czechia, 18–22 November 2016.
23. Mercorelli, P. Biorthogonal wavelet trees in the classification of embedded signal classes for intelligent sensors using machine learning applications. *J. Frankl. Inst.* **2007**, *344*, 813–829. [CrossRef]
24. Mercorelli, P. Denoising and harmonic detection using nonorthogonal wavelet packets in industrial applications. *J. Syst. Sci. Complex.* **2007**, *20*, 325–343. [CrossRef]

Review

A Review of Tracking and Trajectory Prediction Methods for Autonomous Driving

Florin Leon and Marius Gavrilescu *

Faculty of Automatic Control and Computer Engineering, "Gheorghe Asachi" Technical University of Iași, Bd. Mangeron 27, 700050 Iași, Romania; florin.leon@academic.tuiasi.ro
* Correspondence: marius.gavrilescu@academic.tuiasi.ro

Abstract: This paper provides a literature review of some of the most important concepts, techniques, and methodologies used within autonomous car systems. Specifically, we focus on two aspects extensively explored in the related literature: tracking, i.e., identifying pedestrians, cars or obstacles from images, observations or sensor data, and prediction, i.e., anticipating the future trajectories and motion of other vehicles in order to facilitate navigating through various traffic conditions. Approaches based on deep neural networks and others, especially stochastic techniques, are reported.

Keywords: autonomous driving; object tracking; trajectory prediction; deep neural networks; stochastic methods

Citation: Leon, F.; Gavrilescu, M. A Review of Tracking and Trajectory Prediction Methods for Autonomous Driving. *Mathematics* **2021**, *9*, 660. https://doi.org/10.3390/math9060660

Academic Editor: Denis N. Sidorov

Received: 28 January 2021
Accepted: 17 March 2021
Published: 19 March 2021

Publisher's Note: MDPI stays neutral with regard to jurisdictional claims in published maps and institutional affiliations.

Copyright: © 2021 by the authors. Licensee MDPI, Basel, Switzerland. This article is an open access article distributed under the terms and conditions of the Creative Commons Attribution (CC BY) license (https://creativecommons.org/licenses/by/4.0/).

1. Introduction

Autonomous car technology is already being developed by many companies on different types of vehicles. Complete driverless systems are still at an advanced testing phase, but partially automated systems have been around in the automotive industry for the last few years. Autonomous driving technology has been the focus of multiple research and development efforts by various car manufacturers, universities, and research centers, since the middle 1980s.

A famous competition was the DARPA Urban Challenge in 2007. Other examples include the European Land-Robot Trial, which has been held since 2006, the Intelligent Vehicle Future Challenge, between 2009 and 2013, as well as the Autonomous Vehicle Competition, held between 2009 and 2017. Since the early stages of autonomous driving technology development, research in the related fields has been garnering significant interest in universities and industry worldwide.

In this review, we focus on two aspects of an autonomous car system:

- *Tracking:* identifying traffic participants, i.e., cars, pedestrians, and obstacles from sequences of images, sensor data, or observations. It is assumed that some preprocessing of sensor data and/or input images has already been done;
- *Prediction:* assessing the future motion of surrounding vehicles in order to navigate through various traffic scenarios. Beside the prediction of the simple physical behavior of the agents based on a set of past observations, an important issue is to take into account their possible interactions.

The paper is composed of two main parts that focus on these topics.

Section 2 deals with tracking problems as addressed in the related literature. We cover aspects concerning the extraction and use of various features for the detection of pedestrians, vehicles, and obstacles across sequences of images and sensor data. Also, we address the various ways in which authors tackle the problems of ensuring detection consistency, temporal coherence, or occlusion handling. We present methods using deep neural networks, but also alternative, conventional approaches.

Section 3 addresses the problem of motion and behavior prediction in traffic scenarios. We discuss various solutions proposed in the related literature for predicting the trajectory of the ego car with respect to the behavior of other traffic participants. We address methods based on deep neural networks and stochastic models, as well as various mixed approaches.

Section 4 contains some conclusions with regard to the aspects discussed throughout the paper.

2. Tracking Methods

Object tracking is an important part of ensuring accurate and efficient autonomous driving. The identification of objects such as pedestrians, cars, and various obstacles from images and vehicle sensor data is a significant and complex interdisciplinary domain. It involves contributions from computer vision, signal processing, and/or machine learning. Object tracking is an essential part of ensuring safe autonomous driving, since it can aid in obstacle avoidance, motion estimation, the prediction of the intentions of pedestrians and other vehicles, as well as path planning. Most sensor data that have to be processed take the form of point clouds, images, or a combination of the two. Point cloud data may be handled in a multitude of ways, the most common of which is some form of 3D grid, where a voxel engine is used to traverse the point space. Some situations call for a reconstruction of the environment from the point cloud which involves various means of resampling and filtering. In some instances, stereo visual information is available and disparities must be computed from the left-right images. Stereo matching is not a trivial task and has the drawback that the computations required for reasonable accuracy usually have a significant impact on performance. In other cases, multiple types of sensor data are available, thereby requiring registration, point matching, and image/point cloud fusion. The problem is further complicated by the necessity to account for temporal cues and to estimate motion from time-based frames.

The scenes involved in autonomous driving scenarios rarely feature a single individual target. Most commonly, multiple objects must be identified and tracked concurrently, some of which may be in motion relative to the vehicle and to each other. As such, most approaches in the related literature handle more than one object and are therefore aimed at solving multiple object tracking problems (MOT).

The tracking problem can be summarized as follows: a sequence of sensor data is available from one or multiple vehicle-mounted acquisitions devices. Considering that several observations are identified in all or some of the frames from the sequence, how can the observations from each frame be associated with a set of objects (pedestrians, vehicles, and various obstacles) and how can the trajectories of each such object be reconstructed and predicted as accurately as possible?

Most related methods involve assigning an ID or identifying a response for all objects detected within a frame, and then attempting to match the IDs across subsequent frames. This is often a complex task, considering that the tracked objects may enter and leave the frame at different timestamps. They may also be occluded by the environment or may occlude each other. Additional problems may be caused by defects in the acquired images: noise, sampling or compression artifacts, aliasing, or acquisition errors.

Object tracking for automated driving most commonly has to operate on real-time video. As such, the objective is to correlate tracked objects across multiple video frames, in addition to individual object identification. Accounting for variations in motion comes with an additional set of pitfalls, such as when objects are affected by rotation or scaling transformations, or when the movement speed of the objects is high relative to the frame rate.

In the majority of cases, images are the primary modality for perceiving the scene. As such, a lot of efforts from the related literature are in the direction of 2D MOT. These methods are based on a succession of detection and tracking steps: consecutive detections that are similarly classified are linked together to determine trajectories. A significant challenge comes from the inevitable presence of noise in the acquired images, which may adversely change the features of similar objects across multiple frames. Consequently, the computation of robust features is an important aspect of object detection. Features are representative of a wide array of object properties: color, frequency and distribution, shape, geometry, contours, or correlations within segmented objects. Nowadays, the most popular feature detection methods involve supervised learning. Features start out as groups of random values and are progressively refined using machine learning algorithms. Such approaches require appropriate training data and a careful selection of hyperparameters, often through trial-and-error. However, many results from the related literature show that supervised classification and regression methods offer the best results both in terms of accuracy and robustness to affine transformations, occlusion, and noise.

2.1. Methods Using Neural Networks

In terms of classifying objects from images, neural networks have seen a steady rise in popularity in recent years, particularly the more elaborate and complex convolutional and recurrent networks from the field of deep learning. Neural networks have the advantage of being able to learn important and robust features given training data that is relevant and in sufficient quantity. Considering that a significant percentage of automotive sensor data consists of images, convolutional neural networks (CNNs) are seeing widespread use in the related literature, for both classification and tracking problems. The advantage of CNNs over more conventional classifiers lies in the convolutional layers, where various filters and feature maps are obtained during training. CNNs are capable of learning object features by means of multiple complex operations and optimizations. The appropriate choice of network parameters and architecture can ensure that these features contain the most useful correlations that are needed for the robust identification of the targeted objects. While this choice is most often an empirical process, a wide assortment of network configurations exist in the related literature that are aimed at solving classification and tracking problems, with high accuracies claimed by the authors. Where object identification is concerned, in some cases the output of the fully-connected component of the CNN is used, whereas in other situations the values of the convolutional layers are exploited in conjunction with other filtering and refining methods.

2.1.1. Learning Features from Convolutional Layers

Many results from the related literature systematically demonstrate that convolutional features are more useful for tracking than other explicitly computed ones (Haar, Fused Histogram of Oriented Gradients (FHOG), color labeling). An example in this sense is [1], which handles MOT using combinations of values from convolutional layers located at multiple levels. The method is based on the notion that lower-level layers account for a larger portion of the input image and therefore contain more details from the identified objects. This makes them useful, for instance, for handling occlusion. Conversely, top-level layers are more representative of semantics and are useful in distinguishing objects from the background. The proposed CNN architecture uses dual fully-connected components, for higher and lower-level features, which handle instance-level and category-level classification (Figure 2 in [1]). The proper identification of objects, particularly where occlusion events occur, involves the generation of appearance models of the tracked objects. These often result from the appropriate processing of the features learned within convolutional layers.

In [2], the authors note that the output of the fully-connected component of a CNN is not suitable for handling infrared images. Their attempt to directly transfer CNNs pretrained with traditional images for use with infrared sensor data is unsuccessful, since only the information from the convolutional layers seem to be useful for this purpose. Furthermore, the layer data itself require some level of adaptation to the specifics of infrared images. Typically, infrared data offer much less spatial information than visual images. It is much more suited, for example, in depth sensors for gathering distances to objects, albeit at a significantly lower resolution compared to regular image acquisition. As such, convolutional layers from infrared images are used in conjunction with correlation filters to generate a set of weak trackers. This process provides response maps with regard to the targets' locations. The weak trackers are then combined in ensembles which form stronger response maps with a much greater tracking accuracy. The response map of an image is, generally, an intensity image where higher values indicate a change or a desired feature/shape/structure, as the original image is processed by an operator or correlation filter of some kind. By matching or fusing responses from multiple images within a video sequence, one could identify similar objects (i.e., the same pedestrian) across the sequence and subsequently construct their trajectories.

The potential of correlation filters is also exploitable for regular images. These have the potential to boost the information extracted from the activations of convolutional layers. In [3] the authors find that by applying appropriate filters to information drawn from shallow convolutional layers, a level of robustness similar to using deeper layers or a combination of multiple layers can be achieved. In [4], the authors also note the added robustness obtainable by post-filtering convolutional layers. By using particle and correlation filters, basic geometric and spatial features can be deduced for the tracked objects, which, together with a means of adaptively generating variable models, can be made to handle both simple and complex scenes.

An alternative approach can be found in [5], where discriminative correlation filters are used to generate an appearance model from a small number of samples. The overall approach involves feature extraction, post-processing, and the generation of response maps for carrying out better model updates within the neural network. Contrary to other similar results, the correlation filters used throughout the system are learned within a one-layer CNN, which eventually can be used to make predictions based on the response maps. Furthermore, residual learning is employed in order to avoid model degradation, instead of the much more frequently-used method of stacking multiple layers. Other tracking methods learn a similar kind of mapping from samples in the vicinity of the target object using deep regression [6,7], or by estimating and learning depth information [8].

The authors of [9] note that correlation filters have limitations imposed by the feature map resolution. They propose a novel solution where features are learned in a continuous domain, using an appropriate interpolation model. This allows for the more effective resolution-independent compositing of multiple feature maps, resulting in superior classification results.

Methods based on discriminative correlation filters are notoriously prone to excessive complexity and overfitting, and various means are available for optimizing the more traditional methods. The most noteworthy in this sense is [10], who employs efficient convolution operators, a training sample distribution scheme and an optimal update strategy in an attempt to boost performance and reduce the number of parameters. A promising result that demonstrates significant robustness and accuracy is [11], who use a CNN where the first set of layers are shared, as in a standard CNN. These layers then branch into multiple domain-specific ones. This approach has the benefit of splitting the tracking problem into subproblems which are solved separately in their respective layer sets. Each domain has its own training sequences and can be customized to address a specific issue, such as distinguishing a target with specific shape parameters from the background. A similar concept is exploited by [12], i.e., a network with components distinctly trained for a specific problem. In this case, multiple recurrent layers are used to model different structural properties of the tracked objects, which are incorporated into a parent CNN with the same purpose of improving accuracy and robustness. The Recurrent Neural Network (RNN) layers generate what the authors refer to as "structurally-aware feature maps" which, when combined with pooled versions of their non-structurally aware counterparts, significantly improve the classification results.

2.1.2. High-Level Features, Occlusion Handling, and Feature Fusion

Appearance models offer high-level features that are also used to account for occlusion in much simpler and efficient systems. In [13], appearance descriptors are compounded to form an appearance space. With properly-determined metrics, observations having a similar appearance are identified using a nearest-neighbor approach. Switching from image-space to an appearance space seems to effectively handle occlusions, reducing their negative impact at a negligible performance cost.

A possible alternative to appearance-based classification is the use of template-based metrics. Such an approach uses a reference region of interest (ROI) drawn from one or multiple frames and attempts to match it in subsequent frames using an appropriately-constructed metric. Template-based methods perform well for partial detections, thereby accounting for occlusion and/or noise. This is because the template need not be perfectly or completely matched for a successful detection to occur. An example of a template-based method is provided by [14], which involves three CNNs, one for template generation, one dedicated to region searching and one for handling background areas. The method is somewhat similar to what could be achieved by a generative adversarial network (GAN). A "searcher" network attempts to fit multiple subimages within the positive detections provided by the template component while simultaneously attempting to maximize the distance to the negative background component. The candidate subimages generated by the three components are fed through a loss function that is designed to favor candidates closer to template regions than to background ones. Performance-wise, such an approach is claimed to provide impressive framerates and care should be taken when using template or reference-based methods. These are generally suited for situations where there is no significant variation in the overall tone of the frames. Such methods have a much higher failure rate when, for instance, the lighting conditions change during tracking. An example of this phenomenon is when the tracked object moves from a brightly-lit area to a shaded one.

An improvement on the use of appearance and shared tracking information is provided by [15] in the form of a CNN-based single object tracker that generates and adapts the appearance models for multi-frame detection (Figure 3 in [15]). The use of pooling layers and shared features accounts for drift effects caused by occlusion and inter-object dependency. A spatial and temporal attention mechanism is responsible for dynamically discriminating between training candidates based on the level of occlusion. Training samples are weighted based on their occlusion status, which optimizes the training process where both classification accuracy and performance are concerned. Generally speaking, pooling operations have two important effects: on the one hand, the subimage of the feature map is increased, since a pooled feature map contains information from a larger area of the originating image; on the other hand, the reduced size of a pooled map means fewer computational resources are required to process it, which improves performance. The major downside of pooling is that spatial positioning is further diluted with each additional layer. Multiple related papers exploit the so called "ROI pooling", which commonly refers to a pooling operation being applied to the bounding box of an identified object. The resulting reduced representation will hopefully be more robust to noise and geometric variations across multiple frames. ROI pooling is successfully used by [16] to improve the performance of their CNN-based classifier. The authors observe that positioning cues are adversely affected by pooling. A potential solution is to reposition the misaligned ROIs via bilinear interpolation. This reinterpretation of pooling is referred to as "ROI align". The gain in performance is significant, while the authors demonstrate that the positioning of the ROIs is stabilized.

Tracking stabilization is fundamental in automotive application, where effects such as jittering, camera shaking, and spatial/temporal noise commonly occur. Occlusion handling plays an important role in ensuring ROI stability and accuracy. Some authors handle this topic extensively, such as [17], who propose a deep neural network for tracking occluded body parts, by processing features extracted from a VGG19 network. Some authors use different interpretations of the feature concept, adapted to the specifics of autonomous driving. Reference [18] creates custom feature maps by encoding various properties of the detections in raster images (bounding boxes, positions, velocities, accelerations). These images are sent through a CNN that generates raster features that the authors demonstrate to provide more reliable correlations and more accurate trajectories than using features derived directly from raw data.

The idea of tracking robustness and stability is sometimes solvable using image and object fusion. The related methods are referred to as being "instance-aware". This concept means that a targeted object is matched across the image space and across multiple frames by fusing identified objects with similar characteristics. Reference [19] proposes a fusion-based method that uses single-object tracking to identify multiple candidate instances. Subsequently, it builds target models for potential objects by fusing information from detection and background cues. The models are updated using a CNN, which ensures robustness to noise, scaling, and minor variations of the targets' appearance. As with many other related approaches, an online implementation offloads most of the processing to an external server leaving the embedded device from the vehicle to carry out only minor, frequent tasks. Since quick reactions of the system are crucial for safe vehicle operation, performance and a rapid response of the underlying software is essential, which is why the online approach is popular in this field. Fusion methods are also applied for multimodal inputs, such as in [20], who propose a model based on a convolutional autoencoder to obtain features from a combination of multiple sensor sources, in order to account for improved environment perception.

Also in the context of ensuring robustness and stability, some authors apply fusion techniques to information extracted from convolutional layers. It has been previously mentioned that important correlations can be drawn from deep and shallow layers that can be exploited together for identifying robust features in the data. This principle is used for instance in [21]. In order to ensure robustness and performance, various features extracted from layers in different parts of a CNN are fused to form stronger characteristics that are affected to a lesser degree by noise, spatial variations, and perturbations in the acquired images. The identified relationships between CNN layers are exploited in order to account for lost spatial information that occurs in deeper layers. The method is claimed to have improved accuracy over the state-of-the-art of the time, which is consistent with the idea of ensuring robustness and low failure rates. Deeper features are more consistent and allow for stronger classification, while shallow features compensate for the detrimental effects of filtering and pooling. This allows for deep features to be better integrated into the spatial context of the images. On a similar note, in [22] features from multiple layers that individually constitute weak trackers are combined to form a stronger one, by means of a hedging algorithm. The practice of using multiple weak methods into a more effective one has significant potential and is based on the principle that each individual weak component contains some piece of meaningful information on the tracked object, while also having useless data mostly found in the form of noise. By appropriately combining the contributions of each weak component, a stronger one can be generated. As such, methods that exploit compound classifiers typically show robustness to variances of illumination, affine transforms, or camera shaking. The downside of such methods is that multiple groups of weak features are needed, which causes penalties in real-time response. Additionally, the fusion algorithm has its own performance-impacting overhead.

Alternative approaches exist which mitigate this to some extent. For example, the use of multiple sensors directly supplies the necessary data, as opposed to relying on multiple features computed from the same camera or pair of cameras. An example in this direction is provided in [23], where an image gallery from a multi-camera system is fed into a CNN in an attempt to solve multi-target multi-camera tracking and target re-identification problems. For correct and consistent re-identification, an observation in a specific image is matched against several ones from other cameras using correlations as part of a similarity metric. Such correlation among images from multiple cameras are learned during training and subsequently clustered to provide a unified agreement between them. Eventually, after a training process that exploits a custom triplet loss function, features are obtained to be further used in the identification process. In terms of performance, the method boasts substantial accuracy considering the multi-camera setup. The idea of compositing robust features from a multi-faceted architecture is further exploited in works such as [24]. A triple-net setup is used to generate features that account for appearance, spatial cues, and temporal consistency.

2.1.3. Ensuring Temporal Coherence

One of the most significant challenges for autonomous driving is accounting for temporal coherence in tracking. Nearly all automotive scenarios involve video and motion across multiple frames. Consequently, handling image sequence data and accounting for temporal consistency are key factors in ensuring successful predictions, accuracy, and reliability. Essentially, solving temporal tracking is a compound problem. On the one hand, it involves tracking objects in single images considering all the problems induced by noise, geometry and the lack of spatial information. On the other hand, it should ensure that the tracking is consistent across multiple frames. That is, assigning correct IDs to the same objects in a continuous video sequence.

This presents a lot of challenges, for instance when objects become occluded in some frames and are exposed in others. In some cases, the tracked objects suffer affine transformations across frames, of which rotation and shearing are notoriously difficult to handle. Additionally, the objects may change shape due to noise, aliasing and other acquisition-related artifacts that may be present in the images. Video is rarely if ever acquired at "high enough" resolution and is in many cases in some lossy compressed format. As such, the challenge is to identify features that are robust enough to handle proper classification and to ensure temporal consistency considering all pitfalls associated with processing video data. This often involves a "focus and context" approach: key targets are identified based on features learned from current frames and from the context of the tracked object. Processing a key frame in a video sequence provides the focus, while the information from previous frames form the context.

For this type of problem, one popular approach is to integrate recurrent components into the classifier, which inherently account for the context provided by a set of elements from a sequence. Neural networks with recurrent layers, such as long short-term memory (LSTM) and gated recurrent units (GRU), are commonly employed in the related literature for the processing of temporal data. When training and exploiting recurrent layers to classify sequences, the results from one frame carry over to the computations that take place for subsequent frames. As such, when processing the current frame, resulting detections also account for what was found in previous frames. For automotive applications, one advantage of neural networks is that they can be trained off-site, while the resulting model can be ported to the embedded device in the vehicle where predictions and tracking can occur at usable speeds. While training a recurrent network or multiple collaborating networks can be a lengthy process, forward-propagating new data can happen quite fast, making these algorithms a good choice for real-time tracking.

Another concept that consistently appears in the related literature is "historical matching". The idea is to carry over part of the characteristics of tracked objects across multiple frames, by building an affinity model from shape, appearance, positional, and motion cues. This is achieved in [25] using dual CNNs with multistep training, which handle appearance matching using various filtering operations and linearly composing the resulting features across multiple timestamps. The notion of determining and preserving affinity is also exploited in [26] where data consisting of frame pairs several timestamps apart are fed into dual VGG networks (models based on convolutional neural networks with an architecture designed for image recognition tasks). The resulting features are permuted and incorporated into association matrices that are further used to compute object affinities. This approach has the benefit of partially accounting for occlusion using only a limited number of frames, since the affinity of an object that is partially occluded in one frame may be preserved if it appears fully in the pair frame.

Ensuring the continuity of high-level features such as appearance models is not a trivial task, and multiple solutions exist. For example [27] uses a CNN modified with a discriminative component intended to correct for temporal errors that may accumulate in the appearance of tracked objects across multiple frames. Discriminative network behavior is also exploited in [28] where selectively trained dual networks are used to generate and correlate appearance with a motion stream. Also, decomposing the tracking problem into localization and motion using multiple component networks is a frequently-encountered solution, further exploited in works such as [29,30]. As such, using two networks that work in tandem is a popular approach and seems to provide accurate results throughout the available literature (Figure 2 in [30]).

In this context, siamese convolutional networks have the ability to learn similarities by comparing features from dual-stream convolutional layers. One example is provided by [31], where appearance and motion are handled by a combination of CNNs that work together within a unified framework. The motion component uses spotlight filtering over feature maps that result from subtracting features drawn from dual CNNs. A space-invariant feature map is then generated using pooling and fusion operations. The other component handles appearance by filtering and fusing features from a different arrangement of convolutional layers. Data from ROIs in the acquired images are passed to both components. Motion responses from one component are correlated with appearance responses from the other. Both components produce feature maps that are composed together to form space- and motion-invariant characteristics to be further used for target identification. As such, a common functionality of such models is to feed the similarities learned among different inputs to subsequent network components that carry out the classification/detection task [32,33].

Some authors take this concept further by employing several network components [34], each of which contributes features exhibiting specific and limited correlations. When joined together, the features form a complete appearance model of the tracked objects. Other approaches map network components to flow graphs, the traversal of which enables optimal cost-function and feature learning [35]. It is worthy of noting that the more complicated the architecture of the classifier, the more elaborate the training process and the poorer the performance. A careful balance should therefore be reached between the complexity of the classifier, the completeness of the resulting features, and the amount of processing and training data needed to produce high-accuracy results. All this should involve a computational cost consistent with the needs of automotive applications. For instance, in [36] the authors propose a lightweight solution where the feature extractor consists in only two convolutional layers, while a careful selection of motion patterns solves the data association problem.

In [37], the idea of object matching from frame pairs is further explored using a three-component setup: a siamese network configuration handles single object tracking and generates short-term cues in the form of tracklet images, while a modified version of GoogleNet Inception-v4 generates re-identification features from multiple tracklets. The third component is based on the idea that there may be a large overlap in the previously-computed features, which are consequently treated as switcher candidates. As a result, a switcher-aware logic handles the situation where IDs of different objects may be interchanged during frame sequences mainly as a result of partial occlusion.

As the difficulty of the tracking problem increases, so does the need to design systems capable of learning increasingly useful and robust features. In this sense, many solutions consist in models that extract features expressing increasingly abstract concepts, which have the potential for greater generalization. Therefore, a lot of effort is directed toward identifying object features that are higher-level, more abstract representations of how the object fits within the overall context of the acquired video sequence. Examples of such concept are the previously-mentioned "affinity"; another is "attention", where some authors propose neural-network-based solutions for estimating attention and generating attention maps. Reference [15] computes attention features that are spatially and temporally sound using an arrangement of ROI identification and pooling operations. Reference [38] uses attention cues to handle the inherent noise from conventional detection methods, as well as to compensate for frequent interactions and overlaps among tracked targets. A two-component system handles noise and occlusion, and produces spatial attention maps by matching similar regions from pair frames. Temporal coherence is achieved by weighing observations across the trajectory differently, thereby assigning them different levels of attention. This process results is filtering criteria used to successfully account for similar observations while eliminating dissimilar ones. Another noteworthy contribution is [39], where attention maps are generated using reciprocative learning. The input frame is sent back-and-forth through several convolutional layers: in the forward propagation phase classification scores are generated, while the back-propagation produces attention maps from the gradients of the previously-obtained scores. The computed maps are further used as regularization terms within a classifier. The advantage of this approach is its simplicity compared to other similar ones. The authors claim that their method for generating attention features ensures long-term robustness. Other methods that use frame pairs and no recurrent components do not seem to work as well for very long-term sequences. Recently, attention mechanisms have been gaining significant ground for solving temporal consistency problems, since they allow the underlying model the freedom to weigh selective portions of a time-based sequence. Other noteworthy examples of works where attention mechanisms are incorporated into a CNN-based detector are [40,41].

2.1.4. LSTM-Based Methods

Generally, methods that are based on non-recurrent CNN-only approaches are best suited to handle short scenes where quick reactions are required in a brief situation that can be captured in a limited number of frames. Various literature studies show that LSTM-based methods have more potential to ensure the proper handling of long-term dependencies while avoiding various mathematical pitfalls. One example in this sense is the "vanishing gradient" problem, which in practice manifests as a mis-trained network resulting in drift effects and false positives. Furthermore, handling long-term dependencies means having to deal with occlusions to a greater extent than in shorter term scenarios.

Most approaches combine various classifiers that handle spatial and shape-based classification with LSTM components that deal with temporal coherence. An early example of an RNN implementation is [42], which uses an LSTM-based classifier to track objects in time, across multiple frames (Figure 1 in [42]). The authors demonstrate that an LSTM-based approach is better suited to removing and reinserting candidate observations to account for objects that leave/reenter the visible area of the scene. This provides a solution to the track initiation and termination problem based on data associations found in features obtained from the LSTM layers. This concept is exploited further by [43] where various cues are determined to assess long-term dependencies using a dual LSTM network. One LSTM component tracks motion, while the other handles interactions, and the two are combined to compute similarity scores between frames. The results show that using recurrent components to handle lengthy sequences produces more reliable results than other methods based on frame pairs. Some implementations using LSTM layers focus on tracking-while-driving problems, which pose additional challenges compared to most established benchmarks using static cameras. As an alternative to solutions that involve creating models of vehicle behavior, Reference [44] circumvent the need for vehicle modeling by directly inputting sensor measurements into an LSTM network to predict future vehicle positions and to analyze temporal behavior. A more elaborate attempt is [45] where instead of raw sensor data, the authors establish several maneuver classes and feed maneuver sequences to LSTM layers in order to generate probabilities for the occurrence of future maneuver instances. Eventually, multiple such maneuvers can be used to construct the trajectory and/or anticipate the intentions of the vehicles.

Furthermore, increasing the length of the sequence increases accuracy and stability over time, up to a certain limit where the network saturates and no longer improves. A solution to this problem would be to split the features into multiple sub-features, followed by reconnecting them to form more coherent long-term trajectories. This is achieved in [46] where a combined CNN and RNN-based feature extractor generates tracklets over lengthy sequences. The tracklets are split on frames that contain occlusions. A recombination mechanism based on gated recurrent units (GRUs) recombines the tracklet pieces according to their similarities, followed by the reconstruction of the complete trajectory using polynomial curve fitting.

Some authors do further modifications to LSTM layers to produce classifiers that generate abstract high-level features, such as those found in appearance models. A good example in this sense is [47] where LSTM layers are modified to do multiplication operations and use customized gating schemes between the recurrent hidden state and the derived features. The newly-obtained LSTM layers are better at producing appearance-related features than conventional LSTMs, which excel at motion prediction. Where trajectory estimation is concerned, LSTM-based methods exploit the gating that takes place in the recurrent layers, as opposed to regular RNNs, which pass candidate features into the next recurrent iteration without discriminating between them. The filters inherently present in gated LSTMs have the potential to eliminate unwanted feature candidates which may represent unwanted trajectory paths. Candidates which eventually lead to correctly-estimated motion cues are maintained. Furthermore, LSTMs demonstrate an inherent capability to predict trajectories that are interrupted by occlusion events or by reduced acquisition capabilities. This idea is exploited in order to find solutions to the problem of estimating the layout of a full environment from limited sensor data, a concept referred to in the related literature as "seeing beyond seeing" [48]. Given a set of sensors with limited capability, the idea is to perform end-to-end tracking using raw sensor data without the need to explicitly identify high-level features or to have a pre-existing detailed model of the environment. In this sense, recurrent architectures have the potential to predict and reconstruct occluded parts of a particular scene from incomplete or partial raw sensor output. The network is trained with partial data and it is updated through a mapping mechanism that makes associations with an unoccluded scene. Subsequently, the recurrent layers make their own internal associations and become capable of filling in the missing gaps that the sensors have been unable to acquire. Specifically, given a hidden state of the world that is not directly captured by any sensor, an RNN is trained using sequences of partial observations in an attempt to update its belief concerning the hidden parts of the world. The resulting information is used to "unocclude" the scene that was initially only partially perceived through limited sensor data. Upon training, the network is capable of defining its own interpretation of the hidden state of the scene. The previously-mentioned result is elaborated upon by a group that includes the same authors [49]. A similar approach previously applied in basic robot guidance is extended for use in assisted driving. In this case, more complex information can be inferred from raw sensor input, in the form of occupancy maps. Together with a deep network-based architecture, these allow for predicting the probabilities of obstacle presence even in occluded portions within the field of view. In [50], the idea of using LSTM layers to process sensor data is depicted by modeling actor trajectories and activities based on the output of an arrangement on inertial sensors. The proposed neural network learns correlations among sensor outputs and consequently forms an inertial odometry model.

In more recent studies, authors tend to add supplementary processing stages to their LSTM-based models. This additional effort seems to stem from the need to generate and incorporate an increasingly-refined and abstract array of features into the tracking process. As tracking scenarios increase in complexity, the resulting problem space increases in size and dimensionality. This motivates the need for extending an LSTM-centered model by incorporating it into a broader system. An example of such an approach is [51], where sequential dependencies are handled by LSTM layers as in commonly the case in such works. While relying on convolutional feature maps in the initial phases of the tracking pipeline, there is an additional mechanism for preparing selection proposals for the LSTM layers to process. Additionally, the common problem of feature inadequacy and class imbalance in the learning phase is handled by a GAN-based stage where the candidate samples are augmented. It is worth mentioning that, as more and more layers of different types are added to such a system, the reliability of the selected features may increase together with robustness to potential biases in the training data. However, at the same time, there is the risk that training and validating such a system may become a tedious, time consuming task. As the complexity increases, so does the need for extending the training data set and supplement the required computational resources. Other efforts in the direction of producing more usable features involve determining pedestrian intention as suggested by [52]. In this scenario, an LSTM model is used in conjunction with an intention filter to select suitable trajectory offset hypotheses so as to add to the reliability of the predicted result. The use of intention as a defining concept for features is also explored by [53], who enhance LSTM cells by introducing additional speed and correlation components. These components serve to model the more complex interactions required to define intention. In [54], instead of refining feature candidates using additional mechanisms, the authors choose to change the representation of the respective features. Specifically, LSTM layers are reconfigured and repurposed to handle multidimensional hidden states as opposed to the 1D vectors used traditionally. This increases the ability of such layers to accurately model spatial and temporal interactions among pedestrians. Conversely, in [55] the authors adapt an arrangement of LSTM layers to process sparse 3D data structures as opposed to changing internal data representation. The authors choose to model the interactions among pedestrians using graphs and, consequently, graph convolutional networks. Such systems still rely on LSTM layers to encode temporal dependencies. The spatial and sequence-related relationships among the tracked actors (represented as nodes) is modeled by determining connections in the form of graph edges [56,57].

2.1.5. Miscellaneous Neural Network-Based Methods

An interesting alternative to conventional deep learning architectures is the use of GANs, as demonstrated in [58]. GANs train generative models and filter their results using a discriminative component. GANs are notoriously difficult to train, which is one of the reasons why they see seldom use in the related literature. In terms of tracking, GANs alleviate the need to compute expensive appearance features and minimize the fragmentation that typically occurs in more conventional trajectory prediction models. A generative component produces and updates candidate observations, of which the least updated are eliminated. The generative-discriminative model is used in conjunction with an LSTM component to process and classify candidate sequences. This approach has the potential to produce high-accuracy models of human behavior, especially group behavior. At the same time, it is significantly more lightweight than previously-considered CNN-based solutions.

Another "outlier" solution in the related literature is [59], one of the few efforts involving reinforcement learning for MOT applications. The proposed model is split into two parts: a predictive component based on a CNN, which treats pedestrian detections as agents and determines the displacement of a target agent from its initial location; a decision network, which uses the resulting predictions and detections within a deep reinforcement learning network where the actions among the agents and their environment are rewarded so as to maximize their shared utility. Consequently, the collaborative interactions of multiple agents are exploited in order to simultaneously detect and track them more effectively. Other driver-centric reinforcement learning-based solutions determine driving rules for collision avoidance by weighing vehicle paths against potential pedestrian trajectories [60].

2.2. Other Techniques

While the current state-of-the art methods for MOT are mostly neural network-based, there also exist a multitude of other approaches which exploit more traditional, unsupervised means of providing reliable tracking. Neural networks gained popularity in recent years due in no small part to the availability of more powerful hardware, particularly GPUs, which allowed for training models capable of handling realistic scenarios in a reasonable amount of time. Neural networks however have the downside of needing vast amounts of reliable training data. Also, they require a lot of experimentation and trial-and-error before the right design and hyperparameter set is found for a particular scenario. There are, however, situations where training data may not be readily available in sufficient quantity and variety. Such cases call for a more straightforward design and a more intuitive model that can provide reliable tracking without necessarily requiring supervised learning. Neural network models are harder to understand in terms of how they function, and, while as deterministic as their non-neural network-based counterparts, are less intuitive and meant for use in a "black-box" manner. This is where other, more transparent methods come into place.

The tracking problem can be formulated similarly to the neural-network case: given a set of observations/appearances/segmented objects in multiple video frames, the task is to develop a means of determining relationships among these elements across the frames and to come up with a means of predicting their path. Various authors formulate this problem differently, for instance some methods involve determining tracklets in each frame and then assembling object trajectories in a full video sequence by combining tracklets from all or some of the frames [61]. Traditional, non-NN-based approaches, especially non-supervised ones, generally formulate much more straightforward models. Some are based on a graph or flow-oriented interpretation of the tracked scene. Others rely on emitting hypotheses as to the potential trajectories of the tracked targets, or otherwise formulating some probabilistic approach to predicting the evolution of objects in time. It is worth noting that many of the more conventional, unsupervised algorithms from the related literature do not generalize the solution as well as a NN-based method. Consequently, they are usable in a limited number of scenarios, by comparison. Some works attempt to circumvent this problem using evolutionary algorithms as multicriteria optimization methods [62]. However, while capable of covering a significant portion of the problem space, such methods have the downside that the optimal trajectory needs to be periodically recalculated, which can hinder performance especially for on-board-only systems. Also, methods that attempt to account for temporal consistency do not handle time sequences as lengthy as, for instance, an LSTM network. The likely explanation is that an unsupervised method requires far more processing capabilities the more frame elements it is fed. In the case of a properly-trained neural network, the amount of computational resources required does not increase as much with the length of the associated sequence. However, in practice, especially on an embedded device as required in automotive tracking, porting a more conventional method may be more convenient in terms of implementation and platform compatibility than running a pre-trained NN model.

Another important aspect worth mentioning is that conventional methods are much more varied in terms of their underlying algorithms, as opposed to an NN-based architecture which features various arrangements of the same two or three neural network types, with additional processing of layer activations or outputs as the case may be. For this reason, we do not attempt to cover all the approaches ever developed for object tracking, but we rather focus on representative works featuring various successful attempts at MOT.

2.2.1. Traditional Algorithms and Methods Focusing on High-Performance

The Kalman filter is a popular method with many applications in navigation and control, particularly with regard to predicting the future path of an object, associating multiple objects with their trajectories, while demonstrating significant robustness to noise. Generally, Kalman-based methods are used for simpler tracking, particularly in online scenarios where the tracker only accesses a limited number of frames at a time, possibly only the current and previous ones. An example of the use of the Kalman filter is [63], where a combination of the aforementioned filter and the Munkres algorithm as the min-cost estimator is used in a simple setup focusing on performance. The method requires designing a dynamic model of the tracked objects' motion, and is much more sensitive to the type of detector. However the proper parameters are established, the simplicity of the method allows for significant real-time performance.

Similar methods are frequently used in simple scenarios where a limited number of frames are available and the detections are accurate. In such situations, the simplicity of the implementations allows for quick response times even on low-spec embedded client devices. In the same spirit of providing an easy, straightforward method that works well for simple scenarios, Reference [64] provides an approach based on bounding-box regression. Given multiple object bounding boxes from a set of ordered frames, the authors use a regression model to predict the positions of the objects' bounding boxes in following frames. An important restriction of such an approach is that it only successfully detects targets that move only slightly across consecutive frames, making it reliable in scenarios where the frame rate is high enough and relatively stable. Furthermore, a reliable detector is a must in such situations, and crowded scenes with frequent occlusion events are not handled properly. As with the previous approach, this is well suited for easy cases where robust image acquisition is available and performance and implementation simplicity are a priority. Unfortunately, noisy images are fairly common in automotive scenarios where, for efficiency and cost reasons, a compromise may be made in terms of the quality and performance of the cameras and sensors. It is often desirable that the software be robust to noise so as to minimize the hardware costs.

In Reference [65], tracking is done by a particle filter for each track. The authors use the Munkres assignment for bounding boxes within consecutive images for each track. A cost matrix is then generated based on the associations made among bounding boxes from current and previous images. Specifically, the cost of associating two bounding boxes is determined from the Euclidean distance between the centers of the boxes, as well as their size variation. This approach is simple to implement, but the assignment algorithm has an $O(n^3)$ complexity, which is likely too high for real-time tracking.

Various attempts exist for improving noise robustness while maintaining performance, for example in [66]. In this case, the lifetime of tracked objects is modeled using a Markov Decision Process (MDP). The policy of the MDP is determined using reinforcement learning, whose objective is to learn a similarity function for associating tracked objects. The positions and lifetimes of the objects are modeled using transitions between MDP states. Reference [67] also use MDPs in a more generalized scheme, involving multiple sensors and cameras and fusing the results from multiple MDP formulations. Note that Markov models can be limiting when it comes to automotive tracking, since a typical scene with multiple interacting targets does not exhibit the Markov property where the current state only depends on the previous one. In this regard, the related literature features multiple attempts to improve reliability. Reference [68] propose an elaborate pipeline featuring multiview tracking, ground plane projection, maneuver recognition, and trajectory prediction. The method involves an assortment of approaches, which include Hidden Markov Models and Variational Gaussian mixture models. Such efforts show that an improvement over traditional algorithms involves sequencing together multiple different methods, each with its own role. As such, there is the risk that the overall resulting approach may be too fragmented and too cumbersome to implement, interpret, and improve properly.

Works such as [69] attempt to circumvent such limitations by proposing alternatives to tried-and-tested Markov models, in this case in the form of a system that determines behavioral patterns in an effort to ensure global consistency for tracking results. There are multiple ways to exploit behavior in order to guide the tracking process. For instance, a possible solution would rely on learning and minimizing/maximizing an energy function that associates behavioral patterns with potential trajectory candidates. This concept is exemplified by [70], who propose a method based on minimizing a continuous energy function aimed at handling the very large space of potential trajectory solutions. A limited, discrete set of behavior patterns impose limitations on the energy function. While such a limitation offers better guarantees that a global optimum will eventually be reached, it may not allow for a complete representation of the system.

An alternative approach which is also designed to handle occlusions is [71], where the divide-and-conquer paradigm is used to partition the solution space into smaller subsets, thereby optimizing the search for the optimal variant. The authors note that while detections and their respective trajectories can be extracted rather efficiently from crowded scenes, the presence of ambiguities induced by occlusion events may raise significant detection errors. The proposed solution involves subdividing the object assignment problem into subproblems, followed by a selective combination of the best features found within the subdivisions (Figure 3 in [71]). The number and types of the features are variable, thereby accounting for some level of flexibility for this approach. One particular downside is that once the scene changes, the problem itself also changes and the subdivisions need to reoccur and update, therefore making this method unsuitable for scenes acquired from moving cameras.

A similar problem is posed in [61], where it is also noted that complex scenes pose tracking difficulties due to occlusion events and similarities among different objects. This issue is handled by subdividing object trajectories into multiple tracklets and subsequently determining a confidence level for each such tracklet, based on its detectability and continuity. Actual trajectories are then formed from tracklets connected based on their confidence values. One advantage of this method in terms of performance is that tracklets can be added to already-determined trajectories in real-time as they become available without requiring complex processing or additional associations. Additionally, linear discriminant analysis is used to differentiate objects based on appearance criteria. The concept of appearance is more extensively exploited by [72], who use motion dynamics to distinguish between targets with similar features. They approach the problem by determining a dynamics-based similarity between tracklets using generalized linear assignment. As such, targets are identified using motion cues, which are complementary to more well established appearance models. While demonstrating adequate performance and accuracy, it is worth mentioning that motion-based features are sensitive to camera movement and are considerably more difficult to use in automotive situations. Motion assessment metrics that work well for static cameras may be less reliable when the cameras are in motion and image jittering and shaking occur.

The idea of generating appearance models using traditional means is exemplified in [73], who use a combination appearance models learned using a regularized least squares framework and a system for generating potential solution candidates in the form of a set of track hypotheses for each successful detection. The hypotheses are arranged in trees, each of which are scored and selected according to the best fit in terms of providing usable trajectories. An alternative to constructing an elaborate appearance model is proposed by [74], who directly involve the shape and geometry of the detections within the tracking process, therefore using shape-based cost functions instead of ones based on pixel clusters. Furthermore, results focusing on tracking-while-driving problems may opt for a vehicle behavior model, or a kinematic model, as opposed to one that is based on appearance criteria. Examples of such approaches are [75–77], where the authors build models of vehicle behavior from parameters such as steering angles, headings, offset distances, and relative positions. Note that kinematic and motion models are generally more suited to situations where the input consists in data from radar, Light Detection and Ranging (LiDAR) or Global Positioning Systems (GPS), as opposed to image sequences. In particular, attempting to reconstruct visual information from LiDAR point clouds is not a trivial task and may involve elaborate reconstruction, segmentation and registration preprocessing before a suitable detection and tracking pipeline can be designed [78].

Another class of results from related literature follows a different paradigm. Instead of employing complex energy minimization functions and/or statistical modeling, other authors opt for a simpler, faster approach that works with a limited amount of information drawn from the video frames. The motivation is that in some cases the scenarios may be simple enough that a straightforward method that alleviates the need for extended processing may prove just as effective as more complex and elaborate counterparts. An example in this direction is [79] whose method is based on scoring detections by determining overlaps between their bounding boxes across multiple consecutive frames. A scoring system is then developed based on these overlaps and, depending on the resulting scores, trajectories are formed from sets of successive overlaps of the same bounding boxes. Such a method does not directly handle crowded scenes, occlusions or fast moving objects whose positions are far apart in consecutive frames, however it may present a suitable compromise in terms of accuracy in scenarios where performance is detrimental and the embedded hardware may not allow for more complex processing. This is in contrast to high-performance methods that use on-board hardware to provide a lot of the information required for tracking, therefore reducing the reaction time of the underlying system in high-speed scenarios [80]. An additional important consideration for this type of problem is how the tracking method is evaluated.

Most authors use a common, established set of benchmarks which, while having a certain degree of generality, cannot cover every situation that a vehicle might be found in. As such, some authors such as [81] devote their work to developing performance and evaluation metrics and data sets, which allow for covering a wide range of potential problems that may arise in MOT scenarios. As such, the choice in the method used for tracking is as much a consequence of the diversity of situations and events claimed to be covered by the method, as it results from the evaluation performed by the authors. For example, as was the case for NN-based methods, most evaluations are done for scenes with static cameras, which are only partly relevant for automotive applications. The advantage of the methods presented thus far lies in the fact that they generally outperform their counterparts in terms of the required processing power and computational resources, which is a plus for vehicle-based tracking where the client device is usually a low-power solution. Furthermore, some methods can be extended rather easily, as the need may be, for instance by incorporating additional features or criteria when assembling trajectories from individual detections, by finding an optimizer that ensures additional robustness, or, as is already the case with some of the previously-mentioned papers, by incorporating a light-weight supervised classifier in order to boost detection and tracking accuracy. Additionally, the problem of false or malicious information from other traffic participants (for example in a multi-vehicle situation) has the potential to affect the accuracy of such methods. One proposed solution to this issue is to cluster the observations drawn from cooperative tracking according to their reliability and their potential to adversely affect the tracking results [82].

2.2.2. Methods Based on Graphs and Flow Models

A significant number of results from the related literature present the tracking solution as a graph search problem or otherwise model the tracking scene using a dependency graph or flow model. There are multiple advantages to using such an approach: graph-based models tailor well to the multi-tracking problem since, like a graph, it is formed from inter-related nodes each with a distinct set of parameter values. The relationships that can be determined among tracked objects or a set of trajectory candidates can be modeled using edges with edge costs. Graph theory is well understood and graph traversal and search algorithms can be widely found, with implementations readily available on most platforms. Likewise, flow models can be seen as an alternative interpretation of graphs, with node dependencies modeled through operators and dependency functions, forming an interconnected system. Unlike a traditional graph, data from a flow model progresses in an established direction that starts from initial components where acquired data are handled as input; the data then traverse intermediate nodes where they are processed in some manner and end up at terminal nodes where the results are obtained and exploited. Like graphs, flow models allow for loops that implement refinement techniques and in-depth processing via multiple local iterations.

Most methods that exploit graphs and flow models attempt to solve the tracking problem using a minimum path or minimum cost-type approach. An example in this sense is [83], where multi-object tracking is modeled using a network flow model subjected to min-cost optimization. Each path through the flow model represents a potential trajectory, formed by concatenating individual detections from each frame. Occlusion events are modeled as multiple potential directions arising from the occlusion node and the proposed solution handles the resulting ambiguities by incorporating pairwise costs into the flow network.

A more straightforward solution is presented by [84], who solve multi-tracking using dynamic programming and formulate the scenario as a linear program. They subsequently handle the large number of resulting variables and constraints using k-shortest paths. One advantage of this method seems to be that it allows for reliable tracking from only four overlapping low resolution low fps video streams, which is in line with the cost-effectiveness required by automotive applications.

Another related solution is [85], where a cost function is developed from estimating the number of potential trajectories as well as their origins and end frames. Then, the scenario is handled as a shortest-path problem in a graph, which the authors solve using a greedy algorithm. This approach has the advantage that it uses well-established methods, therefore affording some level of simplicity to understanding and implementing the algorithms.

In [86], a similar graph-based solution divides the problem into multiple subproblems by exploring several graph partitioning mechanisms and uses greedy search based on Adaptive Label Iterative Conditional Modes. Partitioning allows for successful disassociation of object identities in circumstances where said identities might be confused with one another. Also, methods based on solution space partitioning have the advantage of being highly scalable, therefore allowing fine tuning of their parameters in order to achieve a trade-off between accuracy and performance. Multiple extensions of the graph-based problem exist in the related literature, for instance, when multiple other criteria are incorporated into the search method. Reference [87] incorporate appearance and motion-based cues into their data association mechanism, which is modeled using a global graph representation and makes use of generalized minimum clique graphs to locate representative tracklets in each frame. Among other advantages, this allows for a longer time span to be handled, albeit for each object individually.

Another related approach is provided in [88], where the solution consists in a collaborative model which makes use of a detector and multiple individual trackers, whose interdependencies are determined by finding associations with key samples from each detected region in the processed frames. These interdependencies are further exploited via a sample selection method to generate and update appearance models for each tracker.

As extensions of the more traditional graph-based models that use greedy algorithms to search for suitable candidate solutions and update the resulting models in subsequent processing steps, some authors handle the problem using hypergraphs. These extend the concept of classical graphs by generalizing the role of graph edges. In a conventional graph an edge joins two nodes, while in a hypergraph edges are sets of arbitrary combinations of nodes. Therefore an edge in a hypergraph connects to multiple nodes, instead of just two as in the traditional case. This structure has the potential to form more extensive and complete models using a singular unified concept and to alleviate the need for costly solution space partitioning or subdivision mechanisms. Another use of the hypergraph concept is provided by [89], who build a hypergraph-based model to generate meaningful data associations capable of handling the problem of targets with similar appearance and in close proximity to one-another, a situation frequently encountered in crowded scenes. The hypergraph model allows for the formulation of higher-order relationships among various detections, which, as mentioned in previous sections, have the potential to ensure robustness against simple transformations, noise, and various other spatial and temporal inaccuracies. The method is based on grouping dense neighborhoods of tracklets hierarchically, forming multiple layers which enable more fine-grained descriptions of the relationships that exists in each such neighborhood. A related but much more recent result [90] is also based on the notion that hypergraphs allow for determining higher order dependencies among tracklets, but in this case the parameters of the hypergraph edges are learned using a structural support vector machine (SSVM), as opposed to being determined empirically. Trajectories are established as a result of determining higher order dependencies by rearranging the edges of the hypergraph so as to conform to several constraints and affinity criteria (Figure 1 in [90]). While demonstrating robustness to affine transforms and noise, such methods still cannot handle complex crowded scenes with multiple occlusions and, compared to previously-mentioned methods, suffer some penalties in terms of performance, since updating the various parameters of hypergraph edges can be computationally costly.

2.3. Discussion

Most of the results from the available literature focus on generating abstract, high-level features of the observations found in the processed images, since, generally, the more abstract the feature, the more robust it should be to transformations, noise, drift, and other undesired artifacts and effects. Most authors rely on an arrangement of CNNs where each component has a distinct role in the system, such as learning appearance models, geometric and spatial patterns, or learning temporal dependencies. It is worth noting that a strictly CNN-based method needs substantial tweaking and careful parameter adjustment before it can accomplish the complex task of consistent detection in space and across multiple frames.

LSTM-based architectures seem to show more promising results for ensuring long-term temporal coherence, since this is what they were designed for, while also being simpler to implement and train. For the purposes of autonomous driving, an LSTM-based method shows promise, considering that training should happen offline and that a heavily-optimized solution is needed to achieve a real-time response. Designing such a system also requires a fair amount of trial-and-error since currently there is no well established manner to predict which network architecture is suited to a particular purpose.

One particularly promising direction for automotive tracking are solutions that make use of limited sensor data and that are able to efficiently predict the surrounding environment without requiring a full representation or reconstruction of the scene. These approaches circumvent the need for lengthy video sequences, heavy image processing and the computation of complicated object features while being especially designed to handle occlusion and objects outside of the immediate field of view. As such, where automotive tracking is concerned, the available results from the state-of-the art seem to suggest that an effective solution would make use of partial data while being able to handle temporal correlations across lengthy sequences using an LSTM component.

Other, unsupervised approaches not reliant on neural networks offer a more straightforward model with an easier implementation. The downside often consists in the lack of generalization that such systems are capable of. The features required for detection and stability often have to be manually established, as opposed to supervised methods that can learn meaningful features on their own. The choice in terms of the most useful and reliant tracking model ultimately rests on many factors, among which we mention: the size and complexity of the problem, the availability of training data, the available computational resources, and, ultimately, the requirements in terms of accuracy, coherence, and stability.

We summarize our findings in Table 1, where we classify the works referenced in this Section according to the main method and the general approach used throughout. We choose to feature distinctive categories for solutions relying mainly on convolutional layers and on recurrent ones. Other methods are grouped into their own category. This choice is motivated by the fact that, as of yet, deep neural networks consistently show the most promise for the problems described throughout this Section. Many authors have found inventive and effective solutions to tracking problems using neural network-based models, since they offer the most robust features while being natively designed to solve focus-and-context problems in data sequences.

Table 1. Classification of tracking solutions from existing works.

Category	Strengths/Weaknesses	Overall Approach	Contributions
Methods based on convolutional neural networks	Strengths: - good at learning spatial, shape and geometric features - reduced computational load compared to regular neural networks - translation invariance Weaknesses: - cannot determine temporal dependencies in sequence data without additional mechanisms	Tracking from features directly learned by simple single-stream convolutional layers	[8,16,18]
		Dual-stream CNNs with data associations performed by additional model components	[6,19,28,37]
		Tracking using responses from convolutional features processed through correlation filters	[2–5,22,23]
		Multistream CNNs that determine similarities between multiple ROIs and target templates	[14]
		Models that determine appearance descriptors or that generate appearance representations from convolutional features	[1,13,25,27,32]
		Convolutional models that account for temporal coherence using a multi-network pipeline	[24]
		Tracking from features learned by fusing responses from dual stream convolutional layers (Siamese CNNs)	[26,30,31,38]
		Models that generate features from convolutional layers and use attention mechanisms for temporal coherence and matching	[15,39–41]
		Multi-stream convolutional layers used for detecting pedestrian poses	[17]
		Siamese networks combining convolutional features with complementary features from image processing	[33]
Methods based on recurrent neural networks	Strengths: - good for processing data series - good for learning temporal features and dependencies, and for ensuring temporal coherence Weaknesses: - cannot deduce interactions without additional mechanisms - generally more difficult to train	Models based on LSTM cell configurations that directly predict vehicle/obstacle occupancy	[44,49]
		Model for motion prediction based on vehicle maneuvers	[45]
		LSTM-based architectures that generate appearance and motion models and learn interaction information over extended sequences	[43,47,51]
		Multi-layer GRU-based architecture which splits and reconnects tracklets generated from convolutional features	[46]
		Basic RNN that encodes information from multiple frame sequences	[42,48]
		LSTM layers that focus on learning and interpreting actor intentions	[52,53]
		LSTM-based object detection and tracking adapted for sequences of higher-dimensional data	[55]
		LSTM models that encode relationships between actors using graph representations	[56,57]
		LSTM model that uses multidimensional internal representations of data sequences	[54]
Methods not relying on neural networks	Strengths: - designing a working model is more straightforward compared to neural networks - most are not training data-dependent Weaknesses: - traditional, classic methods do not model sequence dependencies as effectively as many RNN-based solutions	Models that represent and predict actor relationships using flow-networks and graphs	[83,85,89,90]
		Models relying on geometric representations, kinematics and pose estimations	[74,75,77,91]
		Models that ensure detection coherence using adaptive partitioning of the problem space	[71,86]
		Methods relying on Markov models and Markov decision processes	[66–68]
		Methods that build appearance models and/or use appearance similarity metrics	[72,73,87,88]
		Methods using a multi-stage tracking pipeline incorporating filtering, segmentation, clustering and/or data association	[76,78]
		Methods relying on lightweight filtering and optimization for high-speed high-performance applications	[63,64,80,84]

3. Trajectory Prediction Methods

In order to navigate through complex traffic scenarios safely and efficiently, autonomous cars should be able to predict the way in which the other traffic participants will behave in the near future with a sufficient degree of accuracy. The prediction of their motion is especially difficult because there are usually multiple interacting agents in a scene. Also, driver behavior is multi-modal, e.g., in different situations, from a certain common past trajectory, several different future trajectories may emerge. An autonomous car must also find a balance between the safety of people involved (its own passengers and other human drivers, or pedestrians) and choosing an efficient speed to reach its destination, without any perturbations to existing traffic. Predicting the future state of its environment is particularly important when the autonomous vehicle should act proactively, e.g., when changing lanes, overtaking other traffic participants and managing intersections [45].

Other difficulties come from the requirement that such a system must be prepared to handle rare, exceptional situations. However, because of the great number of possibilities involved, it should take into account only a reasonable subset of possible future scene evolutions [92] and often, it is important to identify the most probable solution [93].

Reasoning about the intentions of other drivers is a particularly helpful ability. Trajectory prediction can be treated on two different levels of abstraction. On the higher level, one can identify the overall intentions regarding a discrete set of possible actions, e.g., changing a lane or moving left or right in an intersection. On the lower level, one can predict the actual continuous trajectories of the road users [94].

Trajectory prediction needs to be precise but also computationally efficient [95]. The latter requirement can be satisfied by recognizing some constraints that reduce the size of the problem space. For example, the current speed of a vehicle affects its stopping time or the allowed curvature of its future trajectory so as to maintain the stability of the vehicle. Even if each driver has his/her own driving style, it is assumed that traffic rules will be obeyed, at least to some extent, and this will constrain the set of possible future trajectories [93].

A recent white paper [96] states that a solution for the prediction and planning tasks of an autonomous car may consider a combination of the following properties:

- *Predicting only a short time into the future.* Given the probabilistic nature of trajectory prediction, the farther one predicts into the future, the less certain the results become. Moreover, the probability distribution of the predicted trajectories disperses and thus becomes less useful altogether;
- *Relying on physics where possible.* Machine learning models, e.g., deep networks, can be used to predict trajectories, but they can suffer from approximation errors. Especially in simple, non-interactive scenarios, e.g., when the vehicles have constant speeds or accelerate/decelerate in a foreseeable manner, using physics-based extrapolations can provide more precise results. Also, each type of vehicle may have its own dynamics, so the identification of the vehicle class before prediction is a necessary initial step;
- *Considering whether road users obey traffic rules.* The autonomous car may plan as if the other traffic participants observed the imposed traffic rules, e.g., cars stopping at red lights or pedestrians not crossing the street in forbidden areas. However, defensive safety measures must be in place to prevent accidents with the so-called "vulnerable" road users;
- *Recognizing particular traffic situations.* For example, the behavior of traffic participants caught in a traffic jam differs from their behavior in flowing traffic.

Further, Reference [96] asserts that the self-driving car system should be prepared not only for the worst-case illegal behavior of the other traffic participants, but also for their worst-case legal behavior. The prediction system should be able to learn what the "reasonable" conduct of the other drivers may be in various circumstances. This may also depend on local conditions, such as different "driving cultures" in different countries.

3.1. Problem Description

To tackle the trajectory prediction task, one needs to have access to real-time data from sensors such as LiDAR, radar, or camera, and to a functioning system that allows detection and tracking of traffic participants in real-time. Examples of pieces of information that describe a traffic participant are: the bounding box, position, velocity, acceleration, heading, and yaw rate, i.e., the change in the heading angle. It may also be needed to have mapping data of the area where the ego car is driving, i.e., road and crosswalk locations, lane directions, and other relevant map-related information. Past and future positions are represented in an ego car-centric coordinate system. Also, one needs to model the static context with road and crosswalk polygons, as well as lane directions and boundaries [97]. An example of available information on which the prediction module can operate is presented in Figure 1.3 in [98].

More formally, prediction can be defined as reasoning about probable outcomes based on past observations [99]. Let X_t^i be a vector with the spatial coordinates of agent i at observation time t, with $t \in \{1, 2, ..., T_{obs}\}$, where T_{obs} is the present time step in the series of observations. The past trajectory of agent i is a sequence $X^i = \{X_1^i, X_2^i, ..., X_{T_{obs}}^i\}$. Based on the past trajectories of all agents, one needs to estimate the future trajectories of all agents, i.e., $\hat{Y}^i = \{\hat{Y}_{T_{obs}+1}^i, \hat{Y}_{T_{obs}+2}^i, ..., \hat{Y}_{T_{pred}}^i\}$.

It is also possible to first generate the trajectories in the Frenet coordinate system along the current lane of the ego vehicle, and then convert it to the Cartesian coordinate system [93]. The Frenet coordinate system is useful to simplify the motion equations when cars travel on curved roads. It consists of longitudinal and lateral axes, denoted as s and d, respectively. The curve that goes through the center of the road determines the s axis and indicates how far along the car is on the road. The d axis indicates the lateral displacement of the car. d is 0 on the center of the road and its absolute value increases with the distance from the center. Also, it can be positive or negative, depending on the side of the road.

3.2. Classification of Methods

There are several classification approaches presented in the literature regarding trajectory planning methods.

An online tutorial [100] distinguishes the following categories:

1. *Model-based approaches.* They identify common behaviors of the vehicle, e.g., changing lane, turning left, turning right, determining the maximum turning speed, etc. A model is created for each possible trajectory the vehicle can follow and then probabilities are computed for all these models. One of the simplest approaches to compute the probabilities is the autonomous multiple modal (AMM) algorithm. First, the states of the vehicle at times $t - 1$ and t are observed. Then the process model is computed at time $t - 1$ resulting in the expected states for time t. Then the likelihood of the expected state with the observed state is compared, and the probability of the model at time t is computed. Finally, the model with the highest probability is selected;

2. *Data-driven approaches.* In these approaches a black box model (usually a neural network) is trained using a large quantity of training data. After training, the model is applied to the observed behavior to make the prediction. The training of the model is usually computationally expensive and is made offline. On the other hand, the prediction of the trajectories, once the model is trained, is quite fast and can be made online, i.e., in real-time. Some of these methods also employ unsupervised clustering of trajectories using, e.g., spectral clustering or agglomerative clustering, and define a trajectory pattern for each cluster. In the prediction stage, the partial trajectory of the vehicle is observed, it is compared with the prototype trajectories, and the trajectory most similar to a prototype is predicted.

A survey [101] proposes a different classification based on three increasingly abstract levels:

1. *Physics-based motion models.* They apply the laws of physics to estimate the trajectory of a vehicle, by considering inputs such as steering, acceleration, weight, and even the coefficient of friction of the pavement in order to predict outputs such as position, speed, and heading. *Challenges* are related to noisy sensors and sensitivity to initial conditions. *Methods* include Kalman filters and Monte Carlo sampling. *Advantages.* Such models are very often used in the context of safety, as classic fail-safe methods when more sophisticated approaches, such as those using machine learning, are utilized for prediction. They can also be employed in situations that lack intricate interactions between road users. These models do not have to be very simple, as they can include detailed representations of vehicle kinematics, road geometry, etc. *Disadvantages.* They are usually appropriate for short-term predictions, e.g., less than a second, because they cannot predict maneuvers that aim to accomplish higher level goals, e.g., slowing down to prepare to turn in an intersection or because the vehicle in front is expecting a pedestrian to cross the street;
2. *Maneuver-based motion models.* They try to estimate the series of maneuvers that the cars perform on their way, but consider each vehicle to be deciding independently from the other traffic participants. These models attempt to identify such maneuvers as early as possible, and then assume that the maneuvers continue into the near future and estimate the corresponding trajectories. They use either prototype trajectories or maneuver intention estimation. *Challenges* are related to occlusions and the complexity of intentions. *Methods* include clustering, hidden Markov models, and reinforcement learning. *Advantages.* The identified maneuvers serve as a priori information or evidence that conditions future motion. Therefore, they tend to be more reliable than the physics-based ones and their predictions remain relevant for longer periods of time. *Disadvantages.* Because of the independence assumption, these models cannot handle the ways in which the maneuvers of a car influence the behavior of its neighbors. The interactions between traffic participants can be strong in scenarios with a high density of agents, e.g., intersections, possibly with specific priority rules. By ignoring the inter-agent interactions, these models tend to provide less accurate interpretations of such situations;
3. *Interaction-aware motion models.* This is the most general class of models, where the maneuvers of the vehicles are considered to be influenced by those of their neighboring road users. These models use prototype trajectories or dynamic Bayesian networks. *Challenges* refer to the ability to detect interactions and to a possible combinatorial explosion. *Methods* include coupled hidden Markov models, dynamically-linked hidden Markov models, and even rule-based systems. *Advantages.* The inclusion of inter-agent dependencies contributes to a better understanding of the situation. On the one hand, they facilitate longer-term predictions compared to physics-based models. On the other hand, they can be more reliable than maneuver-based models. *Disadvantages.* Because they often have to compute all possible trajectories, they can be inefficient from the computational point of view. Therefore, they may not be appropriate for real-time use cases.

The higher the level of abstraction of a prediction model, the more computationally expensive the model tends to become. Therefore, algorithms have been proposed that focus only on the most plausible trajectories. Also, the performance of the prediction methods are highly coupled with risk estimation possibilities. Therefore, the authors of [101] consider that successful approaches in this area should consider both vehicle motion modeling and risk estimation.

A classification somewhat similar with the previous two is mentioned in [102], which distinguishes the following motion prediction categories of methods:

1. *Learning-based motion prediction*: learning from the observation of the past movements of vehicles in order to predict the future motion;
2. *Model-based motion prediction*: using motion models;
3. *Motion prediction with a cognitive architecture*: trying to reproduce human behavior.

In the rest of this section, we present some specific approaches classified by their main prediction "paradigm", namely neural networks and other methods, most of which use some kind of stochastic representation of the agents' behavior in the environment. This is especially useful since some works use the same model to address different abstraction levels of the trajectory prediction task.

3.3. Methods Using Neural Networks

Many of the approaches presented in the literature that are based on neural networks use either recurrent neural networks (RNNs), which explicitly take into account a history composed of the past states of the agents, or simpler convolutional neural networks (CNNs). Other authors use conditional variational autoencoders (CVAEs) or more recent methods such as generative adversarial networks (GANs) and attention mechanisms.

A generative system is DESIRE [99], which has the goal of predicting the future locations of multiple interacting agents in dynamic (driving) scenes. It can handle the multi-modal nature of the prediction, i.e., for the same set of inputs, the predicted outputs may have several distinct values (a one-to-many mapping). It also takes into account the scene context and the interactions between traffic participants. It uses a single end-to-end neural network model, which the authors report to be computationally efficient. Using a deep learning framework, DESIRE can rank and refine the set of generated trajectories by considering the long-term future values, i.e., the sum of discounted rewards.

The corresponding optimization problem tries to maximize the potential future reward of the prediction, using the following mechanisms (Figure 2 in [99]):

1. *Diverse sample generation:* A conditional variational auto-encoder (CVAE) is used to capture the multi-modal nature of future trajectories. It uses stochastic latent variables which can be sampled to generate multiple possible future hypotheses for a single set of past information. The CVAE is combined with an RNN that encodes the past trajectories, to generate hypotheses using another RNN;
2. *Inverse optimal control (IOC)-based ranking and refinement:* After including the context and the interactions, the most likely trajectories are identified using potential future rewards, similar to inverse optimal control (IOC) or inverse reinforcement learning (IRL). The agents maximize long-term values. The authors believe that in this way the generalization capabilities of the model are improved and the model is more reliable for longer-term predictions. Since a reward function is difficult to design for general traffic scenarios, it is learned by means of IOC. The RNN model assigns rewards to each prediction hypothesis and assesses its quality based on the accumulated long-term rewards. In the testing phase, there are multiple iterations in order to obtain more accurate refinements of the future prediction;
3. *Scene context fusion:* This module aggregates the agent interactions and the context encoded by a CNN. Then this information is passed to an RNN scoring module which computes the rewards.

In [103], a method to predict the trajectories of the neighboring traffic participants is proposed using a long short-term memory (LSTM) network, with the goal of taking into account the relationship between the ego car and surrounding vehicles. The LSTM is a type of recurrent neural network (RNN) capable of learning long-term dependencies. Generally, an RNN has a vanishing gradient problem. An LSTM is able to deal with this through a forget gate, designed to control the information between the memory cells in order to store the most relevant previous data. The proposed method considers the ego car and four surrounding vehicles. It is assumed that drivers generally pay attention to the relative distance and speed with respect to the other cars when they intend to change a lane. Based on this assumption, the relative amounts between the target and the four surrounding vehicles are used as the input of the LSTM network. The feature vector x_t at time step t is defined by twelve features: lateral position of target vehicle, longitudinal position of target vehicle, lateral speed of target vehicle, longitudinal speed of target vehicle, relative distance between target and preceding vehicle, relative speed between target and preceding vehicle, relative distance between target and following vehicle, relative speed between target and following vehicle, relative distance between target and lead vehicle, relative speed between target and lead vehicle, relative distance between target and ego vehicle, and relative speed between target and ego vehicle. The input vector of the LSTM network is sequence data with x_t's for past time steps. The output is the feature vector at the next time step $t + 1$. A trajectory is predicted by iteratively using the output result of the network as the input vector for the subsequent time step.

In [44], an efficient trajectory prediction framework is proposed, which is also based on an LSTM. This approach is data-driven and learns complex behaviors of the vehicles from a massive amount of trajectory data. The LSTM receives the coordinates and velocities of the surrounding vehicles as inputs and produces probabilistic information about the future positions of the traffic participants on an occupancy grid map (Figure 1 in [44]). The proposed method is reported to have better prediction accuracy than Kalman filtering.

The occupancy grid map is widely adopted for probabilistic localization and mapping. It reflects the uncertainty of the predicted trajectories. In [44], the occupancy grid map is constructed by partitioning the range under consideration into several grid cells. The grid size is determined such that a grid cell approximately covers a quarter of a lane to recognize the movement of the vehicles on the same lane, as well as the lengths of the vehicles (Figure 3 in [44]).

When predictions are needed for different time ranges, e.g., $\Delta = 0.5, 1, 2$ s, the LSTM is trained independently for each time range. The LSTM produces the probability of occupancy for each grid cell. Let (x, y) be the identifier of a cell in the occupancy grid. Then the softmax layer in LSTM i computes the probability $P_o(i_x, i_y)$ for the grid element (i_x, i_y). Finally, the outputs of the n LSTMs are combined using $P_o(i_x, i_y) = 1 - \prod_{i=1}^{n} \left(1 - P_o^{(i)}(i_x, i_y)\right)$. The probability of occupancy $P_o(i_x, i_y)$ summarizes the prediction of the future trajectory for all n vehicles in the single map.

Alternatively, the same LSTM architecture can be used to directly predict the coordinates of a vehicle as a regression task. Instead of using the softmax layer to compute probabilities, the system can produce two real coordinate values x and y.

In [45], another LSTM model is described for interaction-aware motion prediction. Confidence values are assigned to the maneuvers that are performed by vehicles. Based on them, a multi-modal distribution over future motions is computed. More specifically, the model computes probabilities for each type of maneuver, based on six maneuver classes. The input to the LSTM is represented by the past positions of the ego car and its neighbors, and the geometry of the road lanes.

Social LSTM [104], used for predicting the trajectory of pedestrians, uses LSTM with a social pooling layer which allows neighbors, up to a certain distance, to exchange information. The hidden states of their corresponding LSTMs are pooled together and used as an input for the following prediction step.

Taking into account the time constraints of a real-time system, Reference [97] uses a simple feed-forward CNN architecture for the prediction task. The authors use an RGB image to represent the scene context. However, a vector of velocity, acceleration, and yaw rate can also be included. In this case, this vector is concatenated with the flattened output of the CNN. Then, these aggregated features are sent to a fully connected layer.

A similar approach is used in [18], which predicts multiple possible trajectories together with their probabilities. The context is also encoded as an image that is passed to a CNN. Given the raster image and the state estimates of agents at a time step, the CNN is used to predict a multitude of possible future state sequences, as well as the probability of each sequence.

As part of a complete software stack for autonomous driving, NVIDIA created a system based on a CNN, called PilotNet [105], which outputs steering angles given images of the road ahead. This system is trained using road images paired with the steering angles generated by a human driving a car that collects data. The authors identified the elements of the road image that have the greatest effect on the steering decision. It seems that in addition to learning the obvious features such as lane markings, edges of roads and other cars, the system learns more subtle features that would be hard to anticipate and program by engineers, e.g., bushes lining the edge of the road and atypical vehicle classes, while ignoring structures in the camera images that are not relevant to driving. This capability is derived from data without the need of hand-crafted rules.

In [94], the authors propose a learnable end-to-end model with a deep neural network that reasons about both high level behavior and long-term trajectories. Inspired by how humans perform this task, the network exploits motion and prior knowledge about the road topology in the form of maps containing semantic elements such as lanes, intersections, and traffic lights. The so-called IntentNet is a CNN that outputs three types of variables in a single forward pass: the detection scores for vehicle and background classes, the action probabilities corresponding to the discrete intentions, and bounding box regressions in the current and future time steps representing the intended trajectory. This design enables the system to propagate uncertainty through the different components and is reported to be computationally efficient.

A sequence-to-sequence CNN architecture is also used in [95] for an end-to-end trajectory prediction model. The authors say that the results are comparable to those of other, more complex approaches using LSTMs. Trajectory histories are embedded by means of a fully connected layer. Stacked convolutional layers are used to learn temporal dependencies in a consistent manner. Then, the features from the final convolutional layer are passed through a fully-connected layer to simultaneously generate all predicted positions. The authors report that the results when only one time step at a time is predicted are worse than the results when all future times are predicted at the same time.

The CoverNet model [106] uses a CNN in combination with a trajectory set generated from the input state containing, e.g., speed, acceleration, and yaw rate. The image features pass though some fully-connected layers and produce probabilities for each mode using softmax.

The Y-net model [107] uses the U-Net architecture [108] for the semantic segmentation of the input image. It also computes a distribution for the future trajectories, where the sampled points are clustered using k-means [109].

The TraPHic model [110] uses a hybrid LSTM-CNN network. The trajectory information is passed through LSTMs to construct three maps for the horizon, neighbors, and ego vehicle. The first two are further passed through different CNNs and concatenated with the ego car tensor. The resulting latent representations are passed through another LSTM to predict the ego trajectory.

The EvolveGraph approach [111] uses graphs to model the behavior of heterogeneous agents. It proposes an observation graph, fully connected, to represent the agents in the scene, and an interaction graph for the agent–agent and agent–context interactions. It also employs an encoder–decoder technique, with the encoder using softmax for edge classification and the decoder generating a Gaussian mixture distribution for prediction.

The TNT model [112] uses VectorNet [113], a hierarchical graph neural network, to encode the context of a scene, including road lanes and the position of the traffic signs, beside the trajectories of the agents. The generated set of trajectories is finally filtered to reject similar instances.

Generative approaches are also used. For example, PRECOG [114] follows the idea of identifying high-level goals and condition the predictions based on those. It employs CNNs and RNNs, and also a generative model where the latent variables stand for plausible behavior of the agents in a scene. Reference [115] uses a so-called "conditional flow" variational autoencoder (CF-VAE) that can handle multi-modal conditional distributions. PECNet [116] conditions the predicted trajectories on their endpoints with a conditional variational autoencoder (CVAE) and proposes the "truncation trick", i.e., truncating the sampling distribution with a smaller standard deviation for cases with a few samples to increase the diversity for multi-modal prediction.

Several trajectory prediction models employ Generative Adversarial Networks (GANs) [117]. This architecture has two components: a generator and a discriminator. Instead of training the generator model to directly match the desired data distribution, in this case the generator is trained so that it increases the error rate of the discriminator. In turn, the discriminator tries to distinguish whether a given sample belongs to the true data distribution or is generated by the generator. Both components are engaged in a competition to outsmart the other one, and from this process the generator learns to generate data that resemble the true data distribution. In the domain of trajectory prediction, Social GAN [118] uses a GAN where the generator is composed of an LSTM-based encoder, a context-pooling module, and an LSTM-based decoder. The discriminator uses LSTMs as well.

Other models employ GANs in conjunction with attention mechanisms. AEE-GAN [119] uses attention in order to alleviate the issues given by the complexity of a scene with many heterogeneous interacting agents. For trajectory encoding, it also uses LSTMs. A characteristic feature is the enhanced attention module containing two components: one for recurrent visual attention enforcement (RVAE) and one for social enforcement (SE). The results of the RVAE are visualized with the Grad-Cam method [120], which creates a heatmap with the attention weights of the image pixels.

Another GAN-based architecture is Social Ways [121], which uses three types of losses: discrimination loss for the discriminator, adversarial loss for the generator, and information loss for both. SoPhie [122] uses a GAN module together with a feature extractor module composed of a CNN and several LSTMs encoders, and an attention module with two components: physical attention and social attention.

Attention mechanisms are also used with techniques other than GANs. MHA-JAM [123] uses a CNN for the transformation of the input image and LSTMs for trajectory encoding, whose outputs then pass to several attention heads that provide the data for the LSTM decoders.

Other authors rely on methods to handle graphs explicitly. For example, DAG-NET [124] uses an attention-enhanced graph neural network (GNN) together with a recurrent variational encoder (RVAE) composed of a variational autoencoder (VAE) and a recurrent neural network (RNN). Multiverse [125] is also based on a graph attention network that is used by a convolutional recurrent neural network (ConvRNN). Unlike other approaches, it uses an occupancy grid for a coarse-grained prediction, which is further refined by a fine-grained prediction. Graphs are also employed by Trajectron++ [126], where a scene is represented as a spatio-temporal graph in which nodes denote the agents and edges denote their interactions. A local map is processed by a CNN, trajectories are encoded with LSTMs, multi-modal solutions are handled by means of a CVAE, and the trajectory decoders are based on gated recurrent units (GRUs).

P2T$_{IRL}$ [127] uses similar techniques, i.e., attention, GRU, CNN, but it conditions trajectories by means of a policy learned with inverse reinforcement learning (IRL) on a grid that represents the scene.

3.4. Methods Using Stochastic Techniques

The authors of [128] use Partially Observable Markov Decision Processes (POMDPs) for behavior prediction and nonlinear receding horizon control, or model predictive control, for trajectory planning. The POMDP models the interactions between the ego vehicle and the obstacles. The action space is discretized into: acceleration, deceleration, and maintaining the current speed. For each of the obstacle vehicles, three types of intentions are considered: going straight, turning, and stopping. The reward function is chosen so that the agents make the maximum progress on the road while avoiding collisions. A particle filter is implemented to update the belief of each motion intention for each obstacle vehicle. For the ego car, the bicycle kinematic model is used to update the state.

Article [129] presents a method to predict trajectories in dense city environments. The authors recorded the trajectories of cars comprising over 1000 h of driving in San Francisco and New York. By relating the current position of an observed car to this large dataset of previously exhibited motion in the same area, the prediction of its future position can be directly performed. Under the hypothesis that the car follows the same trajectory pattern as one of the cars in the past at the same location had followed. This non-parametric method improves over time as the amount of samples increases and avoids the need for more complex models.

Paper [93] presents a trajectory prediction method that combines the constant yaw rate and acceleration (CYRA) motion model with maneuver recognition. The maneuver recognition module selects the current maneuver from a predefined set (e.g., keep lane, change lane to the right or to the left, and turn at an intersection) by comparing the center lines of the road lanes to a local curvilinear model of the path of the vehicle. The proposed method combines the short-term accuracy of the former technique and the longer-term accuracy of the latter. The authors use mathematical models that take into account the position, speed, and acceleration of vehicles.

In [130], a method is presented that evaluates the probabilistic prediction of real traffic scenes with varying start conditions. The prediction is based on a particle filter, which estimates the behavior-describing parameters of a microscopic traffic model, i.e., the driving style as a distribution of behavior parameters. This method seems to be applicable for long-term trajectory planning. The driving style parameters of the intelligent driving model (IDM) are continuously estimated, together with the relative motion between objects. By measuring vehicle accelerations, a driving style estimation can be provided from the first detection without the need of a long observation time before performing the prediction. By using a particle filter, it is possible to handle continuous behavior changes with arbitrarily shaped parameter distributions. Forward propagation using Monte Carlo simulation provides an approximate probability density function of the future scene.

In first-order Markov models, a state prediction depends only on the previous observed state, therefore, if the set of past trajectories has common subsequences, the quality of future predictions may be poor. An additional problem is that the data obtained from sensors can be affected by occlusions. The approaches based on Gaussian processes (GPs) overcome this problem by modeling motion patterns as velocity flow fields and provide good performance in the presence of noise. Another advantage is that the predictions have a simple analytical form, and this can be used to assess the risk in traffic scenarios.

As the traffic participants have a mutual influence on one another, their interaction is explicitly considered in [102], which is inspired by an optimization problem. For motion prediction, the collision probability of a vehicle performing a certain maneuver is computed. The prediction is performed based on the safety evaluation and the assumption that drivers avoid collisions. This combination of the intention of each driver and the driver's local risk assessment to perform a maneuver leads to an interaction-aware motion prediction. The authors compute the probability that a collision will occur anywhere in the whole scene, considering that the number of different maneuvers is limited (e.g., lane changes, acceleration, maintaining the speed, deceleration, and combinations), and then the proposed system assesses the danger of possible future trajectories.

The same concept of considering risk is used in [92], which applies a Bayesian approach combined with maneuver-based trajectory prediction. First, a collection of high-level driving maneuvers is assessed for each vehicle with inference in the Bayesian network that models the traffic scene. Then, maneuver-based probabilistic trajectory prediction models are employed to predict the configuration of each vehicle forward in time. The proposed system has three main parts: the maneuver detection, the prediction, and the criticality assessment. In the last part, the individual joint distributions are used together with a parametric free space map-based representation of the environment with probability distribution functions to estimate the probability of a collision between the ego car and any of its neighbors within the prediction horizon via Monte Carlo simulation.

The authors of [68] propose a framework with three interacting modules: a trajectory prediction module based on a motion-based interaction model combined with maneuver-specific variational Gaussian mixture models, a maneuver recognition module based on hidden Markov models (HMMs) for assigning confidence values for maneuvers being performed by surrounding vehicles, and a vehicle interaction module that handles the context of the scene and assigns final predictions by minimizing an energy function based on outputs of the other two modules. The paper defines ten maneuver classes defined by combinations of lane passes, overtakes, cut-ins, and drifts into the ego lane. A corresponding energy minimization problem is set so that the predictions where cars come too close to one another are penalized.

3.5. Mixed Methods

The authors of [131] use a model-based approach relying on vehicle kinematics and an assumption that drivers plan trajectories in such a way as to minimize an unknown cost function. They introduce an IRL algorithm to learn the cost functions of other vehicles in an energy-based generative model. Langevin sampling, a Monte Carlo-based sampling algorithm, is used to directly sample the control sequence. Langevin sampling is shown to generate better predictions with higher stability. It seems that this algorithm is more flexible than standard IRL methods, and can learn higher-level, non-Markovian cost functions defined over entire trajectories. The cost functions are extended with neural networks in order to combine the advantages of both model-based and model-free learning. The study uses both environment structure, in the form of kinematic vehicular constraints, which can be modeled very accurately, and the assumption that human drivers optimize their trajectories according to a subjective cost function.

Multiple deep neural network architectures are designed to learn the cost functions, some of which augment a set of hand-crafted features. The human-crafted cost functions are defined as ten components: the distance to the goal, the distance to the center of the lane, the penalty of collision to other vehicles (inversely proportional to the distance to other vehicles), the L2-norm of acceleration and steering, the L2-norm for the difference of acceleration and steering between two frames, the heading angle to lane, and the difference to the speed limit.

The application of deep learning and mixture models for the prediction of human drivers in traffic is investigated in [132]. The chosen approach is a mixture density network (MDN) where the neural model has LSTM units and the mixture model consists of univariate Gaussian distributions. It applies multi-task learning, in that by sharing the representation between multiple tasks, one enables the model to generalize better. A limitation is that the tasks usually have to be related to some extent. For example, a single neural network can predict both longitudinal and lateral accelerations from the same input, where the first few layers in the network are shared between the two tasks, and then separated into two different layers to produce the final outputs. To capture the intention of the driver, another layer is used in parallel to the motion prediction layer after the LSTM layers. This layer indicates if the driver intends to switch lane and remain there within the next four seconds.

Another algorithm is the Predictron [133]. This architecture is an abstract model based on a Markov reward process, which can be rolled forward for a series of "imagined" planning steps. The predictron is trained end-to-end with the objective that the accumulated values computed in each forward pass should approximate the true value function. It is reported to demonstrate more accurate predictions than conventional deep neural network architectures.

The Monte Carlo Tree Search (MCTS) [134] algorithm can also be used in the context of trajectory planning. It simulates the possible future trajectories starting from the current state, then it evaluates the performance of the leaves using an evaluation function, e.g., a "value network", and finally it uses these evaluations to update the internal values along the trajectory. The architecture presented in [135], called MCTSnet, incorporates the simulation-based search into a neural network, working with vector embeddings. Its advantage is that gradient-based optimization can be used to train the network end-to-end.

However, internal action sequences directing the control flow of the network cannot be differentiated. To address this, an approximate method for credit assignment is proposed that allows to learn this part of the search network from data.

3.6. Discussion

We summarize the works and their specific techniques presented in the trajectory prediction section in Table 2. The table is sorted by publication year in order to give the reader an impression about the overall progress in this field.

The datasets that were used as benchmarks by the papers were also included. We must mention that all authors report experimental results on some kind of datasets, e.g., driving data specifically collected in some areas of the world or synthetic data collected from simulators. However, only the publicly available, real-world datasets were included in the table.

Information about the general capabilities of the methods was included in terms of the ability to provide multi-modal predictions and whether the social context, i.e., the other agents in the scene, was taken into account. Here, we only mention the approaches that handle the interactions and the context explicitly, e.g., with some kind of pooling mechanism or graph representation, not those that just consider an image as the input, which implicitly contains graphical depictions of all agents.

Some of the works also predict the trajectories of pedestrians, not only vehicles. However, we do not distinguish between these case studies, but only mention the main methods which can be used in both situations.

Table 2. Overview of trajectory prediction solutions.

Contribution	Year	Datasets	Multi-Modal	Social Context	Methods
[93]	2013				physics model (CYRA), maneuver recognition
[102]	2013			Yes	interaction-based, risk estimation, discrete maneauvers
[92]	2016		Yes		Bayesian networks, maneuver-based, risk estimation, Monte Carlo simulation
Social LSTM [104]	2016	ETH, UCY		Yes	LSTM
DESIRE [99]	2017	KITTY, Stanford Drone	Yes	Yes	CVAE, GRU, IRL
[44]	2017		Yes (probabilities of occupancy grid cells)	Yes	LSTM, occupancy grid, softmax
PilotNet [105]	2017				CNN, outputs steering angles
[130]	2017		Yes		particle filter, IDM, driving style estimation, Monte Carlo simulation
Predictron [133]	2017				Markov reward process, DNN (fully-connected deep neural network)
[103]	2018	I-80		Yes (only four neighbors)	LSTM
[45]	2018	NGSIM, I-80	Yes	Yes	LSTM, maneuver-based
[97]	2018				CNN
[18]	2018		Yes		CNN

Table 2. Cont.

Contribution	Year	Datasets	Multi-Modal	Social Context	Methods
IntentNet [94]	2018		Yes		CNN, intention-based, discrete intention set
[95]	2018	ETH, UCY			CNN
[128]	2018		Yes		POMDP, particle filter, discrete states and actions
[129]	2018				probabilistic sampling
[68]	2018			Yes	Gaussian mixture models, HMM, discrete maneuvers
[132]	2018	NGSIM			DNN, MDN, LSTM
MCTSNet [135]	2018				DNN, vector embeddings, Monte Carlo tree search
Social GAN [118]	2018	ETH, UCY	Yes	Yes	GAN, LSTM
[131]	2019	NGSIM			IRL, Langevin sampling, DNN
PRECOG [114]	2019	nuScenes	Yes	Yes	GRU, CNN, generative model
Social Ways [121]	2019	ETH, UCY	Yes	Yes	GAN, LSTM, generative model, attention
SoPhie [122]	2019	ETH, UCY, Stanford Drone	Yes	Yes	GAN, attention, CNN, LSTM
TraPHic [110]	2019	NGSIM		Yes	LSTM-CNN hybrid network
MHA-JAM [123]	2020	nuScenes	Yes	Yes	multi-head attention, LSTM, CNN
AEE-GAN [119]	2020	Waymo, Stanford Drone, ETH, UCY	Yes	Yes	attention, GAN, LSTM
CF-VAE [115]	2020	Stanford Drone	Yes	Yes	CF-VAE
CoverNet [106]	2020	nuScenes	Yes		softmax for discrete trajectory set
DAG-NET [124]	2020	Stanford Drone, SportVU	Yes	Yes	VAE, RNN, attention, GNN
EvolveGraph [111]	2020	Honda 3D, Stanford Drone, SportVU	Yes	Yes	graphs, GRU, Gaussian mixture
Multiverse [125]	2020	VIRAT/ActEV	Yes		ConvRNN, occupancy grid, graph attention network
P2T$_{IRL}$ [127]	2020	Stanford Drone	Yes		IRL, attention, GRU, CNN
PECNet [116]	2020	Stanford Drone, ETH, UCY	Yes	Yes	CVAE (Endpoint VAE), truncation trick
TNT [112]	2020	Argoverse, Interaction, Stanford Drone	Yes	Yes	VectorNet, MLP (classic multilayer perceptron)
Trajectron++ [126]	2020	ETH, UCY, nuScenes	Yes	Yes	LSTM, attention, GRU, CVAE, Gaussian mixture model
Y-Net [107]	2020	Stanford Drone, ETH, UCY	Yes		U-Net, k-means

In general, many authors use CNNs to process the graphical inputs, e.g., camera-based images or maps, and LSTMs for trajectory encoding and decoding. Because of the constraints of real-time requirements, some works also use CNN architectures for prediction [97]. They seem to be able to model complex relations and capture spatial correlations in the data [136]. Some papers state that they are also competitive in modeling temporal data [95], with performance comparable to that of the LSTMs, but with a much simpler internal structure. Multi-modal predictions are often made with some kind of generative models such as CVAE. The methods based on CNNs seem to be more lightweight and fast than those containing LSTM and CVAE components. Still, a large number of approaches combine these techniques in some way.

Other works employ more recent techniques such as GANs and graph representations in conjunction with neural networks. Attention mechanisms also seem promising to distinguish the important features in the context of a complex scene with many interacting agents.

The data themselves may cause difficulties, because a network only learns what is present in the data, and hopefully generalizes well, but there may be situations where the humans do not behave according to previous observations. This is one drawback of using neural networks. However, it seems that the advantages of using data-driven approaches outperform the disadvantages.

Many methods that belong to the stochastic paradigm try to estimate the probabilities of discrete maneuvers. When using, e.g., hidden Markov models, the movement of the traffic participants is evaluated independently, an assumption which is true only for simple scenarios. Gaussian Process regression can quantify uncertainty, but it is also limited in its ability to model complex interactions. For this purpose, other techniques such as Bayesian networks can be used instead, with the disadvantage of an increased computation time and thus a difficulty in handling real-time learning tasks [97].

Although it is possible to do multi-step prediction with a Kalman filter, it cannot be extended far into the future with reasonable accuracy. A multi-step prediction done solely by a Kalman filter was found to be accurate up until 10–15 time steps, after which the predictions diverged and ended up being worse than constant velocity inference [132]. This emphasizes the advantages of data-driven approaches, as it is possible to observe almost an infinite number of variables that may all affect the driver, whereas the Kalman filter relies solely on the physical movement of the vehicle.

Another approach is to learn policies in a supervised way, e.g., imitation learning. The cost function of a human driver can be estimated with inverse reinforcement learning and then a policy can be extracted from the cost function [136]. However, this may again be inefficient for real-time applications [97].

In multi-agent contexts such as those defined by traffic scenarios, since an agent's actions depend on the other agents' actions, uncertainty can propagate to future states with the consequence that an agent completely stops because all possible actions are deemed as unacceptably unsafe. This is known as the "freezing-robot" problem. Deadlock avoidance and multi-objective decision making are very common in practice, e.g., in autonomous robotics [137–139].

Finally, it should be mentioned that in this section, we have addressed the trajectory prediction problem. A related, but distinct problem, is trajectory planning, i.e., finding an optimal path from the current location to a given goal location. Its aim is to produce smooth trajectories with small changes in curvature, so as to minimize both the lateral and the longitudinal acceleration of the ego vehicle. For this purpose, there are several methods reported in the literature, e.g., using cubic spline interpolation, trigonometric spline interpolation, Bézier curves, or clothoids, i.e., curves with a complex mathematical definition, which have a linear relation between the curvature and the arc length, and allow smooth transitions from a straight line to a circle arc or vice versa. Deep reinforcement learning methods [140,141] such as policy gradients [142], deep Q-network [143], actor-critic [144], asynchronous advantage actor-critic [145], proximal policy optimization [146], trust region policy optimization [147], imagination-augmented agents [148], or proximal gradient temporal difference learning [149] can also be used to decide the possible maneuvers that the ego car can make in order to optimize criteria related to risk and efficiency.

4. Conclusions

Learning-based approaches have basically become the norm for autonomous driving problems. Although explicit rule-based methods may have an important advantage in the form of explicit knowledge, hand-crafted rules usually take a considerable amount of effort to devise and validate, and usually do not have satisfactory generalization capabilities because of the great variability of situations that may appear in a driving context. Unfortunately, techniques based on learning typically require large quantities of data in order to cover a sufficiently large part of the space of possible driving behaviors.

Because they capture the generative structure of vehicle trajectories, model-based methods can potentially learn more from fewer data than model-free methods. However, good cost functions are challenging to learn, and simple, hand-crafted representations may not generalize well across tasks and contexts. In general, model-based methods can be less flexible and may underperform model-free methods in the limit of infinite data. Model-free methods take a data-driven approach, aiming to learn predictive distributions over trajectories directly from data. These approaches are more flexible and require less knowledge engineering in terms of the type of vehicles, maneuvers, and scenarios, but the amount of data they require may be very large.

The past three decades have seen increasingly rapid progress in driverless vehicle technology. In addition to the advances in computing and perception hardware, this rapid progress has been enabled by major theoretical progress in computational aspects. Autonomous cars are complex systems that can be decomposed into a hierarchy of decision making problems, where the solution of one problem is the input to the next. The breakdown into individual decision making problems has enabled the use of well-developed methods and technologies from a variety of research areas.

This literature review has concentrated only on two aspects: tracking and trajectory prediction. It can serve as a reference for assessing the computational tradeoffs between various choices for algorithm design.

Author Contributions: Writing—Sections 1 and 2, M.G., Sections 3 and 4, F.L.; funding acquisition, F.L. All authors have read and agreed to the published version of the manuscript.

Funding: This research was funded by Continental AG within the *Proreta 5* project.

Institutional Review Board Statement: Not applicable.

Informed Consent Statement: Not applicable.

Data Availability Statement: Not applicable.

Acknowledgments: We kindly thank Continental AG for their great cooperation within *Proreta 5*, which is a joint research project of the Technical University of Darmstadt, University of Bremen, "Gheorghe Asachi" Technical University of Iași and Continental AG.

Conflicts of Interest: The authors declare no conflict of interest.

References

1. Chen, L.; Ai, H.; Shang, C.; Zhuang, Z.; Bai, B. Online multi-object tracking with convolutional neural networks. In Proceedings of the 2017 IEEE International Conference on Image Processing (ICIP), Beijing, China, 17–20 September 2017; pp. 645–649. [CrossRef]
2. Liu, Q.; Lu, X.; He, Z.; Zhang, C.; Chen, W.S. Deep Convolutional Neural Networks for Thermal Infrared Object Tracking. *Knowl.-Based Syst.* **2017**. [CrossRef]
3. Danelljan, M.; Häger, G.; Khan, F.S.; Felsberg, M. Convolutional Features for Correlation Filter Based Visual Tracking. In Proceedings of the 2015 IEEE International Conference on Computer Vision Workshop (ICCVW), Santiago, Chile, 7–13 December 2015; pp. 621–629. [CrossRef]

4. Mozhdehi, R.J.; Reznichenko, Y.; Siddique, A.; Medeiros, H. Deep Convolutional Particle Filter with Adaptive Correlation Maps for Visual Tracking. In Proceedings of the 2018 25th IEEE International Conference on Image Processing (ICIP), Athens, Greece, 7–10 October 2018; pp. 798–802. [CrossRef]
5. Song, Y.; Ma, C.; Gong, L.; Zhang, J.; Lau, R.W.H.; Yang, M.H. CREST: Convolutional Residual Learning for Visual Tracking. In Proceedings of the 2017 IEEE International Conference on Computer Vision (ICCV), Venice, Italy, 22–29 October 2017; pp. 2574–2583.
6. Wang, C.; Galoogahi, H.K.; Lin, C.H.; Lucey, S. Deep-LK for Efficient Adaptive Object Tracking. In Proceedings of the 2018 IEEE International Conference on Robotics and Automation (ICRA), Brisbane, QLD, Australia, 21–25 May 2018; pp. 627–634.
7. Du, M.; Ding, Y.; Meng, X.; Wei, H.L.; Zhao, Y. Distractor-Aware Deep Regression for Visual Tracking. Sensors 2019, 19, 387. [CrossRef] [PubMed]
8. Zhou, H.; Ummenhofer, B.; Brox, T. DeepTAM: Deep Tracking and Mapping. In Computer Vision—ECCV 2018; Ferrari, V., Hebert, M., Sminchisescu, C., Weiss, Y., Eds.; Springer International Publishing: Cham, Switzerland, 2018; pp. 851–868.
9. Danelljan, M.; Robinson, A.; Khan, F.S.; Felsberg, M. Beyond Correlation Filters: Learning Continuous Convolution Operators for Visual Tracking. In Proceedings of the Computer Vision—ECCV 2016—14th European Conference, Amsterdam, The Netherlands, 11–14 October 2016; pp. 472–488. [CrossRef]
10. Danelljan, M.; Bhat, G.; Khan, F.S.; Felsberg, M. ECO: Efficient Convolution Operators for Tracking. In Proceedings of the 2017 IEEE Conference on Computer Vision and Pattern Recognition, CVPR 2017, Honolulu, HI, USA, 21–26 July 2017; pp. 6931–6939. [CrossRef]
11. Nam, H.; Han, B. Learning Multi-domain Convolutional Neural Networks for Visual Tracking. In Proceedings of the 2016 IEEE Conference on Computer Vision and Pattern Recognition (CVPR), Las Vegas, NV, USA, 27–30 June 2016; pp. 4293–4302.
12. Fan, H.; Ling, H. SANet: Structure-Aware Network for Visual Tracking. In Proceedings of the 2017 IEEE Conference on Computer Vision and Pattern Recognition Workshops (CVPRW), Honolulu, HI, USA, 21–26 July 2017; pp. 2217–2224.
13. Wojke, N.; Bewley, A.; Paulus, D. Simple online and realtime tracking with a deep association metric. In Proceedings of the 2017 IEEE International Conference on Image Processing (ICIP), Beijing, China, 17–20 September 2017; pp. 3645–3649. [CrossRef]
14. Li, K.; Kong, Y.; Fu, Y. Multi-stream Deep Similarity Learning Networks for Visual Tracking. In Proceedings of the 26th International Joint Conference on Artificial Intelligence, IJCAI'17, Melbourne, Australia, 19–25 August 2017; pp. 2166–2172.
15. Chu, Q.; Ouyang, W.; Li, H.; Wang, X.; Liu, B.; Yu, N. Online Multi-object Tracking Using CNN-Based Single Object Tracker with Spatial-Temporal Attention Mechanism. In Proceedings of the 2017 IEEE International Conference on Computer Vision (ICCV), Venice, Italy, 22–29 October 2017; pp. 4846–4855.
16. Cui, Z.; Lu, N.; Jing, X.; Shi, X. Fast Dynamic Convolutional Neural Networks for Visual Tracking. In Proceedings of the 10th Asian Conference on Machine Learning, Beijing, China, 14–16 November 2018.
17. Fabbri, M.; Lanzi, F.; Calderara, S.; Palazzi, A.; Vezzani, R.; Cucchiara, R. Learning to Detect and Track Visible and Occluded Body Joints in a Virtual World. In Proceedings of the Computer Vision—ECCV 2018—15th European Conference, Munich, Germany, 8–14 September 2018; pp. 450–466. [CrossRef]
18. Cui, H.; Radosavljevic, V.; Chou, F.; Lin, T.; Nguyen, T.; Huang, T.; Schneider, J.; Djuric, N. Multimodal Trajectory Predictions for Autonomous Driving using Deep Convolutional Networks. In Proceedings of the 2019 International Conference on Robotics and Automation (ICRA), Montreal, QC, Canada, 20–24 May 2019; pp. 2090–2096. [CrossRef]
19. Chu, P.; Fan, H.; Tan, C.C.; Ling, H. Online Multi-Object Tracking With Instance-Aware Tracker and Dynamic Model Refreshment. In Proceedings of the 2019 IEEE Winter Conference on Applications of Computer Vision (WACV), Waikoloa, HI, USA, 7–11 January 2019; pp. 161–170. [CrossRef]
20. Shahian Jahromi, B.; Tulabandhula, T.; Cetin, S. Real-Time Hybrid Multi-Sensor Fusion Framework for Perception in Autonomous Vehicles. Sensors 2019, 19, 4357. [CrossRef] [PubMed]
21. Bhat, G.; Johnander, J.; Danelljan, M.; Khan, F.S.; Felsberg, M. Unveiling the Power of Deep Tracking. In Proceedings of the European Conference on Computer Vision (ECCV), Munich, Germany, 8–14 September 2018.
22. Qi, Y.; Zhang, S.; Qin, L.; Yao, H.; Huang, Q.; Lim, J.; Yang, M. Hedged Deep Tracking. In Proceedings of the 2016 IEEE Conference on Computer Vision and Pattern Recognition (CVPR), Las Vegas, NV, USA, 27–30 June 2016; pp. 4303–4311. [CrossRef]
23. Ristani, E.; Tomasi, C. Features for Multi-target Multi-camera Tracking and Re-identification. In Proceedings of the 2018 IEEE/CVF Conference on Computer Vision and Pattern Recognition, Salt Lake City, UT, USA, 18–23 June 2018; pp. 6036–6046. [CrossRef]
24. Teng, Z.; Xing, J.; Wang, Q.; Lang, C.; Feng, S.; Jin, Y. Robust Object Tracking Based on Temporal and Spatial Deep Networks. In Proceedings of the 2017 IEEE International Conference on Computer Vision (ICCV), Venice, Italy, 22–29 October 2017; pp. 1153–1162. [CrossRef]
25. Yoon, Y.C.; Boragule, A.; Yoon, K.; Jeon, M. Online Multi-Object Tracking with Historical Appearance Matching and Scene Adaptive Detection Filtering. In Proceedings of the 2018 15th IEEE International Conference on Advanced Video and Signal Based Surveillance (AVSS), Auckland, New Zealand, 27–30 November 2018; pp. 1–6.
26. Sun, S.; Akhtar, N.; Song, H.; Mian, A.S.; Shah, M. Deep Affinity Network for Multiple Object Tracking. IEEE Trans. Pattern Anal. Mach. Intell. 2021, 43, 104–119. [CrossRef] [PubMed]
27. Li, S.; Ma, B.; Chang, H.; Shan, S.; Chen, X. Continuity-Discrimination Convolutional Neural Network for Visual Object Tracking. In Proceedings of the 2018 IEEE International Conference on Multimedia and Expo (ICME), San Diego, CA, USA, 23–27 July 2018.

28. Hahn, M.; Chen, S.; Dehghan, A. Deep Tracking: Visual Tracking Using Deep Convolutional Networks. *CoRR* **2015**, arXiv:1512.03993v1 [cs.CV].
29. Yang, L.; Liu, R.; Zhang, D.; Zhang, L. Deep Location-Specific Tracking. In Proceedings of the 25th ACM International Conference on Multimedia, MM '17, Mountain View, CA, USA, 23–27 October 2017; ACM: New York, NY, USA, 2017; pp. 1309–1317. [CrossRef]
30. Feichtenhofer, C.; Pinz, A.; Zisserman, A. Detect to Track and Track to Detect. In Proceedings of the 2017 IEEE International Conference on Computer Vision (ICCV), Venice, Italy, 22–29 October 2017; pp. 3057–3065.
31. Jiang, X.; Zhen, X.; Zhang, B.; Yang, J.; Cao, X. Deep Collaborative Tracking Networks. In Proceedings of the 29th The British Machine Vision Conference, Newcastle, UK, 3–6 September 2018.
32. Chen, L.; Lou, J.; Xu, F.; Ren, M. Grid-based multi-object tracking with Siamese CNN based appearance edge and access region mechanism. *Multimed. Tools Appl.* **2020**, *79*. [CrossRef]
33. Zhang, W.; Du, Y.; Chen, Z.; Deng, J.; Liu, P. Robust adaptive learning with Siamese network architecture for visual tracking. *Vis. Comput.* **2020**. [CrossRef]
34. Son, J.; Baek, M.; Cho, M.; Han, B. Multi-object Tracking with Quadruplet Convolutional Neural Networks. In Proceedings of the 2017 IEEE Conference on Computer Vision and Pattern Recognition (CVPR), Honolulu, HI, USA, 21–26 July 2017; pp. 3786–3795. [CrossRef]
35. Schulter, S.; Vernaza, P.; Choi, W.; Chandraker, M.K. Deep Network Flow for Multi-object Tracking. In Proceedings of the 2017 IEEE Conference on Computer Vision and Pattern Recognition (CVPR), Honolulu, HI, USA, 21–26 July 2017; pp. 2730–2739.
36. Wang, N.; Zou, Q.; Ma, Q.; Huang, Y.; Luan, D. A light tracker for online multiple pedestrian tracking. *J. Real-Time Image Process.* **2021**, *18*, 1–17. [CrossRef]
37. Feng, W.; Hu, Z.; Wu, W.; Yan, J.; Ouyang, W. Multi-Object Tracking with Multiple Cues and Switcher-Aware Classification. *CoRR* **2019**, arXiv:1901.06129v1 [cs.CV].
38. Zhu, J.; Yang, H.; Liu, N.; Kim, M.; Zhang, W.; Yang, M.H. Online Multi-Object Tracking with Dual Matching Attention Networks. In Proceedings of the Computer Vision—ECCV 2018, Munich, Germany, 8–14 September 2018; Springer International Publishing: Cham, Switzerland, 2018; pp. 379–396.
39. Pu, S.; Song, Y.; Ma, C.; Zhang, H.; Yang, M.H. Deep Attentive Tracking via Reciprocative Learning. In Proceedings of the 32nd International Conference on Neural Information Processing Systems, NIPS'18, Montréal, QC, Canada, 3–8 December 2018.
40. Wang, Y.; Zhang, Z.; Zhang, N.; Zeng, D. Attention Modulated Multiple Object Tracking with Motion Enhancement and Dual Correlation. *Symmetry* **2021**, *13*, 266. [CrossRef]
41. Meng, F.; Wang, X.; Wang, D.; Shao, F.; Fu, L. Spatial–Semantic and Temporal Attention Mechanism-Based Online Multi-Object Tracking. *Sensors* **2020**, *20*, 1653. [CrossRef] [PubMed]
42. Milan, A.; Rezatofighi, S.H.; Dick, A.; Reid, I.; Schindler, K. Online Multi-Target Tracking using Recurrent Neural Networks. In Proceedings of the 31st AAAI Conference on Artificial Intelligence, San Francisco, CA, USA, 4–9 February 2017.
43. Sadeghian, A.; Alahi, A.; Savarese, S. Tracking the Untrackable: Learning to Track Multiple Cues with Long-Term Dependencies. In Proceedings of the 2017 IEEE International Conference on Computer Vision (ICCV), Venice, Italy, 22–29 October 2017; pp. 300–311.
44. Kim, B.; Kang, C.M.; Kim, J.; Lee, S.H.; Chung, C.C.; Choi, J.W. Probabilistic vehicle trajectory prediction over occupancy grid map via recurrent neural network. In Proceedings of the 2017 IEEE 20th International Conference on Intelligent Transportation Systems (ITSC), Yokohama, Japan, 16–19 October 2017; pp. 399–404.
45. Deo, N.; Trivedi, M.M. Multi-Modal Trajectory Prediction of Surrounding Vehicles with Maneuver based LSTMs. In Proceedings of the 2018 IEEE Intelligent Vehicles Symposium (IV), Changshu, China, 26–30 June 2018; pp. 1179–1184.
46. Ma, C.; Yang, C.; Yang, F.; Zhuang, Y.; Zhang, Z.; Jia, H.; Xie, X. Trajectory Factory: Tracklet Cleaving and Re-Connection by Deep Siamese Bi-GRU for Multiple Object Tracking. In Proceedings of the 2018 IEEE International Conference on Multimedia and Expo (ICME), San Diego, CA, USA, 23–27 July 2018. [CrossRef]
47. Kim, C.; Li, F.; Rehg, J.M. Multi-object Tracking with Neural Gating Using Bilinear LSTM. In Proceedings of the European Conference on Computer Vision (ECCV), Munich, Germany, 8–14 September 2018.
48. Ondrúška, P.; Posner, I. Deep Tracking: Seeing Beyond Seeing Using Recurrent Neural Networks. In Proceedings of the Thirtieth AAAI Conference on Artificial Intelligence, AAAI'16, Phoenix, AZ, USA, 12–17 February 2016.
49. Dequaire, J.; Ondruska, P.; Rao, D.; Wang, D.Z.; Posner, I. Deep tracking in the wild: End-to-end tracking using recurrent neural networks. *Int. J. Robot. Res.* **2018**, *37*, 492–512. [CrossRef]
50. Chen, C.; Zhao, P.; Lu, C.; Wang, W.; Markham, A.; Trigoni, N. Deep-Learning-Based Pedestrian Inertial Navigation: Methods, Data Set, and On-Device Inference. *IEEE Internet Things J.* **2020**, *7*, 4431–4441. [CrossRef]
51. Du, Y.; Yan, Y.; Chen, S.; Hua, Y. Object-adaptive LSTM network for real-time visual tracking with adversarial data augmentation. *Neurocomputing* **2020**, *384*, 67–83. [CrossRef]
52. Huang, Z.; Hasan, A.; Shin, K.; Li, R.; Driggs-Campbell, K. Long-Term Pedestrian Trajectory Prediction Using Mutable Intention Filter and Warp LSTM. *IEEE Robot. Autom. Lett.* **2021**, *6*, 542–549. [CrossRef]
53. Quan, R.; Zhu, L.; Wu, Y.; Yang, Y. Holistic LSTM for Pedestrian Trajectory Prediction. *IEEE Trans. Image Process.* **2021**, *30*, 3229–3239. [CrossRef] [PubMed]

54. Song, X.; Chen, K.; Li, X.; Sun, J.; Hou, B.; Cui, Y.; Zhang, B.; Xiong, G.; Wang, Z. Pedestrian Trajectory Prediction Based on Deep Convolutional LSTM Network. *IEEE Trans. Intell. Transp. Syst.* **2020**, 1–18. [CrossRef]
55. Huang, R.; Zhang, W.; Kundu, A.; Pantofaru, C.; Ross, D.A.; Funkhouser, T.; Fathi, A. An LSTM Approach to Temporal 3D Object Detection in LiDAR Point Clouds. In *Computer Vision—ECCV 2020*; Vedaldi, A., Bischof, H., Brox, T., Frahm, J.M., Eds.; Springer International Publishing: Cham, Switzerland, 2020; pp. 266–282.
56. Dan, X. Spatial-Temporal Block and LSTM Network for Pedestrian Trajectories Prediction. *arXiv* **2020**, arXiv:2009.10468.
57. Zhou, Y.; Wu, H.; Cheng, H.; Qi, K.; Hu, K.; Kang, C.; Zheng, J. Social graph convolutional LSTM for pedestrian trajectory prediction. *IET Intell. Transp. Syst.* **2021**, *15*, 396–405. [CrossRef]
58. Fernando, T.; Denman, S.; Sridharan, S.; Fookes, C. Tracking by prediction: A deep generative model for multi-person localisation and tracking. In Proceedings of the IEEE Winter Conference on Applications of Computer Vision (WACV 2018), Lake Tahoe, NV, USA, 12–15 March 2018; IEEE: Lake Tahoe, NV, USA, 2018; pp. 1122–1132. [CrossRef]
59. Ren, L.; Lu, J.; Wang, Z.; Tian, Q.; Zhou, J. Collaborative Deep Reinforcement Learning for Multi-Object Tracking. In Proceedings of the European Conference on Computer Vision (ECCV), Munich, Germany, 8–14 September 2018.
60. Li, J.; Yao, L.; Xu, X.; Cheng, B.; Ren, J. Deep reinforcement learning for pedestrian collision avoidance and human-machine cooperative driving. *Inf. Sci.* **2020**, *532*, 110–124. [CrossRef]
61. Bae, S.; Yoon, K. Robust Online Multi-object Tracking Based on Tracklet Confidence and Online Discriminative Appearance Learning. In Proceedings of the 2014 IEEE Conference on Computer Vision and Pattern Recognition, Columbus, OH, USA, 23–28 June 2014; pp. 1218–1225. [CrossRef]
62. Receveur, J.B.; Victor, S.; Melchior, P. Autonomous car decision making and trajectory tracking based on genetic algorithms and fractional potential fields. *Intell. Serv. Robot.* **2020**, *13*. [CrossRef]
63. Bewley, A.; Ge, Z.; Ott, L.; Ramos, F.; Upcroft, B. Simple online and realtime tracking. In Proceedings of the 2016 IEEE International Conference on Image Processing (ICIP), Phoenix, AZ, USA, 25–28 September 2016; pp. 3464–3468. [CrossRef]
64. Bergmann, P.; Meinhardt, T.; Leal-Taixé, L. Tracking without bells and whistles. *arXiv* **2019**, arXiv:1903.05625.
65. Mogelmose, A.; Trivedi, M.M.; Moeslund, T.B. Trajectory analysis and prediction for improved pedestrian safety: Integrated framework and evaluations. In Proceedings of the 2015 IEEE Intelligent Vehicles Symposium (IV), Seoul, Korea, 28 June–1 July 2015; pp. 330–335. [CrossRef]
66. Xiang, Y.; Alahi, A.; Savarese, S. Learning to Track: Online Multi-object Tracking by Decision Making. In Proceedings of the 2015 IEEE International Conference on Computer Vision (ICCV), Santiago, Chile, 7–13 December 2015; pp. 4705–4713. [CrossRef]
67. Rangesh, A.; Trivedi, M.M. No Blind Spots: Full-Surround Multi-Object Tracking for Autonomous Vehicles using Cameras and LiDARs. *arXiv* **2019**, arXiv:1802.08755.
68. Deo, N.; Rangesh, A.; Trivedi, M.M. How Would Surround Vehicles Move? A Unified Framework for Maneuver Classification and Motion Prediction. *IEEE Trans. Intell. Veh.* **2018**, *3*, 129–140. [CrossRef]
69. Maksai, A.; Wang, X.; Fleuret, F.; Fua, P. Non-Markovian Globally Consistent Multi-object Tracking. In Proceedings of the 2017 IEEE International Conference on Computer Vision (ICCV), Venice, Italy, 22–29 October 2017; pp. 2563–2573. [CrossRef]
70. Milan, A.; Roth, S.; Schindler, K. Continuous Energy Minimization for Multitarget Tracking. *IEEE TPAMI* **2014**, *36*, 58–72. [CrossRef] [PubMed]
71. Solera, F.; Calderara, S.; Cucchiara, R. Learning to Divide and Conquer for Online Multi-target Tracking. In Proceedings of the 2015 IEEE International Conference on Computer Vision (ICCV), ICCV '15, Santiago, Chile, 7–13 December 2015; IEEE Computer Society: Washington, DC, USA, 2015; pp. 4373–4381. [CrossRef]
72. Dicle, C.; Camps, O.I.; Sznaier, M. The Way They Move: Tracking Multiple Targets with Similar Appearance. In Proceedings of the 2013 IEEE International Conference on Computer Vision, Sydney, NSW, Australia, 1–8 December 2013; pp. 2304–2311. [CrossRef]
73. Kim, C.; Li, F.; Ciptadi, A.; Rehg, J.M. Multiple Hypothesis Tracking Revisited. In Proceedings of the 2015 IEEE International Conference on Computer Vision (ICCV), Santiago, Chile, 7–13 December 2015; pp. 4696–4704. [CrossRef]
74. Sharma, S.; Ansari, J.A.; Murthy, J.K.; Krishna, K.M. Beyond Pixels: Leveraging Geometry and Shape Cues for Online Multi-Object Tracking. In Proceedings of the 2018 IEEE International Conference on Robotics and Automation (ICRA), Brisbane, QLD, Australia, 21–25 May 2018; pp. 3508–3515.
75. Andersen, H.; Chong, Z.J.; Eng, Y.H.; Pendleton, S.; Ang, M.H. Geometric path tracking algorithm for autonomous driving in pedestrian environment. In Proceedings of the 2016 IEEE International Conference on Advanced Intelligent Mechatronics (AIM), Banff, AB, Canada, 12–15 July 2016; pp. 1669–1674. [CrossRef]
76. Manjunath, A.; Liu, Y.; Henriques, B.; Engstle, A. Radar Based Object Detection and Tracking for Autonomous Driving. In Proceedings of the 2018 IEEE MTT-S International Conference on Microwaves for Intelligent Mobility (ICMIM), Munich, Germany, 15–17 April 2018; pp. 1–4. [CrossRef]
77. Cao, J.; Song, C.; Peng, S.; Song, S.; Zhang, X.; Xiao, F. Trajectory Tracking Control Algorithm for Autonomous Vehicle Considering Cornering Characteristics. *IEEE Access* **2020**, *8*, 59470–59484. [CrossRef]
78. Kampker, A.; Sefati, M.; Rachman, A.S.A.; Kreisköther, K.; Campoy, P. Towards Multi-Object Detection and Tracking in Urban Scenario under Uncertainties. *arXiv* **2018**, arXiv:1801.02686.

79. Bochinski, E.; Eiselein, V.; Sikora, T. High-Speed tracking-by-detection without using image information. In Proceedings of the 2017 14th IEEE International Conference on Advanced Video and Signal Based Surveillance (AVSS), Lecce, Italy, 29 August–1 September 2017; pp. 1–6. [CrossRef]
80. Shan, Y.; Zheng, B.; Chen, L.; Chen, L.; Chen, D. A Reinforcement Learning-Based Adaptive Path Tracking Approach for Autonomous Driving. *IEEE Trans. Veh. Technol.* **2020**, *69*, 10581–10595. [CrossRef]
81. Ristani, E.; Solera, F.; Zou, R.S.; Cucchiara, R.; Tomasi, C. Performance Measures and a Data Set for Multi-Target, Multi-Camera Tracking. *arXiv* **2016**, arXiv:1609.01775.
82. Pi, W.; Yang, P.; Duan, D.; Chen, C.; Cheng, X.; Yang, L.; Li, H. Malicious User Detection for Cooperative Mobility Tracking in Autonomous Driving. *IEEE Internet Things J.* **2020**, *7*, 4922–4936. [CrossRef]
83. Chari, V.; Lacoste-Julien, S.; Laptev, I.; Sivic, J. On pairwise costs for network flow multi-object tracking. In Proceedings of the 2015 IEEE Conference on Computer Vision and Pattern Recognition (CVPR), Boston, MA, USA, 7–12 June 2015; pp. 5537–5545.
84. Berclaz, J.; Fleuret, F.; Turetken, E.; Fua, P. Multiple Object Tracking Using K-Shortest Paths Optimization. *IEEE Trans. Pattern Anal. Mach. Intell.* **2011**, *33*, 1806–1819. [CrossRef] [PubMed]
85. Pirsiavash, H.; Ramanan, D.; Fowlkes, C.C. Globally-optimal greedy algorithms for tracking a variable number of objects. In Proceedings of the CVPR 2011, Colorado Springs, CO, USA, 20–25 June 2011; pp. 1201–1208. [CrossRef]
86. Ristani, E.; Tomasi, C. Tracking Multiple People Online and in Real Time. In Proceedings of the 12th Asian Conference on Computer Vision, Singapore, 1–5 November 2014.
87. Roshan Zamir, A.; Dehghan, A.; Shah, M. GMCP-Tracker: Global Multi-object Tracking Using Generalized Minimum Clique Graphs. In *Computer Vision—ECCV 2012*; Fitzgibbon, A., Lazebnik, S., Perona, P., Sato, Y., Schmid, C., Eds.; Springer: Heidelberg/Berlin, Germany, 2012; pp. 343–356.
88. Naiel, M.A.; Ahmad, M.O.; Swamy, M.; Lim, J.; Yang, M.H. Online multi-object tracking via robust collaborative model and sample selection. *Comput. Vis. Image Underst.* **2017**, *154*, 94–107. [CrossRef]
89. Wen, L.; Li, W.; Yan, J.; Lei, Z.; Yi, D.; Li, S.Z. Multiple Target Tracking Based on Undirected Hierarchical Relation Hypergraph. In Proceedings of the 2014 IEEE Conference on Computer Vision and Pattern Recognition, Columbus, OH, USA, 23–28 June 2014; pp. 1282–1289. [CrossRef]
90. Wen, L.; Du, D.; Li, S.; Bian, X.; Lyu, S. Learning Non-Uniform Hypergraph for Multi-Object Tracking. In Proceedings of the AAAI Conference on Artificial Intelligence, Honolulu, HI, USA, 27 January–1 February 2019; pp. 8981–8988. [CrossRef]
91. Dolatabadi, M.; Elfring, J.; van de Molengraft, R. Multiple-Joint Pedestrian Tracking Using Periodic Models. *Sensors* **2020**, *20*, 6917. [CrossRef] [PubMed]
92. Schreier, M.; Willert, V.; Adamy, J. An Integrated Approach to Maneuver-Based Trajectory Prediction and Criticality Assessment in Arbitrary Road Environments. *IEEE Trans. Intell. Transp. Syst.* **2016**, *17*, 2751–2766. [CrossRef]
93. Houenou, A.; Bonnifait, P.; Cherfaoui, V.; Yao, W. Vehicle trajectory prediction based on motion model and maneuver recognition. In Proceedings of the 2013 IEEE/RSJ International Conference on Intelligent Robots and Systems, Tokyo, Japan, 3–7 November 2013; pp. 4363–4369. [CrossRef]
94. Casas, S.; Luo, W.; Urtasun, R. IntentNet: Learning to Predict Intention from Raw Sensor Data. In Proceedings of the 2nd Annual Conference on Robot Learning, CoRL 2018, Zürich, Switzerland, 29–31 October 2018; pp. 947–956.
95. Nikhil, N.; Morris, B.T. Convolutional Neural Network for Trajectory Prediction. In Proceedings of the Computer Vision—ECCV 2018 Workshops, Munich, Germany, 8–14 September 2018; pp. 186–196. . [CrossRef]
96. Aptiv.; Audi.; Baidu.; BMW.; Continental.; Daimler.; Fiat.; Chrysler Automobiles.; HERE.; Infineon.; Intel.; Volkswagen. Safety First for Automated Driving. Available online: https://www.daimler.com/documents/innovation/other/safety-first-for-automated-driving.pdf (accessed on 2 July 2019).
97. Djuric, N.; Radosavljevic, V.; Cui, H.; Nguyen, T.; Chou, F.; Lin, T.; Schneider, J. Motion Prediction of Traffic Actors for Autonomous Driving using Deep Convolutional Networks. *CoRR* **2018**, arXiv:1808.05819v3 [cs.LG].
98. Ward, E. Models Supporting Trajectory Planning in Autonomous Vehicles. Ph.D. Thesis, KTH Royal Institute of Technology, Stockholm, Sweden, 2018.
99. Lee, N.; Choi, W.; Vernaza, P.; Choy, C.B.; Torr, P.H.S.; Chandraker, M.K. DESIRE: Distant Future Prediction in Dynamic Scenes with Interacting Agents. In Proceedings of the 2017 IEEE Conference on Computer Vision and Pattern Recognition (CVPR), Honolulu, HI, USA, 21–26 July 2017; pp. 2165–2174. [CrossRef]
100. Singh, A. Prediction in Autonomous Vehicle–All You Need to Know. Available online: https://medium.com/m/global-identity?redirectUrl=https%3A%2F%2Ftowardsdatascience.com%2Fprediction-in-autonomous-vehicle-all-you-need-to-know-d8811795fcdc (accessed on 28 January 2021).
101. Lefèvre, S.; Vasquez, D.; Laugier, C. A survey on motion prediction and risk assessment for intelligent vehicles. *Robomech J.* **2014**, *2014*, 1–14. [CrossRef]
102. Lawitzky, A.; Althoff, D.; Passenberg, C.F.; Tanzmeister, G.; Wollherr, D.; Buss, M. Interactive scene prediction for automotive applications. In Proceedings of the 2013 IEEE Intelligent Vehicles Symposium (IV), Gold Coast, QLD, Australia, 23–26 June 2013; pp. 1028–1033. [CrossRef]
103. Woo, H.; Sugimoto, M.; Wu, J.; Tamura, Y.; Yamashita, A.; Asama, H. Trajectory Prediction of Surrounding Vehicles Using LSTM Network. In Proceedings of the 2013 IEEE Intelligent Vehicles Symposium (IV), Gold Coast, QLD, Australia, 26–28 June 2013.

104. Alahi, A.; Goel, K.; Ramanathan, V.; Robicquet, A.; Fei-Fei, L.; Savarese, S. Social LSTM: Human Trajectory Prediction in Crowded Spaces. In Proceedings of the 2016 IEEE Conference on Computer Vision and Pattern Recognition (CVPR), Las Vegas, NV, USA, 27–30 June 2016; pp. 961–971. [CrossRef]
105. Bojarski, M.; Yeres, P.; Choromanska, A.; Choromanski, K.; Firner, B.; Jackel, L.D.; Muller, U. Explaining How a Deep Neural Network Trained with End-to-End Learning Steers a Car. *CoRR* **2017**, arXiv:1704.07911v1 [cs.CV].
106. Phan-Minh, T.; Grigore, E.C.; Boulton, F.A.; Beijbom, O.; Wolff, E.M. CoverNet: Multimodal Behavior Prediction using Trajectory Sets. *arXiv* **2020**, arXiv:1911.10298.
107. Mangalam, K.; An, Y.; Girase, H.; Malik, J. From Goals, Waypoints and Paths to Long Term Human Trajectory Forecasting. *arXiv* **2020**, arXiv:2012.01526.
108. Ronneberger, O.; Fischer, P.; Brox, T. U-Net: Convolutional Networks for Biomedical Image Segmentation. In *Medical Image Computing and Computer-Assisted Intervention—MICCAI 2015*; Navab, N., Hornegger, J., Wells, W.M., Frangi, A.F., Eds.; Springer International Publishing: Cham, Switzerland, 2015; pp. 234–241.
109. MacQueen, J.B. Some methods for classification and analysis of multivariate observations. In *Proceedings of the Fifth Berkeley Symposium on Mathematical Statistics and Probability*; University of California Press: Berkeley, CA, USA, 1967; pp. 281–297.
110. Chandra, R.; Bhattacharya, U.; Bera, A.; Manocha, D. TraPHic: Trajectory Prediction in Dense and Heterogeneous Traffic Using Weighted Interactions. In Proceedings of the 2019 IEEE/CVF Conference on Computer Vision and Pattern Recognition (CVPR), Long Beach, CA, USA, 15–20 June 2019; pp. 8475–8484. [CrossRef]
111. Li, J.; Yang, F.; Tomizuka, M.; Choi, C. EvolveGraph: Multi-Agent Trajectory Prediction with Dynamic Relational Reasoning. *arXiv* **2020**, arXiv:2003.13924.
112. Zhao, H.; Gao, J.; Lan, T.; Sun, C.; Sapp, B.; Varadarajan, B.; Shen, Y.; Shen, Y.; Chai, Y.; Schmid, C.; et al. TNT: Target-driveN Trajectory Prediction. *arXiv* **2020**, arXiv:2008.08294.
113. Gao, J.; Sun, C.; Zhao, H.; Shen, Y.; Anguelov, D.; Li, C.; Schmid, C. VectorNet: Encoding HD Maps and Agent Dynamics from Vectorized Representation. *arXiv* **2020**, arXiv:2005.04259.
114. Rhinehart, N.; McAllister, R.; Kitani, K.; Levine, S. PRECOG: PREdiction Conditioned On Goals in Visual Multi-Agent Settings. In Proceedings of the International Conference on Computer Vision (ICCV), Seoul, Korea, 27 October–2 November 2019.
115. Bhattacharyya, A.; Hanselmann, M.; Fritz, M.; Schiele, B.; Straehle, C.N. Conditional Flow Variational Autoencoders for Structured Sequence Prediction. *arXiv* **2020**, arXiv:1908.09008.
116. Mangalam, K.; Girase, H.; Agarwal, S.; Lee, K.H.; Adeli, E.; Malik, J.; Gaidon, A. It Is Not the Journey But the Destination: Endpoint Conditioned Trajectory Prediction. In *Computer Vision—ECCV 2020*; Vedaldi, A., Bischof, H., Brox, T., Frahm, J.M., Eds.; Springer International Publishing: Cham, Switzerland, 2020; pp. 759–776.
117. Goodfellow, I.; Pouget-Abadie, J.; Mirza, M.; Xu, B.; Warde-Farley, D.; Ozair, S.; Courville, A.; Bengio, Y. Generative Adversarial Nets. In *Advances in Neural Information Processing Systems*; Ghahramani, Z., Welling, M., Cortes, C., Lawrence, N., Weinberger, K.Q., Eds.; Curran Associates, Inc.: Red Hook, NY, USA, 2014; Volume 27.
118. Gupta, A.; Johnson, J.; Li, F.F.; Savarese, S.; Alahi, A. Social GAN: Socially Acceptable Trajectories with Generative Adversarial Networks. In Proceedings of the IEEE Conference on Computer Vision and Pattern Recognition (CVPR 2018), Salt Lake City, UT, USA, 18–23 June 2018.
119. Lai, W.C.; Xia, Z.X.; Lin, H.S.; Hsu, L.F.; Shuai, H.H.; Jhuo, I.H.; Cheng, W.H. Trajectory Prediction in Heterogeneous Environment via Attended Ecology Embedding. In Proceedings of the 28th ACM International Conference on Multimedia, MM '20, Seattle, WA, USA, 12–16 October 2020; Association for Computing Machinery: New York, NY, USA, 2020; pp. 202–210. [CrossRef]
120. Selvaraju, R.R.; Cogswell, M.; Das, A.; Vedantam, R.; Parikh, D.; Batra, D. Grad-CAM: Visual Explanations from Deep Networks via Gradient-Based Localization. In Proceedings of the 2017 IEEE International Conference on Computer Vision (ICCV), Venice, Italy, 22–29 October 2017; pp. 618–626. [CrossRef]
121. Amirian, J.; Hayet, J.B.; Pettre, J. Social Ways: Learning Multi-Modal Distributions of Pedestrian Trajectories with GANs. In Proceedings of the CVPR Workshops, Long Beach, CA, USA, 16–20 June 2019.
122. Sadeghian, A.; Kosaraju, V.; Sadeghian, A.; Hirose, N.; Rezatofighi, H.; Savarese, S. SoPhie: An Attentive GAN for Predicting Paths Compliant to Social and Physical Constraints. In Proceedings of the IEEE Conference on Computer Vision and Pattern Recognition, CVPR 2019, Long Beach, CA, USA, 16–20 June 2019; pp. 1349–1358. [CrossRef]
123. Messaoud, K.; Deo, N.; Trivedi, M.M.; Nashashibi, F. Trajectory Prediction for Autonomous Driving based on Multi-Head Attention with Joint Agent-Map Representation. *arXiv* **2020**, arXiv:2005.02545.
124. Monti, A.; Bertugli, A.; Calderara, S.; Cucchiara, R. DAG-Net: Double Attentive Graph Neural Network for Trajectory Forecasting. *arXiv* **2020**, arXiv:2005.12661.
125. Liang, J.; Jiang, L.; Murphy, K.; Yu, T.; Hauptmann, A. The Garden of Forking Paths: Towards Multi-Future Trajectory Prediction. *arXiv* **2020**, arXiv:1912.06445.
126. Salzmann, T.; Ivanovic, B.; Chakravarty, P.; Pavone, M. Trajectron++: Dynamically-Feasible Trajectory Forecasting With Heterogeneous Data. In Proceedings of the European Conference on Computer Vision (ECCV), Glasgow, UK, 23–28 August 2020.
127. Deo, N.; Trivedi, M.M. Trajectory Forecasts in Unknown Environments Conditioned on Grid-Based Plans. *arXiv* **2020**, arXiv:2001.00735.

128. Zhou, B.; Schwarting, W.; Rus, D.; Alonso-Mora, J. Joint Multi-Policy Behavior Estimation and Receding-Horizon Trajectory Planning for Automated Urban Driving. In Proceedings of the 2018 IEEE International Conference on Robotics and Automation (ICRA), Brisbane, QLD, Australia, 21–25 May 2018; pp. 2388–2394. [CrossRef]
129. Suraj, M.S.; Grimmett, H.; Platinský, L.; Ondrúška, P. Predicting trajectories of vehicles using large-scale motion priors. In Proceedings of the 2018 IEEE Intelligent Vehicles Symposium (IV), Changshu, China, 26–30 June 2018; pp. 1639–1644. [CrossRef]
130. Hoermann, S.; Stumper, D.; Dietmayer, K. Probabilistic long-term prediction for autonomous vehicles. In Proceedings of the 2017 IEEE Intelligent Vehicles Symposium (IV), Los Angeles, CA, USA, 11–14 June 2017; pp. 237–243. [CrossRef]
131. Xu, Y.; Zhao, T.; Baker, C.; Zhao, Y.; Wu, Y.N. Learning Trajectory Prediction with Continuous Inverse Optimal Control via Langevin Sampling of Energy-Based Models. *CoRR* **2019**, arXiv:1904.05453v1 [cs.LG].
132. Andersson, J. Predicting Vehicle Motion and Driver Intent Using Deep Learning. Master's Thesis, Chalmers University of Technology, Göteborg, Sweden, 2018.
133. Silver, D.; van Hasselt, H.; Hessel, M.; Schaul, T.; Guez, A.; Harley, T.; Dulac-Arnold, G.; Reichert, D.; Rabinowitz, N.; Barreto, A.; et al. The Predictron: End-To-End Learning and Planning. In Proceedings of the 34th International Conference on Machine Learning, Sydney, Australi, 6–11 August 2017; pp. 3191–3199.
134. Silver, D.; Huang, A.; Maddison, C.J.; Guez, A.; Sifre, L.; van den Driessche, G.; Schrittwieser, J.; Antonoglou, I.; Panneershelvam, V.; Lanctot, M.; et al. Mastering the Game of Go with Deep Neural Networks and Tree Search. *Nature* **2016**, *529*, 484–489. [CrossRef]
135. Guez, A.; Weber, T.; Antonoglou, I.; Simonyan, K.; Vinyals, O.; Wierstra, D.; Munos, R.; Silver, D. Learning to Search with MCTSnets. *CoRR* **2018**, arXiv:1802.04697v2 [cs.AI].
136. Schwarting, W.; Alonso-Mora, J.; Rus, D. Planning and Decision-Making for Autonomous Vehicles. *Annu. Rev. Control Robot. Auton. Syst.* **2018**, *1*, 187–210. [CrossRef]
137. Zhou, Y.; Hu, H.; Liu, Y.; Lin, S.W.; Ding, Z. A distributed method to avoid higher-order deadlocks in multi-robot systems. *Automatica* **2020**, *112*, 108706. [CrossRef]
138. Foumani, M.; Moeini, A.; Haythorpe, M.; Smith-Miles, K. A cross-entropy method for optimising robotic automated storage and retrieval systems. *Int. J. Prod. Res.* **2018**, *56*, 6450–6472. [CrossRef]
139. Foumani, M.; Gunawan, I.; Smith-Miles, K. Resolution of deadlocks in a robotic cell scheduling problem with post-process inspection system: Avoidance and recovery scenarios. In Proceedings of the 2015 IEEE International Conference on Industrial Engineering and Engineering Management (IEEM), Singapore, 6–9 December 2015; pp. 1107–1111. [CrossRef]
140. Sutton, R.S.; Barto, A.G. *Reinforcement Learning*; MIT Press: Cambridge, MA, USA, 2018.
141. Lapan, M. *Deep Reinforcement Learning Hands-On*; Packt Publishing: Birmingham, UK, 2018.
142. Grondman, I.; Busoniu, L.; Lopes, G.A.D.; Babuska, R. A Survey of Actor-Critic Reinforcement Learning: Standard and Natural Policy Gradients. *IEEE Trans. Syst. Man Cybern. Part (Appl. Rev.)* **2012**, *42*, 1291–1307. [CrossRef]
143. Mnih, V.; Kavukcuoglu, K.; Silver, D.; Rusu, A.A.; Veness, J.; Bellemare, M.G.; Graves, A.; Riedmiller, M.; Fidjeland, A.K.; Ostrovski, G.; et al. Human-level control through deep reinforcement learning. *Nature* **2015**, *518*, 529–533. [CrossRef] [PubMed]
144. Konda, V.R.; Tsitsiklis, J.N. Actor-Critic Algorithms. *SIAM* **2000**, *42*, 1008–1014.
145. Mnih, V.; Badia, A.P.; Mirza, M.; Graves, A.; Lillicrap, T.P.; Harley, T.; Silver, D.; Kavukcuoglu, K. Asynchronous Methods for Deep Reinforcement Learning. *arXiv* **2016**, arXiv:1602.01783.
146. Schulman, J.; Wolski, F.; Dhariwal, P.; Radford, A.; Klimov, O. Proximal Policy Optimization Algorithms. *CoRR* **2017**, arXiv:1707.06347v2 [cs.LG].
147. Schulman, J.; Levine, S.; Moritz, P.; Jordan, M.I.; Abbeel, P. Trust Region Policy Optimization. *CoRR* **2015**, arXiv:1502.05477v5 [cs.LG].
148. Weber, T.; Racanière, S.; Reichert, D.P.; Buesing, L.; Guez, A.; Rezende, D.J.; Badia, A.P.; Vinyals, O.; Heess, N.; Li, Y.; et al. Imagination-Augmented Agents for Deep Reinforcement Learning. In Proceedings of the 31st International Conference on Neural Information Processing, Long Beach, CA, USA, 4–9 December 2017; pp. 5694–5705. [CrossRef]
149. Liu, B.; Ghavamzadeh, M.; Gemp, I.; Liu, J.; Mahadevan, S.; Petrik, M. Proximal Gradient Temporal Difference Learning: Stable Reinforcement Learning with Polynomial Sample Complexity. *J. Artif. Intell. Res.* **2018**, *63*, 461–494. [CrossRef]

Article

Solving Regression Problems with Intelligent Machine Learner for Engineering Informatics

Jui-Sheng Chou [1,*], Dinh-Nhat Truong [1,2] and Chih-Fong Tsai [3]

1. Department of Civil and Construction Engineering, National Taiwan University of Science and Technology, Taipei City 106335, Taiwan; d10605806@mail.ntust.edu.tw
2. Department of Civil Engineering, University of Architecture Ho Chi Minh City (UAH), Ho Chi Minh City 700000, Vietnam
3. Department of Information Management, National Central University, Taoyuan City 320317, Taiwan; cftsai@mgt.ncu.edu.tw
* Correspondence: jschou@mail.ntust.edu.tw

Abstract: Machine learning techniques have been used to develop many regression models to make predictions based on experience and historical data. They might be used singly or in ensembles. Single models are either classification or regression models that use one technique, while ensemble models combine various single models. To construct or find the best model is very complex and time-consuming, so this study develops a new platform, called intelligent Machine Learner (iML), to automatically build popular models and identify the best one. The iML platform is benchmarked with WEKA by analyzing publicly available datasets. After that, four industrial experiments are conducted to evaluate the performance of iML. In all cases, the best models determined by iML are superior to prior studies in terms of accuracy and computation time. Thus, the iML is a powerful and efficient tool for solving regression problems in engineering informatics.

Keywords: applied machine learning; classification and regression; data mining; ensemble model; engineering informatics

1. Introduction

Machine Learning (ML)-based methods for building prediction models have attracted abundant scientific attention and are extensively used in industrial engineering [1–3], design optimization of electromagnetic devices, and other areas [4,5]. The ML-based methods have been confirmed to be effective for solving real-world engineering problems [6–8]. Various supervised ML techniques (e.g., artificial neural network, support vector machine, classification and regression tree, linear (ridge) regression, and logistic regression) are typically used individually to construct single models and ensemble models [9,10]. To construct a series of models and identify the best one among these ML techniques, users need a comprehensive knowledge of ML and spend a significant effort building advanced models.

The primary objective of this research is to develop a user-friendly and powerful ML platform, called intelligent Machine Learner (iML), to help its users to solve real-world engineering problems with a shorter training time and greater accuracy than before. The iML can automatically build and scan all regression models, and then identify the best one. Novice users with no experience of ML can easily use this system. Briefly, the iML (1) helps users to make prediction model easily; (2) provides an overview of the parameter settings for the purpose of making objective choices; and (3) yields clear performance indicators, facilitating reading and understanding of the results, on which decisions can be based.

Four experiments were carried out to evaluate the performance of iML and were compared with previous studies. In the first experiment, empirical data concerning enterprise resource planning (ERP) for software projects by a leading Taiwan software provider over the last five years were collected and analyzed [1]. The datasets in the other three experiments were published on the UCI website [11–13]. Specifically, the purpose of the second experiment was to train a regression model of comparing the performance of CPU processors by using some characteristics as input. The third experiment involved forecasting the demand supporting structured productivity and high levels of customer service, and the fourth experiment involved estimating the total bikes rented per day.

The rest of this paper is organized as follows. Section 2 reviews application of machine learning techniques in various disciplines. Section 3 presents the proposed methodology and iML framework. Section 4 introduces the evaluation metrics to measure accuracy of the developed system. Section 5 demonstrates iML's interface. Section 6 shows benchmarks between iML and WEKA (a free, open source program). Section 7 exhibits the applicability of iML in numerical experiments. Section 8 draws conclusions, and provides managerial implications and suggestions for future research.

2. Literature Review

Numerous researchers in various fields, such as ecology [14,15], materials properties [16–18], water resource [19], energy management [20], and decision support [21,22], use data-mining techniques to solve regression problems, and especially project-related problems [23,24]. Artificial neural network (ANN), support vector machine/regression (SVM/SVR) classification and regression tree (CART), linear ridge regression (LRR), and logistic regression (LgR) are the most commonly used methods for this purpose and are all considered to be among the best machine learning techniques [25–27]. Similarly, four popular ensemble models, including voting, bagging, stacking and tiering [28–30], can be built based on the meta-combination rules of aforementioned single models.

Chou (2009) [31] developed a generalized linear model-based expert system for estimating the cost of transportation projects. Dandikas et al. (2018) [32] assessed the advantages and disadvantages of regression models for predicting potential of biomethane. The results indicated that the regression method could predict variations in the methane yield and could be used to rank substrates for production quality. However, least squares-based regression usually leads to overfitting a model, failure to find unique solutions, and issues dealing with multicollinearity among the predictors [33], so ridge regression, another type of regularized regression, is favorably integrated in this study to avoid the above problems. Additionally, Sentas and Angelis (2006) [34] investigated the possibility of using some machine learning methods for estimating categorical missing values in software cost databases. They concluded that multinomial logistic regression was the best for imputation owing to its superior accuracy.

The general regression neural network was originally designed chiefly to solve regression problems [24,35]. Caputo and Pelagagge (2008) [36] compared the ANN with the parametric methods for estimating the cost of manufacturing large, complex-shaped pressure vessels in engineer-to-order manufacturing systems. Their comparison demonstrated that the ANN was more effective than the parametric models, presumably because of its better mapping capabilities. Rocabruno-Valdés et al. (2015) [37] developed models based on ANN for predicting the density, dynamic viscosity, and cetane number of methyl esters and biodiesel. Similarly, Ganesan et al. (2015) [38] used ANN to predict the performance and exhaust emissions of a diesel electricity generator.

SVM was originally developed by Vapnik (1999) for classification (SVM) and regression (SVR) [39,40]. Jing et al. (2018) [41] used SVM to classify air balancing, which is a key element for heating, ventilating, air-conditioning (HAVC), and variable air volume (VAV) system installation, and is useful for improving the energy efficiency by minimizing unnecessary fresh air to the air-conditioned zones. The results demonstrated that SVM achieved 4.6% of relative error value and is a promising approach for air balancing. García-Floriano et al. (2018) [42] used SVR to model software maintenance (SM) effort prediction. The SVR model was superior to regression, neural networks, association rules and decision trees, with 95% confidence level.

The classification and regression tree method (CART), introduced by Breiman et al. (2017) [43], is an effective method to solve classification and regression problems [42]. Choi and Seo (2018) [44] predicted the fecal coliform in the North Han River, South Korea by CART models, the test results showed the total correct classification rates of the four models ranged from 83.7% to 93.0%. Ru et al. (2016) [45] used the CART model to predict cadmium enrichment levels in reclaimed coastal soils. The results showed that cadmium enrichment levels had an accuracy of 78.0%. Similarly, Li (2006) [16] used CART to predict materials properties and behavior. Chou et al. (2014, 2017) [26,46] utilized the CART method to modeling steel pitting risk and corrosion rate and forecasting project dispute resolutions.

In addition to the aforementioned single models, Elish (2013) [47] used voting ensemble for estimating software development effort. The ensemble model outperformed all the single models in terms of Mean Magnitude of Relative Error (MMRE), and achieved competitive percentage of observations whose Magnitude of Relative Error (MRE) is less than 0.25 (PRED (25)) and recently proposed Evaluation Function (EF) results. Wang at el. (2018) demonstrated that ensemble bagging tree (EBT) model could accurately predict hourly building energy usage with MAPE ranging from 2.97% to 4.63% [48]. Comparing to the conventional single prediction model, EBT is superior in prediction accuracy and stability. However, it requires more computation time and is short of interpretability owing to its sophisticated model structure.

Chen et al. (2019) [49] showed that the stacking model outperformed the individual models, achieving the highest R^2 of 0.85, followed by XGBoost (0.84), AdaBoost (0.84) and random forest (0.82). For the estimation of hourly PM2.5 in China, the stacking model exhibited relatively high stability, with R^2 ranging from 0.79 to 0.92. Basant at el. (2016) [50] proposed a three-tier quantitative structure-activity relationship (QSAR) model. This model can be used for the screening of chemicals for future drug design and development process and safety assessment of the chemicals. In comparison with previously studies, the QSAR models on the same endpoint property showed the encouraging statistical quality of the proposed models.

According to the reviewed literature, various machine learning platforms have been developed for the past decades, such as the Scikit-Learn Python libraries, Google's TensorFlow, WEKA and Microsoft Research's CNTK. Users can find it easy to use a machine learning tool and/or framework to solve numerous problems as per their needs [51]. ML-based approaches have been confirmed to be effective in providing decisive information. Since there is no best model suitable to predict all problems (the "No Free Lunch" theorem [52,53]), a comprehensive comparison of single and ensemble models embedded within an efficient forecasting platform for solving real-world engineering problems is imperatively needed. The iML platform proposed in this study can efficiently address this issue.

3. Applied Machine Learning
3.1. Classification and Regression Model
3.1.1. Artificial Neural Network (ANN)

Neural networks (or artificial neural networks) comprise information-processing units, which are similar to the neurons in the human brain, except that a neural network is composed of artificial neurons (Figure 1) [54]. Particular, back-propagation networks (BPNNs) are widely used, and are known to be the most effective network models [55,56].

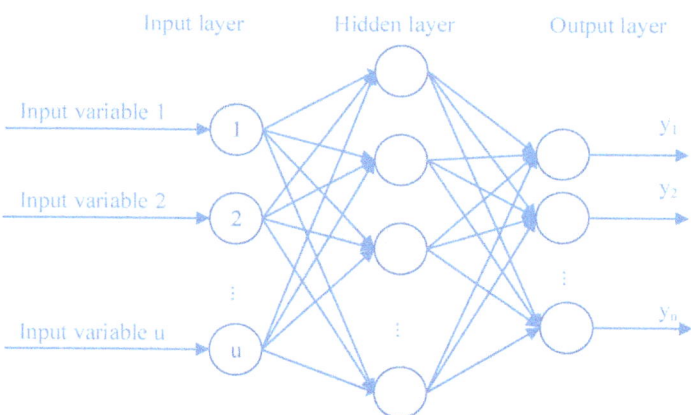

Figure 1. Artificial neural network (ANN) model.

Equation (1) uses sigmoid function to activate each neuron in a hidden output layer, and the Scaled Conjugate Gradient Algorithm is used to calculate the weights of the network. BPNNs will be trained until the stopping criteria is reached by default settings in MATLAB.

$$net_k = \sum w_{kj} O_j \text{ and } y_k = f(net_k) = \frac{1}{1 + e^{-net_k}} \qquad (1)$$

where net_k is the activation of the kth neuron; j is the set of neurons in the preceding layer; w_{kj} is the weight of the connection between neuron k and neuron j; O_j is the output neuron j; and y_k is the sigmoid or logistic transfer function.

3.1.2. Support Vector Machine (SVM) and Support Vector Regression (SVR)

Developed by Cortes and Vapnik (1995) [57], SVM is used for binary classification problems. The SVM was created based on decision hyper-planes that determine decision boundaries in an input space or a high-dimensional feature space [40,58]. Binary classification can only classify samples into negative and positive while multi-class classification problems are complex (Figure 2). In this study, One Against All (OAA) is used to solve multiple classification problems.

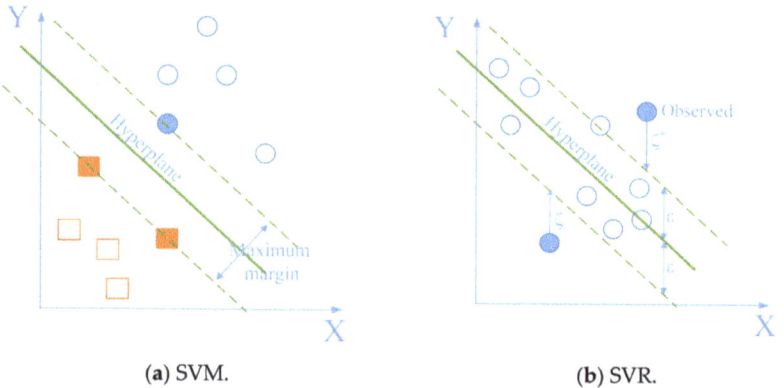

(a) SVM. (b) SVR.

Figure 2. Support Vector Machine (SVM) and Support Vector Regression (SVR) models.

The OAA-SVM constructs m SVM models for m-class classification problems, and the ith SVM model is trained based on the dataset of the ith class which includes a positive class and a negative class. In training, a set of l data points $(x_i, y_i)_{i=1}^{l}$, where $x_i \in R^n$ the input data, and $y_i \in (1, 2, \ldots, m)$ is the class label of x_i; the ith SVM model is solved using the following optimization problem equation [59].

$$\min_{w^i, b, \xi} J\left(w^i, b, \xi\right) = \frac{1}{2}\left(w^i\right)^T w^i + C \sum_{i=1}^{l} \xi_j^i \quad (2)$$

$$\text{subject to:} \begin{cases} \left(w^i\right)^T \varphi(x_j) + b^i \geq 1 - \xi_j^i, y_j = i, \\ \left(w^i\right)^T \varphi(x_j) + b^i \leq -1 + \xi_j^i, y_j \neq i, \\ \xi_j^i \geq 0, j = 1, \ldots, l. \end{cases} \quad (3)$$

When the SVM models have been solved, the class label of example x is predicted as follows:

$$y(x) = \arg \max_{i=1\ldots m} \left(\left(w^i\right)^T \varphi(x) + b^i\right) \quad (4)$$

where i is the ith SVM model; w^i is a vector normal to the hyper-plane; b^i is a bias, $\varphi(x)$ is a nonlinear function that maps x to a high-dimension feature space, ξ^i is the error in misclassification, and $C \geq 0$ is a constant that specifies the trade-off between the classification margin and the cost of misclassification.

To train the SVM model, radial basic function (RBF) kernel maps samples non-linearly into a feature space with more dimensions. In this study, the RBF kernel is used as SVM kernel function.

$$K(x_i, x_j) = \exp\left(\frac{-\|x_i - x_j\|^2}{2\sigma^2}\right) \quad (5)$$

where σ is a positive parameter that controls the radius of RBF kernel function.

Support vector regression (SVR) [40] is one version of SVM. SVR computes a linear regression function for the new higher-dimensional feature space using ε-insensitive loss while simultaneously reducing model complexity of the model by minimizing $\|w\|^2$. This process can be implemented by introducing (non-negative) slack variables ξ_i, ξ_i^* to measure the deviation in training samples outside the ε-insensitive zone. The SVR can be formulated as the minimization of the following equation:

$$\min_{w, b, \xi} J(w, b, \xi) = \frac{1}{2}(w)^T w + C \sum_{i=1}^{l} (\xi_i + \xi_i^*) \quad (6)$$

$$\text{subject to:} \begin{cases} y_i - f(x_i, w) \leq \varepsilon + \xi_i^* \\ f(x_i, w) - y_i \leq \varepsilon + \xi_i \\ \xi_i^*, \xi_i \geq 0, i = 1, \ldots, n \end{cases} \quad (7)$$

When SVR model has been solved, the value of example x is predicted as follows.

$$f(x) = \sum (\alpha_i - \alpha_i^*) K(x_i, x) + b \quad (8)$$

where $K(x_i, x)$ is the kernel function and α_i^*, α_i are Lagrange multipliers in the dual function.

3.1.3. Classification and Regression Tree (CART)

Classification and regression tree technique is described as a tree on which each internal (non-leaf) node represents a test of an attribute, each branch represents the test result, and each leaf (or terminal) node has a class label and class result (Figure 3) [60]. The tree is "trimmed" until total error is minimized to optimize the predictive accuracy of the tree by minimizing the number of branches. The training CART is constructed through the Gini index. The formulas are as follows.

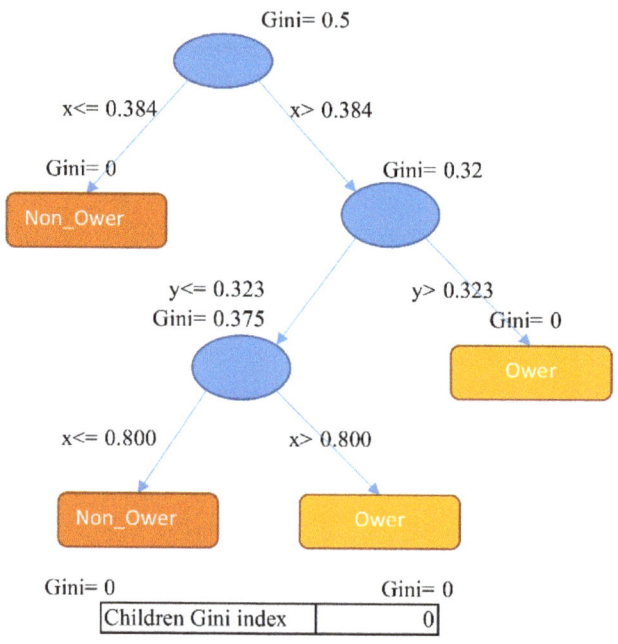

Figure 3. The classification and regression tree (CART) model.

$$g(t) = \sum_{j \neq i} p(j|t)p(i|t) \tag{9}$$

$$p(j|t) = \frac{p(j,t)}{p(t)} \tag{10}$$

$$p(j,t) = \frac{p(j)N_j(t)}{N_j} \tag{11}$$

$$p(t) = \sum_j p(j,t) \tag{12}$$

$$Gini\ index = 1 - \sum p(j,t)^2 \tag{13}$$

where i and j are the categorical variables in each item; $N_j(t)$ is the recorded number of nodes t in category j; and N_j. is the recorded number of the root nodes in category j; and $p(j)$ is the prior probability value for category j.

3.1.4. Linear Ridge Regression (LRR) and Logistic Regression (LgR)

Statistical models of the relationship between dependent variables (response variables) and independent variables (explanatory variables) are developed using linear regression (Figure 4). The general formula for multiple regression models is as follows.

$$y = f(x) = \beta_o + \sum_{j=1}^{n} \beta_j x_j + \varepsilon \tag{14}$$

where y is a dependent variable; β_o is a constant; β_j is a regression coefficient ($j = 1, 2, \ldots, n$), and ε is an error term.

Linear ridge regression (LRR) is a regularization technique that can be used together with generic regression algorithms to model highly correlated data [61,62]. Least squares method is a powerful technique for training the LRR model, which denotes β to minimize the Residual Sum Squares (RSS)-function. Therefore, the cost function is presented as below.

$$Cost(\beta) = RSS(\beta) = \sum_{i=1}^{l}(y-y')^2 + \lambda\left(\sum_{j=1}^{n}\beta_j^2\right) \quad (15)$$

$$y' = \beta_0 + \sum \beta_j x_j \quad (16)$$

where λ is a pre-chosen constant, which is the product of a penalty term and the squared norm in the β vector of regression method, and y' is the predicted values.

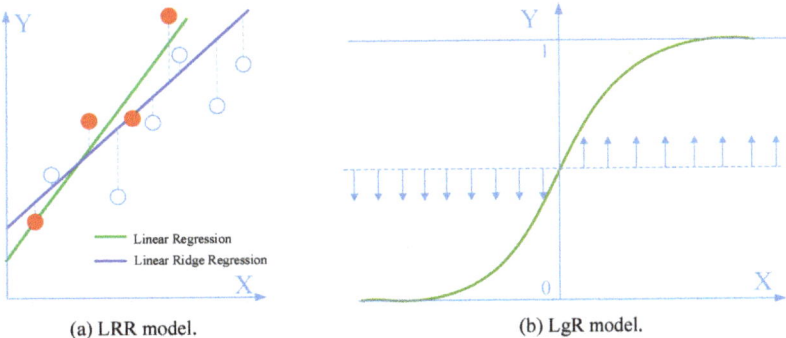

(a) LRR model.   (b) LgR model.

Figure 4. Linear Ridge Regression (LRR) and Logistic Regression (LgR) models.

Statistician David Cox developed logistic regression in 1958 [63]. An explanation of logistic regression begins with an explanation of the standard logistic function. Equation (17) mathematically represents the logistic regression model.

$$p(x) = \frac{1}{1+e^{-(\beta_0+\sum_{j=1}^{n}\beta_j x_j)}} \quad (17)$$

where $p(x)$ is the probability that the dependent variable equals a "success" or "case" rather than a failure or non-case. β_0 and β_j are found by minimizing cost function defined in Equation (18).

$$Cost(\beta) = -\left(\sum_{i=1}^{l}(y_i \ln(p(x_i)) + (1-y_i)\ln(1-p(x_i)))\right) + \frac{\lambda}{2}\sum_{j=1}^{n}\beta_j^2 \quad (18)$$

where y_i is the observed outcome of case x_i, having 0 or 1 as possible values [64]

3.2. Ensemble Regression Model

In this study, several ensemble schemes, including voting, bagging, stacking, and tiering were investigated using the input data and described as below.

- *Voting*: The voting ensemble model combines the outputs of the single models using a meta-rule. The mean of the output values is used in this study. According to the adopted ML models, 11 voting models are trained in this study, including (1) ANN + SVR, (2) ANN + CART, (3) ANN + LRR, (4) SVR + CART, (5) SVR + LRR, (6) CART + LRR, (7) ANN + SVR + CART, (8) ANN + CART + LRR, (9) ANN + CART + LRR, (10) SVR + CART + LRR, (11) ANN + SVR + CART + LRR. Figure 5a presents the voting ensemble model.
- *Bagging*: The bagging ensemble model duplicates samples at random, and each regression model predicts values from the samples independently. The meta-rule is

applied to all of the outputs in this study. Bagging ensemble model is depicted at Figure 5b.
- *Stacking*: The stacking ensemble model is a two-stage model, and Figure 5c describes the principle of the model. In stage 1, each single model predicts one output value. Then, these outputs are used as inputs to train a model by these machine learning techniques again to make a meta-prediction in stage 2. There are four stacking models herein, including ANN (ANN, SVR, CART, LRR); SVR (ANN, SVR, CART, LRR); CART (ANN, SVR, CART, LRR); LRR (ANN, SVR, CART, LRR).

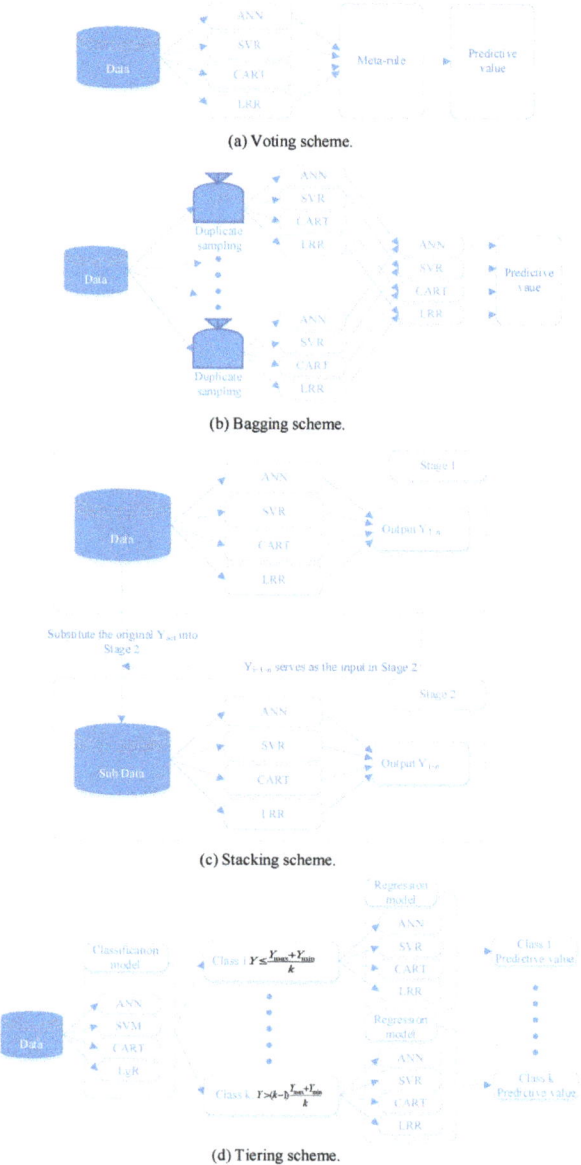

Figure 5. Ensemble models.

- *Tiering*: Figure 5d illustrates the tiering ensemble model. There are two tiers inside a tiering ensemble model in this study. The first tier is to classify data into *k* classes on the basis of T value [18]. Machine learning technique in the first tier for classifying data needs to be identified. After classifying the data, the regression machine learning is used to train each data (Sub Data) of each class (second tier) to predict results. In the iML, we developed three types of models, including 2-class, 3-class, and 4-class. The equation for calculating T value is:

$$T = \frac{y_{max} + y_{min}}{k} \tag{19}$$

where T is standard value, k is the number of classes, and y_{max} and y_{min} are the maximum and minimum of actual values, respectively.

3.3. K-Fold Cross Validation

K-fold cross validation is used to compare two or more prediction models. This method randomly divides a sample into a training sample and a test sample by splitting into K subsets. K-1 subsets are selected to train the model while the other is used to test, and this training process is repeated K times (Figure 6). To compare models, the average of performance results (e.g., RMSE, and MAPE) is computed. Kohavi (1995) stated that K = 10 provides analytical validity, computational efficiency, and optimal deviation [65]. Thus, K = 10 is used in this study. Performance metrics will be explained in details Section 4.

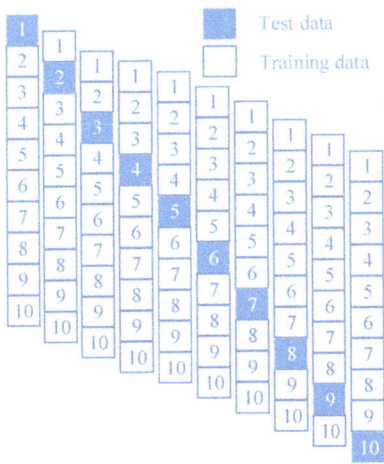

Figure 6. K-fold cross-validation method.

3.4. Intelligent Machine Learner Framework

Figure 7 presents the structure of iML. In stage 1 (data preprocessing), the data is classified distinctly for particular use in the Tiering ensemble model. Meanwhile, all data is divided into two main data groups, namely, learning data and test data, and the learning data is duplicated for training ensemble models.

At the next stage, all retrieved data is automatically used for training models, which include single models (ANN, SVR, LRR, and CART), and ensemble models (voting, bagging, stacking, and tiering). Notably, the tiering ensemble model needs to employ a classification technique to assign a class label to the original input at the first tier. A corresponding regression model for the particular class is then adopted at the second tier to obtain the predictive value [17,26].

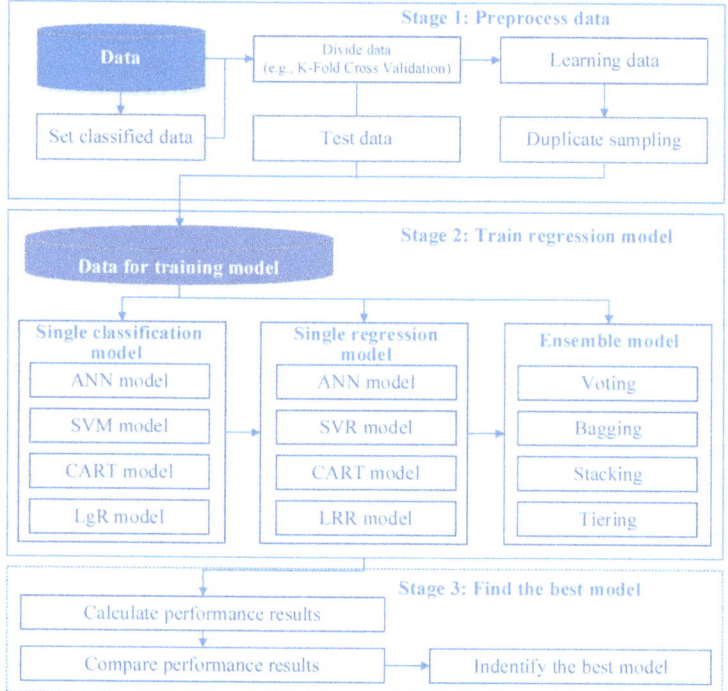

Figure 7. Intelligent machine leaner framework.

Finally, in stage 3 (find the best model), the predictive performances of all the models learned (trained) in stage 2 using test dataset are compared to identify the best models. Section 4 describes the performance evaluation metrics in detail.

4. Mathematical Formulas for Performance Measures

To measure the performance of classification models, the accuracy, precision, sensitivity, specificity and the area under the curve (AUC) are calculated. For the regression models, five-performance measures, (i.e., correlation coefficient (R), mean absolute error (MAE), mean absolute percentage error (MAPE), root mean squared error (RMSE), and total error rate (TER)) are calculated. Table 1 presents a confusion matrix and Table 2 exhibits those performance measures [17,66].

Table 1. Confusion matrix.

		Actual Class	
		Positive	Negative
Predicted class	Positive	True positive	False Negative
	Negative	False positive	True negative

In Table 2, MAE is the mean absolute difference between the prediction and the actual value. MAPE represents the mean percentage error between prediction and actual value, the smaller value of MAPE, the better prediction result achieved by the model. The MAPE is the index typically used to evaluate the accuracy of prediction models. RMSE represents the dispersion of errors by a prediction model. The statistical index that shows the linear correlation between two variables is denoted as R. Lastly, TER is the total difference of predicted and actual values [17].

Table 2. Mathematical formulas for performance measures.

Measure	Formula	Measure	Formula
Accuracy	Accuracy $= \frac{tp+tn}{tp+fp+tn+fn}$	Mean absolute error	MAE $= \frac{1}{n} \sum_{i=1}^{n} \left| y_i - y'_i \right|$
Precision	Precision $= \frac{tp}{tp+fp}$	Mean absolute percentage error	MAPE $= \frac{1}{n} \left| \frac{y_i - y'_i}{y_i} \right|$
Sensitivity	Sensitivity $= \frac{tp}{tp+fn}$	Root mean square error	RMSE $= \sqrt{\frac{1}{n} \sum_{i=1}^{n} (y_i - y'_i)^2}$
Specificity	Specificity $= \frac{tn}{tn+fp}$	Correlation coefficient	R $= \frac{n \sum y_i \cdot y'_i - (\sum y_i)(\sum y'_i)}{\sqrt{n(y_i^2) - (\sum y_i)^2} \sqrt{n(y'_i{}^2) - (\sum y'_i)^2}}$
Area under the curve	AUC $= \frac{1}{2} \left[\left(\frac{tp}{tp+fn} \right) + \left(\frac{tn}{tn+fp} \right) \right]$	Total error rate	TER $= \frac{\left| \sum_{i=1}^{n} y'_i - \sum_{i=1}^{n} y_i \right|}{\sum_{i=1}^{n} y_i}$

tp is the true positives (number of correctly recognized class examples); tn is the true negatives (number of correctly recognized examples that do not belong to the class); fp is the number of false positives (number of examples that were incorrectly assigned to a class); fn is the number of false negatives (number of examples that were not assigned to a class); y_i is actual value; y'_i is predicted value; n is sample size.

The goal is to identify the model that yields the lowest error of test data. To obtain a comprehensive performance measure, the five statistical measures (RMSE, MAE, MAPE, 1-R, and TER) were combined into a synthesis index (SI) using Equation (20). Based on the SI values, the best model is identified.

$$\text{SI} = \frac{1}{m_p} \sum_{i=1}^{m_p} \left(\frac{P_i - P_{min,i}}{P_{max,i} - P_{min,i}} \right) \quad (20)$$

where m_p = number of performance measures; $P_i = i^{th}$ performance measure; and $P_{min,i}$ and $P_{max,i}$ are the maximum and minimum of ith measure. The SI range is 0–1; the SI value close to 0 indicates a better accuracy of the predictive model.

5. Design and Implementation of iML Interface

The iML was developed in MATLAB R2016a on a PC with an Intel Core i5-750 CPU, a clock speed of 3.4 GHz, and 8 GB of RAM, running Windows 10. Figure 8 presents a user-friendly interface for iML. First, users select models on setting-parameters board and set the parameters for the chosen models, which will be trained and analyzed. Next, users choose whether to test with either "K-Fold Validation" or "Percentage Split" before uploading the data. Notably, if "Percentage Split" is selected, the user only has to input percentage value of learning data. Then, users click on the "Run" button to train the model. Finally, the "Make Report" function is to create a report containing performance metrics of all selected models and the identified best model. Figure 9 displays a snapshot of report file in notepad.

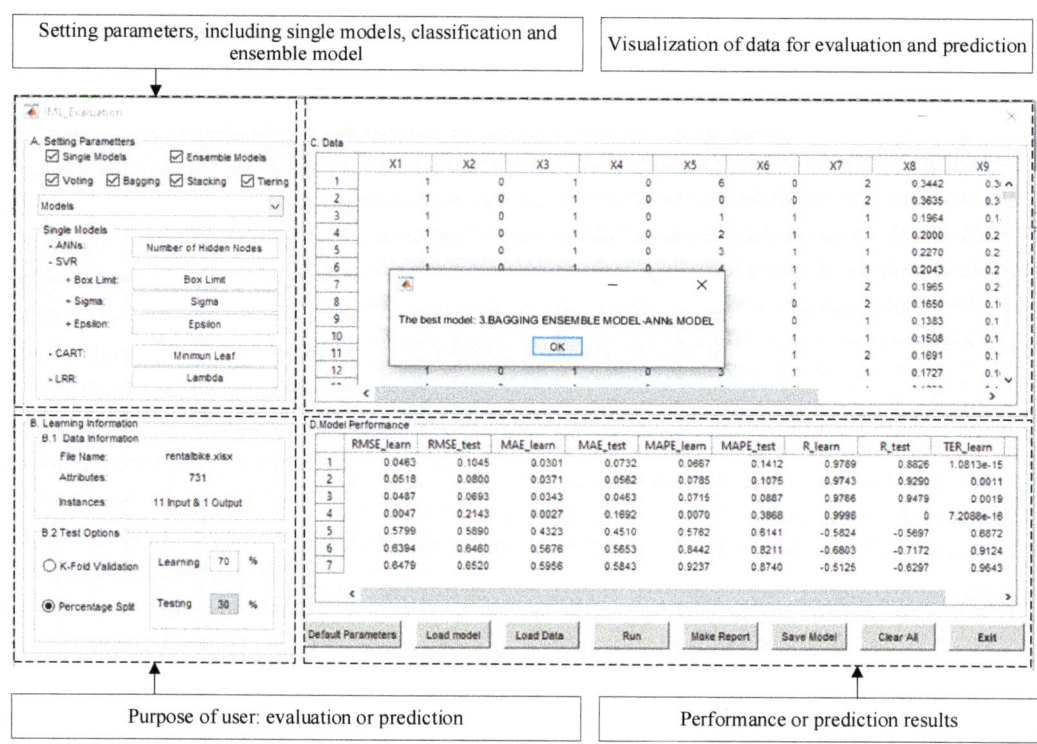

Figure 8. Snapshot of intelligent Machine Learner (iML) interface.

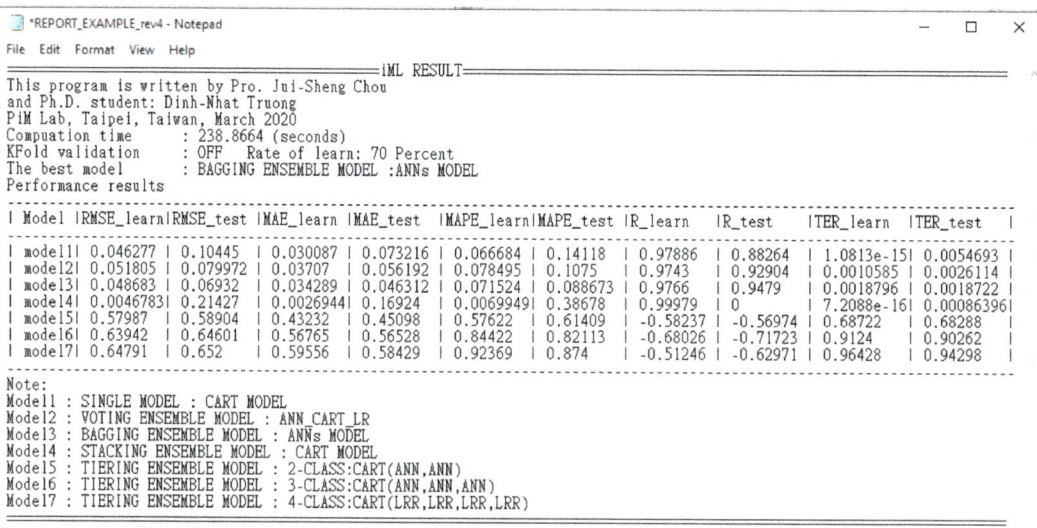

Figure 9. Snapshot of report file.

6. Benchmarks between iML and WEKA

6.1. Publicly Available Datasets

Table 3 shows the publicly available datasets from the UCI Machine Learning Repository (https://archive.ics.uci.edu/mL/index.php; accessed 1 March 2021). The iML is benchmarked with WEKA (a free, open source program) using hold-out validation and K-fold cross-validation on the target datasets. All algorithm parameters are set default for both iML and WEKA platforms.

Table 3. Characteristic of data from UCI Machine Learning Repository.

UCI Data Set	No. of Samples	No. of Attributes	Output Information
Concrete Compressive Strength (Yeh (2006) [67])	1030	8	Concrete compressive strength (MPa)
Real estate valuation (Yeh and Hsu (2018) [68])	414	6	Y = house price of unit area (10,000 New Taiwan Dollar/Ping, where Ping is a local unit, 1 Ping = 3.3 m squared)
Energy efficiency (Tsanas and Xifara (2012) [69])	768	8	y1 Heating Load (kW)
			y2 Cooling Load (kW)
Airfoil Self-Noise (Lau and López (2009) [70])	1503	5	Scaled sound pressure level (dB).

6.2. Hold-Out Validation

In this test, datasets are randomly partitioned into 80% and 20% for learning and test, respectively. Tables 4–8 show the one-time performance results on these five datasets. A model with a normalized SI value of 0.000 is the best prediction model among all the models tested by iML and WEKA. Notably, the best model can be automatically identified by iML with "one-click". To train models with WEKA, the users need to build each model individually. Moreover, iML gives better test results of single, voting and bagging models than those of WEKA. Based on the benchmark results, iML is effective to find the best model in the hold-out validation.

Table 4. Test results by WEKA and iML on concrete compressive strength dataset via hold-out validation.

Model	WEKA					iML				
	R	RMSE (MPa)	MAE (MPa)	MAPE (%)	SI (Ranking)	R	RMSE (MPa)	MAE (MPa)	MAPE (%)	SI (Ranking)
I. Single	0.927	CART 6.546	5.170	18.770	0.142 (7)	0.946	ANN 5.302	3.728	12.673	0.023 (3)
II. Voting	0.936	ANN + CART 6.202	4.930	19.090	0.124 (6)	0.956	ANN + CART 4.771	3.550	12.723	**0.000 (1)**
III. Bagging	0.960	CART 5.044	3.983	15.130	0.032 (4)	0.951	ANN 5.056	3.647	12.249	0.010 (2)
IV. Stacking	0.939	(*) CART 5.986	4.792	17.520	0.104 (5)	0.444	(*) LRR 14.829	11.779	56.775	1.000 (8)

Note: (*) is (ANN + SVR + CART + LRR); bold value denotes the best overall performance.

Table 5. Test results by WEKA and iML on real estate dataset via hold-out validation.

Model	WEKA					iML				
	R	RMSE (U)	MAE (U)	MAPE (%)	SI (Ranking)	R	RMSE (U)	MAE (U)	MAPE (%)	SI (Ranking)
I. Single	0.740	CART 10.762	5.882	13.210	0.321 (6)	0.871	ANN 6.630	4.912	13.591	0.049 (3)
II. Voting	0.745	ANN + CART + LRR 11.054	5.908	12.780	0.327 (7)	0.877	ANN + CART 6.615	4.739	12.867	0.030 (2)
III. Bagging	0.770	CART 10.321	5.281	11.760	0.246 (4)	0.884	CART 6.485	4.381	12.305	**0.000 (1)**
IV. Stacking	0.744	(*) CART 10.748	5.774	12.940	0.311 (5)	0.391	(*) ANN 12.638	10.411	33.807	1.000 (8)

Note: (*) is (ANN + SVR + CART + LRR); bold value denotes the best overall performance; U: house price of unit area (10,000 New Taiwan Dollar/Ping, where Ping is a local unit, 1 Ping = 3.3 m squared).

Table 6. Test results by WEKA and iML on energy efficiency data set (Heating load) via hold-out validation.

Model	WEKA					iML				
	R	RMSE (kW)	MAE (kW)	MAPE (%)	SI (Ranking)	R	RMSE (kW)	MAE (kW)	MAPE (%)	SI (Ranking)
I. Single	0.996	CART 0.914	0.646	3.300	0.418 (6)	0.999	ANN 0.488	0.354	1.700	0.046 (3)
II. Voting	0.996	ANN + CART 0.929	0.729	3.820	0.449 (7)	0.999	ANN + CART 0.495	0.336	1.617	0.045 (2)
III. Bagging	0.997	CART 0.870	0.619	3.210	0.354 (5)	0.999	ANN 0.426	0.311	1.519	**0.000 (1)**
IV. Stacking	0.998	(*) CART 0.754	0.524	2.480	0.231 (4)	0.998	(*) LRR 3.454	3.226	17.658	1.000 (8)

Note: (*) is (ANN + SVR + CART + LRR); bold value denotes the best overall performance.

Table 7. Test results by WEKA and iML on energy efficiency dataset (Cooling load) via hold-out validation.

Model	WEKA					iML				
	R	RMSE (kW)	MAE (kW)	MAPE (%)	SI (Ranking)	R	RMSE (kW)	MAE (kW)	MAPE (%)	SI (Ranking)
I. Single	0.986	CART 1.524	1.006	3.900	0.320 (6)	0.992	ANN 1.231	0.884	3.577	0.038 (2)
II. Voting	0.987	ANN + CART 1.504	1.064	4.330	0.293 (4)	0.988	ANN + CART 1.509	0.982	3.544	0.222 (3)
III. Bagging	0.986	CART 1.565	1.046	4.030	0.358 (7)	0.993	ANN 1.177	0.809	3.165	**0.000 (1)**
IV. Stacking	0.986	(*) SVR 1.537	0.979	3.700	0.314 (5)	0.989	(*) LRR 4.290	3.762	17.305	1.000 (8)

Note: (*) is (ANN + SVR + CART + LRR); bold value denotes the best overall performance.

Table 8. Test results by WEKA and iML on airfoil self-noise dataset via hold-out validation.

Model	WEKA				SI (Ranking)	iML				SI (Ranking)
	R	RMSE (dB)	MAE (dB)	MAPE (%)		R	RMSE (dB)	MAE (dB)	MAPE (%)	
I. Single	0.898	CART 3.185	2.339	1.880	0.502 (6)	0.953	ANN 2.149	1.577	1.259	0.044 (2)
II. Voting	0.893	ANN + CART 3.471	2.649	2.100	0.591 (7)	0.952	ANN + CART 2.163	1.633	1.301	0.058 (3)
III. Bagging	0.922	CART 2.902	2.135	1.710	0.332 (4)	0.958	ANN 2.031	1.494	1.194	**0.000 (1)**
IV. Stacking	0.905	(*) CART 3.082	2.271	1.820	0.450 (5)	0.952	(*) LRR 7.050	5.648	4.613	1.000 (8)

Note: (*) is (ANN + SVR + CART + LRR); bold value denotes the best overall performance.

6.3. K-Fold Cross-Validation

Tenfold cross-validation is used to evaluate the generalized performance of WEKA and iML. Tables 9–13 show the average performance measures of five datasets, respectively. Similarly, iML identifies better models in single, voting, and bagging schemes than those trained by WEKA. The best model for each dataset is automatically determined by iML. Therefore, iML is a powerful tool to find the best model in the cross-fold validation.

Table 9. Performance of WEKA and iML on concrete compressive strength dataset via tenfold cross-validation.

Model	WEKA				SI (Ranking)	iML				SI (Ranking)
	R	RMSE (MPa)	MAE (MPa)	MAPE (%)		R	RMSE (MPa)	MAE (MPa)	MAPE (%)	
I. Single	0.923	CART 6.434	4.810	15.510	0.228 (5)	0.946	ANN 5.411	4.003	13.866	0.154 (3)
II. Voting	0.917	ANN + CART 6.823	5.213	17.230	0.265 (7)	0.955	ANN + CART 4.903	3.506	12.397	0.111 (2)
III. Bagging	0.932	CART 6.082	4.598	15.030	0.205 (4)	0.980	CART 3.359	2.432	8.356	**0.000 (1)**
IV. Stacking	0.924	(*) SVR 6.436	4.852	15.530	0.229 (6)	0.613	(*) ANN 14.381	10.867	44.759	1.000 (8)

Note: (*) is (ANN + SVR + CART + LRR); bold value denotes the best overall performance.

Table 10. Performance of WEKA and iML on real estate valuation dataset via tenfold cross-validation.

Model	WEKA				SI (Ranking)	iML				SI (Ranking)
	R	RMSE (U)	MAE (U)	MAPE (%)		R	RMSE (U)	MAE (U)	MAPE (%)	
I. Single	0.807	CART 8.021	5.197	15.270	0.314 (5)	0.813	ANN 8.011	5.388	14.991	0.315 (6)
II. Voting	0.805	SVR + CART 8.091	5.198	15.090	0.315 (7)	0.821	ANN + CART + LRR 7.878	5.376	15.116	0.308 (4)
III. Bagging	0.828	CART 7.637	5.017	14.930	0.280 (2)	0.925	CART 4.774	3.201	8.974	**0.000 (1)**
IV. Stacking	0.819	(*) SVR 7.823	4.969	14.440	0.284 (3)	0.432	(*) ANN 12.309	9.526	32.267	1.000 (8)

Note: (*) is (ANN + SVR + CART + LRR); bold value denotes the best overall performance; U: house price of unit area (10,000 New Taiwan Dollar/Ping, where Ping is a local unit, 1 Ping = 3.3 m squared).

Table 11. Performance of WEKA and iML on energy efficiency dataset (Heating load) via tenfold cross-validation.

Model	WEKA					iML				
	R	RMSE (kW)	MAE (kW)	MAPE (%)	SI (Ranking)	R	RMSE (kW)	MAE (kW)	MAPE (%)	SI (Ranking)
I. Single		CART					ANN			
	0.995	1.046	0.712	3.200	0.459 (7)	0.999	0.484	0.360	1.722	0.049 (2)
II. Voting		ANN + CART					ANN + CART			
	0.997	0.853	0.641	3.190	0.309 (4)	0.999	0.497	0.352	1.602	0.053 (3)
III. Bagging		CART					ANN			
	0.997	0.915	0.633	2.890	0.324 (5)	0.999	0.384	0.291	1.409	**0.000 (1)**
IV. Stacking		(*) SVR					(*) LRR			
	0.996	0.872	0.639	2.990	0.337 (6)	0.998	3.522	3.226	18.181	1.000 (8)

Note: (*) is (ANN + SVR + CART + LRR); bold value denotes the best overall performance.

Table 12. Performance of WEKA and iML on energy efficiency dataset (Cooling load) via tenfold cross-validation.

Model	WEKA					iML				
	R	RMSE (kW)	MAE (kW)	MAPE (%)	SI (Ranking)	R	RMSE (kW)	MAE (kW)	MAPE (%)	SI (Ranking)
I. Single		CART					ANN			
	0.982	1.812	1.183	4.160	0.460 (5)	0.993	1.140	0.799	3.161	0.150 (2)
II. Voting		ANN + CART					ANN + CART			
	0.982	1.831	1.276	4.770	0.491 (7)	0.989	1.415	0.900	3.206	0.250 (3)
III. Bagging		CART					ANN			
	0.983	1.785	1.160	4.070	0.444 (4)	0.997	0.808	0.556	2.129	**0.000 (1)**
IV. Stacking		(*) SVR					(*) LRR			
	0.982	1.827	1.195	4.210	0.465 (6)	0.989	4.108	3.619	17.253	1.000 (8)

Note: (*) is (ANN + SVR + CART + LRR); bold value denotes the best overall performance.

Table 13. Performance of WEKA and iML on airfoil self-noise dataset via tenfold cross-validation.

Model	WEKA					iML				
	R	RMSE (dB)	MAE (dB)	MAPE (%)	SI (Ranking)	R	RMSE (dB)	MAE (dB)	MAPE (%)	SI (Ranking)
I. Single		CART					ANN			
	0.877	3.314	2.381	1.910	0.497 (5)	0.946	2.239	1.660	1.331	0.152 (2)
II. Voting		ANN + CART					ANN + CART			
	0.851	3.685	2.747	2.220	0.641 (7)	0.946	2.246	1.664	1.334	0.152 (3)
III. Bagging		CART					CART			
	0.911	2.906	2.160	1.730	0.352 (4)	0.971	1.727	1.271	1.023	**0.000 (1)**
IV. Stacking		(*) LRR					(*) LRR			
	0.874	3.374	2.494	1.990	0.525 (6)	0.946	6.894	5.587	4.562	1.000 (8)

Note: (*) is (ANN + SVR + CART + LRR); bold value denotes the best overall performance.

6.4. Discussion

Single, voting, bagging, and stacking models are compared using WEKA and iML, except for the tiering method, which is not available in WEKA. Additionally, unlike manual construction of individual models in WEKA interface, iML can automatically build and identify the best model for the imported datasets. Hold-out validation and tenfold cross-validation are used to evaluate the performance results (R, MAE, RMSE, and MAPE) in each scheme (single, voting, bagging, and stacking). The analytical results of either validation show that most of the models trained by iML are superior to those trained by WEKA using the same datasets. Hence, iML is an effective platform to solve regression problems.

7. Numerical Experiments

This section validates iML by using various industrial datasets, including (1) enterprise resource planning data [1], (2) CPU computer performance data [12], (3) customer data for a logistics company [13], and (4) daily data bike rentals [11]. Table 14 presents the initial parameter settings for these problems.

Table 14. Parameter setting.

Experiment	Model	ANN	SVR and SVM			LRR and LgR	CART
		Number of Hidden Node	C	Sigma	Epsilon	Lambda	Min Leaf
1	Single regression model	30	7.3×10^6	45.67	1.0×10^{-5}	1.0×10^{-8}	1
	Single classification model	30	41,703	3.67	-	1.0×10^{-5}	1
	Ensemble regression model	30	7.3×10^6	45.67	1.0×10^{-5}	1.0×10^{-8}	1
2	Single regression model	30	7.3×10^6	20.03	1.0×10^{-5}	1.0×10^{-8}	1
	Single classification model	30	4200	3.40	-	1.0×10^{-5}	1
	Ensemble regression model	30	7.3×10^6	45.67	1.0×10^{-5}	1.0×10^{-8}	1
3	Single regression model	30	7.3×10^6	30.00	1.0×10^{-5}	1.0×10^{-8}	1
	Single classification model	20	41,703	3.67	-	1.0×10^{-5}	1
	Ensemble regression model	30	7.3×10^6	30.00	1.0×10^{-5}	1.0×10^{-8}	1
4	Single regression model	15	7.3×10^6	45.67	1.0×10^{-5}	1.0×10^{-8}	1
	Single classification model	15	41,703	3.67	-	1.0×10^{-5}	1
	Ensemble regression model	15	7.3×10^6	45.67	1.0×10^{-5}	1.0×10^{-8}	1

7.1. Enterprise Resource Planning Software Development Effort

Enterprise Resource Planning (ERP) data for 182 software projects of a leading Taiwan software provider over the last five years was collected, analyzed, and tested with K-fold cross validation.

7.1.1. Variable Selection

Experienced in-house project managers were interviewed to identify factors that affect the ERP software development effort (SDE). There are 182 samples and 17 attributes, and Table 15 summarizes the descriptive statistical data in details. The input and output attributes are defined by Chou el at. (2012) [1].

7.1.2. iML Results

iML automatically trains the models and calculates the performance values. Then, it compares the SI values (SI_{local} and $SI_{globlal}$) among the selected modeling type (singe, voting ensemble, bagging ensemble, stacking ensemble and tiering ensemble). Table 16 presents the detailed results of iML and Figure 10 plots the RMSE of best models for the studied case. Both SI_{local} and SI_{global} values of bagging ANN ensemble are equal to zero, which indicate that the bagging ANN ensemble is the best model in terms of prediction accuracy.

Table 15. Variables and descriptive statistics for predicting enterprise resource planning (ERP) software development effort.

Variable	Min.	Max.	Mean	Standard Deviation	Data Type
Y: Software development effort (person-hour)	4	2694	258.55	394.69	Numerical
X_1: Program type entry	0	1	Dummy variable		Boolean
X_2: Program type report	0	1	Dummy variable		Boolean
X_3: Program type batch	0	1	Dummy variable		Boolean
X_4: Program type query	0	1	Dummy variable		Boolean
X_5: Program type transaction	0	0	Referential category		Boolean
X_6: Number of programs	1	88	16.73	19.12	Numerical
X_7: Number of zooms	0	2028	100.22	255.40	Numerical
X_8: Number of columns in form	3	3216	397.75	548.06	Numerical
X_9: Number of actions	0	1645	288.44	339.61	Numerical
X_{10}: Number of signature tasks	0	15	0.39	1.77	Numerical
X_{11}: Number of batch serial numbers	0	11	0.31	1.50	Numerical
X_{12}: Number of multi-angle trade tasks	0	22	0.55	2.66	Numerical
X_{13}: Number of multi-unit tasks	0	21	1.10	3.41	Numerical
X_{14}: Number of reference calls	0	528	13.96	49.92	Numerical
X_{15}: Number of confirmed tasks	0	21	1.50	3.99	Numerical
X_{16}: Number of post tasks	0	12	0.23	1.33	Numerical
X_{17}: Number of industry type tasks	0	21	0.80	2.97	Numerical

Table 16. Performances of predictive models for ERP software development effort.

No.	Model	Learn					Test					SI and Ranking	
		RMSE	MAE	MAPE	R	TER	RMSE	MAE	MAPE	R	TER	SI_{local}	SI_{global}
I	Single												
1	ANN	68.81	24.92	19.91%	0.98	1.49%	115.24	61.85	30.65%	0.95	9.86%	0.00 (1)	0.13 (2)
2	SVR	0.00	0.00	0.00%	1.00	0.00%	361.88	255.35	611.63%	Inf	36.81%	0.89 (4)	
3	CART	86.89	40.38	19.56%	0.97	0.00%	196.48	107.02	47.29%	0.85	12.77%	0.15 (2)	
4	LRR	250.21	221.15	617.02%	0.84	48.77%	255.43	227.80	647.08%	0.78	63.59%	0.72 (3)	
II	Voting												
1	(*)	73.30	62.90	157.96%	0.99	12.39%	185.39	139.02	321.77%	0.94	23.90%	0.34 (5)	0.15 (3)
2	ANN + CART + LRR	97.74	83.87	210.62%	0.98	16.52%	139.70	110.32	227.32%	0.94	22.01%	0.21 (2)	
3	ANN + SVR + CART	39.95	19.33	11.22%	0.99	0.50%	181.63	117.20	215.13%	0.94	14.45%	0.22 (3)	
4	ANN + SVR + LRR	87.66	76.90	207.61%	0.99	16.52%	202.91	158.85	419.21%	0.94	30.90%	0.46 (7)	
5	SVR + CART + LRR	93.17	80.47	208.60%	0.99	16.26%	236.10	178.73	427.29%	0.89	32.97%	0.60 (10)	
6	ANN + CART	59.92	28.99	16.83%	0.99	0.75%	123.43	68.46	33.90%	0.94	10.10%	0.01 (1)	
7	ANN + LRR	131.49	115.35	311.42%	0.97	24.78%	152.57	129.46	326.55%	0.94	32.67%	0.34 (6)	
8	ANN + SVR	34.40	12.46	9.96%	1.00	0.75%	199.24	135.42	307.89%	0.95	18.21%	0.31 (4)	
9	SVR + CART	43.45	20.19	9.78%	0.99	0.00%	254.27	166.51	320.26%	0.85	21.09%	0.56 (9)	
10	CART + LRR	139.76	120.71	312.91%	0.96	24.38%	190.91	151.58	337.64%	0.89	32.09%	0.48 (8)	
11	SVR + LRR	125.10	110.57	308.51%	0.97	24.38%	287.41	230.03	626.83%	0.78	47.93%	1.00 (11)	
III	Bagging												
1	ANN	70.28	33.48	21.45%	0.98	2.51%	65.58	40.51	19.50%	0.99	5.59%	0.00 (1)	**0.00 (1)**
2	SVR	174.56	92.31	231.82%	0.91	3.98%	162.38	99.02	87.79%	0.87	11.69%	0.47 (3)	
3	CART	123.05	52.45	25.94%	0.94	2.19%	127.21	79.97	20.11%	0.96	8.58%	0.21 (2)	
4	LRR	249.00	222.81	666.29%	0.88	59.61%	309.18	257.31	269.67%	0.71	10.90%	0.97 (4)	
IV	Stacking												
1	(*) ANN	0.07	0.02	0.03%	1.00	0.00%	361.54	255.19	611.55%	0.71	36.77%	0.80 (2)	1.00 (5)
2	(*) SVR	0.00	0.00	0.00%	1.00	0.00%	361.76	255.53	612.78%	NaN	36.76%	1.00 (4)	
3	(*) CART	52.00	17.40	5.69%	0.99	0.00%	360.24	252.61	593.46%	NaN	34.86%	0.81 (3)	
4	(*) LRR	132.73	97.48	187.26%	0.95	23.62%	289.20	206.32	494.18%	0.62	34.11%	0.03 (1)	
V	Tiering												
1	2-Class (**)	317.04	71.38	24.45%	0.58	20.48%	176.77	65.42	23.57%	0.79	16.13%	0.00 (1)	0.31 (4)
2	3-Class (***)	383.02	115.48	26.68%	0.30	36.54%	278.46	111.74	26.94%	0.51	28.64%	0.54 (2)	
3	4-Class (****)	414.30	151.32	29.96%	0.10	49.11%	347.05	147.22	29.76%	0.26	43.00%	1.00 (3)	

Note: (*) is (ANN + SVR + CART + LRR); (**) SVM-(ANN, SVR); (***) CART-(ANN, SVR, SVR); (****) CART-(CART, SVR, SVR, SVR); (No.): Ranking.

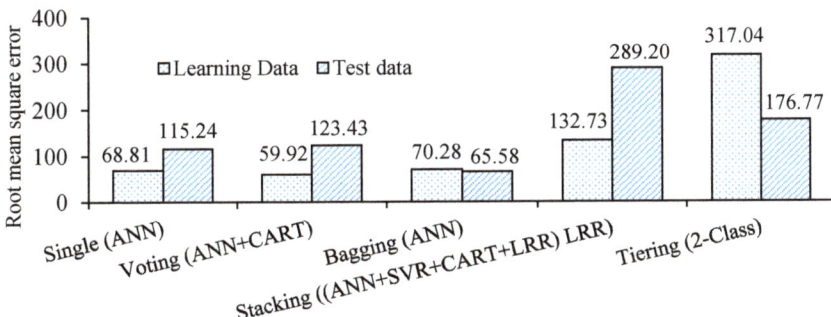

Figure 10. Root mean square errors of best models.

Three models (single, voting, and bagging) provided better results in terms of R (0.94 to 0.99) than the tiering and stacking ensemble models, which had the R values of 0.58 to 0.95. Among these three best models, in terms of MAPE, the bagging model exhibited the best balance of MAPE results from learning and test data (21.45% and 19.50%, respectively). The single and voting models depicted un-balanced MAPEs for training and test data (19.91% and 30.65% for the single model; 16.83% and 33.90% for the voting model). Thus, the bagging model was the best model to predict ERP.

The first experiment indicates that, the iML not only identifies the best model, but also reports the performance values of all the training models. Chou et al. (2012) obtained training and testing MAPEs of 26.8% and 27.3%, and RMSEs of 234.0157 h and 97.2667 h using Evolutionary Support Vector Machine Inference Model (ESIM) [1]. The iML yields the bagging ensemble model with MAPEs of 21.45% and 19.50%, and RMSEs of 70.28hr and 65.58 h for the same training and test data, respectively. As a result, the iML is effective to find the best model among the popular regression models.

7.2. Experiments on Industrial Datasets

Three additional experiments were performed to evaluate iML. To ensure a fair comparison, 70 % of the data was used for learning whereas the remaining 30% was utilized for testing.

7.2.1. Performance of CPU Processors

This experiment is about the comparison of performance of CPU processors. The data for this experiment was taken from Maurya and Gupta (2015) [12]. This dataset contained 209 samples with a total of 6 attributes (Table 17). The descriptions of the attributes are as follows: X_1: Machine cycle time in nanoseconds (integer, input); X_2: Minimum main memory in kilobytes (integer, input); X_3: Maximum main memory in kilobytes (integer, input); X_4: Cache memory in kilobytes (integer, input); X_5: Minimum channels in units (integer, input); X_6: Maximum channels in units (integer, input); and Y: Estimated relative performance (integer, output).

Table 17. Descriptive statistics for CPU processors.

Statistic Value	Input						Output
	X_1	X_2	X_3	X_4	X_5	X_6	Y
Min	17	64	64	0	0	0	15
Max	1500	32,000	64,000	256	52	176	1238
Mean	203.82	2867.98	11,796.2	25.21	4.7	18.27	99.33
Std.	260.26	3878.74	11,726.6	40.63	6.82	26	154.76

7.2.2. Daily Demand Forecasting Orders

This experiment is about the daily demand forecasting orders. The data used in this experiment was taken from Ferreira et al. (2016) [13]. Table 18 shows a statistical analysis of the data. There were 60 samples with 12 attributes, including X_1: Week of the month (first week, second, third or fourth week of month, input); X_2: Day of the week (Monday to Friday, input); X_3: Urgent orders (integer, input); X_4: Non-urgent orders (integer, input); X_5: Type A orders (integer, input); X_6: Type B orders (integer, input); X_7: Orders of type C (integer, input); X_8: Orders from the tax sector (integer, input); X_9: Orders from the traffic controller sector (integer, input); X_{10}: Orders from the banking sector 1 (integer, input); X_{11}: Orders from the banking sector 2 (integer, input); X_{12}: Banking orders 3 (integer, input); and Y: Total orders (integer, output).

Table 18. Variables and descriptive statistics for daily demand forecasting orders.

Statistic Value	Input												Output
	X_1	X_2	X_3	X_4	X_5	X_6	X_7	X_8	X_9	X_{10}	X_{11}	X_{12}	Y
Min	1	2	43.65	77.37	21.83	25.13	74.37	0	11,992	3452	16,411	7679	129.41
Max	5	6	435.30	223.27	118.18	267.34	302.45	865	71,772	210,508	188,411	73,839	616.45
Mean	-	-	172.55	118.92	52.11	109.23	139.53	77.4	44,504.4	46,640.8	79,401.5	23,114.6	300.87
Std.	-	-	69.51	27.17	18.83	50.74	41.44	186.5	12,197.9	45,220.7	40,504.4	13,148	89.6

7.2.3. Total Hourly-Shared Bike Rental per Days

The experiment is about the total hourly-shared bike rental per days. The data was adopted from Fanaee-T and Gama (2014) [11], and statistically analyzed in Table 19. In total, there were 731 samples and 11 attributes, defined as follows: X_1: Season (1: spring, 2: summer, 3: fall, 4: winter, input); X_2: Month (1 to 12, input); X_3: Year (0:2011, 1:2012, input); X_4: Weather day is holiday or not (input); X_5: Day of the week (input); X_6: Working day if day is neither weekend nor holiday is 1, otherwise is 0 (input); X_7: Weather condition (1: Clear, Few clouds, partly cloudy; 2: Mist + Cloudy, Mist + Broken clouds, Mist + Few clouds, Mist; 3: Light Snow, Light Rain + Thunderstorm + Scattered clouds, Light Rain + Scattered clouds; 4: Heavy Rain + Ice Pallets + Thunderstorm + Mist, Snow + Fog, input); X_8: Normalized temperature in Celsius. The values are divided to 41 (max) (input); X_9: Normalized feeling temperature in Celsius. The values are divided to 50 (max) (input); X_{10}: Normalized humidity. The values are divided to 100 (max) (input); X_{11}: Normalized wind speed. The values are divided to 67 (max) (input); and Y: Count of total rental bikes including both casual and registered (output).

Table 19. Variables and descriptive statistics for total hourly-shared bike rental per days.

Statistic Value	Input											Output
	X_1	X_2	X_3	X_4	X_5	X_6	X_7	X_8	X_9	X_{10}	X_{11}	Y
Min	1	0	1	0	0	0	1	0.059	0.079	0	0.022	22
Max	4	1	12	1	6	1	3	0.862	0.841	0.973	0.507	8714
Mean	-	-	-	0.029	2.997	0.684	1.395	0.495	0.474	0.628	0.19	4504.35
Std.	-	-	-	0.167	2.005	0.465	0.545	0.183	0.163	0.142	0.077	1937.21

In this study, to calculate MAPE, the output was normalized and 0.1 was added to prevent a zero value.

$$y_i = \frac{y_i - y_{min}}{y_{max} - y_{min}} + 0.1 \tag{21}$$

where y_i, y_{min}, and y_{max} are actual value, minimum and maximum of actual value, respectively.

7.2.4. Performance Results

Table 20 presents the performance results of all models for the three additional datasets. Using the same dataset in the experiment No. 2, Maurya and Gupta (2015) [12] trained ANN models with the maximum R-learn and R-test values of 0.98146 and 0.98662, respectively. Meanwhile, the iML identifies the ANN single model as the best model with R-learn and R-test values of 0.99990 and 0.99629, respectively. The iML gives out a slightly better model than those of the previous research in this numerical experiment.

Table 20. Performance results of three additional numerical experiments.

No.	Model	Learn					Test				
		RMSE	MAE	MAPE	R	TER	RMSE	MAE	MAPE	R	TER
2	**Single**	2.462	0.569	1.015%	1.000	0.236%	8.738	3.683	3.775%	0.996	2.730%
	Voting	17.509	4.965	2.808%	0.995	0.118%	13.087	5.173	5.685%	0.989	0.921%
	Bagging	40.127	8.893	2.360%	0.981	3.015%	13.484	3.930	2.489%	0.996	3.795%
	Stacking	40.428	9.030	3.389%	0.973	0.000%	64.782	43.615	104.992%	0.842	9.852%
	Tiering-2class	163.338	26.947	3.229%	0.383	24.818%	18.196	5.822	4.629%	0.986	3.220%
	Tiering-3class	167.093	29.784	3.856%	0.340	27.446%	76.406	12.698	5.268%	0.639	13.094%
	Tiering-4class	182.572	44.722	7.970%	0.112	41.332%	91.701	21.089	8.342%	0.422	24.368%
3	Single	0.349	0.080	0.023%	1.000	0.021%	0.317	0.231	0.093%	1.000	0.042%
	Voting	17.417	10.754	3.089%	0.985	0.010%	12.162	10.157	3.993%	0.951	0.867%
	Bagging	0.917	0.399	0.110%	1.000	0.020%	0.296	0.221	0.087%	1.000	0.074%
	Stacking	0.338	0.090	0.026%	1.000	0.014%	0.335	0.251	0.101%	1.000	0.042%
	Tiering-2class	169.296	63.580	14.294%	−0.399	21.483%	214.674	86.711	16.747%	−0.704	27.688%
	Tiering-3class	273.303	212.047	62.384%	−0.664	71.449%	295.065	223.209	62.186%	−0.570	71.054%
	Tiering-4class	329.001	312.397	97.619%	−0.304	99.023%	51.164	45.122	18.684%	0.706	11.339%
4	Single	0.046	0.030	6.670%	0.979	10.450%	0.105	0.073	14.120%	0.883	0.550%
	Voting	0.052	0.037	7.850%	0.974	8.000%	0.080	0.056	10.750%	0.929	0.260%
	Bagging	0.049	0.034	7.150%	0.977	6.930%	0.069	0.046	8.870%	0.948	0.190%
	Stacking	0.005	0.003	0.700%	1.000	21.430%	0.214	0.169	38.680%	0.000	0.086%
	Tiering-2class	0.580	0.432	57.620%	−0.582	58.900%	0.589	0.451	61.410%	−0.570	68.290%
	Tiering-3class	0.639	0.568	84.420%	−0.680	64.600%	0.646	0.565	82.110%	−0.717	90.260%
	Tiering-4class	0.648	0.596	92.370%	−0.513	65.200%	0.652	0.584	87.400%	−0.630	94.300%

Note: No. 2: CPU experiment dataset; No. 3: Customer experiment dataset; No. 4: Rental bike experiment dataset; the bold denotes the best model in each experiment.

In the experiment No. 3, Ferreira et al. (2016) had an analytical result of MAPE 3.45% and iML confirms ANN single model as the best model, with MAPE values for learning and test of 0.023% and 0.093%, respectively [13]. The stacking ANN ensemble also performs well with the MAPEs for the learning and test data by 0.026% and 0.010%, respectively.

Finally, in the experiment No. 4, iML achieves R-learn and R-test values of 0.97660 and 0.94790, with bagging ANN as the best model. In contrast, Fanaee-T and Gama (2014) obtained a maximum R value of 0.91990 [11].

As shown in the above numerical experiments, iML trains and identifies the best models which are better than those in the previous studies.

8. Conclusions and Future Work

This study develops an iML platform to efficiently operate data-mining techniques. The iML is designed to be user-friendly, so users can get the results with only "One-Click". The numerical experiments have demonstrated that iML is a powerful soft computing to identify the best prediction model by automating comparison among diverse machine learning techniques.

To benchmark the effectiveness of iML with WEKA, five datasets collected from the UCI Machine Learning Repository were analyzed via hold-out validation and tenfold cross validation. The performance results indicate that iML can find a more accurate model than that of WEKA in the publicly available datasets. The best prediction model identified by iML is also the best model among all the models trained by iML and WEKA. Notably, iML requires minimal effort from the users to build single, voting, bagging, and stacking models in comparison with WEKA.

Four industrial experiments were carried out to validate the performance of iML. The first experiment involved training a model for prediction of ERP development effort, in which iML yielded an RMSE for learning data with 70.28 h and for testing data with 65.58 h, by using the bagging ANN ensemble (best model). In contrast, Chou et al. (2012) [1] obtained training and testing RMSE values of 234.0157 h and 97.2667 h, respectively.

In the second experiment on performance of CPU processors, iML yielded 0.99990 for R-learning and 0.99629 for R-testing, which are better than those reported in Maurya and Gupta (2015) [12], and confirmed that single ANN was the best model. In the third experiment of daily demand forecasting orders, iML achieved MAPE values of 0.026% (learning) and 0.010% (testing). The results are as excellent as those obtained in Ferreira et al. (2016) [13]. In the fourth experiment for total hourly-shared bike rental, R-learning and R-testing values of 0.97660 and 0.94790 were reached using iML. The test performance was 6% better than that obtained by Fanaee-T and Gama (2014) [11]. In addition to the enhanced prediction performance, the iML possesses ability to determine the best models on the basis of multiple evaluation metrics.

In conclusion, the iML is a powerful and promising prediction platform for solving diverse engineering problems. Since the iML platform can only deal with regression problems, future research should upgrade iML for solving complex classification and time series problems by automatically presenting the alternative models for practical use in engineering applications, as well as adding some other advanced ML methods (such as deep learning models). Moreover, metaheuristic optimization algorithms could be integrated with the iML to help the users finetune the hyperparameters of chosen machine learning models.

Author Contributions: Conceptualization, J.-S.C.; data curation, D.-N.T.; formal analysis, J.-S.C. and D.-N.T.; funding acquisition, J.-S.C.; investigation, J.-S.C., D.-N.T. and C.-F.T.; methodology, J.-S.C. and C.-F.T.; project administration, J.-S.C.; resources, J.-S.C., D.-N.T. and C.-F.T.; software, D.-N.T.; supervision, J.-S.C.; validation, J.-S.C., D.-N.T. and C.-F.T.; visualization, J.-S.C. and D.-N.T.; writing—original draft, J.-S.C., D.-N.T. and C.-F.T.; writing—review and editing, J.-S.C. and D.-N.T. All authors have read and agreed to the published version of the manuscript.

Funding: This research was funded by the Ministry of Science and Technology, Taiwan, under grants 108-2221-E-011-003-MY3 and 107-2221-E-011-035-MY3.

Data Availability Statement: The data that support the findings of this study are available from the UCI Machine Learning Repository or corresponding author upon reasonable request.

Acknowledgments: The authors would like to thank the Ministry of Science and Technology, Taiwan, for financially supporting this research.

Conflicts of Interest: The authors declare that they have no conflict of interest.

References

1. Chou, J.-S.; Cheng, M.-Y.; Wu, Y.-W.; Wu, C.-C. Forecasting enterprise resource planning software effort using evolutionary support vector machine inference model. *Int. J. Proj. Manag.* **2012**, *30*, 967–977. [CrossRef]
2. Pham, A.-D.; Ngo, N.-T.; Nguyen, Q.-T.; Truong, N.-S. Hybrid machine learning for predicting strength of sustainable concrete. *Soft Comput.* **2020**. [CrossRef]
3. Cheng, M.-Y.; Chou, J.-S.; Cao, M.-T. Nature-inspired metaheuristic multivariate adaptive regression splines for predicting refrigeration system performance. *Soft Comput.* **2015**, *21*, 477–489. [CrossRef]
4. Li, Y.; Lei, G.; Bramerdorfer, G.; Peng, S.; Sun, X.; Zhu, J. Machine Learning for Design Optimization of Electromagnetic Devices: Recent Developments and Future Directions. *Appl. Sci.* **2021**, *11*, 1627. [CrossRef]
5. Piersanti, S.; Orlandi, A.; Paulis, F.d. Electromagnetic Absorbing Materials Design by Optimization Using a Machine Learning Approach. *IEEE Trans. Electromagn. Compat.* **2018**, 1–8. [CrossRef]
6. Chou, J.S.; Pham, A.D. Smart artificial firefly colony algorithm-based support vector regression for enhanced forecasting in civil engineering. *Comput.-Aided Civ. Infrastruct. Eng.* **2015**, *30*, 715–732. [CrossRef]

7. Cheng, M.-Y.; Prayogo, D.; Wu, Y.-W. A self-tuning least squares support vector machine for estimating the pavement rutting behavior of asphalt mixtures. *Soft Comput.* **2019**, *23*, 7755–7768. [CrossRef]
8. Al-Ali, H.; Cuzzocrea, A.; Damiani, E.; Mizouni, R.; Tello, G. A composite machine-learning-based framework for supporting low-level event logs to high-level business process model activities mappings enhanced by flexible BPMN model translation. *Soft Comput.* **2019**. [CrossRef]
9. López, J.; Maldonado, S.; Carrasco, M. A novel multi-class SVM model using second-order cone constraints. *Appl. Intell.* **2016**, *44*, 457–469. [CrossRef]
10. Bogawar, P.S.; Bhoyar, K.K. An improved multiclass support vector machine classifier using reduced hyper-plane with skewed binary tree. *Appl. Intell.* **2018**, *48*, 4382–4391. [CrossRef]
11. Fanaee-T, H.; Gama, J. Event labeling combining ensemble detectors and background knowledge. *Prog. Artif. Intell.* **2014**, *2*, 113–127. [CrossRef]
12. Maurya, V.; Gupta, S.C. Comparative Analysis of Processors Performance Using ANN. In Proceedings of the 2015 5th International Conference on IT Convergence and Security (ICITCS), Kuala Lumpur, Malaysia, 24–27 August 2015; pp. 1–5.
13. Ferreira, R.P.; Martiniano, A.; Ferreira, A.; Ferreira, A.; Sassi, R.J. Study on Daily Demand Forecasting Orders using Artificial Neural Network. *IEEE Lat. Am. Trans.* **2016**, *14*, 1519–1525. [CrossRef]
14. De'ath, G.; Fabricius, K.E. Classification and regression trees: A powerful yet simple technique for ecological data analysis. *Ecology* **2000**, *81*, 3178–3192. [CrossRef]
15. Li, H.; Wen, G. Modeling reverse thinking for machine learning. *Soft Comput.* **2020**, *24*, 1483–1496. [CrossRef]
16. Li, Y. Predicting materials properties and behavior using classification and regression trees. *Mater. Sci. Eng. A* **2006**, *433*, 261–268. [CrossRef]
17. Chou, J.-S.; Yang, K.-H.; Lin, J.-Y. Peak Shear Strength of Discrete Fiber-Reinforced Soils Computed by Machine Learning and Metaensemble Methods. *J. Comput. Civ. Eng.* **2016**, *30*, 04016036. [CrossRef]
18. Qi, C.; Tang, X. Slope stability prediction using integrated metaheuristic and machine learning approaches: A comparative study. *Comput. Ind. Eng.* **2018**, *118*, 112–122. [CrossRef]
19. Chou, J.-S.; Ho, C.-C.; Hoang, H.-S. Determining quality of water in reservoir using machine learning. *Ecol. Inform.* **2018**, *44*, 57–75. [CrossRef]
20. Chou, J.-S.; Bui, D.-K. Modeling heating and cooling loads by artificial intelligence for energy-efficient building design. *Energy Build.* **2014**, *82*, 437–446. [CrossRef]
21. Alkahtani, M.; Choudhary, A.; De, A.; Harding, J.A. A decision support system based on ontology and data mining to improve design using warranty data. *Comput. Ind. Eng.* **2018**. [CrossRef]
22. Daras, G.; Agard, B.; Penz, B. A spatial data pre-processing tool to improve the quality of the analysis and to reduce preparation duration. *Comput. Ind. Eng.* **2018**, *119*, 219–232. [CrossRef]
23. Chou, J.-S.; Tsai, C.-F. Preliminary cost estimates for thin-film transistor liquid–crystal display inspection and repair equipment: A hybrid hierarchical approach. *Comput. Ind. Eng.* **2012**, *62*, 661–669. [CrossRef]
24. Chen, T. An ANN approach for modeling the multisource yield learning process with semiconductor manufacturing as an example. *Comput. Ind. Eng.* **2017**, *103*, 98–104. [CrossRef]
25. Wu, X.; Kumar, V.; Quinlan, J.R.; Ghosh, J.; Yang, Q.; Motoda, H.; McLachlan, G.J.; Ng, A.; Liu, B.; Philip, S.Y. Top 10 algorithms in data mining. *Knowl. Inf. Syst.* **2008**, *14*, 1–37. [CrossRef]
26. Chou, J.-S.; Ngo, N.-T.; Chong, W.K. The use of artificial intelligence combiners for modeling steel pitting risk and corrosion rate. *Eng. Appl. Artif. Intell.* **2017**, *65*, 471–483. [CrossRef]
27. Das, D.; Pratihar, D.K.; Roy, G.G.; Pal, A.R. Phenomenological model-based study on electron beam welding process, and input-output modeling using neural networks trained by back-propagation algorithm, genetic algorithms, particle swarm optimization algorithm and bat algorithm. *Appl. Intell.* **2018**, *48*, 2698–2718. [CrossRef]
28. Tewari, S.; Dwivedi, U.D. Ensemble-based big data analytics of lithofacies for automatic development of petroleum reservoirs. *Comput. Ind. Eng.* **2018**. [CrossRef]
29. Priore, P.; Ponte, B.; Puente, J.; Gómez, A. Learning-based scheduling of flexible manufacturing systems using ensemble methods. *Comput. Ind. Eng.* **2018**, *126*, 282–291. [CrossRef]
30. Fang, K.; Jiang, Y.; Song, M. Customer profitability forecasting using Big Data analytics: A case study of the insurance industry. *Comput. Ind. Eng.* **2016**, *101*, 554–564. [CrossRef]
31. Chou, J.-S. Generalized linear model-based expert system for estimating the cost of transportation projects. *Expert Syst. Appl.* **2009**, *36*, 4253–4267. [CrossRef]
32. Dandikas, V.; Heuwinkel, H.; Lichti, F.; Drewes, J.E.; Koch, K. Predicting methane yield by linear regression models: A validation study for grassland biomass. *Bioresour. Technol.* **2018**, *265*, 372–379. [CrossRef] [PubMed]
33. Ngo, S.H.; Kemény, S.; Deák, A. Performance of the ridge regression method as applied to complex linear and nonlinear models. *Chemom. Intell. Lab. Syst.* **2003**, *67*, 69–78. [CrossRef]
34. Sentas, P.; Angelis, L. Categorical missing data imputation for software cost estimation by multinomial logistic regression. *J. Syst. Softw.* **2006**, *79*, 404–414. [CrossRef]

35. Slowik, A. Application of an Adaptive Differential Evolution Algorithm With Multiple Trial Vectors to Artificial Neural Network Training. *IEEE Trans. Ind. Electron.* **2011**, *58*, 3160–3167. [CrossRef]
36. Caputo, A.C.; Pelagagge, P.M. Parametric and neural methods for cost estimation of process vessels. *Int. J. Prod. Econ.* **2008**, *112*, 934–954. [CrossRef]
37. Rocabruno-Valdés, C.I.; Ramírez-Verduzco, L.F.; Hernández, J.A. Artificial neural network models to predict density, dynamic viscosity, and cetane number of biodiesel. *Fuel* **2015**, *147*, 9–17. [CrossRef]
38. Ganesan, P.; Rajakarunakaran, S.; Thirugnanasambandam, M.; Devaraj, D. Artificial neural network model to predict the diesel electric generator performance and exhaust emissions. *Energy* **2015**, *83*, 115–124. [CrossRef]
39. Vapnik, V.N. An overview of statistical learning theory. *IEEE Trans. Neural Netw.* **1999**, *10*, 988–999. [CrossRef]
40. Vapnik, V. *The Nature of Statistical Learning Theory*, 2nd ed.; Springer: New York, NY, USA, 2013.
41. Jing, G.; Cai, W.; Chen, H.; Zhai, D.; Cui, C.; Yin, X. An air balancing method using support vector machine for a ventilation system. *Build. Environ.* **2018**, *143*, 487–495. [CrossRef]
42. García-Floriano, A.; López-Martín, C.; Yáñez-Márquez, C.; Abran, A. Support vector regression for predicting software enhancement effort. *Inf. Softw. Technol.* **2018**, *97*, 99–109. [CrossRef]
43. Breiman, L.; Friedman, J.; Stone, C.J.; Olshen, R.A. *Classification and Regression Trees*; Routledge: New York, NY, USA, 2017; p. 368. [CrossRef]
44. Choi, S.Y.; Seo, I.W. Prediction of fecal coliform using logistic regression and tree-based classification models in the North Han River, South Korea. *J. Hydro-Environ. Res.* **2018**, *21*, 96–108. [CrossRef]
45. Ru, F.; Yin, A.; Jin, J.; Zhang, X.; Yang, X.; Zhang, M.; Gao, C. Prediction of cadmium enrichment in reclaimed coastal soils by classification and regression trees. *Estuar. Coast. Shelf Sci.* **2016**, *177*, 1–7. [CrossRef]
46. Chou, J.-S.; Tsai, C.-F.; Pham, A.-D.; Lu, Y.-H. Machine learning in concrete strength simulations: Multi-nation data analytics. *Constr. Build. Mater.* **2014**, *73*, 771–780. [CrossRef]
47. Elish, M.O. Assessment of voting ensemble for estimating software development effort. In Proceedings of the 2013 IEEE Symposium on Computational Intelligence and Data Mining (CIDM), Singapore, 16–19 April 2013; pp. 316–321.
48. Wang, Z.; Wang, Y.; Srinivasan, R.S. A novel ensemble learning approach to support building energy use prediction. *Energy Build.* **2018**, *159*, 109–122. [CrossRef]
49. Chen, J.; Yin, J.; Zang, L.; Zhang, T.; Zhao, M. Stacking machine learning model for estimating hourly PM2.5 in China based on Himawari 8 aerosol optical depth data. *Sci. Total Environ.* **2019**, *697*, 134021. [CrossRef] [PubMed]
50. Basant, N.; Gupta, S.; Singh, K.P. A three-tier QSAR modeling strategy for estimating eye irritation potential of diverse chemicals in rabbit for regulatory purposes. *Regul. Toxicol. Pharmacol.* **2016**, *77*, 282–291. [CrossRef]
51. Lee, K.M.; Yoo, J.; Kim, S.-W.; Lee, J.-H.; Hong, J. Autonomic machine learning platform. *Int. J. Inf. Manag.* **2019**, *49*, 491–501. [CrossRef]
52. Wolpert, D.H.; Macready, W.G. No free lunch theorems for optimization. *IEEE Trans. Evol. Comput.* **1997**, *1*, 67–82. [CrossRef]
53. Wolpert, D.H.; Macready, W.G. *No Free Lunch Theorems for Search*; Technical Report SFI-TR-95-02-010; Santa Fe Institute: Santa Fe, NM, USA, 1995.
54. Cheng, D.; Shi, Y.; Gwee, B.; Toh, K.; Lin, T. A Hierarchical Multiclassifier System for Automated Analysis of Delayered IC Images. *IEEE Intell. Syst.* **2019**, *34*, 36–43. [CrossRef]
55. Basheer, I.A.; Hajmeer, M. Artificial neural networks: Fundamentals, computing, design, and application. *J. Microbiol. Methods* **2000**, *43*, 3–31. [CrossRef]
56. Jain, A.K.; Jianchang, M.; Mohiuddin, K.M. Artificial neural networks: A tutorial. *Computer* **1996**, *29*, 31–44. [CrossRef]
57. Cortes, C.; Vapnik, V. Support-vector networks. *Mach. Learn.* **1995**, *20*, 273–297. [CrossRef]
58. Chamasemani, F.F.; Singh, Y.P. Multi-class Support Vector Machine (SVM) Classifiers—An Application in Hypothyroid Detection and Classification. In Proceedings of the 2011 Sixth International Conference on Bio-Inspired Computing: Theories and Applications, Penang, Malaysia, 27–29 September 2011; pp. 351–356.
59. Yang, X.; Yu, Q.; He, L.; Guo, T. The one-against-all partition based binary tree support vector machine algorithms for multi-class classification. *Neurocomputing* **2013**, *113*, 1–7. [CrossRef]
60. Tuv, E.; Runger, G.C. Scoring levels of categorical variables with heterogeneous data. *IEEE Intell. Syst.* **2004**, *19*, 14–19. [CrossRef]
61. Chiang, W.; Liu, X.; Zhang, T.; Yang, B. A Study of Exact Ridge Regression for Big Data. In Proceedings of the 2018 IEEE International Conference on Big Data (Big Data), Seattle, WA, USA, 10–13 December 2018; pp. 3821–3830.
62. Marquardt, D.W.; Snee, R.D. Ridge Regression in Practice. *Am. Stat.* **1975**, *29*, 3–20. [CrossRef]
63. Cox, D.R. The regression analysis of binary sequences. *J. R. Stat. Society. Ser. B* **1958**, *20*, 215–242. [CrossRef]
64. Jiang, F.; Guan, Z.; Li, Z.; Wang, X. A method of predicting visual detectability of low-velocity impact damage in composite structures based on logistic regression model. *Chin. J. Aeronaut.* **2021**, *34*, 296–308. [CrossRef]
65. Kohavi, R. A study of cross-validation and bootstrap for accuracy estimation and model selection. In Proceedings of the International Joint Conference on Artificial Intelligence 1995, Montreal, QC, Canada, 20–25 August 1995; pp. 1137–1143.
66. Chou, J.; Truong, D.; Le, T. Interval Forecasting of Financial Time Series by Accelerated Particle Swarm-Optimized Multi-Output Machine Learning System. *IEEE Access* **2020**, *8*, 14798–14808. [CrossRef]
67. Yeh, I.-C. Analysis of Strength of Concrete Using Design of Experiments and Neural Networks. *J. Mater. Civ. Eng.* **2006**, *18*, 597–604. [CrossRef]

68. Yeh, I.C.; Hsu, T.-K. Building real estate valuation models with comparative approach through case-based reasoning. *Appl. Soft Comput.* **2018**, *65*, 260–271. [CrossRef]
69. Tsanas, A.; Xifara, A. Accurate quantitative estimation of energy performance of residential buildings using statistical machine learning tools. *Energy Build.* **2012**, *49*, 560–567. [CrossRef]
70. Lau, K.; López, R. *A Neural Networks Approach to Aerofoil Noise Prediction*; International Center for Numerical Methods in Engineering: Barcelona, Spain, 2009.

Article

Deep Neural Network for Gender-Based Violence Detection on Twitter Messages

Carlos M. Castorena [1,†,‡], Itzel M. Abundez [1,†], Roberto Alejo [1,*,†,‡], Everardo E. Granda-Gutiérrez [2,†], Eréndira Rendón [1,†] and Octavio Villegas [1,†]

1. Division of Postgraduate Studies and Research, National Technological of Mexico, Instituto Tecnológico de Toluca, Metepec 52149, Mexico; ccastorenal@toluca.tecnm.mx (C.M.C.); iabundezb@toluca.tecnm.mx (I.M.A.); erendonl@toluca.tecnm.mx (E.R.); ovillegasc@toluca.tecnm.mx (O.V.)
2. UAEM University Center at Atlacomulco, Autonomous University of the State of Mexico, Toluca 50400, Mexico; eegrandag@uaemex.mx
* Correspondence: ralejoe@toluca.tecnm.mx; Tel.: +52-722-2816463
† Current address: Av. Tecnológico s/n, Agrícola Bellavista, Metepec 52149, Mexico.
‡ These authors contributed equally to this work.

Abstract: The problem of gender-based violence in Mexico has been increased considerably. Many social associations and governmental institutions have addressed this problem in different ways. In the context of computer science, some effort has been developed to deal with this problem through the use of machine learning approaches to strengthen the strategic decision making. In this work, a deep learning neural network application to identify gender-based violence on Twitter messages is presented. A total of 1,857,450 messages (generated in Mexico) were downloaded from Twitter: 61,604 of them were manually tagged by human volunteers as negative, positive or neutral messages, to serve as training and test data sets. Results presented in this paper show the effectiveness of deep neural network (about 80% of the area under the receiver operating characteristic) in detection of gender violence on Twitter messages. The main contribution of this investigation is that the data set was minimally pre-processed (as a difference versus most state-of-the-art approaches). Thus, the original messages were converted into a numerical vector in accordance to the frequency of word's appearance and only adverbs, conjunctions and prepositions were deleted (which occur very frequently in text and we think that these words do not contribute to discriminatory messages on Twitter). Finally, this work contributes to dealing with gender violence in Mexico, which is an issue that needs to be faced immediately.

Keywords: gender-based violence in Mexico; twitter messages; deep neural networks; class imbalance

Citation: Castorena, C.M.; Abundez, I.M.; Alejo, R.; Granda-Gutiérrez, E.E.; Rendón, E.; Villegas, O. Deep Neural Network for Gender-Based Violence Detection on Twitter Messages. *Mathematics* **2021**, *9*, 807. https://doi.org/10.3390/math9080807

Academic Editors: Florin Leon, Mircea Hulea and Marius Gavrilescu

Received: 26 February 2021
Accepted: 6 April 2021
Published: 8 April 2021

Publisher's Note: MDPI stays neutral with regard to jurisdictional claims in published maps and institutional affiliations.

Copyright: © 2021 by the authors. Licensee MDPI, Basel, Switzerland. This article is an open access article distributed under the terms and conditions of the Creative Commons Attribution (CC BY) license (https://creativecommons.org/licenses/by/4.0/).

1. Introduction

Gender-based violence (GBV) is a big concern around the globe [1]. The United Nations (UN) recognized GBV as a problem involving health and development [2]. A UN declaration about GBV, specifically the cause to women, describes it as all those acts of violence that results or potentially could lead into physical, psychological or sexual damage or suffering; it also includes the menacing of doing such acts, coercion to perform them and arbitrary deprivation of liberty, no matter if this is done in public or private circumstances [3].

Mexico has shown an escalation in the number of victims of GBV due to its social, economic and political context [4,5]. Moreover, crisis like the recent novel coronavirus disease (COVID-19) outbreak have exposed critical inequalities in the social and economic environments, as well as the health system, which have negatively contributed to the GBV problem [6].

Efforts of scholars and activists have increasingly turned society and government attention to this problem, warning about how certain conditions of power or privilege tend to reproduce broader relations of inequality, domination, exploitation, victimization and, finally, loss of humanity [1]. In this respect, computer science researchers have developed algorithms and methodologies based on machine learning to address the GBV problem. For example, Ref. [7] presents a camouflaged electronic device to help potential victims of GBV; it allows to send a voice command and Global Positioning System (GPS) location via smartphone to a Control Center, which analyzes the message to properly assist the victim. A similar but more sophisticated work is presented in [8]; it uses two psychological sensors to identify GBV through a robust speaker identification system, based on the evaluation of speech stress conditions by using data augmentation techniques. Rodríguez-Rodríguez et al. [9] used historic open access data to model and forecast GBV through machine learning methods; their methodology produced successful results in three specific Spanish territories with different populations.

GBV has affected many women around the world in online social network environments [10] and several works have been developed to tackle this problem. In Ref. [11], a classification of cyber-bullying detection methods in online social networks was presented; it shows a survey of techniques to automatically identify cyber-bullying through the machine learning algorithms. Another interesting approach is MANDOLA [12]; it is a big-data processing system intended to evaluate the proliferation and effect of online hate-related speech, which is generally inspired by religion beliefs, ethnicity or gender. Gutiérrez-Esparza et al. [13] studied two machine learning algorithms, and the variable importance measures (VIMs) method, to select the best features from the data set, in order to classify situations of cyber-aggression on Facebook for Spanish-language users from Mexico. They collected 2000 Facebook comments, which were manually labeled as racism, violence based on sexual orientation and violence against women, by a group of three machine learning teachers which supported the psychologists who specialized in evaluation and intervention of bullying situation in high schools. Experimental results of these works showed a classification performance greater than 90% in accuracy.

Twitter has been a scenario where violence against women, indigenous, minorities and migrants, is frequent. Consequently, much work has been focused on this problem and the potential use of machine learning has been demonstrated as a methodology in Ref. [14]. In addition, data-mining [15] has been used to detect domestic violence. Other works have been performed for automatic detection of sexual violence [16], cyber-bullying [17], hate expressions [18], offensive or aggressiveness [13,19,20] on the twitter messages' content, in which the feature extraction method, including the appropriate collection of expressions (words), is essential.

On the specific attention to GBV, much research has been performed. Ref. [21] exhibited the use of machine learning methods on Twitter to study about the circumstances implicated in the #MeToo movement (an initiative to denounce GBV), mainly those related to business and marketing activities.

Ref. [22] presented the automatic detection and categorization of misogynous language in Twitter by using different supervised classifiers. Techniques like N-grams, linguistic, syntactic and embedding were used in order to build the feature space of the training data set. One of the main contributions of these work was to make available to the research community a data set of corpus of misogynistic tweets. Ref. [22], and similarly [23] who collect data from Twitter from frequent words in domestic violence, highlighted the importance of building a data set corpus of misogynistic tweets and consider the language regionalization, i.e., the data corpus should be in accordance with regional context [13].

Xue et al. [15] evidenced the viability of employing topic-modeling methods for data-mining on Twitter to identify GBV. An unsupervised algorithm to discover hidden topics in the tweets was used. Twitter messages were converted into a document-term matrix by applying the CountVectorizer method [24], in order to collect words that appear more frequently in domestic violence, which are related to GBV.

In Ref. [16], a deep neural network was applied to identify the risk factor associated with sexual violence on Twitter; however, it did not explain how the messages were pre-processed.

Mohammed et al. [17] recommended an array of unique features obtained from Twitter (based on network, activity type, user as well as the content of the tweet) for the detection of cyber-bullying (which has a direct relationship with GBV). Results showed an AUC of 0.943 indicating that this set of features provides an effective approach for detecting cyber-bullying.

In Ref. [18], an approach to automatic detection of hate expressions on Twitter was shown. Authors collected offensive or hateful expressions for hate speech detection. The pre-processing stage consisted of a cleaning up of the tweets, tokenization, generation of negation expressions (e.g., "not", "never", etc.) and detection of the broadcast of these words. In addition, a feature selection process was done.

Ref. [23] exhibited a technique for detection of xenophobia and misogyny in tweets by using computing methodologies. Authors created a suitable language resource for hate speech recognition in Spanish (Spain), highlighting the importance of language regionalization, i.e., whether it is Spanish from Spain or Mexico.

In [19], an Arabic offensive tweet detector was built. An inherent complexity to classify tweets is noticeable, which is in accordance with the particular language.

In the Mexican Spanish context in Twitter, a few works have been performed for automatic identification of GBV. Most of them have been focused on detection of aggressiveness. Alvarez-Carmona et al. [25] presented an overview of results from MEX-A3T competition (2018), which is addressed to automatic identification of aggressiveness in Mexican Spanish tweets. The competition included two tracks: in the first, author profiling, the aim is to identify the place of residence and occupation of the users; in the second, the goal was detection of aggressiveness in the message. Results showed 76.4% accuracy in the aggressiveness identification task. Results of the deep learning methods used in MEX-A3T did not overcome 68% accuracy [20]. Ref. [20] analyzed the performance of two deep learning models for automatic classification of aggressive Mexican Spanish tweets. It highlighted the low performance of studied deep learning neural networks to identify aggression in Mexican Spanish tweets, i.e., there are still open issues to better understand this topic, thus, they should be addressed.

Based on the previous works, two essential components were identified in the analysis of content in Twitter messages: (a) the suitable collect of expressions (words) related to the topic under study in accordance to regional context, and (b) the extraction features stage by simple techniques like the CountVectorizer method [24], which transforms tweet content into vectors by counting occurrences of each word in each tweet, but also the use of sophisticated methodologies like those presented in [18] or [25].

In relation to the pre-processing stage, it was noted that most of the works need a complex pre-processing or specialized group to manually tag the comments (or use small data sets).

As a relevant concern, it was observed that most recent advances are developed for the English language [25], but the few works performed for other languages agree with the importance of the regional context of the messages in their original tongue [19,23].

In this paper, a simple methodology to identify GBV in Mexican Spanish Twitter messages is studied, which includes three common extraction feature methods: CountVectorizer, TfidfVectorizer and HashingVectorizer. In contrast with other state-of-the-art works, our proposal does not employ a stage to collect expressions related to GBV, but only give to the classifier enough samples previously labeled by human volunteers of tweets containing evidence of GBV or not containing GBV. Thus, the significance of this work can be highlighted as follows:

1. This research contributes to the automatic detection of GBV in Mexican Spanish tweets (specifically contextualized to Mexican language jargon), which is a little faced issue, with the potential use of this work in the early attention of dangerous behaviors in the users.

2. It shows encouraging results in classification of tweets related to GBV. Area under the receiver operating characteristic (AUC) obtained is about 80% by using a deep neural network.
3. Feature extraction method used in this work is very simple, i.e., a minimal preprocessing of the data is needed to classify tweets, which only implies to clean (delete articles, numbers, symbols, conjunction and nonsense words) and tokenize (by means of CountVectorizer, TfidfVectorizer and HashingVectorizer methods) tweets' content.

2. Deep Learning Multilayer Perceptron

Deep learning neural networks are characterized by the increase of the network depth, i.e., the number of hidden layers; then, the multilayer perceptron is a general and intuitive architecture to be transformed to the deep learning multilayer perceptron (DL-MLP) with two or more hidden layers [26].

DL-MLP tries to find a relation between a set of input vectors x and labels id by modifying the parameters linking those sets. The output y_j is a function of x and weight w so that if w is modified, the difference z between the system output and target id could be minimized. DL-MLP uses two or more hidden layers constituted of nodes or neurons. Each neuron is connected with the neurons of the previous layer and the output signal is calculated by combining all the inputs from the preceding layer [27]. The connections between nodes use a neuronal weight (w) to modify the output signal before getting in the neuron; this transformation corresponds to multiply the respective signal (x_i) times the weight (w_i).

The use of multiple layers generates a more complex optimization problem, but gains a reduction in the number of nodes per layer inside the architecture [28]. However, the increase of the computational effort can be overcome by the availability of advanced frameworks like Spark [29] and Tensorflow [30] that provide tools to optimize the cost function of the perceptron. The use of such tools makes possible that the DL-MLP could be used increasingly in big-data problems [31,32], and also increases the capability of abstraction of DL-MLP to complex problems [28].

Usually, DL-MLPs are trained by means the back-propagation algorithm (based on the stochastic gradient descent) [33–35] and initial weights are randomly assigned. One of the most common algorithms of descending gradient optimization is Adam [36], which is based on adaptive estimation of first-order and second-order moments [37]. This algorithm reduces the error between the $f(x, w)$ and $\hat{f}(x, w)$.

Typically, DL-MLP includes different activation functions that modify the linear space to a nonlinear space of the samples x in each hidden layer, namely: Rectified Linear Unit (ReLU) $f(z) = max(0, z)$, tangent function $f(z) = tanh(z)$, Exponential Linear Unit (ELu) $f(z) = z \geq 0 \to z, z < 0 \to (e^z - 1)$ and sigmoid function $f(z) = 1/(1 + e^{-z})$.

3. Deep Learning for Natural Language Processing and Sentiment Analysis

The advent of the world wide web and search engines brought with it the emergence of natural language processing (NLP) [38], which allows a machine to process a natural human language and then translates it into a format that is processable and understandable to a computer [39]. This field has received a lot of attention due to the efficiency in language modeling. Some of the NLP models have been applied in various areas, as they provide great mechanisms to analyze text in real time, in addition to the reliability that they also demonstrate in different tasks [40].

Due to the rapid growth of the Internet, the use of social networks, forums, blogs and other platforms where people from all over the world share their ideas, opinions and comments on multiple topics, has increased. Politics, cinema, sports, music, among others, have given rise to a great deal of unstructured information [41]. For this reason, sentiment analysis has become one of the main challenges addressed by NLP, whose main objective is to extract feelings, opinions, attitudes and emotions from the users [42] through a series of methods, techniques and tools on the detection and extraction of subjective information to detect the polarity of the text, that is, to determine if the given text is positive, negative or neutral [43].

Sentiment analysis has been positioned as one of the essential tools to transform the emotions and attitudes of a text into actionable and understandable information for a machine [44]. It is so important within the NLP that this area has been addressed at 3 different levels [42]: (1) the document level, focused on determining whether an opinion document expresses a positive or negative sentiment, (2) the sentence level, whose task is to check whether each sentence expresses a positive, negative or neutral opinion and (3) the aspect level, responsible for looking directly at the opinion itself.

To address the problems of sentiment analysis, previously, approaches based on machine learning algorithms and the sentiment lexicon have been used. However, these methods have limitations such as limited data, word order and a large number of tagged texts that make them ineffective for NLP tasks [45]. However, for some of these problems, models based on deep learning have been the solution, these methods have been gaining popularity, thus proving to be a better option to face the problem of sentiment analysis and this is attributed to the high performance they show in different tasks of the NLP [46].

For years, the implementation of a deep learning or pattern recognition system in NLP has required careful engineering and extensive experience to design a feature extraction system that can transform raw data into appropriate internal data or in a vector of characteristics that a learning subsystem, generally a classifier, could use to detect patterns [47]. Feature extraction, as a data preprocessing method in the learning algorithm, contributes to performance improvement. The extraction methods used for this task range from simple approaches, such as those based on the bag of words model (like CountVectorizer [24], TfidfVectorizer [48] or HashingVectorizer [49], to more sophisticated approaches, such as transformers [50–53].

3.1. Text Feature Extraction

The CountVectorizer method converts a document d into a numeric vector $d = \{u_1, u_2, \ldots, u_i, \ldots, u_T\}$, where (u_i) is the weight of the word with the number i in the document d. The feature i of the document will be the sum of the times that the word i appears in it; seen in another way, u_i will be made up of the frequency of appearance of each word i in the document d [48].

TfidfVectorizer method uses the CountVectorizer matrix and applies a term frequency-inverse document frequency transformation (TFIDF), which takes a frequency of the word i, and the inverse frequency of occurrences in the document d (Equation (1)), instead of the raw frequencies of occurrence of a token [54].

$$u_i = TF_i * IDF_i, \qquad (1)$$

where the weight (u_i) is a function of TF_i (term frequency), i.e., the appearance frequency of the word i in a document d, and IDF_i (inverse document frequency) which is:

$$IDF_i = log(\text{Total of documents}/DF_i), \qquad (2)$$

being DF_i (document frequency) the quantity of documents in which the word i appears at least once.

By using IDF, the weight of high frequency words that are not significant (like conjunctions, prepositions or common words) is reduced, because these kinds of words will appear in several documents allowing to identify those with specific relevance in certain documents.

HashingVectorizer implementation works in a similar way to CountVectorizer, but it employs the hashing trick to find the token string name to include integer index mapping, normalized as token frequencies. Thus, there is no way to compute the inverse transformation, i.e., it does not consider inverse document frequency. However, it is very efficient for large data sets [49].

The CountVectorizer, HashingVectorizer and TfidfVectorizer methods can use different forms of assigning the number of the words included in a token (this parameter is N_{gram}). In the present work, tokens with 1, 2 or 3 words were used, which can give more relations between the pattern of the data.

4. Methodology

The methodological aspects of the work are exhibited in this section. Details about data collection, pre-processing, classifier parameters and assessment test are explained in order to allow the replication of the experiments. The source code for this work is accessible through https://github.com/ccastore/GenderViolence (accessed on 1 January 2021).

4.1. Data Collection

Data were collected by using the *twlets* (http://twlets.com) tool. Twitter messages were collected from 18–19 May 2019, taking tweets comments in Spanish language and located in Mexico (coordinates $-118.599, 14.388$ to $-86.493, 32.718$). In order to select tweets related to GBV, messages from individual users, companies and organizations that contained words or phrases related to diverse forms of possible GBV were selected. In addition, news pages and political figures were considered.

A total of 1,857,450 messages were retrieved from Twitter. 61,604 of them were manually tagged by human volunteers as follows: messages referring to GBV (those containing possible intention of GBV) and messages not referring to GBV, resulting in 1604 positive and 60,000 negative tweets.

4.2. Data Pre-Processing

Once the messages were retrieved from the Twitter stream, they were pre-processed to transform the input text to a normalized, comprehensible model of numbers sequence, proceeding as follows:

- Cleaning. Deletion of URLs (starting with "http://" or "https://"), tags ("@user"), articles and unrelated expressions (for example words written in languages outside of ANSI coding), exclamation marks, question marks, full stop marks, quotes and others symbols.
- Convert text to uppercase.
- Transform text to matrix of numbers, using CountVectorizer [24], TfidfVectorizer [48] and HashingVectorizer [49] methods, using a N_{gram} token with 1, 2 or 3 words.

Finally, a matrix obtained by CountVectorize, TfidfVectorizer and HashingVectorizer methods were used to build and test the classifier. For this, the hold-out method [27] was applied; it randomly split the original matrix on training (TDS) 70% and testing (TS) 30% data sets, where TDS ∩ TS = ∅.

4.3. Sampling Methods

Oversampling methods are popular and successful techniques to deal with the class imbalance [55]. The most common algorithms are: (a) Random Over Sampling (ROS), that randomly duplicates samples from the minority class to mitigate the class imbalance, and (b) SMOTE, which produces artificial samples in the minority class by interpolation of near occurrences [56]. Specifically, for each minority class, they find the k intra-class nearest neighbors and generate synthetic samples in the direction of those nearest neighbors. In this work, k was set to five in SMOTE (as in Ref. [57]) and ROS and SMOTE were applied to the data set to achieve a relatively balanced class distribution.

In particular, for this work, TDS obtained from CountVectorize, TfidfVectorizer and HashingVectorizer methods contains 1122 GBV and 42,000 non-GBV samples (see Sections 4.1 and 4.2); thus, the resultant over-sampled TDS by SMOTE and ROS is composed of 42,000 GBV and 42,000 non-GBV samples approximately, i.e., those methods balance the class distribution.

4.4. Neural Network Set-Up

DL-MLP was developed on Tensorflow 2.0 and Keras 2.3.1, and Adam algorithm [36] was employed to train it. The Adam algorithm is used to calculate the adaptation of the learning rate for each parameter, storing an exponentially decreasing average of past gradients [30]. The learning rate (η) was established as 0.0006, meanwhile the stopping criterion was 20 epochs with a batch size of 150.

DL-MLP was set-up through of the trial and error method, which is usual in neural network environments. For this, we randomly take from TDS a subset ST (about of 20%), that was split into ST_{train} and ST_{test}, where $ST \subseteq TDS$, and $ST_{train} \cap ST_{test} = \emptyset$. In this process, we use ST_{train} and ST_{test} to assess different configurations of numbers of hidden layers and neurons by layer, and the topology that produced the best classification result was selected. Final architecture was a DL-MLP with six hidden layer and sigmoid activation functions, and the number of hidden nodes for each layer was set as 6, 6, 5, 5, 4 and 3, respectively.

4.5. Classifier Performance

Classification accuracy and error rate are widely used to assess the performance of learning models. Nevertheless, in class imbalanced scenarios these measures are biased to majority classes or more represented classes (for example, in this work, there are much more non-GBV tweets than GBV tweets). Thus, others metrics should be used.

The receiver operating characteristic curve (ROC) is an appropriate instrument to evaluate the classifiers performance on imbalance scenarios, according to the trade-offs between benefits (true positives) and costs (false positives). The quantitative depiction of ROC is the area under the curve (AUC), calculated as $AUC = (sensitivity + specificity)/2$, where *sensitivity* is the percentage of correctly predicted *positive* samples, and *specificity* is the percentage of negative samples predicted correctly [58] (see Table 1). In this work, *sensitivity*, *specificity* and the *AUC* were used to measure the effectiveness of deep learning neural network to identify GBV on Mexican tweets.

Table 1. Confusion matrix for binary classification.

		Predicted Class		
		Positive	Negative	
True class	Positive	True Positive (tp)	False Negative (fn)	sensitivity $\frac{tp}{tp+fn}$
	Negative	False Positive (fp)	True Negative (tn)	specificity $\frac{tn}{tn+fp}$

5. Experimental Results and Discussion

The main experimental results in identifying GBV in Mexican tweets are presented in this section. Table 2 summarizes the results in term of features obtained for extraction methods, classification performance measures *sensitivity*, *specificity* and *AUC*.

The number of features for HashingVectorizer method was calculated as trial-error for this work. Several values were tested and the best value was determined to be 350 features. For CountVectorizer and TfidVectorizer methods the default parameters were used. Thus, the employed algorithms settled on number of features (see Section 3.1).

In Table 2, is noted that the class imbalance severely affects the classifier overall performance. Results obtained without using any sampling method indicate that the classifier does not learn the minority class (GBV tweets). Thus, this approach is not appropriate to identify GBV on Mexican tweets.

Table 2. Classification results obtained after applying feature extraction and two sampling methods, using a deep learning multilayer perceptron (DL-MLP) as classifier. Best results (in bold) and the best AUC for each sampling method (marked with a star) are also indicated.

Sampling	Feature Extraction	N_{gram}	Features	Specificity	Sensitivity	AUC
N/A	CountVectorizer	1	1021	1.0000	0.0000	0.5000
		1, 2	2124	1.0000	0.0000	0.5000
		1, 2, 3	2915	1.0000	0.0000	0.5000
	HashingVectorizer	1	350	1.0000	0.0000	0.5000
		1, 2	350	1.0000	0.0000	0.5000
		1, 2, 3	350	1.0000	0.0000	0.5000
	TfidVectorizer	1	1027	1.0000	0.0000	0.5000
		1, 2	2152	1.0000	0.0000	0.5000
		1, 2, 3	2836	1.0000	0.0000	0.5000
SMOTE	CountVectorizer	1	1016	0.8659	**0.7562**	0.8111 *
		1, 2	2143	0.8906	0.6883	0.7895
		1, 2, 3	2879	**0.8921**	0.7022	0.7972
	HashingVectorizer	1	350	0.7125	**0.8067**	0.7596
		1, 2	350	**0.7235**	0.7490	0.7363
		1, 2, 3	350	0.6773	0.7571	0.7172
	TfidVectorizer	1	1014	0.8714	**0.7449**	0.8082
		1, 2	2130	0.8970	0.6838	0.7904
		1, 2, 3	2865	**0.9012**	0.6712	0.7862
ROS	CountVectorizer	1	1018	0.8926	**0.7241**	0.8083 *
		1, 2	2108	**0.9120**	0.6506	0.7813
		1, 2, 3	2881	0.8980	0.6779	0.7880
	HashingVectorizer	1	350	0.7017	**0.8339**	0.7678
		1, 2	350	**0.7500**	0.7384	0.7442
		1, 2, 3	350	0.7100	0.7108	0.7104
	TfidVectorizer	1	1010	0.8861	**0.7291**	0.8076
		1, 2	2137	**0.9217**	0.6279	0.7748
		1, 2, 3	2933	0.9128	0.6544	0.7836

Results obtained by employing sampling methods (ROS and SMOTE) indicate that the DL-MLP is effective to learn GBV tweets. However, Table 2 shows that when the minority class has a best performance the majority class performance is reduced, as it can be observed from the *sensitivity* and *specificity* values. For example, on ROS with HashingVectorizer, and $N_{gram} = 1$, the high value of *sensitivity* is obtained simultaneously with the worst *specificity* value. A similar performance is observed with SMOTE.

AUC gives a better understanding of the classifier performance for both classes than the *sensitivity* and *specificity* measures. High AUC values imply a best trade-off between benefits (GBV tweets correctly classify) and costs (GBV tweets incorrectly classify). In this respect, it is observed in Table 2 that CountVectorizer with $N_{gram} = 1$ presents the best AUC value. Then, it is suggested that the simplest method obtains the highest score.

A trend in the studied feature extraction methods is that the better values of *specificity* and AUC are obtained when the $N_{gram} = 1$ is used than when applying other values. In other words, experimental results of this work notice that to identify GBV on Mexican tweets, the employment of only the mean of each word is an effective approach.

Table 2 shows that the worst AUC values correspond to the HashingVectorizer method. However, this method was developed to work with big data sets; then, it could explain this behavior because the data set used in this research contains only 61,604 samples.

Finally, with respect to the number of features obtained for the extraction methods (CountVectorizer, HashingVectorizer and TfidVectorizer), there is not evidence in the obtained results about the relationship between the number of features used and the classifier performance.

6. Conclusions

GBV is a problem that exist on the social network Twitter. Many works have been performed to deal with it along with related issues like hate speech, xenophobia, misogyny, domestic violence, among others. A main stage of that research is the collection of a corpus of words related to particular situations and language. In the Mexican Spanish context, few works have been developed to deal with GBV in Twitter messages and the language regionalization has been recognized as critical. In addition, results of the most of those works need to be improved.

Thus, in this paper, a study to identify GBV on Twitter messages in Mexico is presented. Three common feature extraction methods were used (CountVectorizer, TfidfVectorizer and HashingVectorizer) together with a deep learning multilayer perceptron as the classifier. A data set containing 1604 GBV tweets and 60,000 non-GBV tweets from a total of 1,857,450 messages retrieved from Twitter social network were labeled by human volunteers as GBV or non-GBV messages to train and test the proposed scheme.

Experimental results showed that the class imbalance problem significantly affects the classification of GBV messages. In this sense, oversampling methods, mainly ROS and SMOTE, are effective to overcome this problem. Thus, it was noticed that the CountVectorizer method (and a sampling method) allows DL-MLP to identify GBV on Mexican tweets with about 80% AUC. As a remarkable result, it is worth to mention that only a minimal data set pre-processing was applied to obtain important results. TfidfVectorizer and HashingVectorizer methods show competitive results, but CountVectorizer presented a trend to obtain the best results.

Results of this research give evidence that giving enough labeled samples, obtained from Mexican Spanish Twitter messages and transformed by simple feature extraction method like CountVectorizer to DL-MLP, can produce improved classification results.

GBV is an issue that must be immediately addressed. In this sense, this study could potentially contribute to deal with gender violence in Mexico because it provides the analysis of useful tools to identify GVB in online social networks despite the language jargon. However, the classification results should be improved because the rate of GBV tweets that have been predicted correctly (*sensitivity*) is still low. The analysis in specific variants of Spanish of certain tools for the detection of GBV could help to push further research needed to improve the studied strategies on the identification of GBV in Twitter messages in Mexican Spanish.

Thus, future work should be addressed mainly to reduce the human effort to label the GBV texts and to test advanced deep learning models in order to increase the classifier performance, including more sophisticated natural language processing techniques. Currently, we work in an application on streaming to identify GVB, which uses a DL-MLP with a rejection option, i.e., when the classifier has doubts about a tweet's content it is rejected and sent to a human volunteer to be targeted and included in the training data set. We consider that this procedure will allow to improve the classifier performance.

Author Contributions: C.M.C., R.A.: conceptualization, methodology and experiment; I.M.A.: conceptualization and review; E.R.: supervision; R.A., E.E.G.-G.: writing—review and editing. O.V.: Experiment. All authors have read and agreed to the published version of the manuscript.

Funding: This research did not receive external funding.

Institutional Review Board Statement: Not applicable.

Informed Consent Statement: Not applicable.

Data Availability Statement: Not applicable.

Acknowledgments: This work has been partially supported under grants of project 5046/2020CIC from UAEMex.

Conflicts of Interest: The authors declare no conflict of interest.

References

1. Sweeney, B.N. Gender-Based Violence and Rape Culture. In *Companion to Women's and Gender Studies*; John Wiley & Sons, Ltd.: Hoboken, NJ, USA, 2020; Chapter 15, pp. 285–302. [CrossRef]
2. Russo, N.F.; Pirlott, A. Gender-based violence: Concepts, methods, and findings. *Ann. N. Y. Acad. Sci.* **2006**, *1087*, 178–205. [CrossRef]
3. UN. *Declaration on the Elimination of Violence against Women*; UN General Assembly: New York, NY, USA, 1993.
4. Hernández Castillo, R.A. Racialized Geographies and the "War on Drugs": Gender Violence, Militarization, and Criminalization of Indigenous Peoples. *J. Lat. Am. Caribb. Anthropol.* **2019**, *24*, 635–652. [CrossRef]
5. Sanchez, G. Victimization, Offending and Resistance in Mexico: Toward Critical Discourse and Grounded Methodologies in Organized Crime Research. *Vict. Offenders* **2020**, *15*, 390–393. [CrossRef]
6. John, N.; Casey, S.E.; Carino, G.; McGovern, T. Lessons Never Learned: Crisis and gender-based violence. *Dev. World Bioeth.* **2020**. [CrossRef]
7. Domínguez, M.A.; Palomeque, D.; Carrillo, J.M.; Valverde, J.M.; Duque, J.F.; Pérez, B.; Pérez-Aloe, R. Voice-Controlled Assistance Device for Victims of Gender-Based Violence. In *Developments and Advances in Defense and Security*; Rocha, Á., Pereira, R.P., Eds.; Springer: Singapore, 2020; pp. 397–407. [CrossRef]
8. Rituerto-González, E.; Mínguez-Sánchez, A.; Gallardo-Antolín, A.; Peláez-Moreno, C. Data Augmentation for Speaker Identification under Stress Conditions to Combat Gender-Based Violence. *Appl. Sci.* **2019**, *9*, 2298. [CrossRef]
9. Rodríguez-Rodríguez, I.; José-Víctor, R.; Domingo-Javier, P.-Q.; Heras-González, P.; Chatzigiannakis, I. Modeling and Forecasting Gender-Based Violence through Machine Learning Techniques. *Appl. Sci.* **2020**, *10*, 8244. [CrossRef]
10. Andrada, A.V.; Sanchez, J.J.; Sánchez-Serrano, J.L.S. Gender Violence and New Technologies. In *Qualitative and Quantitative Models in Socio-Economic Systems and Social Work*; Springer International Publishing: Cham, Switzerland, 2020; pp. 375–390. [CrossRef]
11. Vyawahare, M.; Chatterjee, M. Taxonomy of Cyberbullying Detection and Prediction Techniques in Online Social Networks. In *Data Communication and Networks*; Jain, L.C., Tsihrintzis, G.A., Balas, V.E., Sharma, D.K., Eds.; Springer: Singapore, 2020; pp. 21–37. [CrossRef]
12. Paschalides, D.; Stephanidis, D.; Andreou, A.; Orphanou, K.; Pallis, G.; Dikaiakos, M.D.; Markatos, E. MANDOLA: A Big-Data Processing and Visualization Platform for Monitoring and Detecting Online Hate Speech. *ACM Trans. Internet Technol.* **2020**, *20*. [CrossRef]
13. Gutiérrez-Esparza, G.O.; Vallejo-Allende, M.; Hernández-Torruco, J. Classification of Cyber-Aggression Cases Applying Machine Learning. *Appl. Sci.* **2019**, *9*, 1828. [CrossRef]
14. Bellmore, A.; Calvin, A.J.; Xu, J.M.; Zhu, X. The five Ws of bullying on Twitter: Who, What, Why, Where, and When. *Comput. Hum. Behav.* **2015**, *44*, 305–314. [CrossRef]
15. Xue, J.; Chen, J.; Gelles, R. Using Data Mining Techniques to Examine Domestic Violence Topics on Twitter. *Violence Gend.* **2019**, *6*, 105–114. [CrossRef]
16. Khatua, A.; Cambria, E.; Khatua, A. Sounds of Silence Breakers: Exploring Sexual Violence on Twitter. In Proceedings of the 2018 IEEE/ACM International Conference on Advances in Social Networks Analysis and Mining (ASONAM), Barcelona, Spain, 28–31 August 2018; pp. 397–400. [CrossRef]
17. Al-garadi, M.A.; Varathan, K.D.; Ravana, S.D. Cybercrime detection in online communications: The experimental case of cyberbullying detection in the Twitter network. *Comput. Hum. Behav.* **2016**, *63*, 433–443. [CrossRef]
18. Watanabe, H.; Bouazizi, M.; Ohtsuki, T. Hate Speech on Twitter: A Pragmatic Approach to Collect Hateful and Offensive Expressions and Perform Hate Speech Detection. *IEEE Access* **2018**, *6*, 13825–13835. [CrossRef]
19. Mubarak, H.; Rashed, A.; Darwish, K.; Samih, Y.; Abdelali, A. Arabic Offensive Language on Twitter: Analysis and Experiments. *arXiv* **2020**, arXiv:2004.02192.
20. Frenda, S.; Banerjee, S. Deep Analysis in Aggressive Mexican Tweets. In Proceedings of the Third Workshop on Evaluation of Human Language Technologies for Iberian Languages (IberEval 2018) Co-Located with 34th Conference of the Spanish Society for Natural Language Processing (SEPLN 2018), Sevilla, Spain, 18 September 2018; Volume 2150, pp. 108–113.
21. Reyes-Menendez, A.; Saura, J.R.; Ferrão, F. Marketing challenges in the #MeToo era: Gaining business insights using an exploratory sentiment analysis. *Heliyon* **2020**, *6*, e03626. [CrossRef] [PubMed]
22. Anzovino, M.; Fersini, E.; Rosso, P. Automatic Identification and Classification of Misogynistic Language on Twitter. In *Natural Language Processing and Information Systems*; Silberztein, M., Atigui, F., Kornyshova, E., Métais, E., Meziane, F., Eds.; Springer International Publishing: Cham, Switzerland, 2018; pp. 57–64. [CrossRef]
23. Plaza-Del-Arco, F.M.; Molina-González, M.D.; Ureña López, L.A.; Martín-Valdivia, M.T. Detecting Misogyny and Xenophobia in Spanish Tweets Using Language Technologies. *ACM Trans. Internet Technol.* **2020**, *20*. [CrossRef]
24. Garreta, R.; Moncecchi, G. *Learning Scikit-Learn: Machine Learning in Python*; Packt Publishing: Birmingham, UK, 2013.

25. Aragón, M.E.; Alvarez, M.A.; Montes-y-Gómez, M.; Escalante, H.J.; Villaseñor, L.; Moctezuma, D. Overview of MEX-A3T at IberLEF 2019: Authorship and Aggressiveness Analysis in Mexican Spanish Tweets. In Proceedings of the Iberian Languages Evaluation Forum co-located with 35th Conference of the Spanish Society for Natural Language Processing, IberLEF@SEPLN 2019, Bilbao, Spain, 24 September 2019; Volume 2421, pp. 478–494.
26. Krig, S. Feature Learning and Deep Learning Architecture Survey. In *Computer Vision Metrics*; Springer International Publishing: Cham, Switzerland, 2016; pp. 375–514. [CrossRef]
27. Haykin, S. *Neural Networks. A Comprehensive Foundation*, 2nd ed.; Pretince Hall: Upper Saddle River, NJ, USA, 1999.
28. Goodfellow, I.; Bengio, Y.; Courville, A. *Deep Learning*; MIT Press: Cambridge, MA, USA, 2016.
29. Zaharia, M.; Xin, R.S.; Wendell, P.; Das, T.; Armbrust, M.; Dave, A.; Meng, X.; Rosen, J.; Venkataraman, S.; Franklin, M.J.; et al. Apache Spark: A Unified Engine for Big Data Processing. *Commun. ACM* **2016**, *59*, 56–65. [CrossRef]
30. Abadi, M.; Barham, P.; Chen, J.; Chen, Z.; Davis, A.; Dean, J.; Devin, M.; Ghemawat, S.; Irving, G.; Isard, M.; et al. TensorFlow: A System for Large-scale Machine Learning. In *OSDI'16: Proceedings of the 12th USENIX Conference on Operating Systems Design and Implementation*; USENIX Association: Berkeley, CA, USA, 2016; pp. 265–283.
31. Guo, Y.; Liu, Y.; Oerlemans, A.; Lao, S.; Wu, S.; Lew, M.S. Deep learning for visual understanding: A review. *Neurocomputing* **2016**, *187*, 27–48. [CrossRef]
32. Reyes-Nava, A.; Sánchez, J.; Alejo, R.; Flores-Fuentes, A.; Rendón-Lara, E. Performance Analysis of Deep Neural Networks for Classification of Gene-Expression microarrays. In *MCPR 2018: Pattern Recognition—10th Mexican Conference*; Springer: Cham, Switzerland, 2018; Volume 10880, pp. 105–115. [CrossRef]
33. Li, D.; Huang, F.; Yan, L.; Cao, Z.; Chen, J.; Ye, Z. Landslide Susceptibility Prediction Using Particle-Swarm-Optimized Multilayer Perceptron: Comparisons with Multilayer-Perceptron-Only, BP Neural Network, and Information Value Models. *Appl. Sci.* **2019**, *9*, 3664. [CrossRef]
34. Pacheco-Sánchez, J.; Alejo, R.; Cruz-Reyes, H.; Álvarez-Ramírez, F. Neural networks to fit potential energy curves from asphaltene-asphaltene interaction data. *Fuel* **2019**, *236*, 1117–1127. [CrossRef]
35. Looney, C.G. *Pattern Recognition Using Neural Networks*, 1st ed.; Oxford University Press: Oxford, UK, 1997.
36. Ruder, S. An overview of gradient descent optimization algorithms. *arXiv* **2016**, arXiv:1609.04747.
37. Kingma, D.P.; Ba, J. Adam: A method for stochastic optimization. *arXiv* **2014**, arXiv:1412.6980.
38. Rao, J.S. A Survey on Sentiment Analysis and Opinion Mining. In *Proceedings of the International Conference on Advances in Information Communication Technology and Computing*; Association for Computing Machinery: New York, NY, USA, 2016; [CrossRef]
39. Rajput, A. Chapter 3—Natural Language Processing, Sentiment Analysis, and Clinical Analytics. In *Innovation in Health Informatics*; Lytras, M.D., Sarirete, A., Eds.; Next Gen Tech Driven Personalized Med and Smart Healthcare, Academic Press: New York, NY, USA, 2020; pp. 79–97. [CrossRef]
40. Devika, M.D.; Sunitha, C.; Ganesh, A. Sentiment Analysis: A Comparative Study on Different Approaches. *Procedia Comput. Sci.* **2016**, *87*, 44–49. [CrossRef]
41. Dashtipour, K.; Ieracitano, C.; Morabito, F.C.; Raza, A.; Hussain, A. An Ensemble Based Classification Approach for Persian Sentiment Analysis. In *Progresses in Artificial Intelligence and Neural Systems*; Springer: Singapore, 2021; pp. 207–215. [CrossRef]
42. Al-Bayati, A.; Al-Araji, A.; Ameen, S. Arabic Sentiment Analysis (ASA) Using Deep Learning Approach. *J. Eng.* **2020**, *26*, 85–93. [CrossRef]
43. Mantyla, M.; Graziotin, D.; Kuutila, M. The Evolution of Sentiment Analyi—A Review of Research Topics, Venues, and Top Cited Papers. *Comput. Sci. Rev.* **2016**, *27*, 16–32. [CrossRef]
44. Mishev, K.; Gjorgjevikj, A.; Vodenska, I.; Chitkushev, L.T.; Trajanov, D. Evaluation of Sentiment Analysis in Finance: From Lexicons to Transformers. *IEEE Access* **2020**, *8*, 131662–131682. [CrossRef]
45. Lin, P.; Luo, X.; Fan, Y. A Survey of Sentiment Analysis Based on Deep Learning. *Int. J. Comput. Inf. Eng.* **2020**, *14*, 473–485.
46. Kapil, P.; Ekbal, A.; Das, D. Investigating Deep Learning Approaches for Hate Speech Detection in Social Media. *arXiv* **2020**, arXiv:2005.14690.
47. Liang, H.; Sun, X.; Sun, Y. Text feature extraction based on deep learning: A review. *J. Wirel. Commun. Netw.* **2017**, *2017*, 211. [CrossRef]
48. Eshan, S.C.; Hasan, M.S. An application of machine learning to detect abusive Bengali text. In Proceedings of the 2017 20th International Conference of Computer and Information Technology (ICCIT), Dhaka, Bangladesh, 22–24 December 2017; pp. 1–6.
49. Hasan, M.; Islam, I.; Hasan, K.M.A. Sentiment Analysis Using Out of Core Learning. In Proceedings of the 2019 International Conference on Electrical, Computer and Communication Engineering (ECCE), Cox'sBazar, Bangladesh, 7–9 February 2019; pp. 1–6.
50. Liu, Y.; Ott, M.; Goyal, N.; Du, J.; Joshi, M.; Chen, D.; Levy, O.; Lewis, M.; Zettlemoyer, L.; Stoyanov, V. RoBERTa: A Robustly Optimized BERT Pretraining Approach. *arXiv* **2019**, arXiv:1907.11692.
51. Yang, Z.; Dai, Z.; Yang, Y.; Carbonell, J.; Salakhutdinov, R.; Le, Q.V. XLNet: Generalized Autoregressive Pretraining for Language Understanding. *arXiv* **2019**, arXiv:1906.08237.
52. Lan, Z.; Chen, M.; Goodman, S.; Gimpel, K.; Sharma, P.; Soricut, R. ALBERT: A Lite BERT for Self-supervised Learning of Language Representations. *arXiv* **2019**, arXiv:1909.11942.
53. Clark, K.; Luong, M.T.; Le, Q.V.; Manning, C.D. ELECTRA: Pre-training Text Encoders as Discriminators Rather Than Generators. *arXiv* **2020**, arXiv:2003.10555.

54. Uğuz, H. A two-stage feature selection method for text categorization by using information gain, principal component analysis and genetic algorithm. *Knowl.-Based Syst.* **2011**, *24*, 1024–1032. [CrossRef]
55. Abdi, L.; Hashemi, S. To Combat Multi-class Imbalanced Problems by Means of Over-sampling Techniques. *IEEE Trans. Knowl. Data Eng.* **2016**, *28*, 1041–4347. [CrossRef]
56. Fernandez, A.; Garcia, S.; Herrera, F.; Chawla, N.V. SMOTE for Learning from Imbalanced Data: Progress and Challenges, Marking the 15-year Anniversary. *J. Artif. Intell. Res.* **2018**, *61*, 863–905. [CrossRef]
57. Chawla, N.V.; Bowyer, K.W.; Hall, L.O.; Kegelmeyer, W.P. SMOTE: Synthetic Minority Over-sampling Technique. *J. Artif. Intell. Res.* **2002**, *16*, 321–357. [CrossRef]
58. Sokolova, M.; Lapalme, G. A systematic analysis of performance measures for classification tasks. *Inf. Process. Manag.* **2009**, *45*, 427–437. [CrossRef]

Article

k-Nearest Neighbor Learning with Graph Neural Networks

Seokho Kang

Department of Industrial Engineering, Sungkyunkwan University, 2066 Seobu-ro, Jangan-gu, Suwon 16419, Korea; s.kang@skku.edu; Tel.: +82-31-290-7596

Abstract: k-nearest neighbor (kNN) is a widely used learning algorithm for supervised learning tasks. In practice, the main challenge when using kNN is its high sensitivity to its hyperparameter setting, including the number of nearest neighbors k, the distance function, and the weighting function. To improve the robustness to hyperparameters, this study presents a novel kNN learning method based on a graph neural network, named kNNGNN. Given training data, the method learns a task-specific kNN rule in an end-to-end fashion by means of a graph neural network that takes the kNN graph of an instance to predict the label of the instance. The distance and weighting functions are implicitly embedded within the graph neural network. For a query instance, the prediction is obtained by performing a kNN search from the training data to create a kNN graph and passing it through the graph neural network. The effectiveness of the proposed method is demonstrated using various benchmark datasets for classification and regression tasks.

Keywords: k-nearest neighbor; instance-based learning; graph neural network; deep learning

Citation: Kang, S. k-Nearest Neighbor Learning with Graph Neural Networks. *Mathematics* **2021**, *9*, 830. https://doi.org/10.3390/math9080830

Academic Editor: Florin Leon

Received: 24 March 2021
Accepted: 9 April 2021
Published: 10 April 2021

Publisher's Note: MDPI stays neutral with regard to jurisdictional claims in published maps and institutional affiliations.

Copyright: © 2021 by the author. Licensee MDPI, Basel, Switzerland. This article is an open access article distributed under the terms and conditions of the Creative Commons Attribution (CC BY) license (https://creativecommons.org/licenses/by/4.0/).

1. Introduction

The k-nearest neighbor (kNN) algorithm is one of the most widely used learning algorithms in machine learning research [1,2]. The main concept of kNN is to predict the label of a query instance based on the labels of k closest instances in the stored data, assuming that the label of an instance is similar to that of its kNN instances. kNN is simple and easy to implement, but is very effective in terms of prediction performance. kNN makes no specific assumptions about the distribution of the data. Because it is an instance-based learning algorithm that requires no training before making predictions, incremental learning can be easily adopted. For these reasons, kNN has been actively applied to a variety of supervised learning tasks including both classification and regression tasks.

The procedure for kNN learning is as follows. Suppose a training dataset $\mathcal{D} = \{(\mathbf{x}_t, \mathbf{y}_t)\}_{t=1}^N$ is given for a supervised learning task, where $\mathbf{x}_t$ and $\mathbf{y}_t$ are the input vector and the corresponding label vector of the t-th instance. $\mathbf{y}_t$ is assumed to be a one-hot vector in the case of a classification task and a scalar value in the case of a regression task. In the training phase, the dataset $\mathcal{D}$ is just stored without any explicit learning from the dataset. In the inference phase, for each query instance $\mathbf{x}$, kNN search is performed to retrieve kNN instances $\mathcal{N}(\mathbf{x}_t) = \{(\mathbf{x}_t^{(i)}, \mathbf{y}_t^{(i)})\}_{i=1}^k$ that are closest to $\mathbf{x}$ based on a distance function d. Then, the predicted label $\hat{\mathbf{y}}$ is obtained as a weighted combination of the labels $\mathbf{y}^{(1)}, \ldots, \mathbf{y}^{(k)}$ based on a weighting function w along with the distance function d as follows:

$$\hat{\mathbf{y}} = f(\mathbf{x}; \mathcal{D}) = \frac{\sum_{i=1}^{k} w(d(\mathbf{x}, \mathbf{x}^{(i)})) \cdot \mathbf{y}^{(i)}}{\sum_{i=1}^{k} w(d(\mathbf{x}, \mathbf{x}^{(i)}))} \quad (1)$$

The difficulty in using kNN is determining the hyperparameters. The three main hyperparameters are the number of neighbors k, the distance function d, and the weighting function w [3]. Firstly, in terms of k, a small k makes it capture a specific local structure in the data, and thus, the outcome can be sensitive to noise, whereas a large k makes it more concentrate on the global structure of the data and suppresses the effect of noise. Secondly, the distance function d determines how to calculate the distance between the input vectors of a pair of instances with nearby instances having high relevance. Popular examples of this function for kNN are the Manhattan, Euclidean, and Mahalanobis distances. Thirdly, the weighting function w determines how much each kNN instance contributes to the prediction. The standard kNN assigns the same weight to each kNN instance (i.e., $w(d) = 1/k$). It is known to be better to assign larger/smaller weights to closer/farther kNN instances based on their distances to the query instance x using a non-uniform weighting function (e.g., $w(d) = 1/d$). Thus, a kNN instance with a larger weight will contribute more to the prediction for the instance.

The performance of kNN is known to be highly sensitive to hyperparameters, the best setting of which depends on the characteristics of the data [3,4]. Thus, the hyperparameters must be chosen appropriately to improve the prediction performance. Since this is a challenging issue, considerable research efforts have been devoted to hyperparameter optimization for kNN, which are introduced briefly in Section 2. Compared to related work, the main aim of this study is end-to-end kNN learning toward improved robustness to the hyperparameter setting and to make predictions for new data without additional optimization procedures.

This study presents a novel end-to-end kNN learning method, named kNN graph neural network (kNNGNN), which learns a task-specific kNN rule from the training dataset in an end-to-end fashion based on a graph neural network. For each instance in the training dataset and its kNN instances, a kNN graph is constructed with nodes representing the label information of the instances and edges representing the distance information between the instances. Then, a graph neural network is built to consider the kNN graph of an instance to predict the label for the instance. The graph neural network can be regarded as a data-driven implementation of implicit weight and distance functions. By doing so, the prediction performance of kNN can be improved without careful consideration of its hyperparameter setting. The proposed method is applicable to any type of supervised learning task, including classification and regression. Furthermore, the proposed method does not require any additional optimization procedure when making predictions for new data, which is advantageous in terms of computational efficiency. To investigate the effectiveness of the proposed method, experiments are conducted using various benchmark datasets for classification and regression tasks.

2. Related Work

This section discusses related work on hyperparameter optimization for the kNN algorithm, which has been actively studied by many researchers. As previously mentioned, kNN learning involves three main hyperparameters: the number of neighbors k, the distance function d, and the weighting function w. A different dataset requires a different hyperparameter setting, and no specific setting can universally be the best for every application, as indicted by the no-free-lunch theorem [5]. Thus, the proper choice of these hyperparameters is critical for obtaining a high prediction performance. In practice, the best hyperparameter setting for a given dataset is usually determined by performing a cross-validation procedure that searches over possible hyperparameter candidates. Various search strategies are applicable, such as grid search, random search [6], and Bayesian optimization [7]. They are time consuming and costly, especially for large-scale datasets. Previous research efforts have focused on choosing the hyperparameters of kNN in more intelligent ways based on heuristics or extra optimization procedures for each query instance.

There are two main research approaches regarding the number of neighbors k. The first approach is to assign different k values to different query instances based on their local neighborhood information instead of a fixed k value [8–12]. The second approach is to employ non-uniform weighting functions to reduce the effect of k on the prediction performance.

For the distance function d, one research approach is to learn task-specific distance functions directly from data to improve the prediction performance, which is referred to as distance metric learning [13,14]. Many methods for this approach were developed for use in the classification settings [15–19], while some were developed for use in the regression settings [20–22]. Another approach is to adjust the distance function in an adaptive manner for each query instance [23–27]. This requires an extra optimization procedure, as well as a kNN search when making a prediction for each query instance.

For the weighting function w, existing methods have focused on designing non-uniform weighting functions that decay smoothly as the distance increases [4]. One main research approach is to assign adaptive weights to the kNN instances of each query instance by performing an extra optimization procedure [23,25–28], which also helps to reduce the effect of k. Another approach is to develop fuzzy versions of the kNN algorithm [29–31].

The three hyperparameters affect each other, which means that the optimal choice of one hyperparameter is dependent on the other hyperparameters. Therefore, they must be considered simultaneously rather than independently. Moreover, the methods involving costly extra optimization procedures when making predictions for query instances are computationally expensive, which is undesirable in practice. In addition, the majority of existing methods focus on specific settings, primarily classification tasks. Developing a universal method that is efficient and applicable to various tasks is beneficial. To address these concerns, this study proposes to jointly learn a distance function and a weighting function using a graph neural network in an end-to-end manner, which aims to make it robust to the choice of k in the prediction performance and is applicable to both classification and regression tasks.

3. Method

3.1. Graph Representation of Data

Suppose that a training set $\mathcal{D} = \{(\mathbf{x}_t, \mathbf{y}_t)\}_{t=1}^N$ is given, where $\mathbf{x}_t \in \mathbb{R}^p$ is the t-th input vector for the input variables and $\mathbf{y}_t$ is the corresponding label vector for the output variable. For a classification task with regard to c classes, $\mathbf{y}_t$ is a c-dimensional one-hot vector where the element corresponding to the target class is set to 1 and all the remaining elements are set to 0. For a regression task with a single output, $\mathbf{y}_t$ is a scalar representing the target value.

The proposed method uses a transformation function g that transforms each input vector $\mathbf{x}_t$ into a graph $\mathcal{G}_t$ such that $\mathcal{G}_t = g(\mathbf{x}_t; \mathcal{D})$. Two hyperparameters need to be determined: the number of nearest neighbors k and the distance function d. They are used only to operate the transformation function g for kNN search from $\mathcal{D}$; however, they are not used explicitly in the learning procedure in Section 3.2. For each $\mathbf{x}_t$, its kNN instances are searched from $\mathcal{D} \setminus \{(\mathbf{x}_t, \mathbf{y}_t)\}$ based on the distance function d, denoted by $\mathcal{N}(\mathbf{x}_t) = \{(\mathbf{x}_t^{(i)}, \mathbf{y}_t^{(i)})\}_{i=1}^k$. Then, the kNN graph $\mathcal{G}_t = (\mathcal{V}_t, \mathcal{E}_t)$ is constructed as a fully connected undirected graph with $k+1$ nodes and $k(k+1)/2$ edges as follows:

$$\mathcal{V}_t = \{\mathbf{v}_t^i | i \in \{0, \ldots, k\}\};$$
$$\mathcal{E}_t = \{\mathbf{e}_t^{i,j} | i \in \{0, \ldots, k\}, j \in \{0, \ldots, k\}, i \neq j\}, \quad (2)$$

where each node feature vector $\mathbf{v}_t^i \in \mathbb{R}^{c+1}$ and edge feature vector $\mathbf{e}_t^{i,j} \in \mathbb{R}^p$ are represented as:

$$\mathbf{v}_t^i = \begin{cases} (\mathbf{0}, 1), & \text{if } i = 0 \\ (\mathbf{y}_t^{(i)}, 0), & \text{otherwise} \end{cases};$$
$$\mathbf{e}_t^{i,j} = |\mathbf{x}_t^{(i)} - \mathbf{x}_t^{(j)}|, \quad (3)$$

where the t-th input vector $\mathbf{x}_t$ is denoted by $\mathbf{x}_t^{(0)}$ for the simplicity of description. The number c is set to the number of classes in the case of classification and is 1 in the case of regression.

In the graph $\mathcal{G}_t$, the 0-th node corresponds to $\mathbf{x}_t$, and the other nodes correspond to the kNN instances of $\mathbf{x}_t$. Each node feature vector $\mathbf{v}_t^i$ represents the label information with the last element set to zero, except that $\mathbf{v}_t^0$ does not contain the label information and has the last element set to one. Each edge feature vector $\mathbf{e}_t^{i,j}$ consists of the absolute difference between each of the input variables $\mathbf{x}_t^{(i)}$ and $\mathbf{x}_t^{(j)}$. Thus, $\mathcal{G}_t$ represents the labels of the kNN instances and pairwise distances between the instances. It should be noted that $\mathcal{G}_t$ does not contain $\mathbf{y}_t$ because it needs to be unknown when making a prediction in a supervised learning setting.

3.2. k-Nearest Neighbor Graph Neural Network

Here, the proposed method named kNNGNN is introduced, which implements kNN learning in an end-to-end manner. It adapts the message-passing neural network architecture [32], which can handle general node and edge features with isomorphic invariance, to build a graph neural network for kNN learning. To learn a kNN rule from the training dataset $\mathcal{D}$, it builds a graph neural network that operates on the graph representation $\mathcal{G} = g(\mathbf{x}; \mathcal{D})$ for an input vector $\mathbf{x}$ given the training dataset $\mathcal{D}$ to predict the corresponding label vector $\mathbf{y}$ as $\hat{\mathbf{y}} = f(\mathcal{G}) = f(g(\mathbf{x}; \mathcal{D}))$.

The model architecture used in this study is as follows. It first embeds each $\mathbf{v}^i$ into a p-dimensional initial node representation vector using an embedding function ϕ as $\mathbf{h}^{(0),i} = \phi(\mathbf{v}^i), i = 0, \ldots, k$. A message-passing step for the graph $\mathcal{G}$ is then performed using two main functions: message function M and update function U. The node representation vectors $\mathbf{h}^{(l),i}$ are updated as below:

$$\mathbf{m}^{(l),i} = \sum_{j | \mathbf{v}^j \in \mathcal{V} \setminus \mathbf{v}^i} M(\mathbf{e}^{i,j}) \mathbf{h}^{(l-1),j}, \forall i,$$
$$\mathbf{h}^{(l),i} = U(\mathbf{h}^{(l-1),i}, \mathbf{m}^{(l),i}), \forall i. \quad (4)$$

After L time steps of message passing, a set of node representation vectors $\{\mathbf{h}^{(l),i}\}_{l=0}^{L}$ per node is obtained. The set for the 0-th node $\{\mathbf{h}^{(l),0}\}_{l=0}^{L}$ is then processed with the readout function r to obtain the final prediction of the label $\mathbf{y}$ as:

$$\hat{\mathbf{y}} = r(\{\mathbf{h}^{(l),0}\}_{l=0}^{L}). \quad (5)$$

The component functions ϕ, M, U, and r are parameterized as neural networks, mostly based on the idea presented in Gilmer et al. [32]. The function ϕ is a two-layer fully connected neural network with p tanh units in each layer. The function M is a two-layer fully connected neural network where the first layer consists of $2m$ tanh units and the second layer outputs a $m \times m$ matrix. The function U is modeled as a recurrent neural network with gated recurrent units (GRUs) [33], which pass the previous hidden state $\mathbf{h}^{(l-1),i}$ and the current input $\mathbf{m}^{(l),i}$ to derive the current hidden state $\mathbf{h}^{(l),i}$ at each time step l. The function r is a two-layer fully connected neural network where the first layer consists of p tanh units and the second layer outputs $\hat{\mathbf{y}}$ by softmax and linear units in the case of classification and regression tasks, respectively. Different types of supervised learning tasks can be addressed using different types of units in the last layer of r.

The model defined above is denoted as the function f. The model makes a prediction from the input vector $\mathbf{x}$ and its kNN instances in $\mathcal{D}$, i.e., $\hat{\mathbf{y}} = f(g(\mathbf{x}; \mathcal{D}))$. The model differs from conventional neural networks in that it does not directly learn the relationship between input and output variables. In terms of kNN learning, the weight and distance functions are embedded implicitly into the function f. Therefore, the function f can be regarded as an implicit representation of a kNN rule, in which the functions M and U work as implicit distance and weighting functions, respectively.

3.3. Learning from Training Data

Given the training dataset $\mathcal{D} = \{(\mathbf{x}_t, \mathbf{y}_t)\}_{t=1}^{N}$, the proposed method learns a task-specific kNN rule from $\mathcal{D}$ in the form of $\hat{\mathbf{y}} = f(g(\mathbf{x}; \mathcal{D}))$. The prediction model f is trained based on the graph representation g using the following objective function $\mathcal{J}$:

$$\mathcal{J} = \frac{1}{N} \sum_{(\mathbf{x}_t, \mathbf{y}_t) \in \mathcal{D}} \mathcal{L}(\mathbf{y}_t, \hat{\mathbf{y}}_t) = \frac{1}{N} \sum_{(\mathbf{x}_t, \mathbf{y}_t) \in \mathcal{D}} \mathcal{L}(\mathbf{y}_t, f(g(\mathbf{x}_t; \mathcal{D}))), \tag{6}$$

where $\mathcal{L}$ is the loss function, the choice of which depends on the target task. The typical choices of the loss function are cross-entropy and squared error for the classification and regression tasks, respectively.

3.4. Prediction for New Data

Once the prediction model f is trained, it can be used to predict unknown labels for new data. The prediction procedure is illustrated in Figure 1. Given a query instance $\mathbf{x}_*$ whose label $\mathbf{y}_*$ is unknown, its kNN instances $\mathcal{N}(\mathbf{x}_*) = \{(\mathbf{x}_*^{(i)}, \mathbf{y}_*^{(i)})\}_{i=1}^{k}$ are searched from the training dataset $\mathcal{D}$ based on the distance function d. Then, the corresponding graph $\mathcal{G}_* = g(\mathbf{x}_*; \mathcal{D})$ is generated. The prediction of $\mathbf{y}_*$, which is denoted by $\hat{\mathbf{y}}_*$, is computed using the model f as:

$$\hat{\mathbf{y}}_* = f(\mathcal{G}_*) = f(g(\mathbf{x}_*; \mathcal{D})). \tag{7}$$

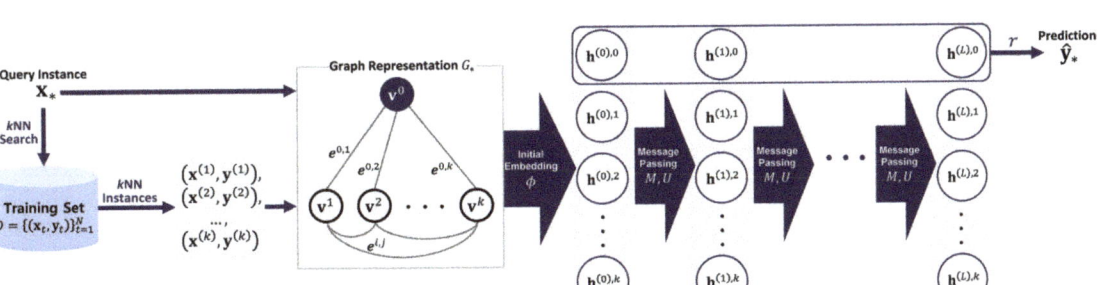

Figure 1. Schematic of the kNN graph neural network (kNNGNN) prediction procedure.

The proposed method does not require additional optimization procedures when making predictions. The prediction for a query instance is simply conducted by performing a kNN search to identify the kNN instances and then processing these instances with the model. This is advantageous in terms of computational efficiency.

As the proposed method learns the kNN rule, incremental learning can be implemented efficiently. This is the main advantage of the kNN algorithm compared to other learning algorithms, especially when additional training data are collected over time after the model is trained. When new labeled data are added to the training dataset $\mathcal{D}$, the prediction performance will be improved without updating the model.

4. Experimental Investigation

4.1. Datasets

The effectiveness of the proposed method was investigated through experiments on various benchmark datasets. They contained 20 classification datasets, and twenty regression datasets were collected from the UCI machine learning repository (http://archive.ics.uci.edu/ml/ (accessed on 10 January 2021) and the StatLib datasets archive (http://lib.stat.cmu.edu/datasets/ (accessed on 10 January 2021)). The datasets used for classification tasks were *annealing, balance, breastcancer, carevaluation, ecoli, glass, heart, ionosphere, iris, landcover, movement, parkinsons, seed, segment, sonar, vehicle, vowel, wine, yeast,* and *zoo*. The datasets used for regression tasks were *abalone, airfoil, appliances, autompg, bikesharing, bodyfat, cadata, concretecs, cpusmall, efficiency, housing, mg, motorcycle, newspopularity, skillcraft, spacega, superconductivity, telemonitoring, wine-red,* and *wine-white*. Each dataset had a different number of instances with a different dimensionality. For each dataset, one-thousand instances were randomly sampled if the size of the dataset was greater than 1000. All numeric variables were normalized into the range of $[-1, 1]$. The details of the datasets used are listed in Tables 1 and 2.

Table 1. Summary statistics of the error rate over different hyperparameter settings on the classification datasets.

Dataset [Size × Dim.]	No. of Classes	Uniform kNN			Weighted kNN			kNNGNN (Proposed)		
		Average	Std. Dev.	Best	Average	Std. Dev.	Best	Average	Std. Dev.	Best
annealing [898 × 38]	5	0.0724	0.0356	0.0175	0.0364	0.0146	**0.0174**	0.0254	**0.0046**	0.0189
balance [625 × 4]	3	0.1453	0.0356	0.1130	0.1436	0.0377	0.1098	**0.1327**	**0.0235**	0.1046
breastcancer [683 × 9]	2	0.0382	0.0051	0.0306	0.0381	0.0051	0.0305	0.0369	0.0051	**0.0290**
carevaluation [1000 × 6]	4	0.2035	**0.0306**	0.1619	0.1858	0.0326	0.1461	**0.1110**	0.0354	**0.0717**
ecoli [336 × 7]	8	0.1853	0.0350	0.1427	**0.1703**	0.0256	**0.1422**	0.1849	0.0238	0.1555
glass [214 × 9]	6	0.3808	0.0425	0.3196	0.3441	0.0229	0.3074	0.3703	0.0259	0.3244
heart [298 × 13]	2	**0.2081**	**0.0311**	0.1792	0.2093	0.0349	**0.1684**	0.2126	0.0366	0.1804
ionosphere [351 × 34]	2	0.1808	0.0448	0.1120	0.1790	0.0428	0.1120	**0.0838**	**0.0122**	**0.0681**
iris [150 × 4]	3	0.0886	0.0522	0.0407	0.0766	0.0417	0.0427	**0.0573**	**0.0157**	0.0413
landcover [675 × 147]	9	0.3550	0.2050	0.1950	0.3450	0.2023	0.1847	**0.2187**	**0.0513**	0.1767
movement [360 × 90]	15	0.4808	0.1436	**0.2125**	0.3679	0.1125	**0.2125**	**0.3300**	0.0375	0.2569
parkinsons [195 × 22]	2	0.1634	0.0458	**0.0775**	0.1373	0.0368	**0.0775**	0.1536	0.0253	0.0918
seed [210 × 7]	3	0.0809	0.0114	**0.0590**	0.0779	0.0090	0.0610	0.0840	**0.0078**	0.0619
segment [1000 × 19]	7	0.1021	0.0281	0.0594	0.0840	0.0184	0.0586	**0.0749**	**0.0079**	**0.0586**
sonar [208 × 60]	2	0.2929	0.0621	**0.1706**	0.2684	0.0556	**0.1706**	0.2387	0.0349	0.1837
vehicle [846 × 18]	4	0.2951	0.0469	0.2276	0.2880	0.0432	0.2239	**0.2775**	0.0391	**0.2191**
vowel [990 × 10]	11	0.2868	0.1367	0.1450	0.1450	0.0553	**0.0516**	0.1504	0.0404	0.0561
wine [178 × 13]	3	0.1035	0.1010	0.0316	0.0861	0.0759	0.0304	**0.0522**	**0.0329**	**0.0270**
yeast [1000 × 8]	10	0.4495	**0.0218**	0.4305	0.4369	0.0292	**0.4116**	0.4514	0.0284	0.4222
zoo [101 × 16]	7	0.2269	0.1656	0.0505	0.1145	0.0776	**0.0484**	**0.1012**	0.0462	0.0709

The lowest values for each dataset are presented in bold.

Table 2. Summary statistics of the RMSE over different hyperparameter settings on the regression datasets.

Dataset [Size × Dim.]	Uniform kNN			Weighted kNN			kNNGNN (Proposed)		
	Average	Std. Dev.	Best	Average	Std. Dev.	Best	Average	Std. Dev.	Best
abalone [1000 × 8]	0.2018	0.0183	0.1812	0.2006	0.0185	**0.1804**	0.1982	**0.0103**	0.1829
airfoil [1000 × 5]	0.2573	0.0185	0.2209	0.2270	**0.0115**	0.2060	**0.1884**	0.0125	**0.1698**
appliances [1000 × 25]	0.2663	0.0173	0.2533	0.2617	0.0190	0.2493	0.2601	**0.0024**	**0.2563**
autompg [392 × 7]	0.1846	0.0137	0.1642	**0.1768**	0.0139	0.1576	0.1821	**0.0092**	0.1715
bikesharing [1000 × 14]	0.1886	0.0315	0.1386	0.1813	**0.0291**	0.1348	**0.0752**	0.0481	**0.0310**
bodyfat [252 × 14]	0.1887	**0.0345**	0.1434	0.1855	0.0350	0.1416	**0.1296**	0.0459	**0.0830**
cadata [1000 × 8]	0.3384	0.0281	0.2999	0.3329	**0.0285**	0.2968	0.3153	0.0243	**0.2848**
concretecs [1000 × 8]	0.2566	0.0197	0.2237	0.2361	**0.0197**	0.2065	**0.2036**	0.0236	**0.1804**
cpusmall [1000 × 12]	0.1370	0.0503	0.0828	0.1247	0.0414	0.0810	**0.0876**	**0.0143**	**0.0708**
efficiency [768 × 8]	0.1468	**0.0326**	0.1077	0.1392	0.0346	0.0980	**0.0700**	0.0326	**0.0483**
housing [506 × 13]	0.2503	0.0197	0.2179	0.2333	**0.0168**	0.2038	**0.2180**	0.0199	**0.1847**
mg [1000 × 6]	0.2947	0.0249	0.2782	**0.2891**	0.0275	**0.2697**	0.2923	**0.0181**	0.2802
motorcycle [133 × 1]	0.2845	0.0473	**0.2415**	0.2862	**0.0170**	0.2730	**0.2694**	0.0195	0.2491
newspopularity [1000 × 58]	0.1242	0.0117	**0.1156**	0.1244	0.0117	0.1158	**0.1218**	**0.0014**	0.1200
skillcraft [1000 × 18]	0.3741	0.0435	0.3365	0.3734	0.0437	0.3354	**0.3603**	0.0241	**0.3317**
spacega [1000 × 6]	0.1576	0.0124	0.1415	**0.1552**	**0.0131**	0.1392	0.1556	0.0097	0.1421
superconductivity [1000 × 81]	0.2962	0.0449	0.2574	0.2790	0.0393	**0.2454**	**0.2655**	**0.0105**	0.2491
telemonitoring [1000 × 16]	0.4274	0.0363	0.3943	0.4235	0.0377	**0.3905**	0.4121	**0.0125**	0.3914
wine-red [1000 × 11]	0.2844	0.0234	0.2668	**0.2751**	0.0272	**0.2547**	0.2812	**0.0088**	0.2708
wine-white [1000 × 11]	0.2895	0.0232	0.2756	**0.2832**	0.0259	**0.2671**	0.2854	**0.0077**	0.2758

The lowest values for each dataset are presented in bold.

4.2. Compared Methods

Three kNN methods that use different weighting schemes w were compared in the experiments: uniform kNN, weighted kNN, and the proposed kNNGNN. The uniform kNN and weighted kNN respectively used the following weighting functions:

$$w_U(d(\mathbf{x}, \mathbf{x}')) = 1/k;$$
$$w_W(d(\mathbf{x}, \mathbf{x}')) = 1/d(\mathbf{x}, \mathbf{x}'). \tag{8}$$

For kNNGNN, the weighting function is embedded implicitly.

For each method, the hyperparameter settings were varied to examine their effects. The candidates for the distance function d were as follows:

$$\text{Manhattan } d_{L1}(\mathbf{x}, \mathbf{x}') = ||\mathbf{x} - \mathbf{x}'||_1;$$
$$\text{Euclidean } d_{L2}(\mathbf{x}, \mathbf{x}') = ||\mathbf{x} - \mathbf{x}'||_2 = \sqrt{(\mathbf{x} - \mathbf{x}')^T(\mathbf{x} - \mathbf{x}')}; \tag{9}$$
$$\text{Mahalanobis } d_M(\mathbf{x}, \mathbf{x}') = \sqrt{(\mathbf{x} - \mathbf{x}')^T S^{-1}(\mathbf{x} - \mathbf{x}')},$$

where S is the covariance matrix of the input variables calculated from the training dataset.

Accordingly, there were a total of nine combinations of distance and weighting functions compared in the experiments, as summarized in Table 3. None of the methods used any additional optimization procedures when making predictions. For kNNGNN, the distance function was only explicitly used for the kNN search to generate graph representations of the data. For each combination, the effect of k was investigated on the prediction performance by varying its value from 1, 3, 5, 7, 10, 15, 20, and 30.

Table 3. Methods compared in the experiments.

		kNN Method (Weighting Function w)		
		Uniform kNN	Weighted kNN	kNNGNN (Proposed)
distance function d	Manhattan (L1)	kNN_L1	WkNN_L1	kNNGNN_L1
	Euclidean (L2)	kNN_L2	WkNN_L2	kNNGNN_L2
	Mahalanobis (M)	kNN_M	WkNN_M	kNNGNN_M

4.3. Experimental Settings

In the experiments, the performance of each method was evaluated using a two-fold cross-validation procedure. In this procedure, the original dataset was divided into five disjoint subsets. Then, two iterations were conducted, each of which used one subset and the other subset as the training and test sets, respectively. As performance measures, the misclassification error rate and root mean squared error (RMSE) were used for the classification and regression tasks, respectively. Given a test set denoted by $\mathcal{D}' = \{(\mathbf{x_t}, \mathbf{y_t})\}_{t=1}^{N'}$, the performance measures are calculated as:

$$\text{ErrorRate} = \frac{1}{N'} \sum_{(\mathbf{x}_t, \mathbf{y}_t) \in \mathcal{D}'} I(\text{argmax}(\mathbf{y}_t) \neq \text{argmax}(\hat{\mathbf{y}}_t));$$
$$\text{RMSE} = \frac{1}{N'} \sum_{(\mathbf{x}_t, \mathbf{y}_t) \in \mathcal{D}'} (\mathbf{y}_t - \hat{\mathbf{y}}_t)^2. \tag{10}$$

For the proposed method, each prediction model was built based on the following configurations. In the objective function $\mathcal{J}$, the loss function $\mathcal{L}$ used for the classification and regression tasks was set to cross-entropy and squared error, respectively. For the model, the hyperparameter L was set to 3, as Gilmer et al. [32] demonstrated any $L \geq 3$ would work. The hyperparameter p was explored on $\{10, 20, 50\}$ by holdout validation. In the training phase, dropout was applied to the function r with a dropout rate of 0.1 for regularization [34]. During the training, eighty percent and 20% of the training set were used to train and validate the model, respectively. The model parameters were updated using the Adam optimizer with a batch size of 20. The learning rate was set to 10^{-3} at the first training epoch and was reduced by a factor of 0.1 if no improvement in the validation loss was observed for 10 consecutive epochs. The training was terminated when the learning rate was decreased to 10^{-7} or the number of epochs reached 500. In the inference phase, for each query instance, thirty different outputs were obtained by performing stochastic forward passes through the trained model with the dropout turned on [35]. The average of these outputs was then used to obtain the predicted label for the instance.

All baseline methods were implemented using the scikit-learn package in Python. The proposed method was implemented based on GPU-accelerated TensorFlow in Python. All experiments were performed 10 times independently with different random seeds. For the results, the average performance over the repetitions was compared. Then, for each of the three weighting functions w, the summary statistics of the performance over different settings of distance functions d and the number of neighbors k are reported.

4.4. Results and Discussion

Figure 2 shows the error rate comparison results of the baseline and proposed methods with varying the hyperparameter settings on 20 classification datasets. Compared to the baseline methods, kNNGNN overall yielded lower error rates at various values of k for most datasets. For the results with different hyperparameters, the average, standard deviation, and best error rate for each dataset are summarized in Table 1. kNNGNN yielded the lowest average and standard deviation of the error rate over different hyperparameters on most datasets, which indicated that the performance of kNNGNN was less sensitive to its hyperparameter settings. In particular, kNNGNN was superior to the baseline method when the hyperparameter k was larger.

Figure 3 compares the baseline and proposed methods in terms of the RMSE with varying hyperparameter settings on 20 regression datasets. As shown in this figure, the performance curves of kNNGNN flattened as k increased on most datasets, whereas the RMSE of the baseline methods tended to increase at large k for some datasets. Table 2 shows the average, standard deviation, and best RMSE for different hyperparameter settings for each dataset. The behavior of kNNGNN was similar to that of the classification tasks. kNNGNN showed stable performance against changes in the hyperparameter settings. kNNGNN yielded the lowest average and standard deviation of the RMSE for the majority of datasets.

In summary, the experimental results successfully demonstrated the effectiveness of kNNGNN in improving the prediction performance for both classification and regression tasks. Although kNNGNN failed to yield the lowest error for some datasets, kNNGNN yielded high robustness to its hyperparameters. This indicated that kNNGNN would provide comparable performance without carefully tuning its hyperparameters; thus, it can be preferred in practice considering the difficulty of choosing the optimal hyperparameter setting. Because the performance curve of kNNGNN flattened at large k values on most datasets, setting a moderate k value around 15~20 would be reasonable considering the trade-off between the performance and computational cost.

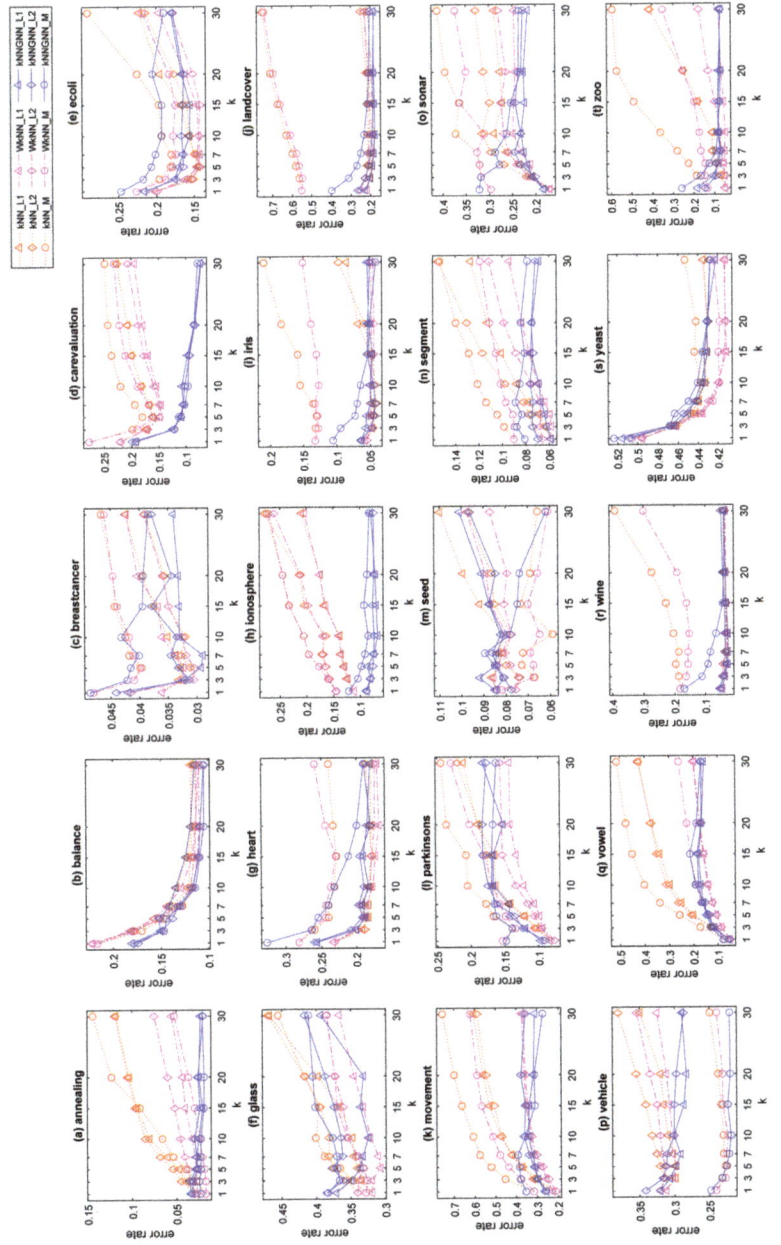

Figure 2. Error rate comparison with varying hyperparameter settings on classification datasets.

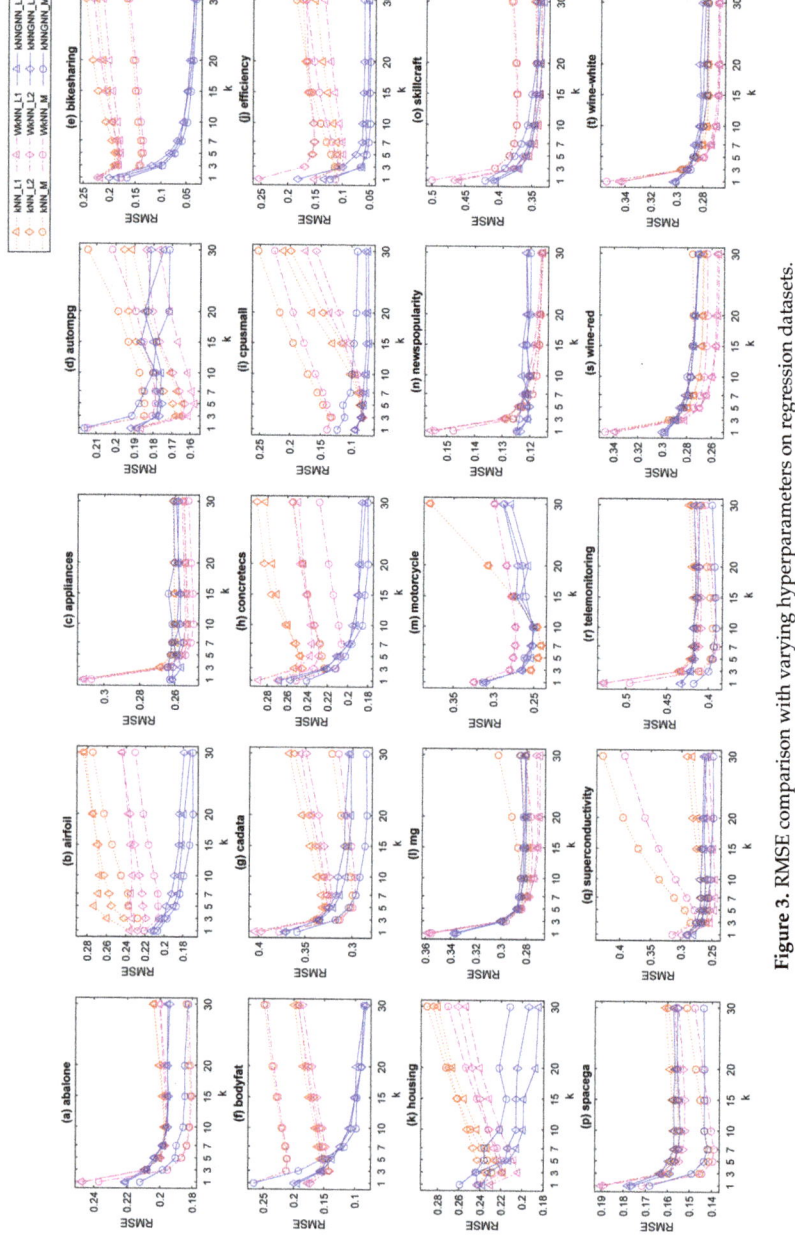

Figure 3. RMSE comparison with varying hyperparameters on regression datasets.

5. Conclusions

This study presented kNNGNN, which learns a task-specific kNN rule from data in an end-to-end fashion. The proposed method constructed the kNN rule in the form of a graph neural network, in which the distance and weighting functions were embedded implicitly. The graph neural network considered the kNN graph of an instance as the input to predict the label of the instance. Owing to the flexibility of neural networks, the method can be applied to any form of supervised learning tasks including classification and regression. It does not require any extra optimization procedure when making predictions for new data, which is beneficial in terms of computational efficiency. Moreover, as the method learns the kNN rule instead of the explicit relationship between the input and output variables, incremental learning can be implemented efficiently.

The effectiveness of the proposed method was demonstrated through experiments on benchmark classification and regression datasets. The results showed that the proposed method can yield comparable prediction performance with less sensitivity to the choice of its hyperparameters. The proposed method allows more robust kNN learning without carefully tuning the hyperparameters. The use of a graph neural network for kNN learning may still have room for improvement and thus merits further investigation. One practical concern is the high complexity of a graph neural network in terms of time and space, which increases with k. A graph neural network cannot be trained in a reasonable amount of time without using a GPU. Alleviation of complexity to improve learning efficiency will be an avenue for future work.

Funding: This work was supported by the National Research Foundation of Korea (NRF) grant funded by the Korea government (MSIT; Ministry of Science and ICT) (Nos. NRF-2019R1A4A1024732 and NRF-2020R1C1C1003232).

Institutional Review Board Statement: Not applicable.

Informed Consent Statement: Not applicable.

Data Availability Statement: Not applicable.

Conflicts of Interest: The authors declare no conflict of interest.

References

1. Wu, X.; Kumar, V.; Quinlan, J.R.; Ghosh, J.; Yang, Q.; Motoda, H.; McLachlan, G.J.; Ng, A.; Liu, B.; Philip, S.Y.; et al. Top 10 algorithms in data mining. *Knowl. Inf. Syst.* **2008**, *14*, 1–37. [CrossRef]
2. Cover, T.; Hart, P. Nearest neighbor pattern classification. *IEEE Trans. Inf. Theory* **1967**, *13*, 21–27. [CrossRef]
3. Jiang, L.; Cai, Z.; Wang, D.; Jiang, S. Survey of improving k-nearest-neighbor for classification. In Proceedings of the International Conference on Fuzzy Systems and Knowledge Discovery, Haikou, China, 24–27 August 2007; pp. 679–683.
4. Atkeson, C.G.; Moore, A.W.; Schaal, S. Locally Weighted Learning. *Artif. Intell. Rev.* **1997**, *11*, 11–73. [CrossRef]
5. Wolpert, D.H. The lack of a priori distinctions between learning algorithms. *Neural Comput.* **1996**, *8*, 1341–1390. [CrossRef]
6. Bergstra, J.; Bengio, Y. Random search for hyper-parameter optimization. *J. Mach. Learn. Res.* **2012**, *13*, 281–305.
7. Snoek, J.; Larochelle, H.; Adams, R.P. Practical Bayesian optimization of machine learning algorithms. In Proceedings of the Advances in Neural Information Processing Systems, Lake Tahoe, NV, USA, 3–6 December 2012; pp. 2951–2959.
8. Zhang, S.; Li, X.; Zong, M.; Zhu, X.; Wang, R. Efficient kNN classification with different numbers of nearest neighbors. *IEEE Trans. Neural Netw. Learn. Syst.* **2017**, *29*, 1774–1785. [CrossRef]
9. Zhang, S.; Cheng, D.; Deng, Z.; Zong, M.; Deng, X. A novel kNN algorithm with data-driven k parameter computation. *Pattern Recognit. Lett.* **2018**, *109*, 44–54. [CrossRef]
10. Wang, J.; Neskovic, P.; Cooper, L.N. Neighborhood size selection in the k-nearest-neighbor rule using statistical confidence. *Pattern Recognit.* **2006**, *39*, 417–423. [CrossRef]
11. García-Pedrajas, N.; del Castillo, J.A.R.; Cerruela-García, G. A proposal for local k values for k-nearest neighbor rule. *IEEE Trans. Neural Netw. Learn. Syst.* **2015**, *28*, 470–475. [CrossRef]
12. Zhang, S.; Li, X.; Zong, M.; Zhu, X.; Cheng, D. Learning k for kNN classification. *ACM Trans. Intell. Syst. Technol.* **2017**, *8*, 43. [CrossRef]
13. Li, D.; Tian, Y. Survey and experimental study on metric learning methods. *Neural Netw.* **2018**, *105*, 447–462. [CrossRef] [PubMed]
14. Kulis, B. Metric Learning: A Survey. *Found. Trends Mach. Learn.* **2013**, *5*, 287–364. [CrossRef]

15. Weinberger, K.Q.; Saul, L.K. Distance metric learning for large margin nearest neighbor classification. *J. Mach. Learn. Res.* **2009**, *10*, 207–244.
16. Goldberger, J.; Roweis, S.; Hinton, G.; Salakhutdinov, R. Neighbourhood components analysis. In *Advances in Neural Information Processing Systems*; MIT Press: Cambridge, MA, USA, 2004; pp. 513–520.
17. Sugiyama, M. Dimensionality reduction of multimodal labeled data by local fisher discriminant analysis. *J. Mach. Learn. Res.* **2007**, *8*, 1027–1061.
18. Davis, J.V.; Kulis, B.; Jain, P.; Sra, S.; Dhillon, I.S. Information-theoretic metric learning. In Proceedings of the International Conference on Machine Learning, Cincinnati, OH, USA, 13–15 December 2007; pp. 209–216.
19. Wang, W.; Hu, B.G.; Wang, Z.F. Globality and locality incorporation in distance metric learning. *Neurocomputing* **2014**, *129*, 185–198. [CrossRef]
20. Assi, K.C.; Labelle, H.; Cheriet, F. Modified large margin nearest neighbor metric learning for regression. *IEEE Signal Process. Lett.* **2014**, *21*, 292–296. [CrossRef]
21. Weinberger, K.Q.; Tesauro, G. Metric Learning for Kernel Regression. In Proceedings of the International Conference on Artificial Intelligence and Statistics, San Juan, Puerto Rico, 21–24 March 2007; pp. 612–619.
22. Nguyen, B.; Morell, C.; De Baets, B. Large-scale distance metric learning for k-nearest neighbors regression. *Neurocomputing* **2016**, *214*, 805–814. [CrossRef]
23. Wang, J.; Neskovic, P.; Cooper, L.N. Improving nearest neighbor rule with a simple adaptive distance measure. *Pattern Recognit. Lett.* **2007**, *28*, 207–213. [CrossRef]
24. Zhou, C.Y.; Chen, Y.Q. Improving nearest neighbor classification with cam weighted distance. *Pattern Recognit.* **2006**, *39*, 635–645. [CrossRef]
25. Jahromi, M.Z.; Parvinnia, E.; John, R. A method of learning weighted similarity function to improve the performance of nearest neighbor. *Inf. Sci.* **2009**, *179*, 2964–2973. [CrossRef]
26. Paredes, R.; Vidal, E. Learning weighted metrics to minimize nearest-neighbor classification error. *IEEE Trans. Pattern Anal. Mach. Intell.* **2006**, *28*, 1100–1110. [CrossRef]
27. Domeniconi, C.; Peng, J.; Gunopulos, D. Locally adaptive metric nearest-neighbor classification. *IEEE Trans. Pattern Anal. Mach. Intell.* **2002**, *24*, 1281–1285. [CrossRef]
28. Kang, P.; Cho, S. Locally linear reconstruction for instance-based learning. *Pattern Recognit.* **2008**, *41*, 3507–3518. [CrossRef]
29. Keller, J.M.; Gray, M.R.; Givens, J.A. A fuzzy k-nearest neighbor algorithm. *IEEE Trans. Syst. Man, Cybern.* **1985**, *SMC-15*, 580–585.
30. Biswas, N.; Chakraborty, S.; Mullick, S.S.; Das, S. A parameter independent fuzzy weighted k-nearest neighbor classifier. *Pattern Recognit. Lett.* **2018**, *101*, 80–87. [CrossRef]
31. Maillo, J.; García, S.; Luengo, J.; Herrera, F.; Triguero, I. Fast and scalable approaches to accelerate the fuzzy k-Nearest neighbors classifier for big data. *IEEE Trans. Fuzzy Syst.* **2019**, *28*, 874–886. [CrossRef]
32. Gilmer, J.; Schoenholz, S.S.; Riley, P.F.; Vinyals, O.; Dahl, G.E. Neural message passing for quantum chemistry. In Proceedings of the International Conference on Machine Learning, Sydney, Australia, 11–15 August 2017; pp. 1263–1272.
33. Cho, K.; Van Merriënboer, B.; Gulcehre, C.; Bahdanau, D.; Bougares, F.; Schwenk, H.; Bengio, Y. Learning phrase representations using RNN encoder-decoder for statistical machine translation. In Proceedings of the Conference on Empirical Methods in Natural Language Processing, Doha, Qatar, 25–29 October 2014; pp. 1724–1734.
34. Srivastava, N.; Hinton, G.; Krizhevsky, A.; Sutskever, I.; Salakhutdinov, R. Dropout: A simple way to prevent neural networks from overfitting. *J. Mach. Learn. Res.* **2014**, *15*, 1929–1958.
35. Gal, Y.; Ghahramani, Z. Dropout as a Bayesian approximation: Representing model uncertainty in deep learning. In Proceedings of the International Conference on Machine Learning, New York, NY, USA, 19–24 June 2016; pp. 1050–1059.

Article

RHOASo: An Early Stop Hyper-Parameter Optimization Algorithm

Ángel Luis Muñoz Castañeda [1,2,*], Noemí DeCastro-García [1,2] and David Escudero García [2]

1. Department of Mathematics, Universidad de León, 24007 León, Spain; ncasg@unileon.es
2. Research Institute of Applied Sciences in Cybersecurity (RIASC), Universidad de León, 24007 León, Spain; descg@unileon.es
* Correspondence: amunc@unileon.es

Abstract: This work proposes a new algorithm for optimizing hyper-parameters of a machine learning algorithm, RHOASo, based on conditional optimization of concave asymptotic functions. A comparative analysis of the algorithm is presented, giving particular emphasis to two important properties: the capability of the algorithm to work efficiently with a small part of a dataset and to finish the tuning process automatically, that is, without making explicit, by the user, the number of iterations that the algorithm must perform. Statistical analyses over 16 public benchmark datasets comparing the performance of seven hyper-parameter optimization algorithms with RHOASo were carried out. The efficiency of RHOASo presents the positive statistically significant differences concerning the other hyper-parameter optimization algorithms considered in the experiments. Furthermore, it is shown that, on average, the algorithm needs around 70% of the iterations needed by other algorithms to achieve competitive performance. The results show that the algorithm presents significant stability regarding the size of the used dataset partition.

Keywords: hyperparameters; machine learning; optimization; inference

1. Introduction

Tuning the hyper-parameter configuration of a machine learning (ML) algorithm is a recommended procedure to obtain a successful ML model for a given problem. Different ML algorithms have specific hyper-parameters whose configuration requires a deep understanding of both the model and the task. Since the hyper-parameter configuration greatly impacts the models' performance, the research in automatic hyper-parameter optimization (HPO) is focused on developing techniques that efficiently find optimal values for the hyper-parameters, maximizing accuracy while avoiding complex and expensive operations. However, this process remains a challenge because not all optimization methods are always suitable for a given problem.

Although there are several methods for tuning both continuous and discrete hyper-parameters, they do not perform equally for all ML algorithms, displaying different consumption of computational resources and stability. The process becomes computationally expensive if too many function evaluations of hyper-parameter values must be carried out to obtain a suitable accuracy. Since the size of the dataset in the HPO phase influences the dynamic complexity of the classifier but not its accuracy [1], another possible limitation is that a HPO algorithm may require a large dataset to work efficiently. Finally, most HPO algorithms are iterative, which suggests that stopping the algorithm when the expected improvement of testing new configurations is low can be a good option [2]. Nevertheless, the ML user does not have information about the rate of convergence and the loss function values. Therefore, the user usually tends to leave the default parameters (more than 50 iterations) or set a high number of iterations to assure good performance ([3]). This fact implies that the algorithm may perform more iterations than needed to obtain an adequate accuracy, with the consequent increased computational cost.

This article proposes a novel early stop HPO algorithm, RHOASo (*RIASC hyperoptimization automatic software*). The work aims to analyze RHOASo, compare its behavior with different state-of-the-art HPO algorithms, and measure its early-stop feature, which entails minimal human intervention.

The research questions that will be studied regarding this new algorithm are the following:

RQ1: Given a dataset, how good is the performance (accuracy, time complexity, sensibility, and specificity) of a ML algorithm when RHOASo is applied?

RQ2: How many iterations does the algorithm need until it stops? How much faster or slower is RHOASo, compared with the other HPO algorithms?

RQ3: Are there statistically meaningful differences between the performance of RHOASo and other HPO algorithms?

RQ4: Are the above results consistent? That is, do they hold for different HPO algorithms and different datasets with different characteristics (size, number of features, etc.)?

In order to test the behavior of RHOASo and answer the above questions, we have evaluated the efficiency of the algorithm combined with three well-known classifier algorithms: random forest (RF), gradient boosting (GB), and multi-layer perceptron (MLP). We choose these three supervised models because each follows a different learning paradigm: RF and GB are ensemble models (bagging and boosting models), and MLP is a type of neural network. Therefore, it is possible to study whether a particular type of model works better with the application of RHOASo.

We have measured the efficiency of RHOASo, by itself and carrying out statistical inference to evaluate how well it performs compared with the other seven HPO algorithms from several families (decision theoretic approaches, Bayesian optimization, evolutionary strategies, etc). The variables considered in the article are the accuracy, the MCC (Matthews correlation coefficient), the time complexity of the whole optimization (from initialization to termination), the sensibility, and the specificity of the obtained models. These variables are collected by applying HPOs over 16 well-known public datasets. Furthermore, the algorithm's performances are studied by working with four different size partitions of each dataset to study their impact on the optimization performance. All of these comparisons are studied through two experiments: (a) Experiment 1 in which RHOASo carries out the number of iterations that it needs until it stops, and the other HPOs have a default input (50 iterations); and (b) Experiment 2 in which the other HPOs perform the same number of trials that RHOASo has conducted. This means that we avoid biased comparisons at the same time that we evaluate RHOASo under the default number of evaluations that a user could specify ([3]).

The results show that RHOASo works efficiently in low-dimension hyper-parameter spaces, and it is competitive in terms of accuracy and computational cost. On average, RHOASo achieves a statistically significant positive difference in terms of efficiency over 70% of the times that the algorithms are applied and obtains good results when it uses small parts of the datasets. Sensitivity and specificity also show positive results in general, although in some unbalanced datasets, there is a bias toward the majority class. Lastly, the automatic early-stop feature of RHOASo lets it finish the tuning process before reaching the fixed number of iterations given as input in the other hyper-parameter optimization methods ($M_e = 34$ vs. 50), with a 30% reduction. When fixing an equal number of iterations for all algorithms, RHOASo loses the advantage decreasing its gain to 50% on average, but remains competitive in two of the three evaluated models. It shows weakness when it is applied with MLP in some datasets.

The article is organized as follows. In Section 2, we state the hyper-parameters optimization problem when the size of the dataset is left as a variable, and we provide an overview of the state-of-the-art methods. In Section 3, we describe the proposed algorithm. First, we set and solve a conditional optimization problem for the logistic function. The solution is a simple iterative algorithm whose discrete analogous is used as the base to define RHOASo. In Section 4, we describe the experimental details of the study. Finally, in Section 5, we develop the obtained results. We analyze the performance of RHOASo when it is run together with the three ML algorithms mentioned above, and it is compared with other HPO algorithms. This is done in two different ways. On one hand, we let the number of iterations of the HPO algorithms have their default values. On the other hand, we allow the HPO algorithms to be run for the same number of iterations that RHOASo needs until it stops.

Additionally, we have included an appendix with the results concerning the first experiment in which case the ML algorithms are decision tree (DT and K-nearest neighbor (KNN). It is shown that the performance of RHOASo compared with the other HPO algorithms is substantially better. Due to the superiority of the performance of RHOASo with respect to the rest of the HPO algorithms when they are run with DT and KNN, we have carried out the complete analysis only with the RF, GB and MLP algorithms.

In order to facilitate the reading of the article, we include below a scheme describing the experimentation and validation phases that were carried out.

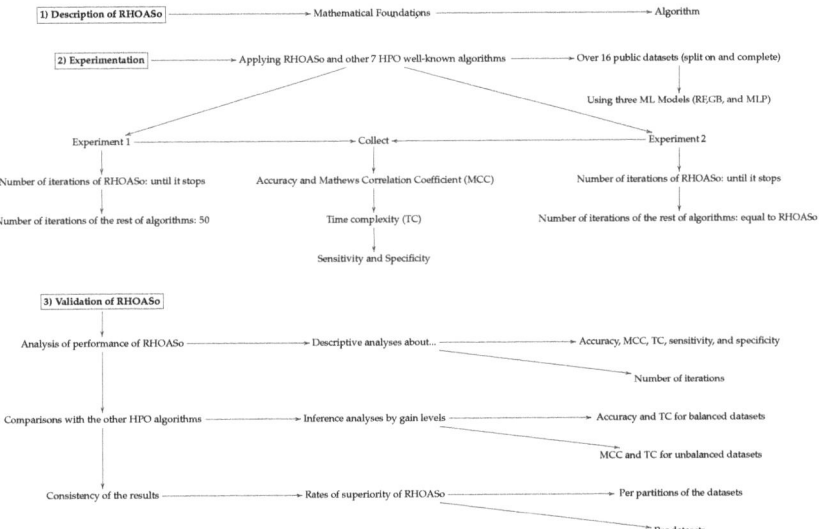

2. Related Work

A hyper-parameter of a ML algorithm is a hidden component that directly influences the algorithm's behavior. Tuning it allows the user to control the performance of the algorithm.

2.1. Problem Statement

Definition 1. *Let $\mathcal{X}$ be a tuple of random variables. Let $\mathcal{Y}$ be the space labels. Let $D^{train} \in \mathcal{X} \times \mathcal{Y}$ be an i. i. d. sample whose distribution function is $\mathcal{P}$.*

A machine learning algorithm $\mathcal{A}$ is a functional as follows:

$$\mathcal{A} : \cup_{n \in \mathbb{N}} (\mathcal{X} \times \mathcal{Y})^n \longrightarrow \mathcal{H}$$

$$D^{train} \mapsto \mathcal{A}(D^{train}) := h_{\mathcal{A}, D^{train}} : \mathcal{X} \longrightarrow \mathcal{Y} \qquad (1)$$

$$x \mapsto h_{\mathcal{A}, D^{train}}(x) = y$$

The model $h_{\mathcal{A}, D^{train}}(x)$ predicts the label of (unseen) instance x minimizing the expected loss function, $\mathcal{L}(D^{train}, h_{\mathcal{A}, D^{train}})$.

This loss function measures the discrepancy between a hypothesis $h \in \mathcal{H}$ and an ideal predictor. The target space of functions of the algorithm, $\mathcal{H}$, depends on specific parameters, $\lambda = (\lambda_1, \ldots, \lambda_n) \in \Lambda$, that might take discrete or continuous values and have to be fixed before applying the algorithm. We use the notation A_λ to refer to the algorithm with the hyper-parameter configuration λ.

In this scenario, another independent data set, $D := D^{train} \cup D^{test}$, serves to evaluate the loss function, $\mathcal{L}(D^{test}, h_{A_\lambda, D^{train}})$, provided by the algorithm $\mathcal{A}_\lambda(D^{train})$. Let the hyper-parameters $\lambda = (\lambda_1, \ldots, \lambda_n)$ remain free in $\mathcal{L}(D^{test}, h_{A_\lambda, D^{train}})$.

In the case of a classification ML problem, we can take the loss function $\mathcal{L}$ as the error rate, that is, one minus the cross-validation value. In this situation, one can define the following function:

$$\Phi_{A,D} : \Lambda \longrightarrow [0,1]$$
$$\lambda \mapsto \text{mean}_{D^{test}} \mathcal{L}(D^{test}, h_{A_\lambda, D^{train}}) \quad (2)$$

The hyper-parameter optimization (HPO) problem consists of trying to reach $\lambda^* := \min_\lambda (\text{mean}_{D^{test}} \mathcal{L}(D^{test}, h_{A_\lambda, D^{train}}))$.

2.2. Overview of the State-of-the-Art Methods

Since the hyper-parameter configuration has a significant effect on the performance of a ML model, the main goal in the HPO research is to find optimal values for the hyper-parameters that maximize the accuracy of the model while minimizing the costs and avoiding manual tuning. In the case where hyper-parameters are continuous, HPO algorithms usually work using gradient descent-based methods ([4–6]) in which the search direction in the hyper-parameter space is determined by the gradient of a model selection criterion at any step.

The discrete case has several approaches that perform differently depending on the ML algorithm and the dataset. Bayesian HPO is a type of surrogate-based optimization ([7]) that tunes the hyper-parameters by keeping the assumed prior distribution of the loss function updated, taking into account the new observations that are selected by the acquisition function. The construction of this surrogate model and the hyper-parameter selection criteria result in several types of sequential model-based optimization (SMBO). The main methods model the error distribution with a Gaussian process ([8]) or tree-based algorithms, such as sequential model-based algorithm configuration (SMAC) or the Tree Parzen Estimators (TPE) method ([9,10]). Another perspective is the radial basis function optimization (RBFOpt) that proposes a deterministic surrogate model to approximate the error function of the hyper-parameters through dynamic coordinate search. These methods require fewer evaluations, improving the associated costs of Gaussian process methods ([11]). Regarding the selection function to choose the next promising hyper-parameter configuration to test in the surrogate-based optimization, the typical approach is to use the expected improvement ([8]). There are other alternatives, such as the predictive entropy search ([12]. Other variants of SMBO can be found in [13,14], where different datasets and tasks are characterized by several measurements that allow to predict a ranking of several combinations of hyper-parameter values.

Another important HPO approach is the decision-theoretic method, where the algorithm obtains the hyper-parameter setting by searching the hyper-parameter space directly following some particular strategy. As examples, we have grid search, which uses brute force, and the simple and effective random search (RS) that tests randomly sampled configurations from the hyper-parameter space [15,16]).

Other optimization algorithms are applied to the problem of discrete hyper-parameter values selection. This is the case, for instance, of the evolutionary algorithms, such as the covariance matrix adaptation evolutionary (CMA-ES) method [17], the simplex Nelder–Mead (NM) method ([18,19]) or the application of continuous techniques over the discrete case such as the particle swarm (PS) ([20,21]).

Although there are several options, these methods provide different results and consumption of computational resources, and they do not perform equally well with all

ML algorithms. Then, we need to consider the costs to choose the HPO method, the size of the data required to run the optimization process effectively, and the human interaction needed. These issues arise in several open research challenges that we have summarized in Table 1.

Table 1. Open research challenge in HPO. Content extracted from [3].

Research Challenge	Description
HPO vs. CASH tools	Research conducted to specialized tools and algorithms for HPO or to CASH (Combined algorithm selection and hyper-parameter optimization) problem
Monitoring HPO	Tools that let the user follow the progress in an interactive way
Less computational costs	HPO remains computationally extremely expensive for certain tasks
Overtuning HPO	Control resampling in an efficient way
Closed black-box	The user can not take decisions about the optimization process and cannot analyze the HPO procedure.
Not supervised learning	Developing HPO algorithms for more types of machine learning models, not only for supervised ones.
Users do not make use of advanced HPO approaches	Potential users have a poor understanding of HPO methods. Missing guidance makes difficult the choice and configuration of of HPO methods
Finishing an HPO method	There are several ways to configure the end of an HPO method, not all of them are easily configurable.

RHOASo is an HPO algorithm that is designed in order for the potential user not to have to configure the end of the process. Currently, the termination of a general HPO algorithm can be carried out in several ways ([3]): (1) an amount of runtime fixed by the user based on intuition; (2) a lower bound of the generalization error specified by the user; and (3) considering the convergence of HPO if no progress is identified. All of these procedures can lead to over-optimistic bounds or excessive runtime that increases the computational cost. In this scenario, RHOASo is able to stop automatically, without losing accuracy and with minimal intervention of the user.

Additionally, many of the state-of-the-art HPO algorithms have, in turn, parameters that must be set up before running them. For instance, when a user wants to tune an (unbounded) integer-valued hyper-parameter of a given ML algorithm, the HPO algorithm requires the user to pre-configure a grid over which it is to be run. In many cases, the higher the size of the grid, the higher the execution cost of the HPO algorithm. A natural way to proceed in these cases is to accelerate the hyper-parameter running process, using early-stopping techniques ([2,22–24]). However, these algorithms still have other parameters that must be set up. Therefore, in some sense, HPO algorithms move the hyper-parameter tuning problem from ML algorithms to themselves, which increases the complexity and cost of the whole process. Table 2 below shows the parameters on which the HPO algorithms used in this work depend.

Table 2. Hyper-parameters of the HPO algorithms considered in this work.

Name	Hyper-Parameters	Library
Particle Swarm	6	[21]
Tree Parzen Estimators	2	[25]
CMA-ES	3	[26]
Nelder–Mead	2	[26]
Random Search	1	[26]
SMAC	30	[27]
RBFOpt	46	[28]

Thus, the natural question is how to tune hyper-parameters of HPO algorithms without increasing the complexity. Since using HPO algorithms over themselves does not solve the problem, it is natural to ask for HPO algorithms depending on as few hyper-parameters as possible and achieving good performance, compared with state-of-the-art HPO algorithms.

Our aim is to present a novel HPO algorithm with only one parameter to be tuned and to analyze its performance, compared with other state-of-the-art HPO algorithms.

3. The Proposed Algorithm: RHOASo

RHOASo is an approach to the HPO problem, whose underlying idea is the reversible gradient-based HPO method proposed for the continuous case ([5]).

Open source code for RHOASo is hosted in GitHub (https://github.com/amunc/RHOASo, accessed on: 25 July 2021), and it is available under the GPL license (version 3). Users do not need to install software separately, save for the Python language. Additionally, this is included in a ML intelligent system, RADSSo (RIASC automated decision support software), and it was used in several research works ([29,30]).

3.1. The Setup

Recall from Section 2.1 that the space of functions in which a learning algorithm takes values is assumed to depend on certain parameters $\lambda = (\lambda_1, \ldots, \lambda_n)$, and this space of functions is denoted by $\mathcal{H}_\lambda$. We make the following assumptions:

1. The hyper-parameters λ_i are discrete.
2. If $\lambda^* = (\lambda_1, \ldots, \lambda_i + 1, \ldots, \lambda_n)$, then $\mathcal{H}_\lambda \subset \mathcal{H}_{\lambda^*}$.

Typical examples of hyper-parameters satisfying such assumptions are the maximum depth in any tree-based machine learning model, the number of trees if the output of the model is a weighted average of the outputs of all the trees, or the number of neurons in hidden layers of a multilayer perceptron (the weights of the inputs of a given neuron may be zero).

Let $\Phi_{A,D}$ be the functional given in Equation (2) for a given machine learning model defined by a dataset D and a model $\mathcal{H}_\lambda$. If we plot the functional $\Phi_{\lambda,D}$ considering, for example, the model random forest with hyper-parameters maximum depth (x-axis) and number of trees (y-axis), we can obtain a figure like those shown in Figure 1 (the plots are obtained from two different datasets).

From the expression $\Phi_{A,D}(\lambda) = \text{mean}_{D^{test}} \mathcal{L}(D^{test}, f_{A_\lambda, D^{train}})$ it follows that if we let both the size of the dataset and the number of iterations in the cross-validation go to infinity, the surface given in Figure 1 becomes smoother, and takes the form shown in Figure 2, which is a concave surface with an asymptote in the plane $z = z_0 \leq 1$.

At this point, one can ask for an algorithm to find a value of $\lambda = (\lambda_1, \lambda_2)$ at which $\Phi_{A,D}$ attains a sufficiently high value while keeping λ_1 and λ_2 as small as possible.

3.2. Motivation of the Algorithm

In order to motivate the algorithm, consider the logistic function $f(x) = 1/(1 + e^{-x})$ restricted to $\mathbb{R}_{>0}$. This is a concave function with an asymptote in $y = 1$, thus it has no maximum. Although the maximization problem of this function has no sense, we can ask for the point $x_0 \in \mathbb{R}$ with higher value $f(x)$ subject to the condition that making x_0 smaller makes the decrease in f important. One way to formalize this question is by considering the maximization problem of the function $\text{Stb}(x) = f(x)f'(x) = e^{-x}f(x)^3$. In a certain sense, maximizing $\text{Stb}(x)$ consists of choosing a point x_0 with a sufficiently large image $f(x_0)$ but whose slope at that point is not too low. Note that $\text{Stb}(x)' = e^{-x}f(x)^3(-1 + 3e^{-x}f(x))$, and therefore $\text{Stb}(x)' = 0$, has only one solution, $x = \ln(2)$, which is a maximum.

Consider now the function $\text{Stb}(x,n) = x^n f(x) f'(x)$, n being a natural number. Then, taking $n > \ln(2)$, the equation $\text{Stb}'(x,n) = 0$ has only one solution x_0, which is greater than $\ln(2)$, and becomes larger as we make n increase. Thus, we can control how small the slope is at the solution x_0 by making n vary.

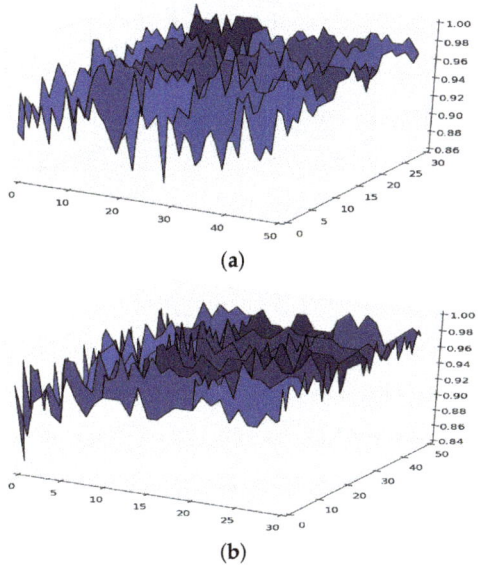

(a)

(b)

Figure 1. Surface of $\Phi_{\lambda,D}$ for random forest. (**a**) Φ in RF. (**b**) Φ in RF.

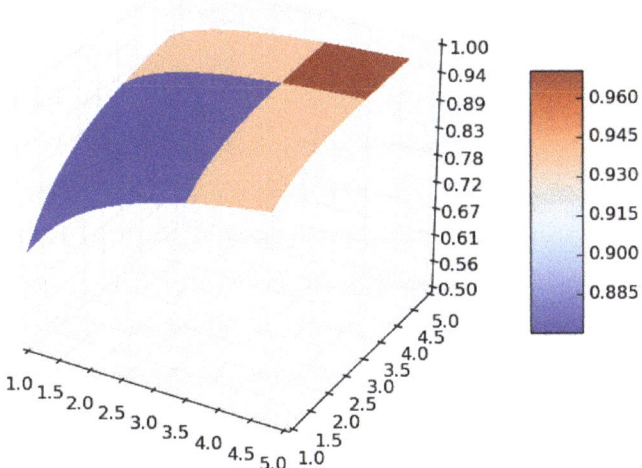

Figure 2. Smoothing of $\Phi_{A,D}$.

In order to explain how RHOASo works, and to link directly with the form it is presented below, let us denote the function f by $\Phi_{A,D}$ and let us restrict its domain of definition to the set of natural numbers. The variable is now denoted by λ. Then, instead of the derivative, we may consider the following function:

$$\frac{\Phi_{A,D}(\lambda + h) - \Phi_{A,D}(\lambda)}{h},$$

where h is a natural number. In order to simplify the notation, we set $h = 1$. Thus, we may consider the optimization problem defined by the following:

$$\max_\lambda \{\text{Stb}(\lambda, n)\},$$

where

$$\text{Stb}(\lambda, n) := \lambda^n \Phi_{A,D}(\lambda)(\Phi_{A,D}(\lambda+1) - \Phi_{A,D}(\lambda)).$$

Now, we can give a simple iterative algorithm to find the value λ close to that at which $\Phi_{A,D}$ attains a sufficiently high value while keeping the magnitude of such coordinate as low as possible; at the iteration i, do: if $\text{Stb}(\lambda_i + 1) > \text{Stb}(\lambda_i)$, then $\lambda_{i+1} := \lambda_i + 1$ and stop otherwise. This is just the most basic algorithm to solve the optimization problem $\max_\lambda \{\text{Stb}(\lambda, n)\}$. Observe that the convergence is always ensured because of the properties of the function $\text{Stb}(\lambda)$.

See Figure 3 to show how the stabilizer function, Stb, behaves in two particular cases.

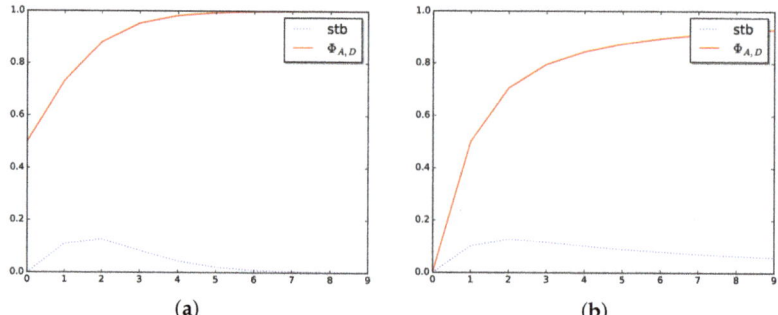

Figure 3. Stabilizer $\text{Stb}(\lambda)$ in dimension one. (**a**) Stabilizer for logistic function. (**b**) Stabilizer for arctangent function.

3.3. The Algorithm

There are two important features we want for the algorithm to have. On one hand, we want to avoid meta-parameters, that is, we want for the algorithm not to depend (strongly) on extra parameters. In state-of-the-art HPO algorithms, the user has to set as input the exact number of iterations that the algorithm must perform. On the other hand, we want the algorithm to give a good result when it is run, giving as input not the whole dataset but a small part of it. The proposed HPO algorithm exploits the consequences derived from assumptions 1 and 2 to reach the objective in a simple way.

Since the hyper-parameters we will work with are discrete, we may assume that the hyper-parameter space is $\Gamma = \mathbb{N}^n$. Suppose that we are at the point $\lambda \in \Gamma = \mathbb{N}^n$ of the hyper-parameter space. The decision about to which next point the algorithm must jump is based on two basic rules:

1. Fix a natural number h. Let Shifts $= \{0, h\}^n \setminus \{(0, \ldots, 0)\}$. We look at the points $\lambda + $ Shifts. These are the *next* points at each possible direction in the space Γ. Since $\mathcal{H}_\lambda \subset \mathcal{H}_{\lambda+\eta}$, for each $\eta \in$ Shifts, it is likely that for each η, $\lambda + \eta$ will give a better result than λ.

2. Given $\lambda = (\lambda_1, \ldots, \lambda_n) \in \Gamma$, define its stabilizer as follows:

$$\text{stb}(\lambda) = \max\{\lambda_i\} \cdot \Phi_{A,D}(\lambda) \cdot \sum_{\lambda' \in \lambda + \text{Shifts}} (\Phi_{A,D}(\lambda') - \Phi_{A,D}(\lambda)) \tag{3}$$

If we are at point λ, we will transit at point $\lambda^* \in \lambda + $ Shifts if the inequality $\text{stb}(\lambda^*) > \text{stb}(\lambda)$ holds true and $\text{stb}(\lambda^*) = \max\{\text{stb}(\lambda') \mid \lambda' \in \lambda + \text{Shifts}\}$.

3. Once the algorithm stops at some point $\lambda^* \in \Omega$, there is a final step at which the point $\lambda' \in \lambda^* + $ Shifts with maximum $\Phi_{A,D}$ is found and given as the final output.

The pseudo-code of the algorithm is included in Algorithms 1–4.

Algorithm 1 Computing stabilizer.

1: **Input:** data, train_data, test_data, features, target
2: **Output:** stb $\in \mathbb{R}$ used for stopping criterion
3: **procedure** GETSTAB(Input,$\lambda_{current}$)
4:     Values $= \emptyset$
5:     Shifts $= \{0,1\}^N$
6:     **for** η in Shifts **do**
7:         $\lambda \leftarrow \lambda_{current} + \eta$
8:         Values $\leftarrow$ Values $\bigcup \{ac(\lambda) - ac(\lambda_{current})\}$
9:         ▷ $ac(\lambda)$ means the accuracy of the model on D_{test_data} trained with D_{train_data} with hyper-parameter configuration λ
10:     **end for**
11:     tot $\leftarrow sum$(Values)
12:     stb $\leftarrow$ tot $\cdot ac(\lambda_{current}) \cdot max(\lambda_{current})$
13:     **return** stb
14: **end procedure**

Algorithm 2 Computing best neighbor.

1: **Input:** data, train_data, test_data, features, target
2: **Output:** best next hyper-parameters
3: **procedure** GETBESTNEIGH(Input,$\lambda_{current}$)
4:     Stb $= \emptyset$
5:     Shifts $= \{0,1\}^N$
6:     **for** η in Shifts **do**
7:         $\lambda \leftarrow \lambda_{current} + \eta$
8:         Stb $\leftarrow$ Stb $\bigcup \{\text{GETSTB}(\text{Input}, \lambda)\}$
9:     **end for**
10:     maximum $= max_\lambda$Stb
11:     **return** maximum, max_Stb
12: **end procedure**

4. Materials and Methods

Some experiments were carried out in order to answer the research questions formulated in the introduction:

1. RQ1: Given a dataset, how good is the performance (accuracy, time complexity, sensibility, and specificity) of a ML algorithm when RHOASo is applied?

Algorithm 3 Last phase.

1: **Input:** data, train_data, test_data, features, target
2: **Output:** next hyper-parameters giving best accuracy
3: **procedure** LASTPHASE($\lambda_{current}$)
4:     Acc $= \emptyset$
5:     Shifts $= \{0,1\}^N$
6:     **for** η in Shifts **do**
7:         $\lambda \leftarrow \lambda_{current} + \eta$
8:         Acc $\leftarrow$ Acc $\bigcup \{ac(\lambda)\}$
9:     **end for**
10:     maximum $= max_\lambda$Acc
11:     **return** maximum
12: **end procedure**

Algorithm 4 RHOASo algorithm.
―――
1: **Input:** data, train_data, test_data, features, target, max_pars
2: **Output:** best hyper-parameters
3: Initialize $\lambda_{current} = (1, \ldots, 1)$
4: **while True do**
5:     $stb_{current} \leftarrow \text{GETSTAB}(\lambda_{current})$
6:     $\lambda, stb \leftarrow \text{GETBESTNEIGH}(\lambda_{current})$
7:     **if** $stb > stb_{current}$ **then**
8:        $\lambda_{current} \leftarrow \lambda$
9:     **else**
10:        **break**
11:     **end if**
12: **end while**
13: $\lambda_{final} = \text{LASTPHASE}(\lambda_{current})$
14: **return** λ_{final}
―――

2. RQ2: How many iterations does the algorithm need until it stops? How much faster or slower is RHOASo, compared with the other HPO algorithms?
3. RQ3: Are there statistically meaningful differences between the performance of RHOASo and other HPO algorithms?
4. RQ4: Are the above results consistent? That is, do they hold for different HPO algorithms and different datasets with different characteristics (size, number of features, etc.)?

The analyses aim to measure the quality of the proposed algorithm and decide whether there are statistically meaningful differences between the performance of the selected HPO methods and RHOASo.

4.1. ML and HPO Algorithms

We have evaluated the efficiency of three well-known ML algorithms:

1. RF is an ensemble classifier consisting of a set of decision trees. Each tree is constructed by applying bootstrap re-sampling (bagging) to the training set, which extracts a subset of samples for training each tree. Therefore, the trees will have a weak correlation and give independent results. In the case of RF, we have two main hyper-parameters: the number of decision trees to be used and the maximum depth for each of them ([31]).
2. GB is another ensemble technique in which the predictors are made sequentially, learning from the previous predictor's mistakes to optimize the subsequent learner. It usually takes fewer iterations to reach close-to-actual predictions, but the stopping criteria have to be chosen carefully. This technique reduces bias and variance but can induce overfitting if too much importance is assigned to the previous errors. We tune two discrete hyper-parameters: the number of predictors and their maximum depth ([32,33]).
3. A MLP is a graph-type model that is organized in ordered layers (input layer, output layer, and hidden layers). Each layer consists of a set of nodes with no connections between them, so the connections occur between nodes belonging to different and contiguous layers. In this study, we set two hidden layers, and we tune the number of neurons in each of these hidden layers.

In Table 3, we give a summary of the ML algorithms we have used together with the hyper-parameters we have tuned. The search space for all hyper-parameters is in the interval [1, 50]. All hyper-parameters not being tuned are set to their default values as per scikit-learn implementation ([34]). We have used 10-fold cross-validation to assess the performance of all ML models combined with the HPO algorithms.

On the other hand, in Table 4 a summary of the HPO algorithms selected for this study is given.

Table 3. ML algorithms together with hyper-parameters we have considered.

Name	Hyper-Parameter 1	Hyper-Parameter 2
RF	Num. Trees	Max. Depth
GB	Num. Trees	Max. Depth
MLP (2 hidden layers)	Num. Neurons Layer 1	Num. Neurons Layer 2

Table 4. HPO algorithms used. Colors explanation: Bayesian methods in blue, decision-theoretic techniques in pink, evolutionary algorithms in brown, and other optimization algorithms in green.

Name	Reference	Python Version	Automatic Early Stop	Library
Particle Swarm	[20,35]	2.7 y 3	X	[21]
Tree Parzen Estimators	[10]	2.7, 3	X	[25]
CMA-ES	[17]	2.7, 3	X	[26]
Nelder–Mead	[18]	2.7, 3	✓	[26]
Random Search	[15]	2.7, 3	X	[26]
SMAC	[9]	3	X	[27]
RBFOpt	[11,36]	2.7, 3	X	[28]

4.2. Datasets

The datasets selected for the experiments are described in Table 5. The choice was motivated by different reasons: availability in public servers to verify the results, the number of instances and classes, and the type of features. There are 16 public benchmark datasets with a different number of variables (8–300), rows (4601–284,807), and classes (2–10). Since the size of the dataset in the HPO phase influences the performance of the classifier ([1]), the performances of the algorithms are analyzed with four different-sized partitions of each dataset ($\mathcal{P} = \{P_1 = 8.3\%, P_2 = 16.6\%, P_3 = 50\%, P_4 = D_i\}$). These proportions are chosen because, in this way, the number of instances is in different orders of magnitude. In addition, each dataset is divided into train data (80%) and validation data (20%) $D_i^{train} \cup D_i^{valid} = D_i$, and the partitioning scheme above is applied to obtain train and validation subsets with the corresponding proportions $P_j(D_i)$.

The transformation of features to obtain treatable datasets to input directly to each model is manually implemented with Python.

Table 5. Descriptions of datasets.

Dataset = D_i	Topic	Classes	Majority Class Proportion	Features	Instances = P_4	P_1	P_2	P_3	Reference
D_1	First order proving	2	0.82	51	4589	382	764	2294	[37]
D_2	Spambase	2	0.6	57	4601	383	766	2300	[38]
D_3	Polish companies Bankruptcy	2	0.978	64	4885	407	814	2442	[39]
D_4	Opto digits	10	0.1	64	5620	468	936	2810	[40]
D_5	Grammatical Facial Expressions	2	0.64	300	27,936	2328	4656	13,968	[41]
D_6	Credit card Fraud Detection	2	0.99	30	284,807	23,733	47,467	142,403	[42]
D_7	Magic Telescope	2	0.64	10	19,020	1585	3170	9519	[43]
D_8	Electricity	2	0.57	8	45,312	3776	7552	22,656	[44,45]
D_9	Wall Robot	4	0.4	24	5456	454	909	2728	[46]
D_{10}	Eye	2	0.55	14	14,980	1248	2496	7490	[47]
D_{11}	Connect 4	3	0.65	43	67,557	5629	11,259	33,778	[48]
D_{12}	Amazon	2	0.94	9	32,769	2730	5461	16,384	[49]
D_{13}	Phishing websites	2	0.55	30	11,055	921	1842	5527	[50]
D_{14}	Higgs	2	0.52	28	98,049	8170	16,341	49,025	[51]
D_{15}	NSL-KDD	6	0.51	42	148,517	12,376	24,752	74,258	[52,53]
D_{16}	Robots in RTLS	3	0.73	12	6422	535	1070	3211	[54]

4.3. Construction of the Response Variables

In order to build response variables to measure the performance of the HPO algorithms, we apply them to each ML model $\mathcal{H}_k$ over each partition, $P_j(D_i)$, obtaining a hyper-parameter configuration $\lambda_{i,j}^k$ for $\mathcal{H}_k$. Then, the learning algorithm with the obtained hyper-parameter configuration is run over D_i^{train} to construct a classifier that is validated over D_i^{valid}. At this step, we collect the obtained accuracy. This scenario is repeated a number of times (trials), giving rise to two different experiments.

1. Experiment 1: the experimentation is repeated 50 times (trials) for all HPOs, except for RHOASo, which automatically stops when it considers that it has obtained an optimal hyper-parameter configuration.
2. Experiment 2: the experimentation is repeated for all HPOs as many times (trials) as RHOASo has carried out until stopping.

The time complexity of the whole process is stored as well. The time complexity, measured in seconds, is the sum of the time needed by the HPO algorithm to find the optimal hyper-parameter configuration, and the time that the ML algorithm uses for training. Then, we create two response variables. We denote by $Acc_{i,j}^k$ the number of trials $\times$ 1 array where the m-th component is the accuracy of the predictive model tested on D_i^{valid} that was trained over D_i^{train} with the hyper-parameters $(\lambda_{i,j}^k)_m$ at the m-th trial. $TC_{i,j}^k$ is the notation for time complexity.

We have also collected the sensitivity and the specificity of each iteration for RHOASo to measure its performance more accurately.

Additionally, we have collected the MCC (Matthew correlation coefficient), which is defined as follows:

$$MCC = \frac{TP \cdot TN - FP \cdot FN}{\sqrt{(TP+FP)(TP+FN)(TN+FP)(TN+FN)}} \quad (4)$$

where TP, TN, FP, and FN denote true positives, true negatives, false positives, and false negatives, respectively. This coefficient works as a substitute metric of the accuracy for unbalanced datasets [55,56]. Since some datasets contain a certain degree of imbalance, we present our results with both indicators, accuracy and MCC.

4.4. Statistical Analyses

Our main objective is to analyze the quality of the RHOASo algorithm and compare it with other HPO algorithms. In order to analyze whether there are meaningful statistical differences among the obtained results by the HPO algorithms and RHOASo, we perform the following statistical analysis:

1. We have conducted descriptive and exploratory analyses of $Acc_{i,j}^{RHOASo}$, $TC_{i,j}^{RHOASo}$, the MCC, the sensibility, and the specificity of RHOASo in order to check how well RHOASo performs the tasks.
2. We have computed the average of iterations that RHOASo executes until it automatically stops.

The following inference tests are carried out for both Experiments 1 and 2.

3. To compare RHOASo's efficiency with that of the other HPO algorithms, we carry out non-parametric tests, due to the non-normality of the data $Acc_{i,j}^k$ and $TC_{i,j}^k$. Wilcoxon's tests for two paired samples are conducted comparing each $Acc_{i,j}^k$ with $Acc_{i,j}^{RHOASo}$ and $TC_{i,j}^k$ with $TC_{i,j}^{RHOASo}$ for all datasets and for the three selected ML algorithms. The choice of Wilcoxon's test is because the response variables that we compare are obtained by the application of the ML algorithms over the same dataset but with different settings $(\lambda_{i,j}^k)$. We obtain the results at a significance level of $\alpha = 0.05$.
4. Once we apply the inference described above, we obtain the p-values of 7 comparisons along 16 datasets with 4 partitions in each one, providing a total of 448 deci-

sions on statistical difference for each ML algorithm and for each response variable, obtaining, thus, 2688 p-values. From these results, we have computed how many times we obtain positive difference ($validity(RHOASo) > validity(H_k)$), negative difference ($validity(RHOASo) < validity(H_k)$) or equality ($validity(RHOASo) = validity(H_k)$), see Table 6.

Table 6. Conditions of validity. The symbols $=, >, <$ denote statistically meaningful equality and difference, and Me denotes the median.

Validity	P_j	$Me(TC_{i,j}^{RHOASo}) > Me(TC_{i,j}^k)$	$Me(TC_{i,j}^{RHOASo}) = Me(TC_{i,j}^k)$	$Me(TC_{i,j}^{RHOASo}) < Me(TC_{i,j}^k)$
		| D_1 | ... | D_{16} |	| D_1 | ... | D_{16} |	| D_1 | ... | D_{16} |
$Me(Acc_{i,j}^{RHOASo}) > Me(Acc_{i,j}^k)$	P_1 P_2 P_3 P_4	$\mathcal{V}^+$	$\mathcal{V}^+$	$\mathcal{V}^*$
$Me(Acc_{i,j}^{RHOASo}) = Me(Acc_{i,j}^k)$	P_1 P_2 P_3 P_4	$\mathcal{V}^-$	$\mathcal{V}^-$	$\mathcal{V}^-$
$Me(Acc_{i,j}^{RHOASo}) < Me(Acc_{i,j}^k)$	P_1 P_2 P_3 P_4	$\mathcal{V}^-$	$\mathcal{V}^-$	$\mathcal{V}^*$

5. Since the blue cells may be understood as being both a positive difference or negative difference, depending on the improvement that we obtain, we have reclassified the results, creating a new table, correcting these cases by the rule described in Equation (5).

$$\mathcal{V}^* = \begin{cases} \mathcal{V}^+ & \text{if } \Delta(Acc_{i,j}(k)) > \Delta(TC_{i,j}(k)) \\ \mathcal{V}^- & \text{if } \Delta(Acc_{i,j}(k)) < \Delta(TC_{i,j}(k)) \end{cases} \quad (5)$$

where

$$\Delta(Acc_{i,j}(k)) = \frac{|Me(Acc_{i,j}^{RHOASo}) - Me(Acc_{i,j}^k)|}{min(Me(Acc_{i,j}^{RHOASo}), Me(Acc_{i,j}^k))} \quad (6)$$

and

$$\Delta(TC_{i,j}(k)) = \frac{|Me(TC_{i,j}^{RHOASo}) - Me(TC_{i,j}^k)|}{min(Me(TC_{i,j}^{RHOASo}), Me(TC_{i,j}^k))} \quad (7)$$

6. We have completed the analysis by computing the rate of each type of validity (red, yellow and green cells) as follows:

$$\mathcal{R}_{\mathcal{V}^\bullet} = \frac{\text{number of cases in the class } \mathcal{V}^\bullet}{N}. \quad (8)$$

where N denotes the number of total possible comparisons. Since we have performed the computations for each ML algorithm, we have $N = 448$.

7. Finally, we compute $\mathcal{R}_{\mathcal{V}^\bullet}$ per partition and per dataset to analyze the consistency of the results.

Note that for studying the cases of unbalanced datasets, we have carried out the analyses described above by changing the accuracy for the MCC.

4.5. Technical Details

The analyses are carried out at high-performance computing over HP ProLiant SL270s Gen8 SE, with two processors, Intel Xeon CPU E5-2670 v2 @ 2.50GHz, with 10 cores each and 128 GB of RAM and one hard disk of 1 TB. The analysis script is implemented in Python language.

5. Results and Discussion

This section is organized according to the research questions we have formulated in the introduction.

5.1. Research Question 1: Given a Dataset, How Good Are the ML Models When RHOASo Is Applied?

Since the behavior of RHOASo is similar in both Experiments 1 and 2, in this section, we detail the performance of RHOASo in Experiment 1, and we include a summary of the results for Experiment 2.

5.1.1. Performance in Experiment 1

We can see in Figure 4 the median of the accuracy when RHOASo is applied. The median for RF is 0.92, for GB it is 0.9058, and for MLP it is 0.8182. We can see that in all of these cases, it is greater than 0.80. We can observe that the achieved accuracy by RHOASo presents great stability in terms of the partitions of the dataset, except for dataset D_6 with model RF, and dataset D_{16} with model GB. In the case of D_6, the variation appears when we change from P_3 to P_4. This dataset is the largest one in the study, with the highest rate of unbalanced data. Then, the most appropriate metric is the MCC. As is discussed later, the MCC remains stable for D_6 in all the partitions. The case of D_{16} is more involved. The most frequent hyper-parameter configurations that RHOASo computes for each partition are *max. depth*: 5, *number trees*: 9 for P_1 (14 times out of 50), *max. depth*: 9, *number trees*: 3 for P_2 (50 times out of 50), *max. depth* : 3, *number trees*: 9 for P_3 (31 times out of 50), and *max. depth*: 9, *number trees*: 3 for P_4 (50 times out of 50). Since the number of features in D_{16} is 12, there are few instances, and the dataset is unbalanced, the most probable explanation is that the model is overfitting the training data. This behavior appears also in the rest of the metrics with the combination D_{16} and GB. The stability is not as evident when the dataset changes, although the general trend is maintained through the ML models. For instance, the obtained models with D_{10} provide the worst accuracy for the three ML algorithms. Although it may seem that the accuracy is not very high, the rest of the HPO does not achieve better results, as is outlined below. For this reason, the fit achieved by RHOASo is considered to be sufficient.

We can see in Figure 5 the median of the MCC ($\in [-1,1]$) when RHOASo is applied. The median for RF is 0.51, for GB it is 0.40, and for MLP it is 0.32. The MCC value considers class imbalance, so the results worsen for accuracy, particularly in datasets 3 and 12, which are highly unbalanced: the minority classes contain less than 1% of total instances. Apart from unbalanced datasets, the trends are similar to those presented when evaluating the accuracy.

As far as the time complexity is concerned, the median results are included in Figure 6. The median value for RF is 1.3028 s, and for GB it is 4.2567 s. The MLP stands out for its high computational cost, with 259.8138 s. Regarding the stability, as expected, the larger the partition size that is used, the larger the time complexity, independently of the ML algorithm used together with RHOASo. Nevertheless, this behavior is different for each ML algorithm. It is worth noting that in the case of RF, the increase in time complexity as the size of the partitions increases is much smoother for most of the datasets, compared to GB and MLP.

Sensitivity and specificity are plotted in Figures 7 and 8. The results are stable across partitions and datasets, which achieve similar results, even when using different models, except for D_{16} and GB. The median of sensitivity is 0.9 for GB, 0.91 for MLP, and 0.88 for RF. In contrast, specificity has a median of 0.7 for GB, 0.69 for MLP, and 0.733 for RF. The lower specificity could be caused by the imbalance between classes in specific datasets (see Table 5), which causes the models to be biased toward the majority class. However, D_{11} has both low specificity and low sensitivity. Overall, the trends are similar to those found in the evaluation of accuracy.

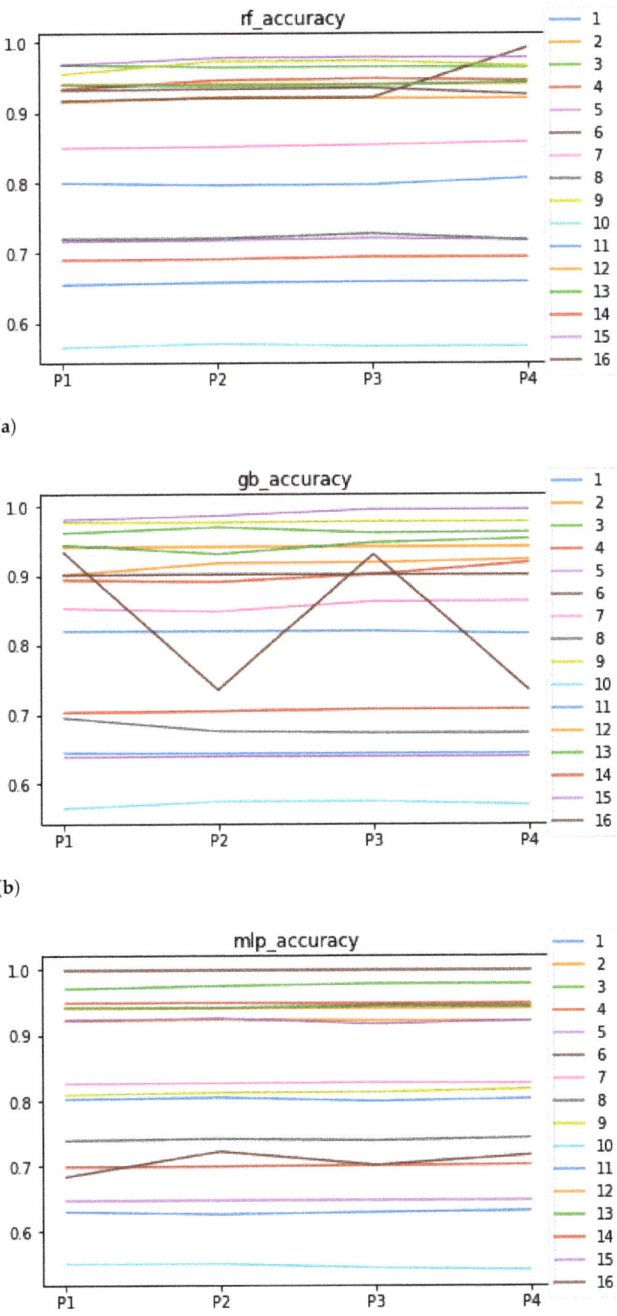

Figure 4. Behavior of RHOASo: accuracy. Each line represents a dataset. (**a**) Accuracy for RF. Axis X: the partition of the dataset. Axis Y: the obtained accuracy. (**b**) Accuracy for GB. Axis X: the partition of the dataset. Axis Y: the obtained accuracy. (**c**) Accuracy for MLP. Axis X: the partition of the dataset. Axis Y: the obtained accuracy.

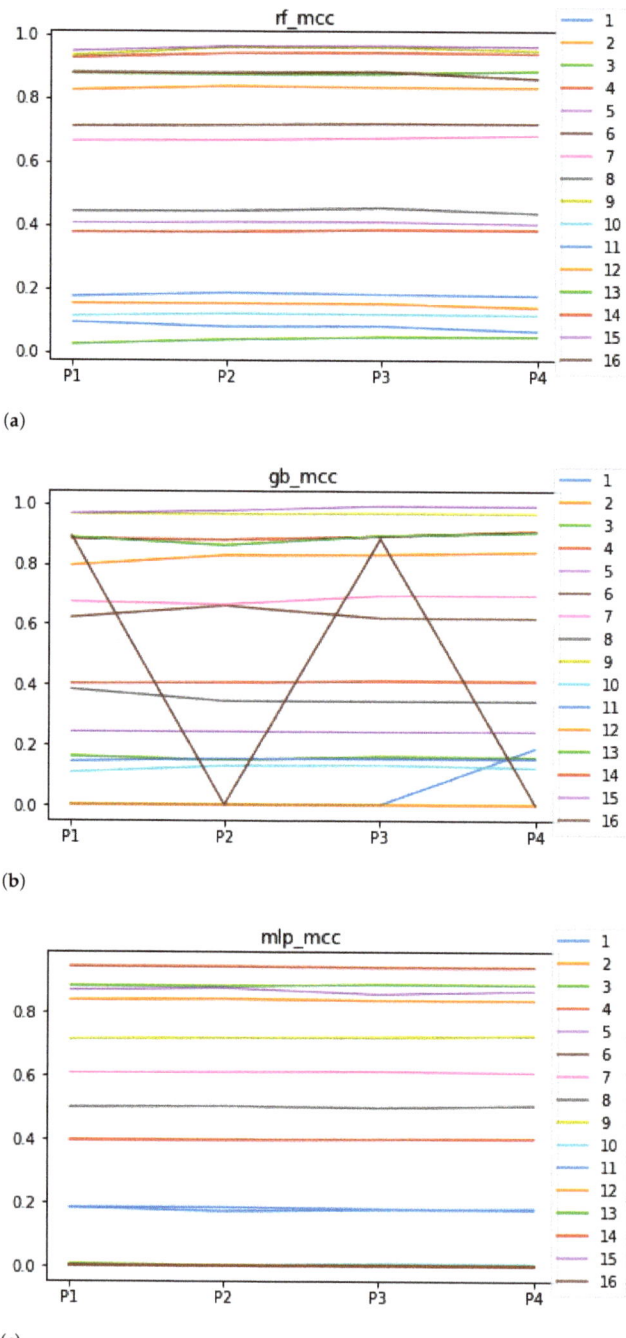

Figure 5. Behavior of RHOASo: MCC. Each line represents a dataset. (**a**) MCC for RF. Axis X: the partition of the dataset. Axis Y: the obtained MCC. (**b**) MCC for GB. Axis X: the partition of the dataset. Axis Y: the obtained MCC. (**c**) MCC for MLP. Axis X: the partition of the dataset. Axis Y: the obtained MCC.

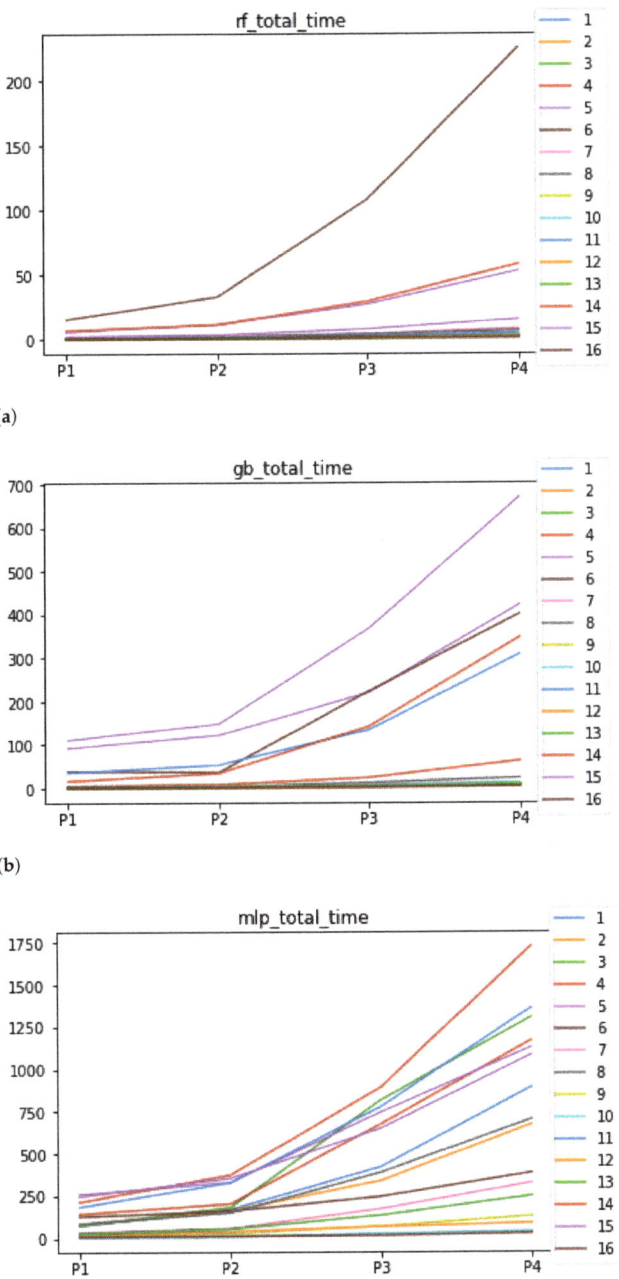

Figure 6. Behavior of RHOASo: time complexity. Each line represents a dataset. (**a**) Time complexity for RF. Axis X: the partition of the dataset. Axis Y: the obtained time complexity in seconds. (**b**) Time complexity for GB. Axis X: the partition of the dataset. Axis Y: the obtained time complexity in seconds. (**c**) Time complexity for MLP. Axis X: the partition of the dataset. Axis Y: the obtained time complexity in seconds.

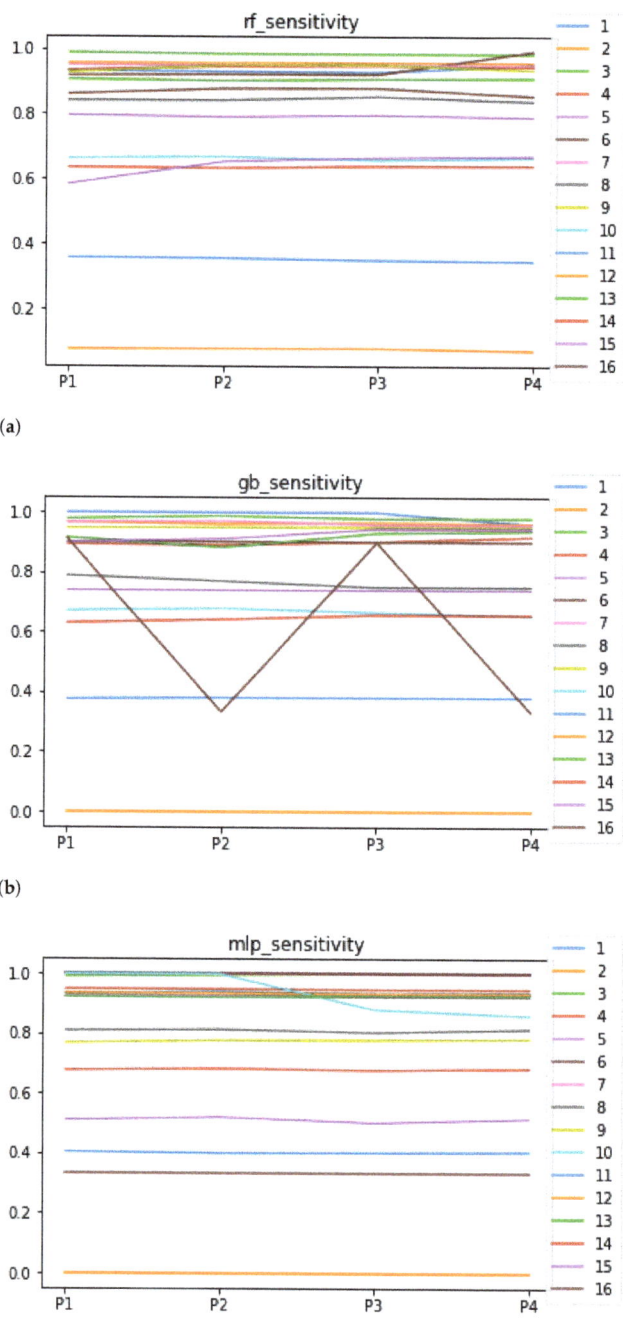

Figure 7. Behavior RHOASo: sensitivity. The datasets are represented with a bar chart for each partition. (**a**) Sensitivity for RF. Axis X: the partition of the dataset. Axis Y: the obtained sensitivity. (**b**) Sensitivity for GB. Axis X: the partition of the dataset. Axis Y: the obtained sensitivity. (**c**) Sensitivity for MLP. Axis X: the partition of the dataset. Axis Y: the obtained sensitivity.

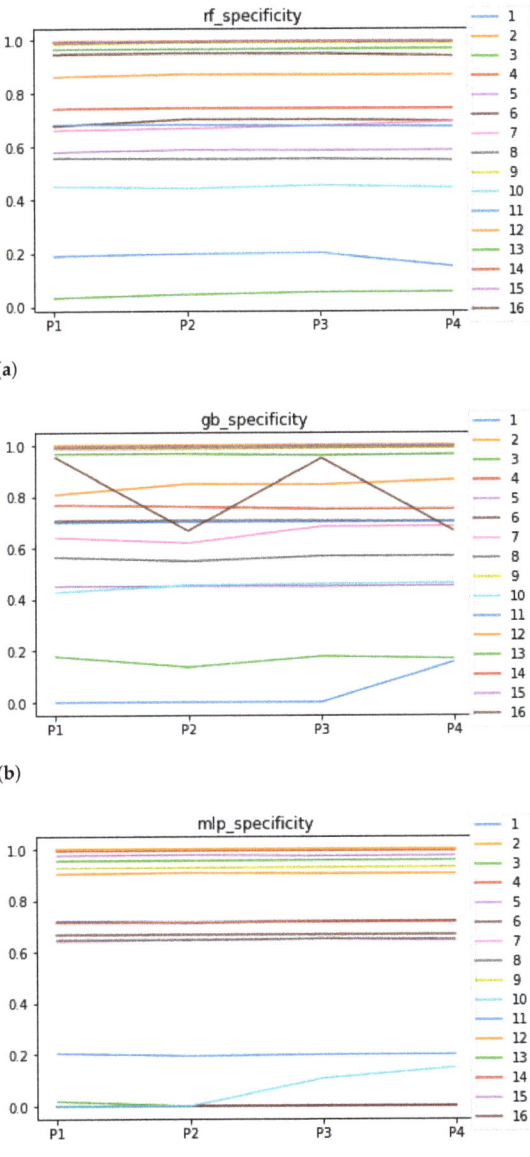

Figure 8. Behavior RHOASo: specificity. The datasets are represented with a bar chart for each partition. (**a**) Specificity for RF. Axis X: the partition of the dataset. Axis Y: the obtained specificity. (**b**) Specificity for GB. Axis X: the partition of the dataset. Axis Y: the obtained specificity. (**c**) Specificity for MLP. Axis X: the partition of the dataset. Axis Y: the obtained specificity.

5.1.2. Performance in Experiment 2

Since the behavior of RHOASo in both experiments is the same, we include in Table 7 a summary with the median of all metrics (without taking into account partitions) that RHOASo has obtained in Experiment 2.

Table 7. Medians of response variables for RHOASo in Experiment 2.

Dataset		Accuracy	MCC	Time Complexity	Sensitivity	Specificity	Iterations
D1	RF	0.8024	0.1776	0.904	0.9419	0.1743	32.0
	GB	0.8194	0.0002	1.2169	0.9999	0.0	30.5
	MLP	0.8018	0.1811	256.0706	0.9367	0.1971	35.0
D2	RF	0.923	0.8394	0.7208	0.9563	0.8732	37.0
	GB	0.9108	0.8141	1.0615	0.9675	0.8274	28.5
	MLP	0.9244	0.8432	237.3997	0.9362	0.9079	32.5
D3	RF	0.9632	0.0496	0.9074	0.9835	0.0573	27.0
	GB	0.9681	0.1249	1.29	0.9866	0.1127	29.0
	MLP	0.9771	0.0	280.6122	0.999	0.0	27.0
D4	RF	0.9499	0.9445	0.8897	0.95	0.9944	37.0
	GB	0.8994	0.8886	14.3729	0.8994	0.9888	33.0
	MLP	0.934	0.9269	339.9584	0.934	0.9927	34.5
D5	RF	0.7175	0.4067	16.6164	0.7885	0.589	37.0
	GB	0.6384	0.2443	189.5368	0.7401	0.4527	38.0
	MLP	0.6464	0.0	321.7485	1.0	0.0	27.0
D6	RF	0.9876	0.7332	35.5843	0.9916	0.6794	33.0
	GB	0.9009	0.6211	127.1753	0.9013	0.6999	33.0
	MLP	0.9922	0.0028	218.3069	1.0	0.0036	27.0
D7	RF	0.8588	0.6845	2.4416	0.9483	0.6965	35.0
	GB	0.8609	0.6916	5.1165	0.9629	0.673	37.0
	MLP	0.8266	0.6106	109.3366	0.9256	0.6444	33.5
D8	RF	0.7325	0.4688	2.4364	0.8881	0.5518	34.5
	GB	0.6726	0.3437	9.2316	0.7504	0.567	38.0
	MLP	0.7374	0.4963	225.6947	0.8138	0.6372	31.0
D9	RF	0.9756	0.9635	1.1602	0.9549	0.9908	36.0
	GB	0.9775	0.9664	3.9133	0.9508	0.9907	32.0
	MLP	0.8014	0.7016	48.2327	0.765	0.9251	35.0
D10	RF	0.5654	0.1156	1.2617	0.6625	0.445	37.0
	GB	0.5703	0.1244	3.665	0.6664	0.4559	37.5
	MLP	0.5457	0.0021	10.986	0.9297	0.0709	27.0
D11	RF	0.6583	0.0276	1.3408	0.3358	0.6683	33.0
	GB	0.6434	0.1519	77.7497	0.381	0.7023	38.0
	MLP	0.6309	0.1779	534.4388	0.4034	0.7164	34.0
D12	RF	0.9406	0.1435	1.1293	0.068	0.9938	29.0
	GB	0.9421	0.0	0.9947	0.0	1.0	27.0
	MLP	0.942	0.0	50.6516	0.0	0.9999	27.0
D13	RF	0.9436	0.886	0.6895	0.9124	0.9682	37.0
	GB	0.9474	0.8936	1.303	0.9273	0.9649	34.0
	MLP	0.9427	0.8841	94.5033	0.9245	0.957	34.5
D14	RF	0.6897	0.3766	20.5842	0.6307	0.7431	38.0
	GB	0.7068	0.4104	81.4707	0.6568	0.7517	38.0
	MLP	0.7003	0.3993	515.4725	0.6811	0.7177	37.0
D15	RF	0.9782	0.9608	4.5483	0.6555	0.9933	33.5
	GB	0.9919	0.9885	305.4672	0.9417	0.9987	35.5
	MLP	0.9217	0.8595	474.1431	0.5171	0.9739	34.0
D16	RF	0.9312	0.8643	0.5872	0.8498	0.9385	35.0
	GB	0.8687	0.6586	0.9878	0.7524	0.8705	30.0
	MLP	0.7348	0.0	11.8172	0.3333	0.6667	27.0

5.2. Research Question 2: How Many Iterations Does the Algorithm Need Until It Stops? How Faster or Slower Is RHOASo Compared with the Other HPO Algorithms?

The number of iterations that RHOASo has needed until stopping for Experiment 1 is included in Figure 9. In the case of Experiment 2, this information can be observed in Table 7.

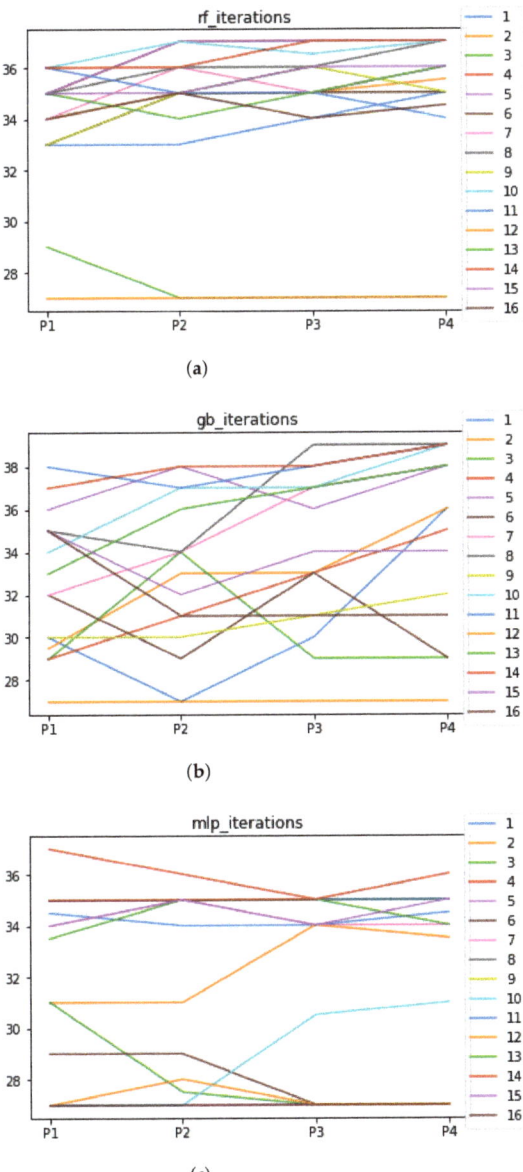

Figure 9. Number of iterations per partition. Each line represents a dataset (medians). (**a**) Number of iterations RF. In axis X, the partition of the dataset. In axis Y, the number of the iterations that RHOASo has carried out. (**b**) Number of iterations GB. In axis X, the partition of the dataset. In axis Y, the number of the iterations that RHOASo has carried out. (**c**) Number of iterations MLP. In axis X, the partition of the dataset. In axis Y, the number of the iterations that RHOASo has carried out.

In Experiment 1, we can see the median of the number of iterations needed by RHOASo for each ML algorithm, each dataset and each partition. As general result, the median of the number of iterations for each partition (computed over all datasets) is 35 for RF, 33.5 for GB, and 34 iterations for MLP. Taking into account that the number of iterations given as input (by default) in the rest of HPO algorithms is equal to 50, it implies that, on average, RHOASo needs approximately 70% of the iterations required by the other algorithms. Additionally, it stops the process by itself. As a consequence of this fact, less time is required by RHOASo to obtain a good enough accuracy and is able to be more competitive than other algorithms. This is most significant in the case of MLP, where each iteration is highly resource consuming. There is not a clear trend relating the partition size and number of iterations, especially in the case of GB, in which there is a greater variability in the number of iterations. It could be expected that a greater amount of data would contribute to a faster convergence, but this is not case. Therefore, it is likely that the functional $\Phi_{A,D}$ may not be as concave, as it would be desirable to perform an effective early stopping.

We have not compared whether RHOASo is faster or slower, compared with the other HPO algorithms for Experiment 2 by the very nature of the design of the experiment.

5.3. Research Question 3: Are There Statistically Meaningful Differences between the Performance of RHOASo and the Other HPO Algorithms?

We remind that we have carried out two experiments:

1. Experiment 1: the experimentation is repeated 50 times (trials) for all HPOs except for RHOASo, which automatically stops when it considers that it has obtained an optimal hyper-parameter configuration.
2. Experiment 2: the experimentation is repeated for all HPOs as many times (trials) as RHOASo has carried out until stopping.

5.3.1. Experiment 1

In Figures 10 and 11, the performances (accuracies and time complexities) that are achieved by the HPO algorithms over each dataset are shown.

However, if we want to compare whether RHOASo obtains any gain against the other HPO algorithms, we need to carry out more detailed analyses. This is the study of the validity of RHOASo.

The rates of validity that are obtained by RHOASo, compared to the rest of the HPO algorithms are included in Table 8. Note that these computations are carried out with the accuracies and time complexities by the analyses that are explained in Section 4.4.

Table 8. Rate of validity of RHOASo vs. HPO across all D_i (% with accuracy and time complexity).

Validity	RF	GB	MLP	Average
$\mathcal{V}$	13.39%	12.27%	15.17%	13.61%
$\mathcal{V}$	5.58%	0.08%	3.12%	2.92%
$\mathcal{V}$	56.25%	45.08%	60.04%	53.79%
$\mathcal{V}$	24.77%	41.74%	21.65%	29.38%

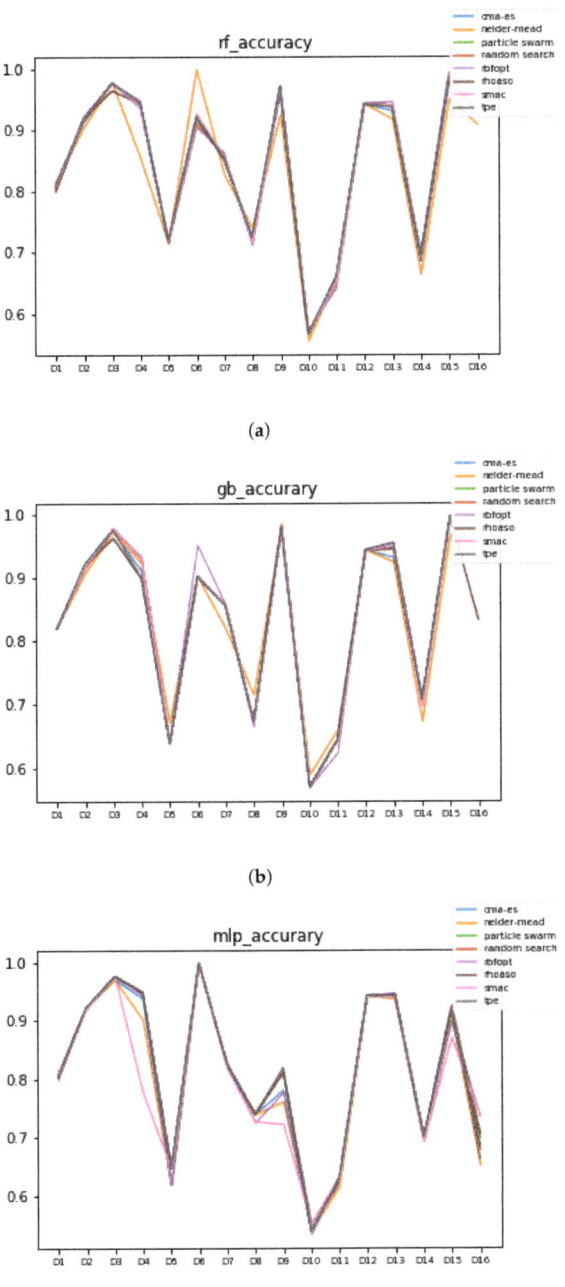

Figure 10. Accuracy complexity. (**a**) Accuracy in RF. In axis X, the dataset. In axis Y, the accuracy obtained. Each line represents a HPO algorithm. (**b**) Accuracy in GB. In axis X, the dataset. In axis Y, the accuracy obtained. Each line represents a HPO algorithm. (**c**) Accuracy in MLP. In axis X, the dataset. In axis Y, the accuracy obtained. Each line represents a HPO algorithm.

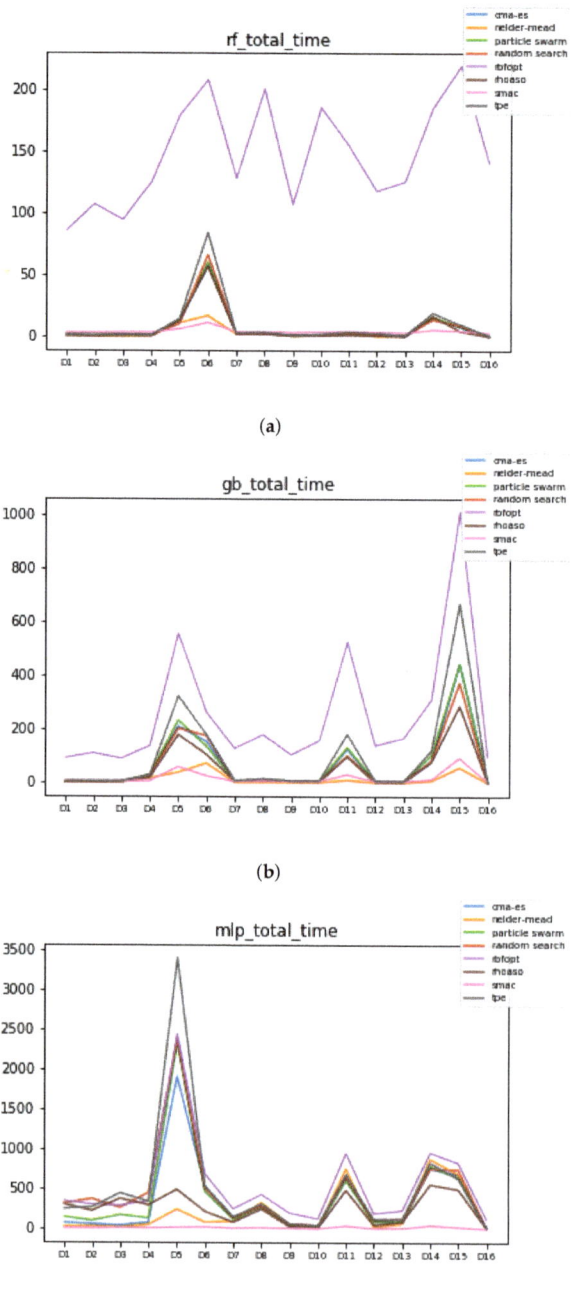

Figure 11. Time complexity. (**a**) Time complexity in RF. In axis X, the dataset. In axis Y, the total time obtained measured in seconds. Each line represents a HPO algorithm. (**b**) Time complexity in GB. In axis X, the dataset. In axis Y, the total time obtained measured in seconds. Each line represents a HPO algorithm. (**c**) Time complexity in MLP. In axis X, the dataset. In axis Y, the total time obtained measured in seconds. Each line represents a HPO algorithm.

On average, the class corresponding to positive statistically significant differences (green class) is higher than 50%. This can be considered a good result since RHOASo achieves better results than the rest of the algorithms in more than half of the cases analyzed. However, there is still a high rate in the blue class. Once we have transformed the blue class (see Section 4.4), we can analyze whether RHOASo is more effective than the rest of the HPO algorithms. The results are included in Table 9, which show that, on average, the class corresponding to positive statistically significant differences is higher than 70%.

Table 9. Rate of validity of RHOASo vs. HPO across all D_i (% with accuracy and time complexity).

Validity	RF	GB	MLP	Average
$\mathcal{V}$	22.76%	21.65%	30.13%	24.74%
$\mathcal{V}$	5.58%	0.89%	3.12%	3.19%
$\mathcal{V}$	71.65%	77.45%	66.74%	71.94%

After confirming that RHOASo is 70% more efficient than the rest of the HPO algorithms, the question that arises is whether there is a pattern in the 30% of the cases in which it does not succeed. For example, it might be possible for RHOASo to fail for datasets with a certain dimensionality, or for a specific ML algorithm. Another possibility is that RHOASo always loses against the same HPO algorithm. For this reason, we are going to study in depth the consistency of the previous results.

Since we have dealt with unbalanced datasets, such as $\mathcal{D}_1, \mathcal{D}_3 \mathcal{D}_4, \mathcal{D}_6$ or $\mathcal{D}_{12}$, we have repeated the analyses, changing the metric of accuracy by MCC so as to avoid over-optimistic scores. In Figure 12, the MCCs that are achieved by the HPO algorithms over each dataset are shown.

The rates of validity that are obtained by RHOASo, compared to the rest of the HPO algorithms, are included in Table 10. Note that these computations are carried out with the MCCs and time complexities by the analyses that are explained in Section 4.4.

Table 10. Rate of validity of RHOASo vs. HPO across all D_i (% with MCC and time complexity).

Validity	RF	GB	MLP	Average
$\mathcal{V}$	22.09%	22.76%	31.02%	25.29%
$\mathcal{V}$	4.68%	0.66%	2.90%	2.75%
$\mathcal{V}$	73.21%	76.56%	66.07%	71.94%

We can observe that RHOASo maintains its rate of gain, overcoming 70% of the cases. In Section 5.4, we study the consistency of these results as well as those situations in which RHOASo does not obtain a gain.

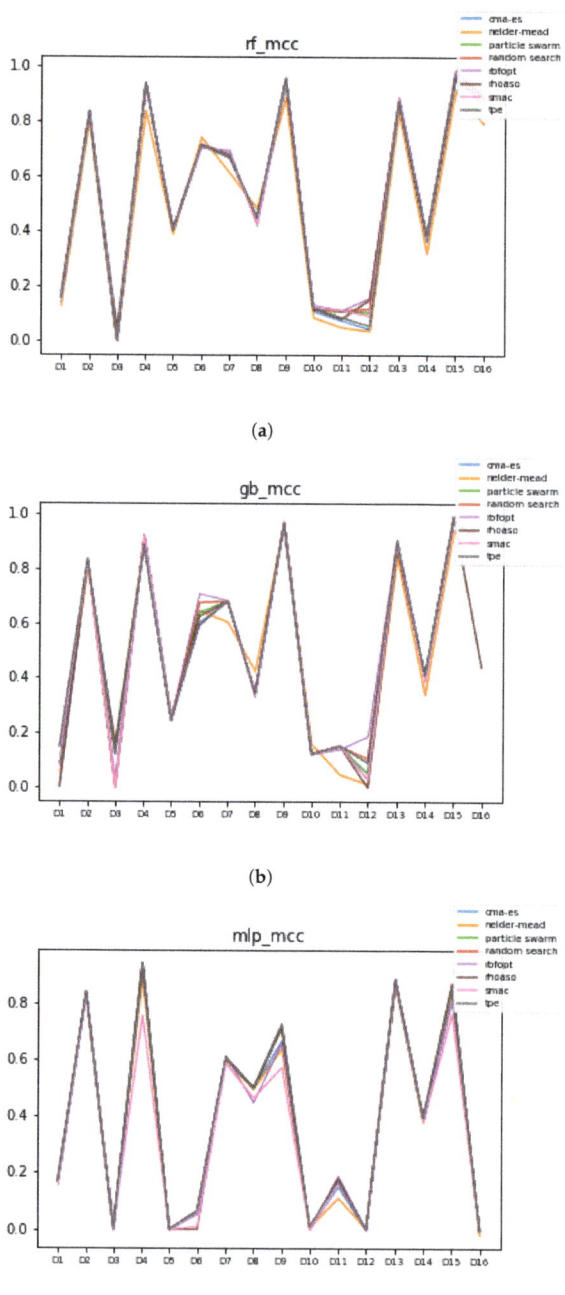

Figure 12. MCC of all HPO algorithms. (**a**) MCC in RF. In axis X, the dataset. In axis Y, the MCC obtained. Each line represents a HPO algorithm. (**b**) MCC in GB. In axis X, the dataset. In axis Y, the MCC obtained. Each line represents a HPO algorithm. (**c**) MCC in MLP. In axis X, the dataset. In axis Y, the MCC obtained. Each line represents a HPO algorithm.

5.3.2. Experiment 2

In Figures 13 and 14, the performances (accuracies and time complexities) that are achieved by the HPO algorithms over each dataset are shown.

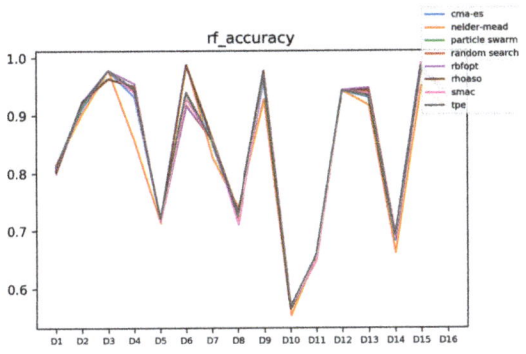

(a)

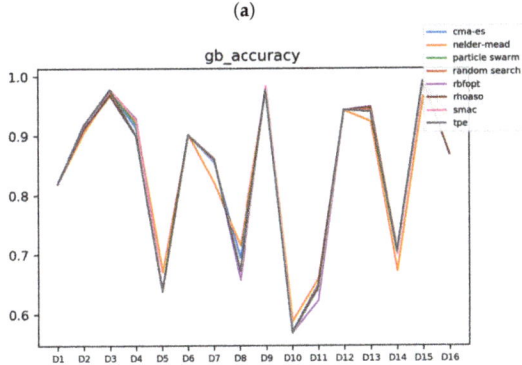

(b)

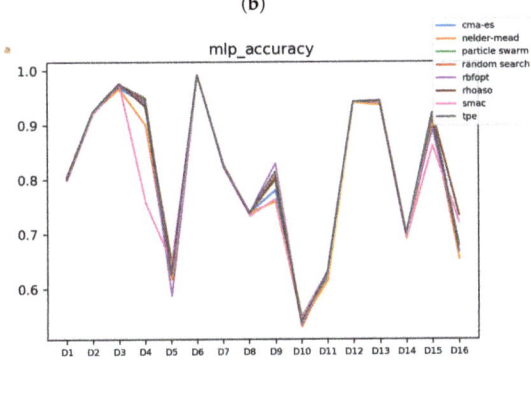

(c)

Figure 13. Experiment 2: Accuracy complexity. (**a**) Accuracy in RF. In axis X, the dataset. In axis Y, the accuracy obtained. Each line represents a HPO algorithm. (**b**) Accuracy in GB. In axis X, the dataset. In axis Y, the accuracy obtained. Each line represents a HPO algorithm. (**c**) Accuracy in MLP. In axis X, the dataset. In axis Y, the accuracy obtained. Each line represents a HPO algorithm.

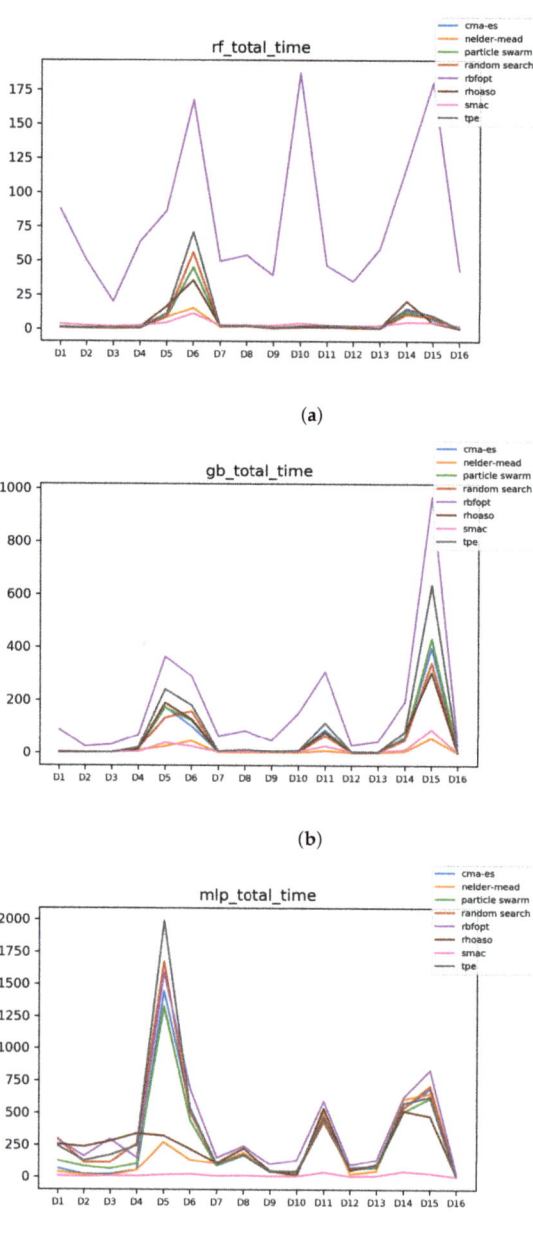

Figure 14. Experiment 2: Time complexity. (**a**) Time complexity in RF. In axis X, the dataset. In axis Y, the total time obtained measured in seconds. Each line represents a HPO algorithm. (**b**) Time complexity in GB. In axis X, the dataset. In axis Y, the total time obtained measured in seconds. Each line represents a HPO algorithm. (**c**) Time complexity in MLP. In axis X, the dataset. In axis Y, the total time obtained measured in seconds. Each line represents a HPO algorithm.

The rates of validity that have been obtained by RHOASo compared to the rest of HPO of algorithms are included in Table 11. Note that these computations are carried out with the accuracies and time complexities by the analyses that are explained in Section 4.4 for Experiment 2.

Table 11. Experiment 2: Rate of validity of RHOASo vs. HPO across all D_i (% with accuracy and time complexity).

Validity	RF	GB	MLP	Average
$\mathcal{V}$	46.87%	30.80%	54.91%	44.19%
$\mathcal{V}$	1.33%	1.33%	3.12%	1.93%
$\mathcal{V}$	51.78%	67.85%	41.96%	53.86%

In this experiment, RHOASo loses some of its advantage over other HPO algorithms, mainly because the improvement in execution time is lower since the number of iterations of the other algorithms is fixed to be equal to that of RHOASo. The average gain is of 53.86%, which is lower than the 71.96% obtained when evaluating the accuracy. Nevertheless, RHOASo performs better for GB, has a slight advantage for RF and is outperformed in MLP. This fact may be related to the search strategy of RHOASo. MLP has worse performance than other models across all datasets and configurations, which increases the number of neurons that tend to perform better, but the search strategy of RHOASo favors configurations with low magnitude of hyper-parameter values. Therefore, the performance when tuning MLP can be expected to be less satisfactory.

For the unbalanced datasets, we have included the analyses changing the metric of accuracy by MCC. In Figure 15, the MCCs that are achieved by the HPO algorithms over each dataset are shown.

The rates of validity that are obtained by RHOASo, compared to the rest of HPO of algorithms are included in Table 12. Note that these computations are carried out with the MCCs and time complexities by the analyses that are explained in Section 4.4.

Table 12. Experiment 2: rate of validity of RHOASo Vs HPO across all D_i (% with MCC and time complexity).

Validity	RF	GB	MLP	Average
$\mathcal{V}$	43.97%	32.58%	57.58%	44.71%
$\mathcal{V}$	2.23%	0.89%	2.90%	2.00%
$\mathcal{V}$	53.79%	66.51%	39.50%	53.27%

When using MCC as the reference metric, the results follow similar trends to that of accuracy. RHOASo gains a slight advantage for RF and incurs a slight loss for MLP, but there are no significant differences.

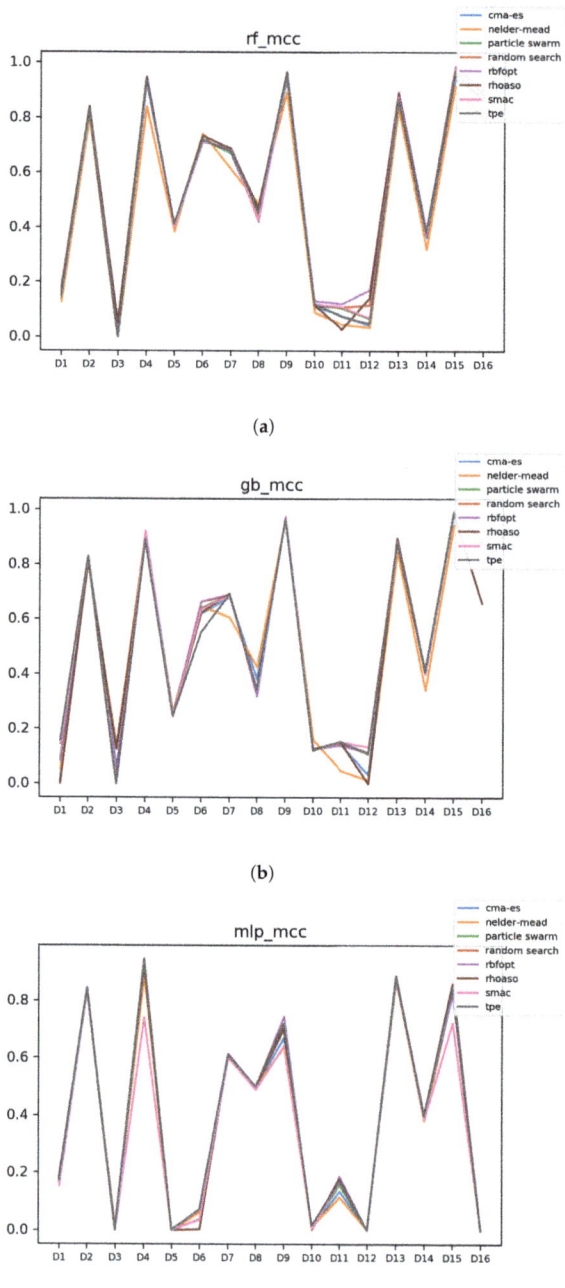

Figure 15. Experiment 2: MCC of all HPO algorithms. (**a**) MCC in RF. In axis X, the dataset. In axis Y, the MCC obtained. Each line represents a HPO algorithm. (**b**) MCC in GB. In axis X, the dataset. In axis Y, the MCC obtained. Each line represents a HPO algorithm. (**c**) MCC in MLP. In axis X, the dataset. In axis Y, the MCC obtained. Each line represents a HPO algorithm.

5.4. Research Question 4: Are the above Results Consistent?

In this section, we analyze whether RHOASo achieves a significant performance improvement consistently across datasets and partitions.

5.4.1. Experiment 1

The rates of validity for each partition computed with the accuracy and time complexity are included in Figure 16. The consistency of the green class is clear when we discriminate them by partitions. That is to say, if we only consider the partitions, RHOASo always outperforms the rest of the HPO algorithms. This may be due to the early stop of RHOASo, consuming less time but achieving good accuracy.

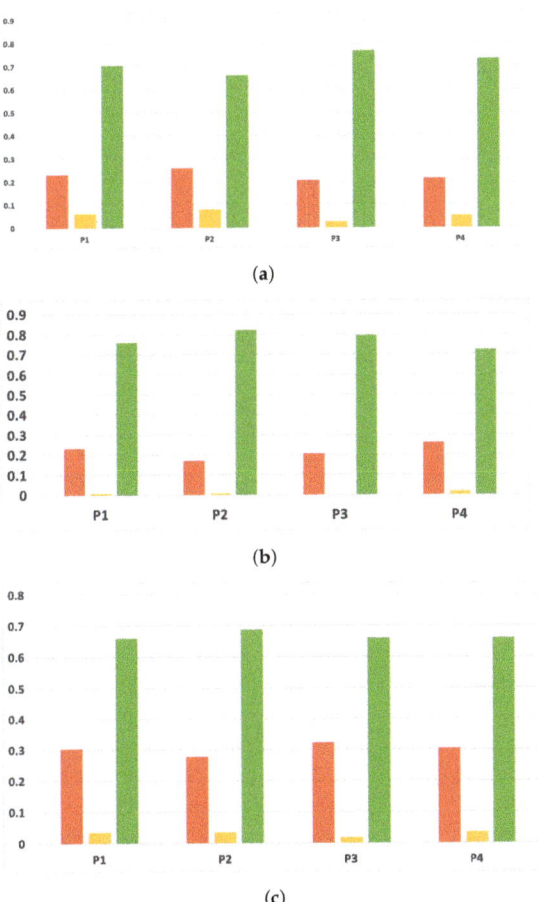

Figure 16. Experiment 1: consistency per partition (accuracy and time complexity). (**a**) Rate of validity per partition in RF. In axis X, the partition of the dataset. In axis Y, the rate of RHOASo for each class of validity. (**b**) Rate of validity per partition in GB. In axis X, the partition of the dataset. In axis Y, the rate of RHOASo for each class of validity. (**c**) Rate of validity per partition in MLP. In axis X, the partition of the dataset. In axis Y, the rate of RHOASo for each class of validity.

As we can see in Figure 17, the above conclusion is not so general when we discriminate them by datasets, except for the case of GB. Nonetheless, the results concerning RF and MLP are not so far from being consistent (RF being closer than MLP), and a deeper analysis should be performed regarding this fact.

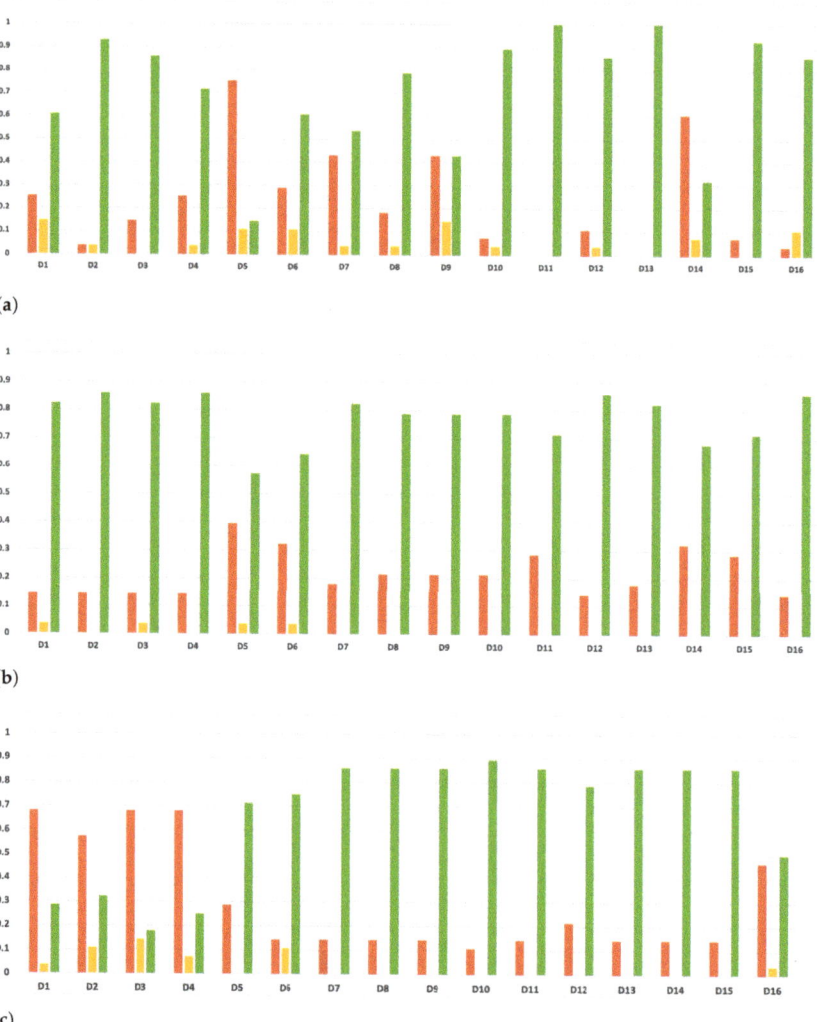

Figure 17. Experiment 1: consistency per dataset. Rate of validity with the accuracy and time complexity. (**a**) Rate of vx per dataset in RF. In axis X, the dataset. In axis Y, the rate of RHOASo for each class of validity. (**b**) Rate of validity per dataset in GB. In axis X, the dataset. In axis Y, the rate of RHOASo for each class of validity. (**c**) Rate of validity per dataset in MLP. In axis X, the dataset. In axis Y, the rate of RHOASo for each class of validity.

In the case of MLP, there is a clear trend for the datasets in which RHOASo losses are D_1, D_2, D_3, and D_4. That is, datasets with a low number of instances (<10,000) but with a high number of features > 50. This failure in the large-dimension small-sized datasets can be due to the early-stop characteristic of the algorithm. Datasets with high dimensionality and low number of instances tend to increase the variance of the results, which contributes to create an irregular surface in $\Phi_{A,D}$, possibly trapping RHOASo in a local maximum. A possible solution could be to substitute the function that maps elements from the hyper-parameter space to the performance of the trained ML model on a validation dataset with an approximated probabilistic model of such function. This suggests that combining the underlying ideas of Bayesian hyper-parameter optimization algorithms with those presented in this paper could yield an early-stop algorithm that works efficiently in high dimensions.

In the case of RF, the gain of RHOASo is not enough for datasets D_5, D_9, and D_{14}. Unlike in the case of MLP, these datasets have no clear commonalities, so we can only hypothesize. We believe that the problem is the same as in the case of MLP: high variance in the results diminishes the effectiveness of RHOASo. However, in this case, this variance could be ascribed to the inherent randomness in the training of RF or in the cross validation sampling.

As we can see in Figure 18, the behavior of the rates of validity for each partition remains consistent when these are computed with the MCC and time complexity.

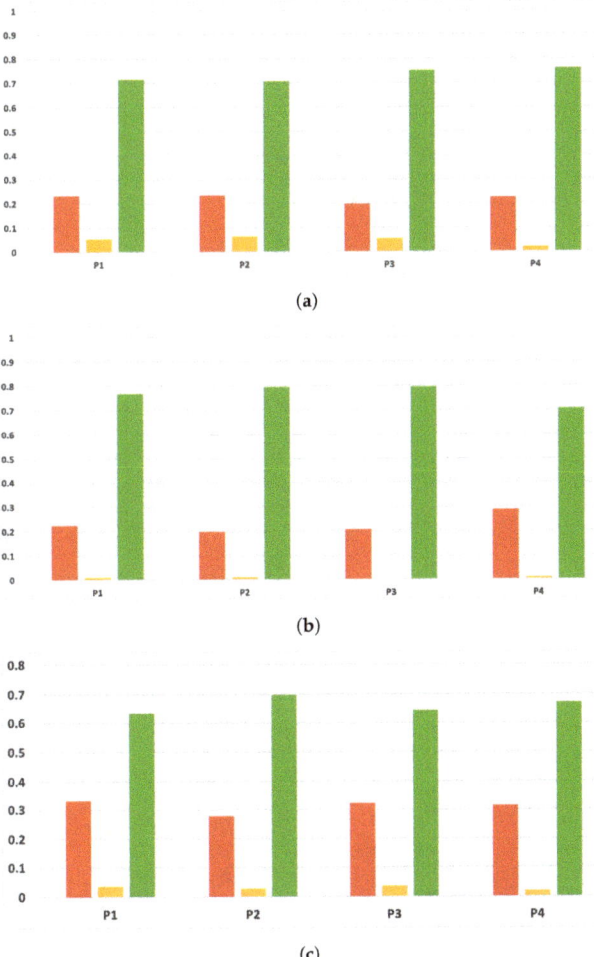

Figure 18. Experiment 1: consistency per partition (MCC and time complexity). (**a**) Rate of validity per partition in RF. In axis X, the partition of the dataset. In axis Y, the rate of RHOASo for each class of validity. (**b**) Rate of validity per partition in GB. In axis X, the partition of the dataset. In axis Y, the rate of RHOASo for each class of validity. (**c**) Rate of validity per partition in MLP. In axis X, the partition of the dataset. In axis Y, the rate of RHOASo for each class of validity.

As we can see in Figure 19, the same conclusions as in the case of accuracy are obtained when we discriminate them by datasets. Nonetheless, the results concerning RF and MLP are not so far from being consistent (RF being closer than MLP), and a deeper analysis should be performed regarding this fact. We note that, in some cases, the green class is increased; this has not occurred for the red class in the unbalanced datasets.

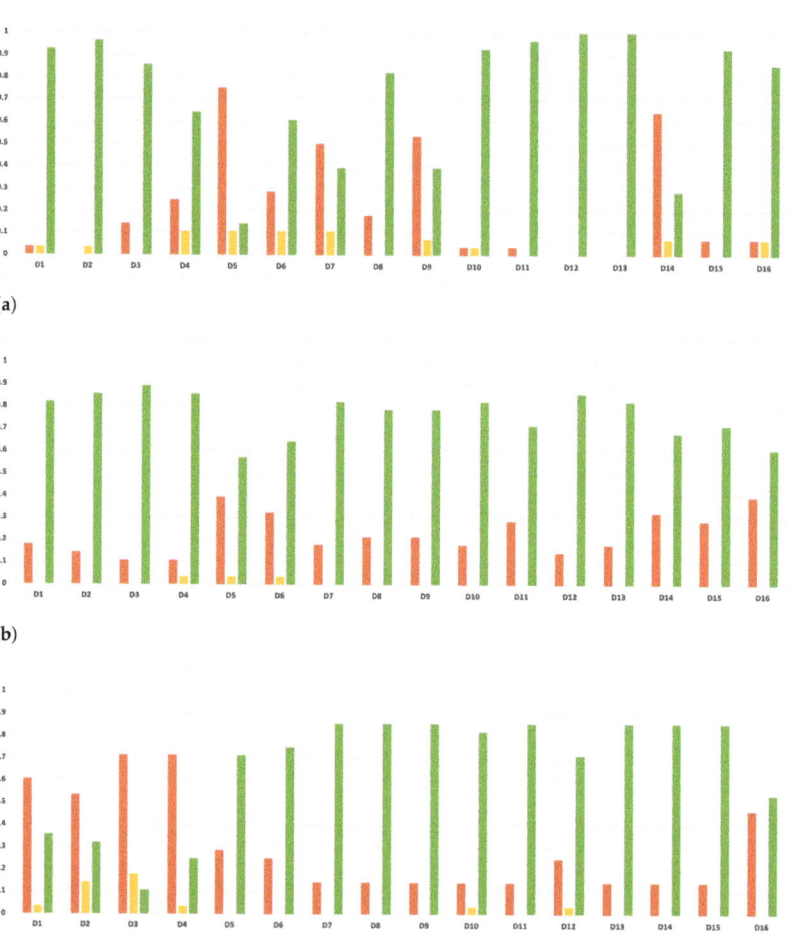

(a)

(b)

(c)

Figure 19. Consistency per dataset: rate of validity with the MCC and time complexity. (**a**) Rate of validity per dataset in RF. In axis X, the dataset. In axis Y, the rate of RHOASo for each class of validity. (**b**) Rate of validity per dataset in GB. In axis X, the dataset. In axis Y, the rate of RHOASo for each class of validity. (**c**) Rate of validity per dataset in MLP. In axis X, the dataset. In axis Y, the rate of RHOASo for each class of validity.

5.4.2. Experiment 2

The rates of validity for each partition computed with the accuracy and time complexity are included in Figure 20. In this experiment, RHOASo is more inconsistent since it loses the advantage of execution time that it had over the other HPOs. RHOASo has a consistent advantage for all partition sizes in GB, but for RF it only improves other algorithms for partitions 3 and 4 and is inferior in MLP in all partitions, although the results improve when the size of the partition increases. This trend is also present for RF and GB. The reasons for this trend are probably the same as those we discussed in Section 5.4.1: the variance in the results trapping RHOASo in the local maxima.

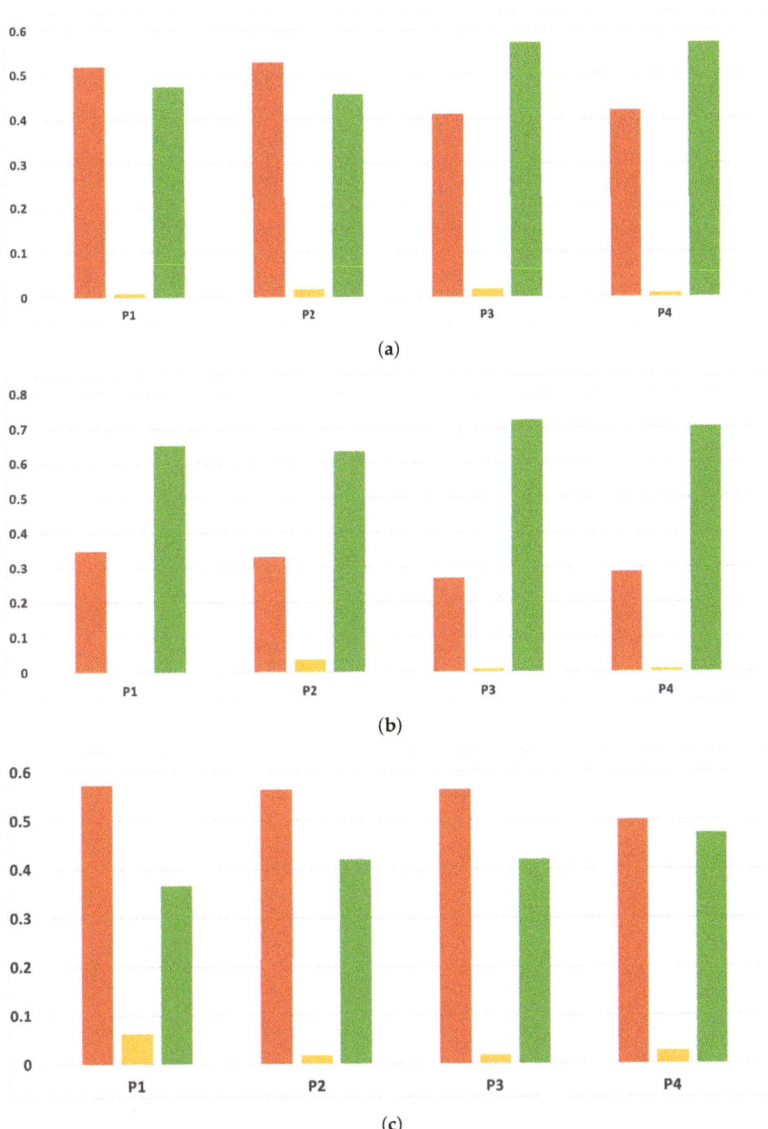

Figure 20. Experiment 2: consistency per partition (accuracy and time complexity). (**a**) Rate of validity per partition in RF. In axis X, the partition of the dataset. In axis Y, the rate of RHOASo for each class of validity. (**b**) Rate of validity per partition in GB. In axis X, the partition of the dataset. In axis Y, the rate of RHOASo for each class of validity. (**c**) Rate of validity per partition in MLP. In axis X, the partition of the dataset. In axis Y, the rate of RHOASo for each class of validity.

As we can see in Figure 21, the validity per dataset is negatively affected. In RF, there is a general decrease, with only datasets 10, 12, 12 and 15 showing a clear advantage for RHOASo. For GB, RHOASo maintains better results than other algorithms for all datasets, except 5 and 6. However, the consistency is lower than in experiment 1. Finally, in MLP, the validity of RHOASo is the lowest among all models, being consistently outperformed in five datasets: 1, 2, 3, 4 and 13.

The datasets with the worst results for each model have no clear commonalities, so it is possible that the neutralization of the execution time advantage of RHOASo is a significant factor in the deterioration of the results.

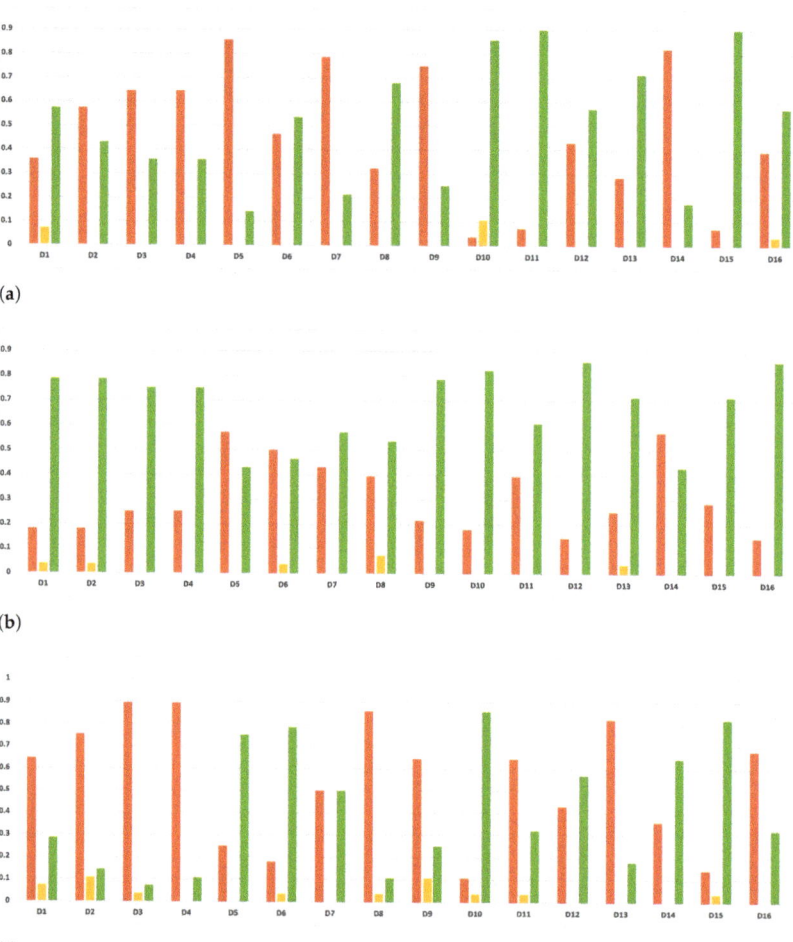

Figure 21. Experiment 2: consistency per dataset. Rate of validity with the accuracy and time complexity. (**a**) Rate of validity per dataset in RF. In axis X, the dataset. In axis Y, the rate of RHOASo for each class of validity. (**b**) Rate of validity per dataset in GB. In axis X, the dataset. In axis Y, the rate of RHOASo for each class of validity. (**c**) Rate of validity per dataset in MLP. In axis X, the dataset. In axis Y, the rate of RHOASo for each class of validity.

In Figure 22, we show the results of the above analysis with the MCC and the time time complexity. The results are very similar to those achieved with accuracy. The main difference is that the validity for RF is improved for partitions 1 and 2, increasing the consistency of the results.

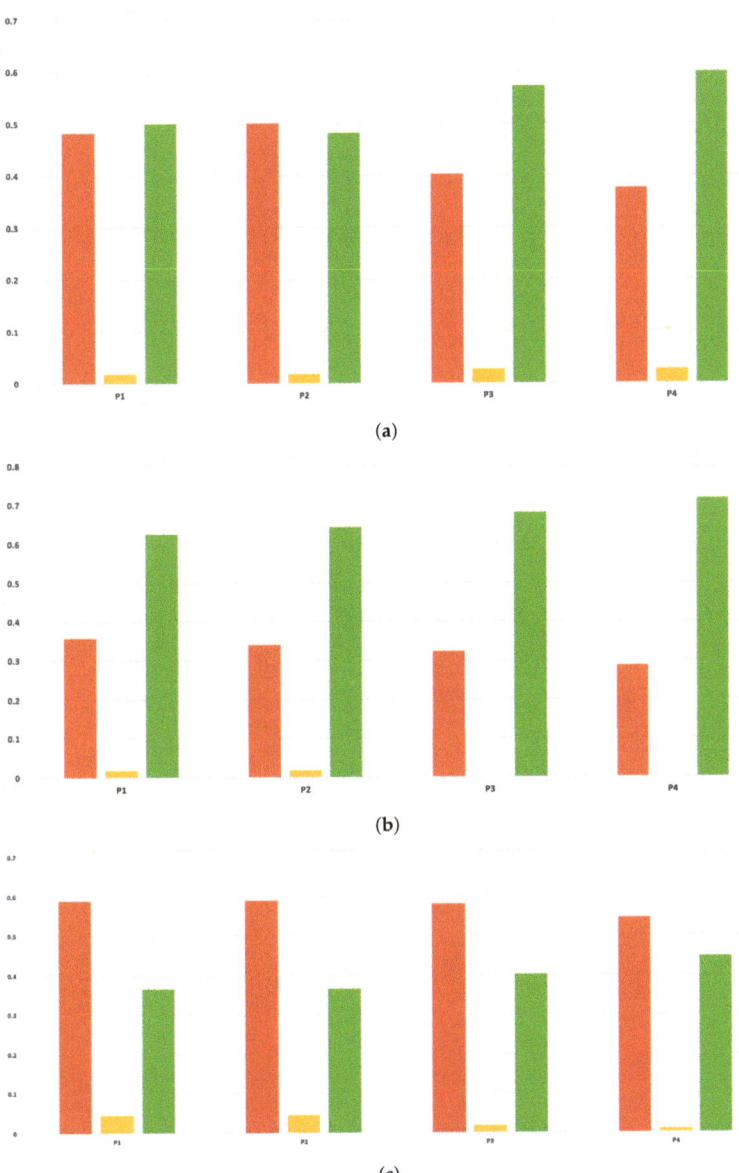

Figure 22. Experiment 2: consistency per partition (MCC and time complexity). (**a**) Rate of validity per partition in RF. In axis X, the partition of the dataset. In axis Y, the rate of RHOASo for each class of validity. (**b**) Rate of validity per partition in GB. In axis X, the partition of the dataset. In axis Y, the rate of RHOASo for each class of validity. (**c**) Rate of validity per partition in MLP. In axis X, the partition of the dataset. In axis Y, the rate of RHOASo for each class of validity.

In Figure 23, we show the rate of validity per dataset, using the MCC as the reference metric. As happened in Experiment 1, the trends are mostly the same as those present when evaluating the accuracy.

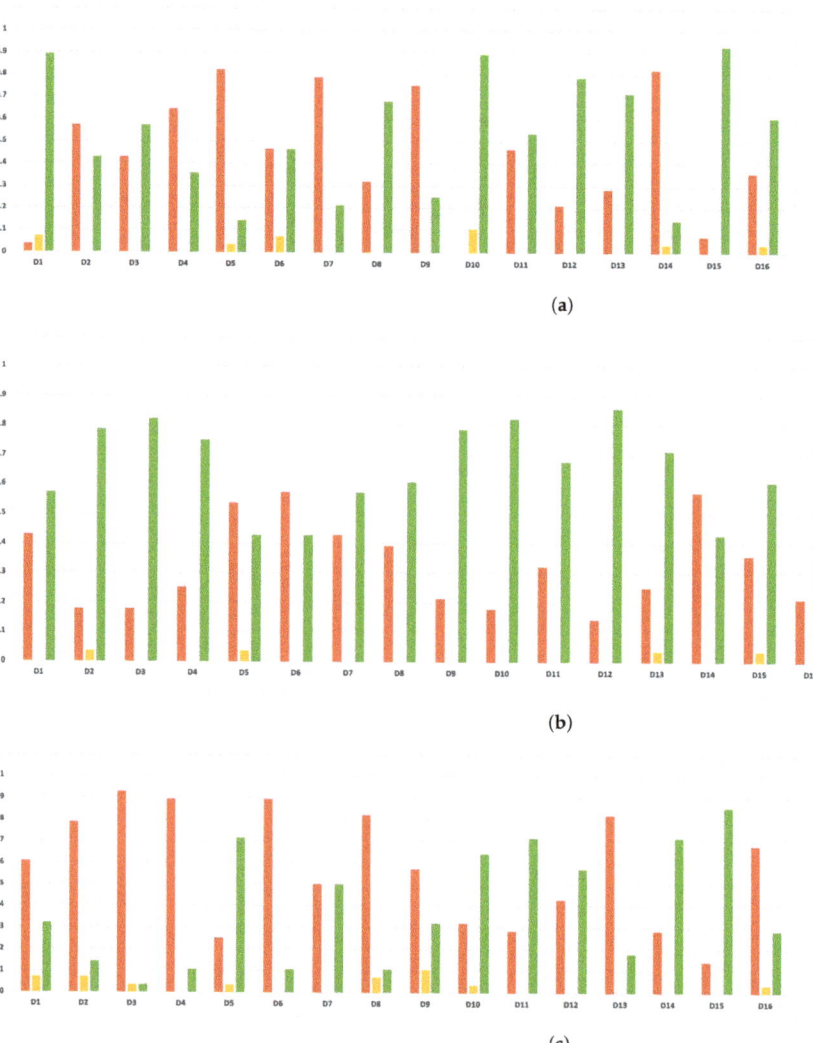

Figure 23. Experiment 2: consistency per dataset. Rate of validity with the MCC and time complexity. (**a**) Rate of validity per dataset in RF. In axis X, the dataset. In axis Y, the rate of RHOASo for each class of validity. (**b**) Rate of validity per dataset in GB. In axis X, the dataset. In axis Y, the rate of RHOASo for each class of validity. (**c**) Rate of validity per dataset in MLP. In axis X, the dataset. In axis Y, the rate of RHOASo for each class of validity.

6. Conclusions

ML provides several powerful tools for data processing that find applications in different fields. Most of the existent models have several hyper-parameters that need to be tuned and have a noticeable impact on their performance. Therefore, HPO algorithms are essential to achieve the highest possible accuracy with minimal human intervention.

In this work, a new HPO algorithm is described as a generalization of the discrete analog of a basic iterative algorithm to obtain the solutions to certain conditional optimization problems for the logistic function. It is shown that its performance is weakly disturbed by changing the size of the data subset with which it is run. The algorithm shows positive statistically meaningful differences in efficiency, regarding the other HPO algorithms considered in this study. The algorithm can finish the tuning process by itself and only requires an upper bound on the number of iterations to perform. Furthermore, it is shown that, on average, it needs around 70% of the iterations needed by the other hyper-parameter optimization algorithms to achieve competitive results.

The results show that the algorithm achieves high accuracy, with similar results for all classifiers on each dataset. In addition, RHOASo can effectively use a small partition size to accelerate the HPO process without sacrificing the final accuracy of the model. Lastly, the automatic early stop ends the tuning process before reaching the fixed number of iterations ($M_e = 34$), further increasing its efficiency.

Future work can be aimed at several lines:

- Test the RHOASO's performance with more machine learning algorithms, such as decision trees or k-nearest neighbors.
- Include more HPO algorithms in the comparison of the effectivity of RHOASo.
- Optimize RHOASo so it can be effective on search spaces of greater dimensions and can better deal with an extremely indented surface of $\Phi_{A,D}$, possibly using a surrogate of the target function, such as in Bayesian optimization.
- Factor the possible inclusion and effect of parallelization.
- Assess RHOASo on data streams.

Author Contributions: Conceptualization : Á.L.M.C.; methodology: Á.L.M.C. and N.D.-G.; software: Á.L.M.C. and D.E.G.; validation: all authors have contributed equally; formal analysis: N.D.-G. and D.E.G.; investigation: Á.L.M.C. and N.D.-G.; data curation: D.E.G.; writing—original draft preparation, review and editing: all authors have contributed equally; project administration and funding acquisition: N.D.-G. All authors have read and agreed to the published version of the manuscript.

Funding: This work was partially supported by the Spanish National Cybersecurity Institute (IN-CIBE) under contract Art.83, key: X54.

Data Availability Statement: The datasets supporting this work are from previously reported studies and datasets, which are cited. The processed data are available from the corresponding author upon request.

Acknowledgments: The authors would like to thank the Spanish National Cybersecurity Institute (INCIBE), who partially supported this work. Additionally, in this research, the resources of the Center of Supercomputation of Castilla y León (SCAYLE) were used.

Conflicts of Interest: The authors declare no conflict of interest. The funders had no role in the design of the study; in the collection, analyses, or interpretation of data; in the writing of the manuscript, or in the decision to publish the results.

Abbreviations

The following abbreviations are used in this manuscript:

CMA-ES	Covariance Matrix Adaptation Evolutionary
GB	Gradient Boosting
HPO	Hyper-Parameters Optimization
ML	Machine Learning
MLP	Multi-Layer Perceptron
NM	Nelder-Mead
PS	Particle Swarm
RBFOpt	Radial Basis Function Optimization
RF	Random Forest
RS	Random Search
SMAC	Sequential Model Automatic Configuration
SMBO	Sequential Model-Based Optimization
TPE	Tree Parzen Estimators

Appendix A. Additional Results

In this appendix, we show the results concerning Experiment 1 in the case that the ML algorithms are DT and KNN. Due to the great difference in the performance of RHOASo with respect to the rest of the HPO algorithms when they are run with these ML algorithms, we have excluded the analyses from the body of the article. However, we believe that the obtained results may be of interest.

We give a brief description of DT and KNN below.

1. DT is a tree-like model, where the internal nodes and their edges encode possibilities and the ending nodes (leafs) encode decisions. Te maximum length of the paths joining the root node and a leaf is called the depth of the tree. There are a number of DT training algorithms, among which we can point out ID3, ID4, ID5 and CART. In this study, we have chosen CART.
2. KNN is a non-parametric classification model the may be also used in regression problems. The training examples are simply vectors in the feature space, carrying their class label. The training phase does not consist of constructing an internal mathematical model but simply allocating training data instances in the feature space. The classification phase is done by looking at the majority label of the k-nearest neighbors of each point. This implies the choice of a metric on the feature space, which by default is usually taken as the $p = 2$ Minkowski distance.

In Table A1, we include the hyper-parameters that we have tuned. We have chosen them because of their influence on the corresponding ML algorithms (see [3], Appendixes A1, A4).Concerning the hyper-parameters of DT to be tuned, we have chosen the minimum number of samples required to split an internal node (min_samples_split) and the minimum number of samples required to be at a leaf node (min_samples_leaf). Regarding the hyper-parameters of KNN, we have considered the number of neighbors to use for queries and p, the Minkowski's distance type.

The search space for all hyper-parameters is in the interval [1, 50]. All hyper-parameters not being tuned are set to their default values in *scikit-learn* implementation ([34]). We have used 10-fold cross-validation to assess the performance of all ML models combined with the HPO algorithms.

Table A1. ML algorithms together with hyper-parameters we have considered.

Name	Hyper-Parameter 1	Hyper-Parameter 2
DT	min_smaples_split	min_samples_leaf
KNN	n_neighbors	p

Appendix A.1. Performance of RHOASo

We can see in Figures A1–A6 that we have plotted the median of the accuracy, MCC, total time, sensitivity, specificity, and the number of iterations when RHOASo is applied together with DT and KNN. The median of accuracy for DT over all datasets is 0.85 and for KNN, it is 0.78. We can observe that the achieved accuracy by RHOASo presents great stability in terms of partitions of the dataset, except for partition P4 and some datasets. The median of MCC for DT is 0.58, and for KNN it is 0.41. Again, RHOASo presents a similar behavior to the accuracy. It can be due to P_4 does not contribute to improve the fit of the models, see [1]. Regarding the time complexity, the median value for DT is 1.11 s, and for KNN it is 22.53 seconds. It can be shown that the total time registered for D_4 in KNN is higher in P_3 than in P_4. This can be caused because the stabilizer of RHOASo achieves an optimum value before in the total dataset compared to P_3. The median of sensitivity is 0.88 for DT, and 0.87 for KNN. In contrast, specificity has a median of 0.87 for DT, and 0.63 for KNN. In addition, these plots inherit the same trends as the graphics of the accuracy and MCC. The median of the number of iterations for DT is 14, and for KNN it is 19. It is worth pointing out the number of iterations for P_3 is larger than for P_4 when we work with KNN in D_4. This is probably caused by the same reason as in the plot of the total time.

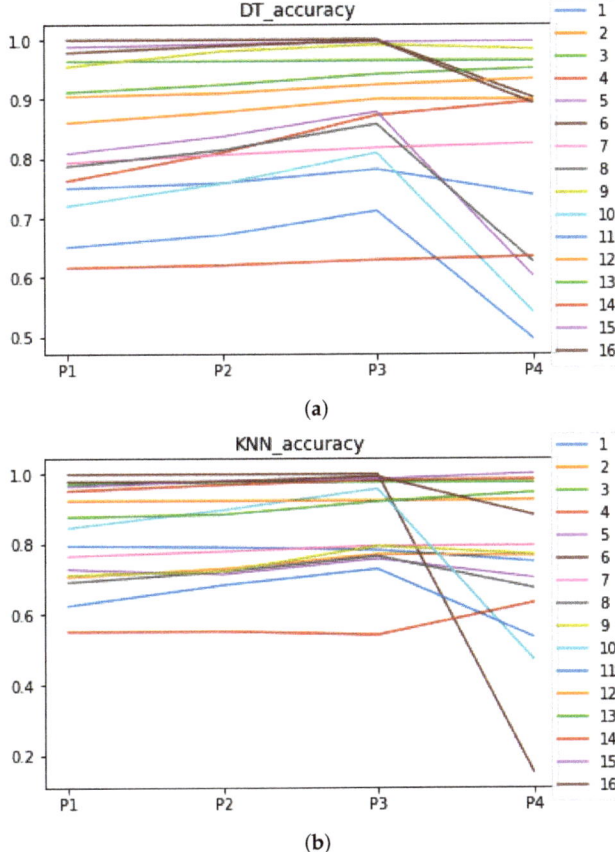

Figure A1. Behavior of RHOASo: accuracy. Each line represents a dataset. (**a**) Accuracy for DT. (**b**) Accuracy for KNN.

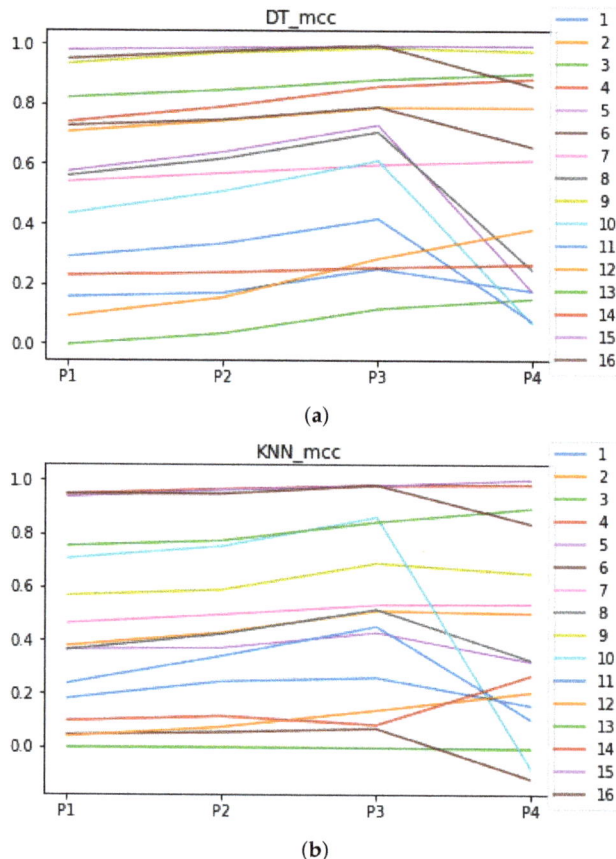

Figure A2. Behavior of RHOASo: MCC. Each line represents a dataset. (**a**) MCC for DT. (**b**) MCC for KNN.

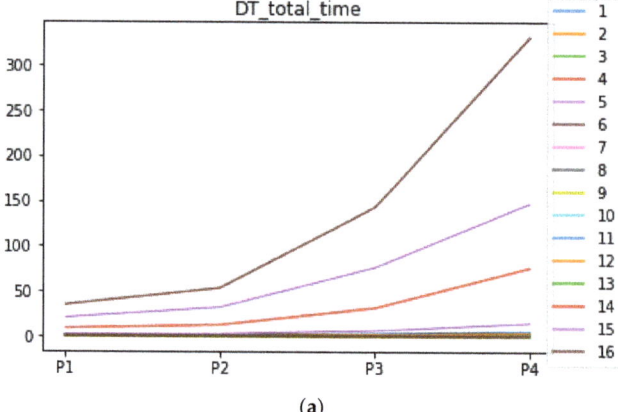

Figure A3. *Cont.*

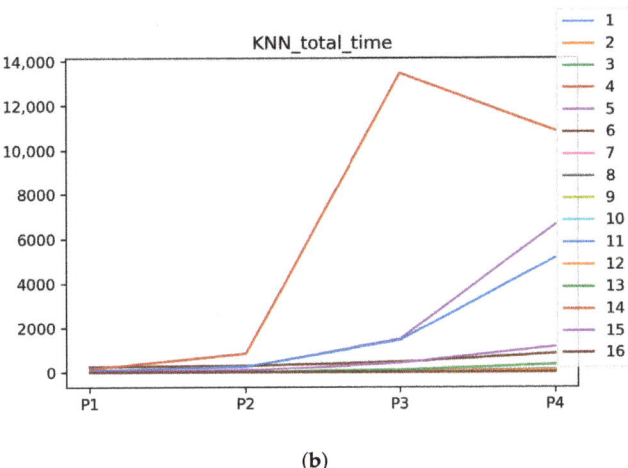

(b)

Figure A3. Behavior of RHOASo: time complexity. Each line represents a dataset. (**a**) Time complexity for DT. (**b**) Time complexity for KNN.

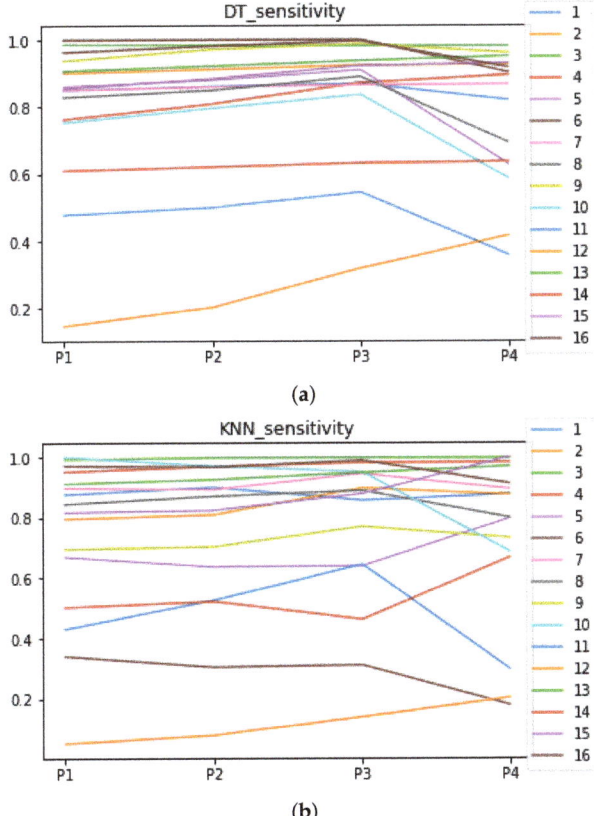

Figure A4. Behavior RHOASo: sensitivity. The datasets are represented with a bar chart for each partition. (**a**) Sensitivity for DT. (**b**) Sensitivity for KNN.

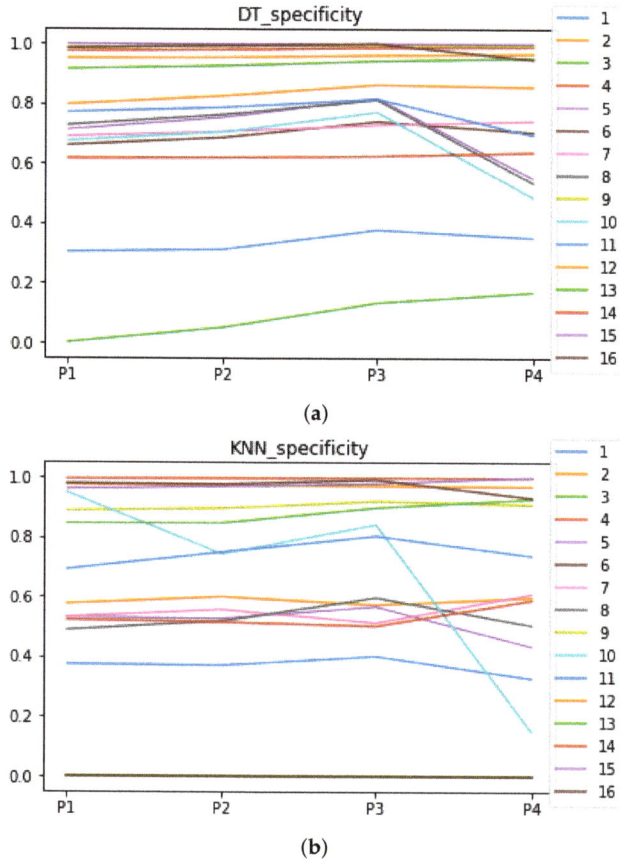

Figure A5. Behavior RHOASo: specificity. The datasets are represented with a bar chart for each partition. (**a**) Specificity for DT. (**b**) Specificity for KNN.

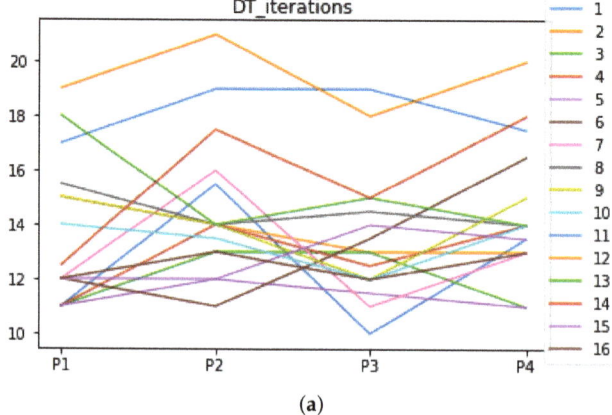

Figure A6. *Cont.*

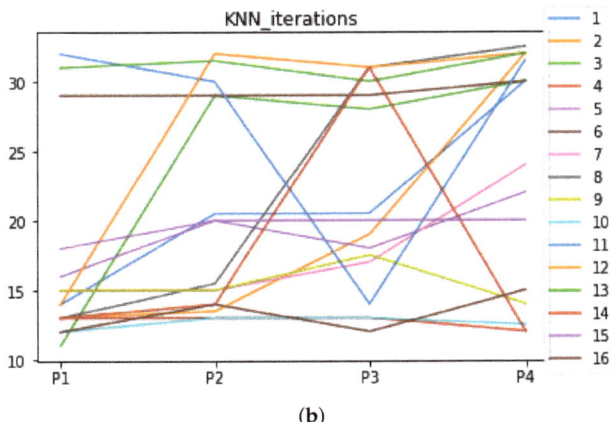

(b)

Figure A6. Number of iterations per partition. Each line represents a dataset (medians). (**a**) Number of iterations DT. (**b**) Number of iterations KNN.

Appendix A.2. Are There Statistically Meaningful Differences between the Performance of RHOASo and the Other HPO Algorithms under DT and KNN?

In Figures A7 and A8, the performances (accuracies and time complexities) that are achieved by the HPO algorithms over each dataset are shown.

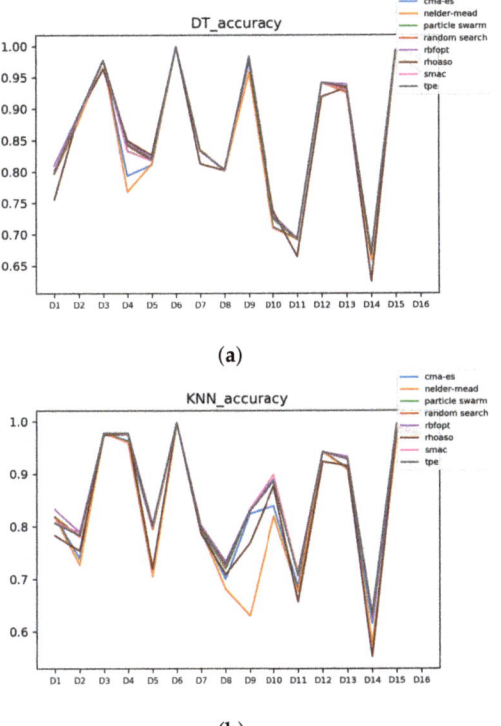

(a)

(b)

Figure A7. Accuracy complexity. (**a**) Accuracy in DT. (**b**) Accuracy in KNN.

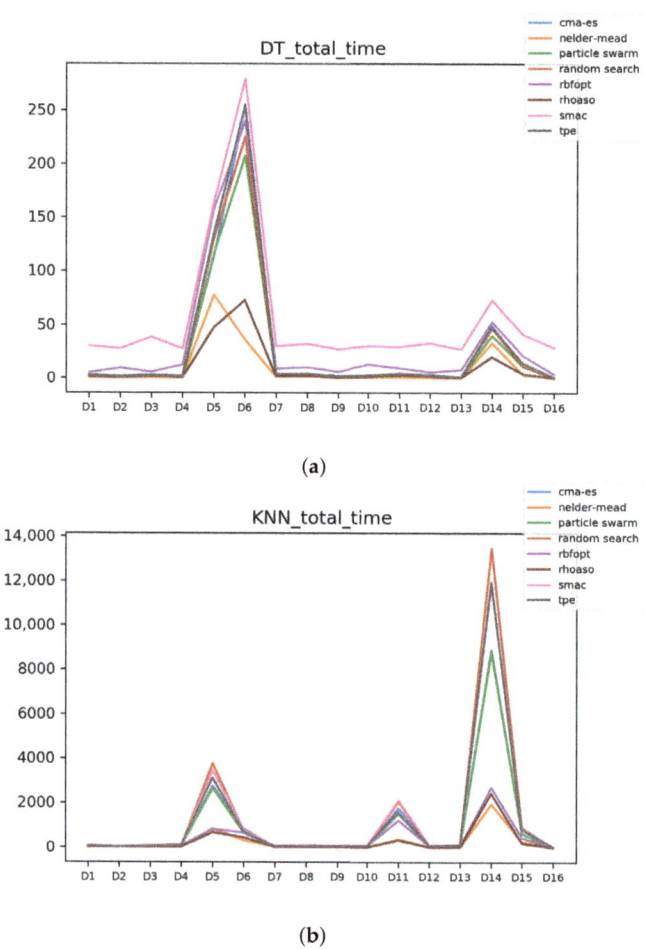

Figure A8. Time complexity. (**a**) Time complexity in DT. (**b**) Time complexity in KNN.

The results of validity of RHOASo faced on the rest of HPO algorithms are included in Table A2, which show that, on average, the class corresponding to positive statistically significant differences is higher than 90%. Note that these computations are carried out with the accuracies and time complexities by the analyses that are explained in Section 4.4.

Table A2. Rate of validity of RHOASo vs. HPO across all D_i (% with accuracy and time complexity).

Validity	DT	KNN	Average
$\mathcal{V}$	10.93%	8.25%	9.59%
$\mathcal{V}$	0%	0%	0%
$\mathcal{V}$	89.06%	91.74%	90.4%

Since we have dealt with unbalanced datasets, we have repeated the analyses, changing the metric of accuracy by MCC so as to avoid over-optimistic scores. In Figure A9, the MCCs that are achieved by the HPO algorithms over each dataset are shown.

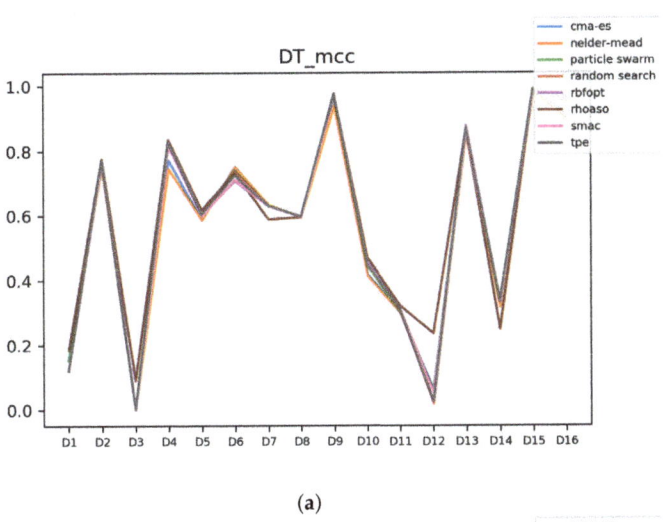

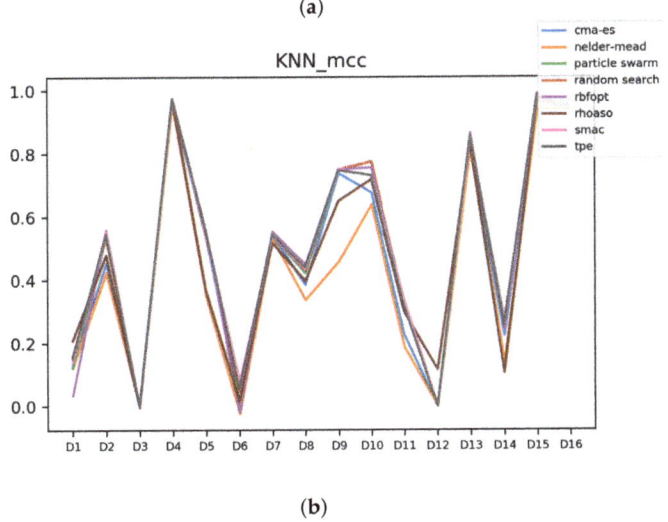

Figure A9. MCC of all HPO algorithms. (**a**) MCC in DT. (**b**) MCC in KNN.

The rates of validity that are obtained by RHOASo, compared to the rest of HPO of algorithms with the MCCs and time complexities, are included in Table A3.

Table A3. Rate of validity of RHOASo vs. HPO across all D_i (% with MCC and time complexity).

Validity	DT	KNN	Average
$\mathcal{V}$	10.49%	10.04%	10.26%
$\mathcal{V}$	0%	0.44%	0.22%
$\mathcal{V}$	89.50%	89.50%	89.507%

We can observe that the rate of gain of RHOASo overcomes the 89% of the cases.

Appendix A.3. Are the above Results Consistent?

In this section, we analyze whether RHOASo achieves a significant performance improvement consistently across datasets and partitions.

The rates of validity for each partition computed with the accuracy and time complexity are included in Figure A10. As can be seen, RHOASo presents a much higher performance than any other HPO algorithm taken into account in this work, and this behavior is independent of the partition.

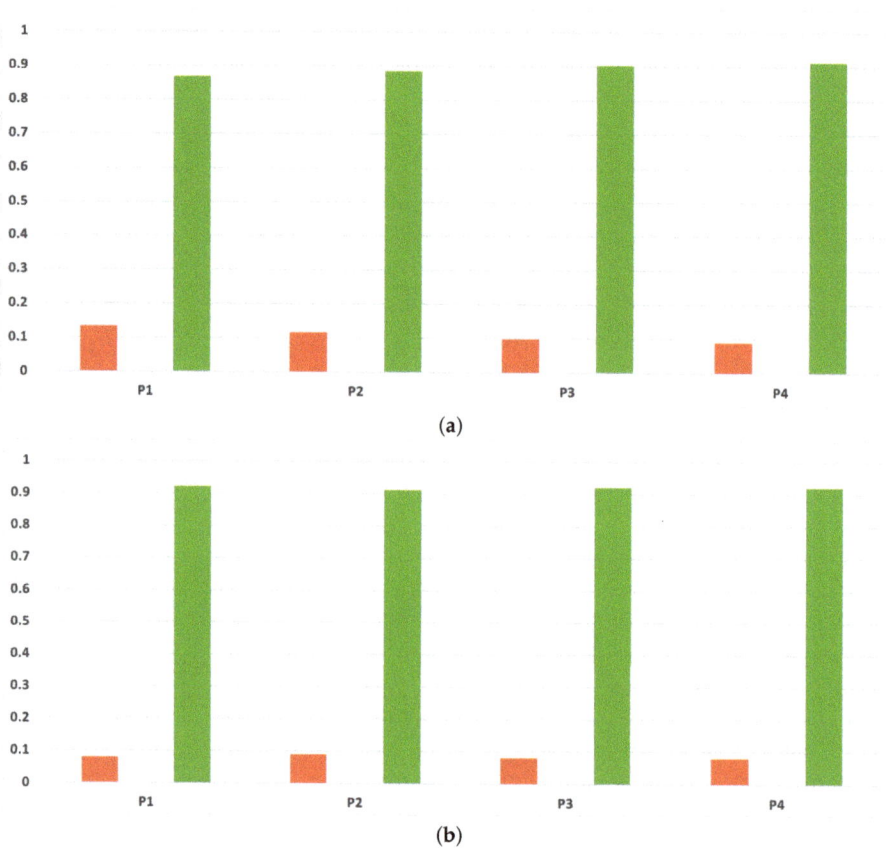

Figure A10. Experiment 1: consistency per partition (accuracy and time complexity). (**a**) Rate of validity per partition in DT. (**b**) Rate of validity per partition in KNN.

As we can see in Figure A11, the above conclusion is the same when we discriminate rates of validity by datasets.

As we can see in Figure A12, the behavior of the rates of validity for each partition remains consistent when these are computed with the MCC and time complexity. As we can seen in Figure A13, the same conclusion as in the case of accuracy is obtained when we discriminate them by datasets.

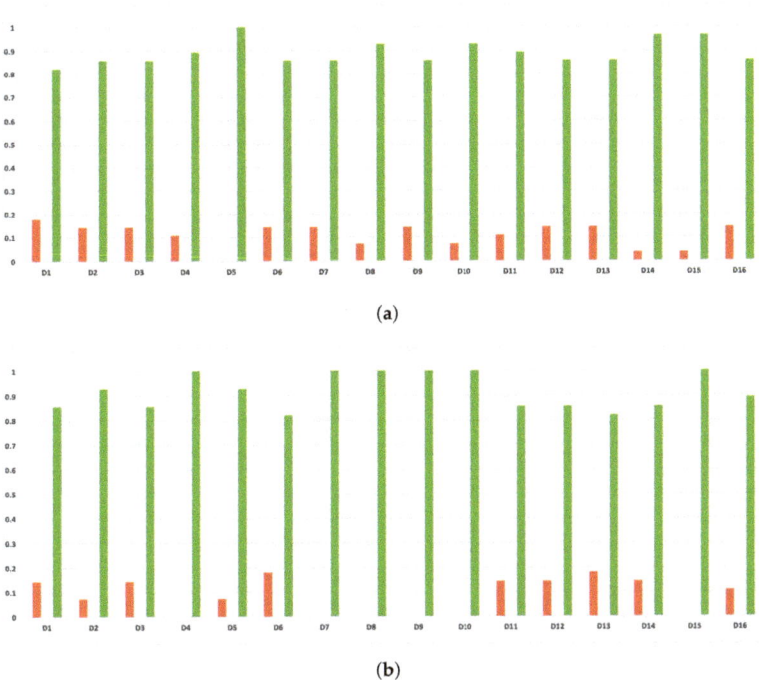

Figure A11. Experiment 1: consistency per dataset. Rate of validity with the accuracy and time complexity. (**a**) Rate of validity per dataset in DT. (**b**) Rate of validity per dataset in KNN.

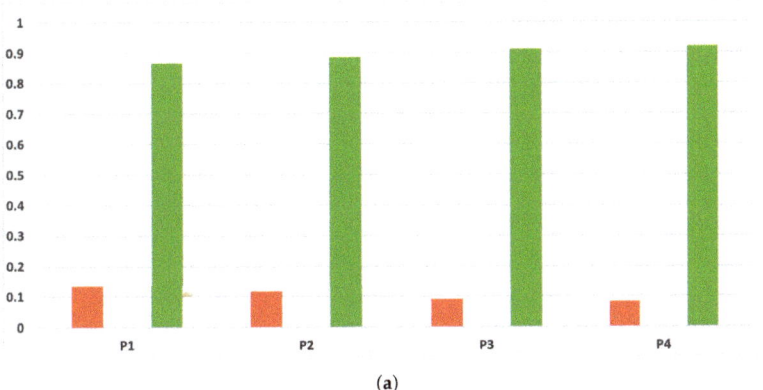

Figure A12. *Cont.*

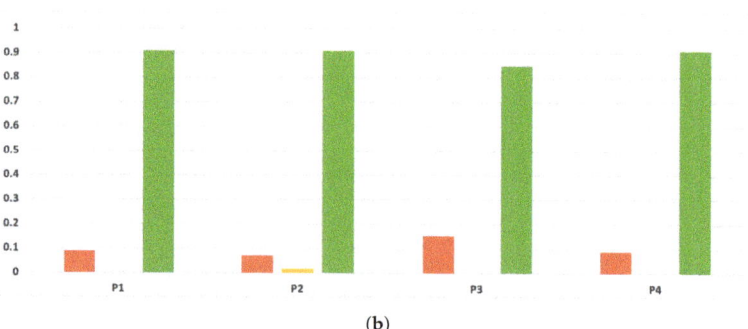

(b)

Figure A12. Experiment 1: consistency per partition (MCC and time complexity). (**a**) Rate of validity per partition in DT. (**b**) Rate of validity per partition in KNN.

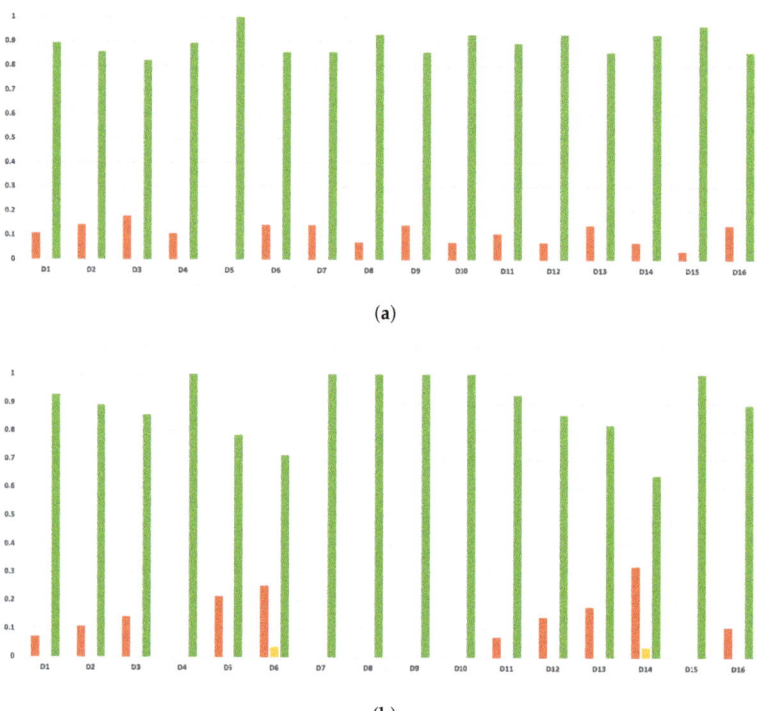

Figure A13. Consistency per dataset: rate of validity with the MCC and time complexity. (**a**) Rate of validity per dataset in DT. (**b**) Rate of validity per dataset in KNN.

References

1. DeCastro-García, N.; Muñoz Castañeda, A.L.; Escudero García, D.; Carriegos, M. Effect of the Sampling of a Dataset in the Hyperparameter Optimization Phase over the Efficiency of a Machine Learning Algorithm. *Complexity* **2019**, *2019*, 16. [CrossRef]
2. Jamieson, K.; Talwalkar, A. Non-stochastic best arm identification and hyperparameter optimization. In Proceedings of the 19th International Conference on Artificial Intelligence and Statistics, AISTATS 2016, Cadiz, Spain, 9–11 May 2016; pp. 240–248.
3. Bischl, B.; Binder, M.; Lang, M.; Pielok, T.; Richter, J.; Coors, S.; Thomas, J.; Ullmann, T.; Becker, M.; Boulesteix, A.L.; et al. Hyperparameter Optimization: Foundations, Algorithms, Best Practices and Open Challenges. *arXiv* **2021**, arXiv:stat.ML/2107.05847.
4. Bengio, Y. Gradient-Based Optimization of Hyperparameters. *Neural Comput.* **2000**, *12*, 1889–1900. [CrossRef]

5. Maclaurin, D.; Duvenaud, D.; Adams, R. Gradient-based hyperparameter optimization through reversible learning. In Proceedings of the 32nd International Conference on Machine Learning (ICML'15). IMLS, Lille, France, 6–11 July 2015; Volume 37, pp. 2113–2122.
6. Franceschi, L.; Donini, M.; Frasconi, P.; Pontil, M. Forward and Reverse Gradient-Based Hyperparameter Optimization. In Proceedings of the 34th International Conference on Machine Learning, Sydney, NSW, Australia, 6–11 August 2017; Precup, D., Teh, Y.W., Eds.; PMLR: International Convention Centre: Sydney, Australia, 2017; Volume 70, pp. 1165–1173.
7. Mockus, J. On Bayesian Methods for Seeking the Extremum. In *Proceedings of the IFIP Technical Conference*; Springer: London, UK, 1974; pp. 400–404.
8. Snoek, J.; Larochelle, H.; Adams, R.P. Practical Bayesian Optimization of Machine Learning Algorithms. In Proceedings of the 25th International Conference on Neural Information Processing Systems (NIPS'12), Lake Tahoe, NV, USA, 3–6 December 2012; Curran Associates Inc.: New York, NY, USA, 2012; Volume 2, pp. 2951–2959.
9. Hutter, F.; Hoos, H.H.; Leyton-Brown, K. Sequential Model-based Optimization for General Algorithm Configuration. In Proceedings of the 5th International Conference on Learning and Intelligent Optimization, Rome, Italy, 17–21 January 2011; Springer: Berlin, Heidelberg, 2011; LION'05, pp. 507–523. [CrossRef]
10. Bergstra, J.; Bardenet, R.; Bengio, Y.; Kégl, B. Algorithms for Hyper-parameter Optimization. In Proceedings of the 24th International Conference on Neural Information Processing Systems, Granada, Spain, 12–15 December 2011; Curran Associates Inc.: New York, USA, 2011; NIPS'11, pp. 2546–2554.
11. Illievski, I.; Akhtar, T.; Feng, J.; Shoemaker, C.A. Efficient hyperparameter optimization for deep learning algorithms using deterministic RBF surrogates. In Proceedings of the Thirty-First AAAI Conference on Artificial Intelligence, San Francisco, CA, USA, 4–9 February 2017; pp. 822–829.
12. Hernández-Lobato, J.M.; Hoffman, M.W.; Ghahramani, Z. Predictive Entropy Search for Efficient Global Optimization of Black-box Functions. In Proceedings of the 27th International Conference on Neural Information Processing Systems (NIPS'14), Montreal, QC, Canada, 8–13 December 2017; MIT Press: Cambridge, MA, USA, 2014; Volume 1, pp. 918–926.
13. Bardenet, R.; Brendel, M.; Kégl, B.; Sebag, M. Collaborative Hyperparameter Tuning. In Proceedings of the 30th International Conference on Machine Learning (ICML'13), Atlanta, GA, USA, 16–21 June 2013; Volume 28, pp. 858–866.
14. Swersky, K.; Snoek, J.; Adams, R.P. Multi-task Bayesian Optimization. In Proceedings of the 26th International Conference on Neural Information Processing Systems (NIPS'13), Lake Tahoe, NV, USA, 5–10 December 2013; Curran Associates Inc.: New York, USA, 2013; Volume 2, pp. 2004–2012.
15. Bergstra, J.; Bengio, Y. Random search for hyper-parameter optimization. *J. Mach. Learn. Res.* **2012**, *13*, 281–305.
16. Nuñez, L.; Regis, R.G.; Varela, K. Accelerated Random Search for constrained global optimization assisted by Radial Basis Function surrogates. *J. Comput. Appl. Math.* **2018**, *340*, 276–295. [CrossRef]
17. Hansen, N.; Ostermeier, A. Completely Derandomized Self-Adaption in Evolution Strategies. *Evol. Comput.* **2001**, *9*, 159–195. [CrossRef]
18. Nelder, J.; Mead, R. A simplex method for function minimization. *Comput. J.* **1965**, *7*, 308–313. [CrossRef]
19. Ozaki, Y.; Yano, M.; Onishi, M. Effective hyperparameter optimization using Nelder-Mead method in deep learning. *Ipsj Trans. Comput. Vis. Appl.* **2017**, *9*, 20. [CrossRef]
20. Clerc, M.; Kennedy, J. The particle swarm-explosion, stability, and convergence in a multidimensional complex space. *IEEE Trans. Evol. Comput.* **2002**, *6*, 58–73. [CrossRef]
21. Fortin, F.; De Rainville, F.; Gardner, M. DEAP: Evolutionary Algorithms Made Easy. *J. Mach. Learn. Res.* **2012**, *13*, 2171–2175.
22. Li, L.; Jamieson, K.; DeSalvo, G.; Rostamizadeh, A.; Talwalkar, A. Hyperband: A novel bandit-based approach to hyperparameter optimization. *J. Mach. Learn. Res.* **2018**, *18*, 1–52.
23. Li, L.; Jamieson, K.; Rostamizadeh, A.; Gonina, E.; Ben-Tzur, J.; Hardt, M.; Recht, B.; Tal-Walkar, A. A System for Massively Parallel Hyperparameter Tuning. In Proceedings of the Machine Learning and Systems 2020, Austin, TX, USA, 2–4 March 2020; pp. 230–246.
24. Falkner, S.; Klein, A.; Hutter, F. BOHB: Robust and Efficient Hyperparameter Optimization at Scale. In Proceedings of the 35th International Conference on Machine Learning. PMLR, Stockholm, Sweden, 10–15 July 2018; Volume 80, pp. 1437–1446.
25. Bergstra, J.; Yamins, D.; Cox, D. Hyperopt: A python library for optimizing the hyperparameters of machine learning algorithms. In Proceedings of the 12th Python in Science Conference (SCIPY 2013), Austin, TX, USA, 24–28 June 2013; pp. 13–20. [CrossRef]
26. Claesen, M.; Simm, J.; Popovic, D.; Moreau, Y.; De Moor, B. Easy Hyperparameter Search Using Optunity. *arXiv* **2014**, arXiv:1412.1114.
27. Lindauer, M.; Eggensperger, K.; Feurer, M.; Falkner, S.; Biedenkapp, A.; Hutter, F. SMAC v3: Algorithm Configuration in Python. 2017. Available online: https://github.com/automl/SMAC3 (accessed on 25 July 2021)
28. Costa, A.; Nannicini, G. RBFOpt: an open-source library for black-box optimization with costly function evaluations. *Math. Program. Comput.* **2018**, *10*, 597–629. [CrossRef]
29. DeCastro-García, N.; Castañeda, Á.L.M.; Fernández-Rodríguez, M. RADSSo: An Automated Tool for the multi-CASH Machine Learning Problem. In *Hybrid Artificial Intelligent Systems*; de la Cal, E.A., Villar Flecha, J.R., Quintián, H., Corchado, E., Eds.; Springer International Publishing: Cham, Switzerland, 2020; pp. 183–194.
30. DeCastro-García, N.; Castañeda, Á.L.M.; Fernández-Rodríguez, M. Machine learning for automatic assignment of the severity of cybersecurity events. *Comput. Math. Methods* **2020**, *2*, e1072. [CrossRef]

31. Breiman, L. Random Forests. *Mach. Learn.* **2001**, *45*, 5–32. [CrossRef]
32. Friedman, J. Greedy function approximation: A gradient boosting machine. *Ann. Statist.* **2001**, *29*, 1189–1232. [CrossRef]
33. Friedman, J. Stochastic gradient boosting. *Comput. Stat. Data Anal.* **2002**, *38*, 367–378. [CrossRef]
34. Pedregosa, F.; Varoquaux, G.; Gramfort, A.; Michel, V.; Thirion, B.; Grisel, O.; Blondel, M.; Prettenhofer, P.; Weiss, R.; Dubourg, V.; et al. Scikit-learn: Machine Learning in Python. *J. Mach. Learn. Res.* **2011**, *12*, 2825–2830.
35. Guo, X.; Yang, J.; Wu, C.; Wang, C.; Liang, Y. A novel LS-SVMs hyper-parameter selection based on particle swarm optimization. *Neurocomputing* **2008**, *71*, 3211–3215. [CrossRef]
36. Diaz, G.I.; Fokoue-Nkoutche, A.; Nannicini, G.; Samulowitz, H. An effective algorithm for hyperparameter optimization of neural networks. *Ibm J. Res. Dev.* **2017**, *61*, 9:1–9:11. [CrossRef]
37. Bridge, J.P.; Holden, S.B.; Paulson, L.C. Machine Learning for First-Order Theorem Proving. *J. Autom. Reason.* **2014**, *53*, 141–172. [CrossRef]
38. Hopkins, E.M.; Reeber, G.F. *Datataset Spambase*. UCI Machine Learning Repository. 1998. Available online: https://archive.ics.uci.edu/ml/datasets/spambase (accessed on 27 August 2019).
39. Zieba, M.; Tomczak, S.; Tomczak, J. Ensemble boosted trees with synthetic features generation in application to bankruptcy prediction. *Expert Syst. Appl.* **2016**, *58*, 93–101. [CrossRef]
40. Alpaydin, E.; Kaynak, C. *Optical Recognition of Handwritten Digits Dataset*. UCI Machine Learning Repository. 1995. Available online: https://archive.ics.uci.edu/ml/datasets/Optical+Recognition+of+Handwritten+Digits (accessed on 27 August 2019).
41. De Almeida Freitas, F.; Peres, S.M.; De Moraes Lima, C.A.; Barbosa, F.V. Grammatical Facial Expressions recognition with Machine Learning. In Proceedings of the 27th International Florida Artificial Intelligence Research Society Conference, FLAIRS 2014, Pensacola Beach, FL, USA, 21–23 May 2014; pp. 180–185.
42. Pozzolo, A.D.; Caelen, O.; Johnson, R.A.; Bontempi, G. Calibrating Probability with Undersampling for Unbalanced Classification. In Proceedings of the 2015 IEEE Symposium Series on Computational Intelligence, Cape Town, South Africa, 8–10 December 2015; pp. 159–166. [CrossRef]
43. Bock, R.; Chilingarian, A.; Gaug, M. Methods for multidimensional event classification: a case study using images from a Cherenkov gamma-ray telescope. *Nucl. Instr. Methods Phys. Res. Sect. Accel. Spectrom. Detect. Assoc. Equip.* **2004**, *516*, 511–528. [CrossRef]
44. Harries, M. SPLICE-2 Comparative Evaluation: Electricity Pricing. In *Technical Report*; The University of South Wales: Cardiff, UK, 1999.
45. Gama, J.; Medas, P.; Castillo, G.; Rodrigues, P. Learning with Drift Detection. In *Advances in Artificial Intelligence—SBIA 2004*; Bazzan, A.L.C., Labidi, S., Eds.; Springer: Berlin/Heidelberg, Germany, 2004; pp. 286–295.
46. Freire, A.L.; Barreto, G.A.; Veloso, M.; Varela, A.T. Short-term memory mechanisms in neural network learning of robot navigation tasks: A case study. In Proceedings of the 6th Latin American Robotics Symposium (LARS 2009), Valparaíso, Chile, 29–30 October 2009; pp. 1–6. [CrossRef]
47. Roesler, O. *Eye dataset*. UCI Machine Learning Repository. 2013. Available online: https://archive.ics.uci.edu/ml/datasets/EEG+Eye+State (accessed on 27 August 2019).
48. Tromp, J. *Connect4 dataset*. UCI Machine Learning Repository. 1995. Available online: https://archive.ics.uci.edu/ml/datasets/Connect-4 (accessed on 27 August 2019).
49. Security, A.I. *Amazon Employee Access Challenge*. Kaggle. 2013. Available online: https://www.kaggle.com/c/amazon-employee-access-challenge (accessed on 27 August 2019).
50. Mohammad, R.M.; Thabtah, F.; McCluskey, L. Predicting phishing websites based on self-structuring neural network. *Neural Comput. Appl.* **2014**, *25*, 443–458. [CrossRef]
51. Baldi, P.; Sadowski, P.; Whiteson, D. Searching for exotic particles in high-energy physics with deep learning. *Nat. Commun.* **2014**, *5*. [CrossRef]
52. Dhanabal, L.; Shantharajah, S. A Study on NSL-KDD Dataset for Intrusion Detection System Based on Classification Algorithms. *Int. J. Adv. Res. Comput. Commun. Eng.* **2015**, *4*, 446–452.
53. DEFCOM. NSL—KDD Dataset. Github. 2015. Available online: https://github.com/defcom17/NSL_KDD (accessed on 27 August 2019).
54. Guerrero-Higueras, A.; DeCastro-García, N.; Matellán, V. Detection of Cyber-attacks to indoor real time localization systems for autonomous robots. *Robot. Auton. Syst.* **2018**, *99*, 75–83. [CrossRef]
55. Matthews, B. Comparison of the predicted and observed secondary structure of T4 phage lysozyme. *Biochim. Biophys. Acta Protein Struct.* **1975**, *405*, 442–451. [CrossRef]
56. Gorodkin, J. Comparing two K-category assignments by a K-category correlation coefficient. *Comput. Biol. Chem.* **2004**, *28*, 367—374. [CrossRef] [PubMed]

Review

Review of Metaheuristics Inspired from the Animal Kingdom

Elena Niculina Dragoi [1,2,*] **and Vlad Dafinescu** [2,3]

[1] Faculty of Automatic Control and Computer Engineering, "Gheorghe Asachi" Technical University, Bld. Dimitrie Mangeron, No. 27, 700050 Iași, Romania
[2] Faculty of Chemical Engineering and Environmental Protection "Cristofor Simionescu", "Gheorghe Asachi" Technical University, Bld. Dimitrie Mangeron, No. 73, 700050 Iași, Romania; vdafinescu@gmail.com
[3] Emergency Hospital "Prof. Dr. N. Oblu", Str. Ateneului No. 2, 700309 Iași, Romania
* Correspondence: elena-niculina.dragoi@academic.tuiasi.ro; Tel.: +40-232-278683

Abstract: The search for powerful optimizers has led to the development of a multitude of metaheuristic algorithms inspired from all areas. This work focuses on the animal kingdom as a source of inspiration and performs an extensive, yet not exhaustive, review of the animal inspired metaheuristics proposed in the 2006–2021 period. The review is organized considering the biological classification of living things, with a breakdown of the simulated behavior mechanisms. The centralized data indicated that 61.6% of the animal-based algorithms are inspired from vertebrates and 38.4% from invertebrates. In addition, an analysis of the mechanisms used to ensure diversity was performed. The results obtained showed that the most frequently used mechanisms belong to the niching category.

Keywords: metaheuristics; optimization; animal-inspired; exploration; exploitation

1. Introduction

A metaheuristic is a high level, problem-independent framework that provides a series of steps and guidelines used to develop heuristic optimizers [1]. Nowadays, the tendency is to use the term for both the general framework and for the algorithms built based on its rules [1]. In the latest years, the literature has shown an increase in the number of proposals of new optimization metaheuristics and their improvements through step alterations, local search procedures or hybridizations [2]. For a few well-known metaheuristics, the numerical evolution of the number of papers (journal and conferences) from the IEEE library is provided in [3]. The work of Hussain et al. in [2] presents a detailed distribution of types of research (basic, improvement, applications) focusing on all metaheuristics and, in [4], a timeline of the history of a set of representative techniques is provided.

This increase has been fueled by the need to efficiently find good solutions for difficult problems, especially for those where classical techniques fail to provide acceptable results within a reasonable amount of time and resources consumed.

All of these new optimizers (as well as the existing ones) follow the principles of the No Free Lunch Theorem (NFL), which states that if "an algorithm gains in performance on one class of problems it necessarily pays for on the remaining problems" [5]. In a simplistic view, this can be interpreted as meaning that one specific algorithm cannot outperform its counterparts on all problems but only on specific types or classes of problems, and it was shown that it is theoretically impossible to have a best general-purpose optimization strategy [6] (more details about NFL and its detailed analysis can be found in [7]). Consequently, researchers will probably never be satisfied with the existing metaheuristics [6], and this gives room for the development of new algorithms, improvements and strategies.

The oldest type of metaheuristic optimizers (from the 60s and 70s) is represented by genetic algorithms, based on the evolutionary processes; however, in their quest for better optimization of metaheuristics, researchers have turned to new sources of inspiration. Nowadays, the world of metaheuristics is large and covers ideas varying from the behavior of the very small, e.g., viruses and bacteria, to the mechanisms of galaxies. This multitude of algorithms can be somewhat overwhelming and, therefore, the objective of this paper is to identify the main directions of research, in terms of sources of inspiration, and to shed some light on the mechanisms used to generate powerful optimizers.

2. Classification and Categorization

When trying to identify the main classes of metaheuristics, various criteria can be applied. Examples include: search path, memory use, neighborhood exploration, number of solutions transferred from one iteration to the next and parallelization ability [8,9]. An extensive discussion related to the issue of classification and categorization for metaheuristics and the different schemes used can be found in [10].

In terms of categorizations, different aspects can be considered. For example, the type of candidate solutions is one of the most used criteria, and it splits the metaheuristic into: (i) individual-based, also known as single solutions, trajectory methods [1] or individualist algorithm [11] and (ii) population-based or collective algorithms [11]. In the individual-based group, a single solution is evolved. The main advantages of these methods consist in simplicity, lower computational costs and a lower number of function evaluation [11]. However, in their basic form, they can become trapped in the local optima and, since there is no information sharing, as there is just one solution, issues such as isolation of optima, deceptiveness and bias of the search space need to be dealt with [12]. Examples of algorithms that belong to this class are: Simulated Annealing (SA) [13]; Tabu Search (TS) [14]; Variable Neighborhood Search (VNH) [15]; Iterated Local Search (ILS) [16], proposed before 2006; Vortex Search (VS) [9], proposed after 2006. In the case of the population-based algorithms, multiple solutions are generated and improved. Distinctive from the individual-based algorithms, the population-based approaches allow some information exchange between the candidate solutions and thus can handle aspects that the individual-based approaches struggle with [12]. However, the cost of the improved performance is higher complexity and a larger number of function evaluations. The majority of metaheuristics are population-based and can themselves be classified into approaches that [17]: (i) increase the population diversity through the variation of control parameters; (ii) maintain population diversity through the replacement of individuals in the current population; (iii) include memory to store promising solutions; (iv) divide the population into subpopulations; (v) combine multiple methods, i.e., hybrid approaches.

When type of search is considered, metaheuristics can be local or global. The local search approaches tend to be more exploitative while the global algorithms are more explorative in nature [2]. On the other hand, in the latest years, the trend is to create hybrids that combine the two types of searches. The best-known examples of local search algorithms are TS, ILS and Greedy Randomized Adaptive Search Procedure (GRASP). Differential Evolution (DE), Particle Swarm Optimization (PSO) and Genetic Algorithms (GA) are examples of global search algorithms. Although the global search approaches can be hybridized to also include local procedures as a means to improve a previously proposed version, the literature presents algorithms that include this global–local search combination from the first version, e.g., the Bat Algorithm (BA) [18], the Shuffle Frog Leap Algorithm (SFL) [19] or Water Wave Optimization (WWO) [20].

When considering the source of inspiration, the majority of authors identify the metaheuristics as evolutionary and swarm intelligence techniques [11]. An extended categorization can be considered, such as the one in [21], where two more groups are included: stochastic and physical. In the latest years, various sources of inspiration have been used for metaheuristics. Therefore, this categorization must be extended to include all of the new methods. As a result, this work performs an extensive literature review of the proposed approaches covering the years 2006–2021. The review is organized around the biological classification of living things (kingdom-phylum-class), and its aim is to determine the main directions of research followed in the last 15 years and to identify new potential directions. The work [22] has a similar aim but focuses on all types of sources of inspirations for metaheuristics. Taking into account the variety of aspects that can be analyzed and the number of algorithms, this work only considers the metaheuristics with a biological base.

Regarding classification of metaheuristics, the work of Stegherr et al. [10] presents a seven-layer classification system. It focuses on structure (with criteria that include discontinuances, population, local search and memory), behavior (with criteria that include the strategy to create new solutions, groups and sub-populations), search (with criteria dealing with the intensification and diversification capabilities), algorithm (with criteria including the basic components incorporated), specific features (dealing with capabilities, i.e., use of adaptive parameters), evaluation (concerning the efficiency on various types of problems) and metaheuristics (which contains the specific algorithm that corresponds to the characteristics form the previous levels). If the first six levels are viewed from a framework perspective, the metaheuristic level deals with algorithms.

3. Source of Inspiration

In order to perform the current review, the main databases searched were: ScienceDirect (https://www.sciencedirect.com/, accessed on 6 August 2021), Web of Science (https://apps.webofknowledge.com/, accessed on 6 August 2021), Google Scholar (https://scholar.google.ro/, accessed on 6 August 2021), Springer Link (https://link.springer.com/, accessed on 6 August 2021) and IEEE Xplore Digital Library (https://ieeexplore.ieee.org/Xplore/home.jsp, accessed on 6 August 2021). The terms used in the search process were "metaheuristics", "nature-inspired optimizers" and "bio-inspired algorithms". The strategy to use both nature-inspired and bio-inspired terms is related to the fact that, in many works, there is not a clear distinction between the two and they are used to describe a variety of metaheuristics. Based on the identified sources, a drill down (study of the references used) and drill up approach (study of the papers citing a specific work) were applied in order to determine additional appropriate manuscripts. For the covered period, 283 algorithms were identified. Their distribution, based on the inspiration source, is presented in Figure 1.

By analyzing the identified categories, two main groups can be distinguished: biological and non-biological sources. The biological sources include animals, plants and humans, while the non-biological sources are represented by the chemical and physical laws of nature. Therefore, broadly speaking, the optimization metaheuristics can be grouped into: (i) biologically-inspired and (ii) nature-inspired.

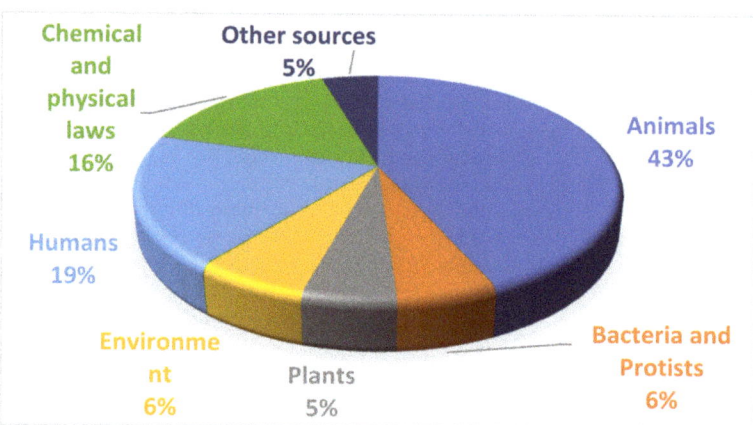

Figure 1. Distribution of newly proposed algorithms in the period 2006–2021, based on their inspiration source.

As it can be observed from Figure 1, the largest group of newly proposed metaheuristics in the considered period have animals as a source of inspiration, and this group is the main focus of the current work. Although humans, from a biological point of view, belong to the animal group, Chordata (vertebrates) phylum, in this work they were not included because they deserve a separate discussion, considering the unique ways of thinking, behaving and interacting with the environment.

In the latest years, various reviews have tried to shed light on the novel approaches that are constantly developed. Examples include: (i) a comprehensive list of algorithms and the steps of a few selected approaches [23]; (ii) a detailed discussion about the main research aspects specific to the field of nature-inspired metaheuristic optimizers [2]. In most works, researchers present the names of the best-known metaheuristics and a few details about the general ideas. This work aims to provide a series of details (such as: source code availability, improvement, applications, mechanisms for controlling the exploration-exploitation balance) in a systematic manner, for each algorithm considered.

3.1. Vertebrates

Most algorithms inspired by animals simulate two main general behaviors: (i) food search (foraging) and (ii) mating. For foraging, there are a number of theoretical models developed to predict the behavior of living things: the optimal foraging theory, the ideal free distribution, game theory and predator-prey models [24]. The theory of optimal foraging was developed to explain the dietary patterns and the resource use, and it states that the individuals using their energy more efficiently for finding food are favored by natural selection [25]. Foraging for food can be an individual activity (solitary foraging—where each individual searches for its food) or can be a social activity (social foraging—where foraging is a group behavior) [26]. The topics of social foraging include: (i) the mechanisms used by the members to find food; (ii) the manner in which the food locations are communicated to other members; (iii) the division of food between group members. The majority of foraging inspired optimization algorithms focus on the first two topics [26]. A taxonomy of foraging inspired algorithms is proposed in [27], where three main categories are identified: vertebrates (with backbone), invertebrates (without backbone) and non-neuronal (organisms that do not possess a central nervous system or brain).

Concerning the mating behavior, different theoretical models that simulate the mating mechanisms of specific species exist. For example, in birds, different strategies are used to display the quality of genes to the potential mates by showing the main physical features: color, shape of specific body parts, etc. In terms of partner combinations, five strategies are encountered: monogamy, polygyny, polyandry, parthenogenesis and promiscuity [28]. These mechanisms are included, in different forms, in the metaheuristic optimizers with the objective of improving diversity and thus increasing performance.

From the total of algorithms inspired from animals, the ones based on vertebrates represent 61.6%, with the most represented sub-groups being birds (Section 3.1.1) and mammals (Section 3.1.2).

3.1.1. Birds

Mating Behavior

One of the best-known algorithms inspired from bird behavior is Cuckoo Search (CS) [29]. It simulates the brood parasitic behavior of some species of cuckoo and, in order to search for new solutions, it uses Levy flight random walk (mutation based on the best solution found so far) and biases/selective random walk (crossover between a current solution and its mutation) [30]. The Levy flight is a random process from the non-Gaussian class, a step which is based on the Levy distribution [31]. The steps of the CS algorithm express three idealized rules: (i) each cuckoo lays an egg and places it into a randomly chosen nest; (ii) the nests with high-quality eggs will be further used in the next generations; (iii) there is a fixed number of nests and there is a probability that the host will discover the foreign egg [32]. The main disadvantage of this algorithm is the fixed value of the scaling factor (that controls the step size) [30] and, in order to improve its performance, various strategies have been applied. A list of different modifications of CS and its applications can be found in [32,33]. Inspired by the same cuckoo breeding behavior, the Cuckoo Optimization Algorithm (COA) was proposed in [34]. When comparing COA and CS, it can be observed that COA is more complex, in the sense that it contains an additional behavioral aspect represented by the immigration process. Also, it uses the k-means clustering algorithm to identify the group that a cuckoo belongs to.

Bird Mating Optimizer (BMO) [28] uses the birds mating process as a framework. Throughout generations, the birds (the population of solutions) apply a probabilistic method to improve the quality of their offspring. The population is divided into males and females. The males can be monogamous, polygamous or promiscuous. On the other hand, the females can be parthenogenetic and polyandrous. In BMO, five species are simulated, and each one has its specific updating pattern.

Developed to adjust the parameters of adaptive neuro-fuzzy inference systems, the Satin Bowerbird Optimizer (SBO) [35] simulates the mating behavior of bowerbirds (a close relative species of the birds-of-paradise). In each iteration, a target individual is determined through roulette wheel selection. The other individuals try to follow it, i.e., they change their position accordingly, and try to improve their strategies by mutation.

Food Search

The flight of eagles (as does, in fact, the flight behavior of many animals and insects) has the typical characteristics of the Levy flight [36]. Based on this idea, the Eagle Strategy (ES) was proposed in [36]. It simulates an idealized two stage strategy (find and chase) [37]. The find step represents the exploration phase realized by the Levy walk and the chase step is an exploitation phase where an intensive local search is performed by the Firefly Algorithm (FA). ES represents a strategy and not an algorithm, and its authors indicate that different algorithms can be used at different stages of the iterations [38]. For example, in [38–40], Differential Evolution (DE) [41] performs the local search procedure that was initially solved by FA.

Chicken Swarm Optimization (CSO) is inspired by swarm behavior [42]. It mimics the flock and foraging behavior of chickens and is based on four simplified rules: (i) the swarm is comprised of several groups and each group has a dominant rooster, some hens and chicks; (ii) the selection of individuals representing each type of bird is based on the fitness value; (iii) the dominance and hen–chick relationship remains unchanged for several iterations; (iv) the search for food is performed around the dominant entity. CSO is a multi-swarm algorithm and its performance analysis showed that it can be efficiently applied to solve benchmarks and real-world problems.

The Crow Search Algorithm (CSA) [43] simulates the behavior of crows when it comes to food, i.e., storing excess food and thievery. Its main principles are: (i) crows live in flocks; (ii) each crow memorizes their hiding places; (iii) crows follow each other to steal food; (iv) crows try to protect their stashes. The algorithm includes a memory of good solutions and, since the movement of crows is performed regardless of whether a newly generated position is worse than the current one, distinctively from many metaheuristics, CSA is not a greedy algorithm.

Simulating the auditory-based hunting mechanism employed by owls, the Owl Search Algorithm (OSA) [44] assumes that the fitness value of each individual is correlated to the intensity of the received sound and that the search space has one global optimum.

The hummingbird's optimization algorithm (HOA) [45] focuses on the foraging processes of hummingbirds. It includes a self-searching phase, based on the individual accumulated experience (using a Levy flight mechanism), and a guide-searching phase that includes information from dominant individuals.

Simulating the social roosting and foraging behavior of ravens, in [26], the Raven Roosting Optimization (RRO) is proposed. The algorithm includes four main components: (i) the perception capability of each individual to find food; (ii) a memory related to the position of previous foraging locations; (iii) transmitting and receiving information about food locations; (iv) probabilistic movement when searching for new resources.

Based on the cooperative hunting behavior of Harris hawks, the Harris Hawks Optimization (HHO) [46] algorithm simulates a series of aspects such as prey exploration, surprise pounce, and attack strategies. The algorithm complexity is O(population_size $\times$ (iterations + iterations $\times$ dimensionality + 1)) and, in exploring or exploiting the search space, a series of strategies such as diversification mechanism, progressive selection scheme and adaptive and time-varying parameters were used.

Aquila is a very successful bird of prey located in the Northern hemisphere that represents the source of inspiration for the Aquila Optimizer (AO) [47]. The model that the AO is based on simulates four hunting methods: (i) high soar with a vertical stoop (corresponding to an expanded exploration step); (ii) contour flight with glide attack (narrowed exploration step); (iii) low flight and slow descent (expanded exploitation step); (iv) walking and prey grabbing (narrowed exploitation). The AO computational complexity is O(solution_number $\times$ (iterations x dimensionality + 1)).

The hunting mechanisms of golden eagles (spiral trajectory for searching food and straight path when attacking, a tendency of cruising at the beginning of the search and attacking at the end, capability to easily change between cruising and attaching) is simulated in the Golden Eagle Optimizer (GEO) [48]. The attack phase corresponds to exploitation and cruising to exploration. To extend applicability, two variants were proposed: the GEO version (single objective) and the MOGEO (multi-objective) version. The GEO computational complexity is O(population_size $\times$ dimensionality $\times$ iterations) and that for MOGEO is O(population_size $\times$ dimensionality $\times$ iterations $\times$ objectives $\times$ archive).

Movement

Based on the characteristics of geese's flight and the general PSO model, a geese-inspired hybrid PSO (Geese-PSO) was proposed in [49]. Although it does not use a completely novel metaphor and the algorithm is a hybridization, it is considered in this work because, to the authors' knowledge, the principle of following the particle ahead and the application of a unidirectional flow of information was not used prior to the proposal of the Geese-PSO approach. In the same direction of research, Migrating Bird

Optimization (MBO) [50] is inspired by the "V" flight formation of migrating birds. MBO is a neighborhood search approach where each solution is improved based on its neighbors.

The swarming behavior of passenger pigeons (the common name given to Blue Pigeons, Merne Rouck Pigeons, wandering long tailed Doves and Wood pigeons) represents the inspiration of the Pigeon Optimization Algorithm (POA) [51]. Pigeons have a specific behavior that can be simplified into a series of rules: (i) in order to enhance the search and reduce the probability of being prey for other animals, flight is performed in flocks; (ii) different flocks have their own solution for movement, which influences the shape of the group; (iii) there is communication between birds through "keck" and "tweet" calls; (iv) the behavior of other pigeons can be imitated; (v) the pigeons responding to the calls are the closest to the source of the call; (vi) in order to have a better probability of survival, each pigeon must lead. Another algorithm inspired by pigeons is Pigeon Inspired Optimization (PIO) [52]. However, in PIO the main idea is to simulate the homing behavior and the mechanisms used by a pigeon to move from a point A to a point B, i.e., orientation through magnetoreception and the sun, and recall of known landmarks close to the destination.

The migratory and attack behavior of seagulls is imitated in the Seagull Optimization Algorithm (SeOA) [53]. Several simplified rules are considered: (i) the migration is performed in groups; (ii) all the individuals travel towards the one with the best fitness; (iii) the attack follows a spiral-like movement. In addition, during the migration process, a mechanism for collision avoidance is included. The Sooty Tern Optimization Algorithm (STOA) [54] has a similar source of inspiration as the SeOA, but based on Sooty Tern seabirds. In this case, the migration behavior represents the exploration phase and the attacking behavior the exploitation one. The complexity of the STOA is O(problem_dimension × iteration × objective_number × population_size × objective_function) and the space complexity is O(objective_number × population_size).

In their fight to survive the harsh conditions of the polar regions, the emperor penguins use a specific strategy of huddling. This represents the main source of inspiration for the Emperor Penguin Optimizer (EPO) [55], where operations such as huddle boundary determination, temperature computation, distance determination and identification of the effective move are mathematically modeled and simulated in order to perform the optimization. The time complexity of EPO is O(k × individual_length × iterations × dimensionality × population_size), where the algorithm termination criteria requires O(k) time. The same mechanisms are also simulated in Emperor Penguins Colony (EPC) [56]. The main difference between the EPO and EPC consists in the manner in which the movement is realized, i.e., in EPC the individuals perform a spiral-like movement.

The foraging and navigation behaviors of African vultures is modeled in the African Vultures Optimization Algorithm (AVOA) [57], where multiple mechanisms to improve the exploration–exploitation balance were proposed: the use of a coefficient vector to change between these phases, use of phase-shift to precent premature convergence and local optimum escape and inclusion of Levy Flight. The AVOA computational complexity is based on initialization, fitness evaluation and vulture update and is O(iteration × population_size) + O(iteration × population_size × dimensionality).

Table 1 summarizes the algorithms briefly presented in this section and shows a series of examples for improvements and applications. In Table 1, where a link to the source code exists, if not specifically indicated, the implementation is provided by a third party. In the application column, due to the fact that it is common to test the performance of a newly proposed algorithm on a set of problems with known characteristics, the standard benchmarks were not specified.

Table 1. Improvements and applications for bird-inspired metaheuristics (alphabetically sorted).

Algorithm	Source Code	Modifications and Improvements	Applications
Bird Mating Optimizer (BMO) [28]		– adaptive population size [58] – hybridization with Differential Evolution [59,60], Teaching Learning-based Optimization [61]	– image segmentation [59] – optimal expansion planning [58] – structural damage [62] – structural design [62] – engineering design [60] – electrochemical discharging machining [63] – photovoltaic modules [64]
Chicken Swarm Optimization (CSO) [42]	(MATLAB—author source) https://www.mathworks.com/matlabcentral/fileexchange/48204-cso, accessed on 7 January 2020	– multi-objective mechanism [65] – inclusion of penalty [66] – learning mechanism [67]	– crude oil price prediction (in combination with ANNs) [68] – projection pursuit evaluation [69] – error control [70]
Crow Search Algorithm (CSA) [43]	(MATLAB—author source) https://www.mathworks.com/matlabcentral/fileexchange/57867-crow-search-algorithm-for-constrained-optimization, accessed on 19 June 2021	– constraint handling [71] – inclusion of Levy flight [72] – multi-objective adaptation [73] – chaotic systems [73,74] – parameter control through diversity population information [75]	– structural design [71] – Parkinson diagnosis [76] – energy problems [72] – image processing [77]
Cuckoo Search (CS) [29]	(MATLAB—author source) https://www.mathworks.com/matlabcentral/fileexchange/29809-cuckoo-search-cs-algorithm, accessed on 4 February 2019	– hybridization with PSO [78] – Shuffle Frog Optimization Algorithm [79] – improvement of specific steps [32] – varying the control parameters [30,80]	– linear antenna array optimization [81] – operating schedule of battery, thermal energy storage, and heat source in a building energy system [82] – power load dispatch [83] – synchronization of bilateral teleoperation systems [84] – 0-1 knapsack problem [79]
Cuckoo Optimization Algorithm (COA) [34]	(MATLAB—author source) https://www.mathworks.com/matlabcentral/fileexchange/35635-cuckoo-optimization-algorithm, accessed on 4 February 2019	– hybridization with HS [85] – adaptation to discrete spaces [86]	– water allocation and crop planning [87] – load frequency control [85] – bilateral teleoperation system [84] – inverse kinematic problem [88] – PID design [34]
Emperor Penguin Optimizer (EPO) [55]		– binary version [89] – multi-objective variant [90,91] – hybridization with Salp Swarm algorithm [92], Social Engineering Optimization [93]	– ranking of cloud service providers [90] – color image segmentation [94] – medical data classification (in combination with Support Vector Machines) [93]
Emperor Penguins Colony (EPC) [56]		– introduction of mutation and crossover operators [95]	– inventory control problem [96] – neuro-fuzzy system [97]

Table 1. *Cont.*

Algorithm	Source Code	Modifications and Improvements	Applications
Harris Hawks Optimization (HHO) [46]	(MATLAB—author source) https://github.com/aliasgharheidaricom/Harris-Hawks-Optimization-Algorithm-and-Applications, accessed on 10 December 2020	– use of chaos [98] – binary version [99] – hybridization with Differential Evolution [100], Salp Swarm Algorithm [99]	– parameter identification photovoltaic cells [98] – productivity prediction of solar still (in combination with Artificial Neural Networks) [101]
Migrating Bird Optimization (MBO) [50]	(Java) http://mbo.dogus.edu.tr, accessed on 15 November 2020	– hybridization with Harmony Search [102], Differential Evolution [103] – new mechanism for leader selection [104], neighborhood search strategy [105], age mechanism [106], crossover mechanism [107], Glover generator in the initialization phase [108] – use of parallel micro-swarms [106]	– scheduling [102,104–106,108] – manufacturing [107]
Owl Search Algorithm (OSA) [109]		– inclusion of opposition-based learning [110], chaos [111] – binary version [112]	– image segmentation [110] – bilateral negotiations [111] – feature selection [112]
Pigeon Inspired Optimization (PIO) [52]	(MATLAB) http://read.pudn.com/downloads713/sourcecode/math/2859919/Code%20of%20Basic%20PIO/Code%20of%20Basic%20PIO.txt__htm, accessed on 15 December 2020	– discretization [113] – inclusion of the heterogeneity principle [114] – use of Cauchy distribution [115], probability factors to adapt the velocity [116] – multi-objective [117] – predator-prey concept [118]	– travelling salesman problem [113] – prediction of bulk commodity futures prices (in combination with extreme learning machine) [119] – automatic carrier landing [115,116] – current motor parameter design [117]
Raven Roosting Optimization (RRO) [26]		– subpopulations with different behavior [120] – hybridization with CSO [121]	– task scheduling [121]
Satin Bowerbird Optimizer (SBO) [35]		– encoding based on complex values [122]	– solid oxide fuel cells [123] – neuro-fuzzy inference systems [35]
Seagull Optimization Algorithm (SeOA) [53]	(Matlab—author code) https://www.mathworks.com/matlabcentral/fileexchange/75180-seagull-optimization-algorithm-soa, accessed on 12 February 2021	– multi-objective [124] – hybridization with Whale Optimization [125], Cuckoo Search [126], Thermal Exchange Optimization [127]	– feature selection [127]
Sooty Tern Optimization Algorithm (STOA) [54]	(Matlab—author code) https://jp.mathworks.com/matlabcentral/fileexchange/76667-sooty-tern-optimization-algorithm-stoa, accessed on 20 June 2021		– model predictive control [128] – industrial engineering problems [54]

3.1.2. Mammals

Food search

Based on the principles of echolocation used by bats to find food, the Bat Algorithm (BA) [18] employs several idealized rules: (i) echolocation is used for distance sensing and prey identification; (ii) the flying pattern is random, with characteristics such as velocity, pulse rate and loudness; (iii) the variation of loudness is assumed to move from a large value to a minimum (constant). The role of the pulse rate and loudness is to balance exploration and exploitation [129]. Another algorithm simulating bats is the Directed Artificial Bat Algorithm (DABA) [130]. DABA considers the individual flight of bats with no interaction between individuals, while in BA, the bat behavior is similar to the PSO particles. Although, compared to BA, the DABA model is closer to the natural behavior, in terms of optimization performance, BA is better [131]. On the other hand, the Dynamic Virtual Bats Algorithm (DVBA) [131] has a population comprised of only two individuals, i.e., explorer and exploiter bats, that dynamically exchange their roles based on their locations.

The echolocation mechanism is not specific to bats; other animals also using it to navigate and to find food, e.g., Dolphin Echolocation [132]. The abbreviation given by its authors is DE; however, so as not to confuse it with Differential Evolution, Dolphin Echolocation will be denoted by DEO in this work. It mimics the manner in which sound, in the form of clicks, is used to track and aim objects. Distinctively from the bat sonar system, which has a short range of 3–4 m, the range of the dolphin sonar varies from a few tens of meters to over a hundred meters. This aspect and the differences in the environmental characteristics lead to the development of totally different sonar systems, and a direct comparison between the two may be difficult.

In their search for food, sperm whales go as deep as 2000–3000 m and can stay underwater without breathing for about 90 min [133]. They are social animals, travel in groups and only the weaker specimens are attacked by predators such as orcas. This behavior was modeled in the Sperm Whale Algorithm (SWA) [133], where the population is divided into subgroups. In each cycle of breathing and feeding, the individual experiences two opposite poles (surface and bottom of the sea); however, because computing the mirror place is expensive and its influence on the search process is limited, it is applied only to the worst solutions. In order to simulate the hunting behavior of humpback whales, i.e., the bubble net feeding method, the Whale Optimization Algorithm (WOA) [134] searches for prey (the exploration phase) and then uses the shrinking encircling mechanism and the spiral updating position (the exploitation phase). A detailed review covering the multiple aspects of WOA is presented in [135].

The Grey Wolf Optimizer (GWO) [136] models a strict social dominant hierarchy and the group hunting mechanisms—tracking, chasing, approaching and attacking the prey—of grey wolfs (Canis lupus). The complexity of GWO is O(problem_dimension × iteration × objective_number × population_size × objective_function) [54]. Similar to other bio-inspired approaches, the GWO suffers from premature convergence. The prey weight and astrophysics concepts were applied in the Astrophysics Inspired Grey Wolf Optimizer (AGWO) [137] to simultaneously improve exploration and exploitation. Although wolves live in packs and communicate over long distances by howling, they have developed unique semi-cooperative characteristics [138]. By focusing on the independent hunting ability, as opposed to the GWO, which uses a single leader to direct the search in a cooperative manner, the Wolf Search Algorithm (WSA) [138] functions with multiple leaders swarming from multiple directions towards the optimal solution.

Spider monkeys are specific to South America and their behavior falls in the category of fission–fusion social structure [139], i.e., based on the scarcity or availability of food, they split from large to smaller groups and vice versa. The algorithm that simulates this structure is called Spider Monkey Optimization (SMO) [139]. It consists of six phases and, unlike the natural system, the position of leader (local or global) is not fixed, depending instead on its ability to search for food. In addition, the optimization procedure does not include the communication tactics specific to spider monkeys. Distinctively, the individual intelligence of chimps used for group hunting is modelled into the Chimp Optimization Algorithm (ChOA), where four types of hunting are included: driving, chasing, blocking and attacking [140]. Another type of ape is represented by gorillas and, in the Artificial Gorilla Troops Optimizer (GTO), their collective life is mathematically modeled to include exploration–exploitation mechanisms [141].

Spotted hyenas have a behavior similar to that of wolves and whales, which uses collective behavior to encircle the prey and attack. Their model is used in the Spotted Hyena Optimizer (SHO) [142], which saves the best-so-far solution, simulates the encircling prey through a circle-shaped neighborhood that can be extended to higher dimensions and controls the exploration–exploitation balance through control parameters. The time complexiy of SHO is O(problem_dimension $\times$ G $\times$ iteration $\times$ objective_number $\times$ population_size $\times$ objective_function), where the time to define the groups of individuals is O(G).

In the Squirrel Search Algorithm (SSA) [109], the gliding behavior of flying squirrels when exploring different areas of a forest in search for food is simulated by considering some simplifications of the natural mechanisms: (i) a squirrel is assumed to be on one tree; (ii) in the forest there are only three types of trees: normal, oak and hickory; (iii) the region under consideration contains three oaks and one hickory tree. It is considered that the squirrel with the best fitness is positioned on a hickory tree and the next three individuals with the best fitness are on oak trees. The other individuals in the population move towards the oak or the hickory, depending on their daily energy requirements. In SSA, the seasonal changes are modeled through control parameters and influence the behavior of the individuals in the population.

Social Behavior

The Lion's Algorithm (LA) [143] is based on the social behavior of lions. It simulates the process of pride forming through mating, removing weak cubs, territorial defense and takeover. The population is formed of males and females and the cub population is subjected to gender grouping (through the application of k-means clustering). LA is not the only approach inspired by lions; the Lion Optimization Algorithm (LOA) [144] is also an example. Distinctively from LA, LOA includes the hunting and migration mechanisms and the mating process is based on differentiation rather than on crossover and mutation. Another lion-inspired approach is the Lion Pride Optimization (LPOA) [145].

Similar to honey bees or ant colonies, blind naked mole rats (a species specific to Africa) have a complex social behavior: (i) they live in large colonies; (ii) a queen and a reduced number of males are responsible for offspring generation; (iii) there are individuals specialized in food search and domestic activities, i.e., taking care of the nest and of the young and in protection against invaders [146]. These mechanisms, in a simplified form, are simulated in the Blind Naked Mole Rats (BNMR) algorithm [146].

Elephants are the largest walking mammals and their successful survival is influenced, among other things, by their social and behavioral structures. The adult males solitarily roam into the wild, they do not commit to any family and can potentially mate over thirty times a year, while the female elephants form matriarchal societies that allow better protection and safe rearing of young calves. The Elephant Search Algorithm (ESA) [147] and Elephant Herding Optimization [148,149] simulate these mechanisms and perform the search.

Table 2 summarizes the algorithms briefly presented in this section and shows a series of examples for improvements and applications. The same structure and idea as in Table 1 are applied.

Table 2. Improvements and applications for some mammal-inspired metaheuristics (alphabetically sorted).

Algorithm	Source Code	Modifications and Improvements	Applications
Bat Algorithm (BA) [18]	(Python) https://github.com/buma/BatAlgorithm, accessed on 20 December 2019	- discrete version [150] - introducing directional echolocation [151] - multi-population, chaotic sequences [129] - inclusion of Doppler effect [152] - hybridization with Invasive Weed Optimization [153], Differential Evolution [154], - binary version [155]	- drugs distribution problem [150] - forecasting motion of floating platforms (in combination with Support Vector Machines -SVM-) [156] - flood susceptibility assessment (in combination with adaptive network-based fuzzy inference system) [157] - job shop scheduling [158] - travelling salesman problem [159] - battery energy storage [160] - constraint [161] and structural optimization [162]
Blind Naked Mole Rats (BNMR) [146]			- data clustering [163]
Chimp optimization algorithm (ChOA), [140]	(MATLAB—author source) https://www.mathworks.com/matlabcentral/fileexchange/76763, accessed on 20 August 2021	- use of sine-cosine functions to update the search process of ChOA [164]	- combination with ANNs for underwater acoustical classification [165] - high level synthesis of data paths in digital filters [164]
Directed Artificial Bat Algorithm (DABA) [130]			- travelling salesman problem [130]
Dolphin Echolocation (DEO) [132]		- exploration improvement [166]	- plastic analysis of moment frames [167] - design of steel frame structure [168] - reactive power dispatch [169]
Dynamic Virtual Bats Algorithm (DVBA) [131]		- parameter setting [170]	
Elephant Herding Optimization (EHO) [148,149]	(MATLAB—author source) http://www.mathworks.com/matlabcentral/fileexchange/53486, accessed on 17 January 2020	- alpha tuning, cultural-based algorithm, biased initialization [171] - hybridization with Cultural Algorithm [172] - multi-objective and discrete [173] - introduction of chaotic maps [174]	- structural design [171] - batch fermented for penicillin production [171] - network detection intrusion (in combination with SVM) [175] - image processing [176] - SVM parameter tuning [177]
Elephant Search Algorithm (ESA) [147]		- chromosome representation, elephant deep search, and baby elephant birth [178]	- data clustering [179,180] - snack food distribution [178] - travelling salesman problem [181]
Grey Wolf Optimizer (GWO) [136]	(MATLAB—author source) http://www.alimirjalili.com/Projects.html, accessed on 6 June 2021	- two phase mutation [182] - introduction of random walk [183] - introduction of cellular topological structure [184] - modification of parameter behavior [185] - binary [186] - multi-objective [187]	- fluid dynamic problems [188] - power systems [185,189] - combination with ANNs [190] - structural engineering [137] - maximum power tracking [191]

Table 2. Cont.

Algorithm	Source Code	Modifications and Improvements	Applications
Lion's Algorithm (LA) [143]		- fertility evaluation, a modified crossover operator and gender clustering [192] - hybridization with a heuristic specific to job shop scheduling [193]	- system identification [192] - rescheduling based congestion management [194] - job shop scheduling [193]
Lion Optimization Algorithm (LOA) [144]			- clustering mixed data [195]
Lion Pride Optimization Algorithm (LPOA) [145]			- double layer barrel vault structures [196] - structural design [145]
Sperm Whale Algorithm (SWA) [133]			- natural gas production optimization [133] - ANN parameter identification [197]
Spider Monkey Optimization (SMO) [139]	(MATLAB, C++, Python–author sources) http://smo.scrs.in, accessed on 10 January 2021	- inclusion of chaos [198], levy flight [199], quadratic approximation [200], age principle for population [201] - hybridization with Limacon curve [202], Nelder–Mead [203] - binary [204]	- load frequency control [205] - irrigation [206] - diabetes classification [207] - capacitor optimal placement [202] - antenna array [204]
Spotted Hyena Optimizer (SHO) [142]		- multi-objective [208]	- structural design [209] - neural network training [210] - airfoil design [211]
Squirrel Search Algorithm (SSA) [109]			- heat flow [109]
Whale Optimization Algorithm (WOA) [134]	(MATLAB—author source) http://www.alimirjalili.com/Projects.html, accessed on 6 June 2021	- hybridization with Nawaz–Enscore–Ham [212], Simulated Annealing [213], Differential Evolution [214] - mechanism of exploration phase [215] - introduction of chaotic maps [216]	- power system [217] - optimal control [218] - structural engineering [215] - drug toxicity [219] - parameter optimization for Elman Networks applied to polymerization process [216] - robot path planning [220] - handwritten binarization [221]

3.1.3. Other Vertebrates

This category includes other sources of inspiration from the vertebrate group that do not belong to the bird and mammal classes.

The SailFish Optimizer (SFO) [222] is inspired by the group hunting of sailfish (*Istiophorus platypterus*), one of the fastest fish in the ocean. This mechanism of alternating attacks on schools of sardines is modeled through the use of energy-based approaches, where, at the beginning of the hunt, both the predator and the prey are energetic and not injured; however, as the hunt continues, the power of the sailfish will decrease and the sardines will become tired and have reduced awareness.

The food catching behavior of Agama lizards is modelled in the Artificial Lizard Search Optimization (ALSO) [223]. The algorithm focuses on new discoveries regarding the mechanisms for movement control through the tail during prey hunting.

The manner in which chameleons catch prey using their long and sticky tongue represents the basis for the Chameleon Swarm Algorithm [224]. The notation given by the authors of this algorithm is CSA, however, since the same notation is used to represent the Crow Search Algorithm, in this work, the Chameleon Swarm Algorithm will be indicated by the ChSA notation. The ChSA follows three main strategies for catching prey: tracking (modelled as a position update step), eye pursuing (modeled as position update in accordance with the position of the prey) and attacking (based on tongue velocity). Distinctively from the majority of metaheuristics, which tend to have less than three parameters, ChSA has five parameters that help in controlling the exploration–exploitation balance.

3.1.4. General

Unlike the other algorithms mentioned in this work that have a source of inspiration represented by a single animal, in the case of the general class, the metaheuristics are based on a general aspect that can be specific to multiple animals or types of animals. Examples proposed prior to 2008 include algorithms such as Genetic Algorithms (where the genetic principles of mutation and crossover are applicable to all species) and Extremal Optimization [225], based on the Bak–Sneppen mechanism, a model of co-evolution between interacting species which reproduces nontrivial features of paleontological data.

Inspired from the encircling mechanisms used by group hunters such as lions, wolves and dolphins, the Hunting Search (HuS) [226] simulates the cooperation of members to catch food. As a perfect correlation between nature and an optimization process cannot be achieved, a set of differences from the real world are taken into account: (i) in the majority of cases, the location of the optimum of a problem is not known, while, in the real world, the hunters can see the prey or sense its presence; (ii) the optimum is set, however, in the real world, the prey dynamically changes its position. Unlike the DEO and GWO, which emulate the specific hunting approaches used by dolphins and wolves, HuS is focused on the cooperation aspect and the repositioning during the hunt. Other approaches which simulate the food searching mechanisms include: the Backtracking Search Algorithm Optimization (BSA) [227], Optimal Foraging Algorithm (OFA) [25], Fish Electrolocation Optimization (FEO) [228] and Marine Predators Algorithm (MPA) [229]. BSA is based on the return of a living creature to previously found fruitful areas. At its core, it is an evolutionary approach that, although it has a very similar structure to the other EAs, differs as follows: (i) mutation is applied to a single individual; (ii) there is a more complex crossover strategy compared with DE; (iii) it is a dual population algorithm; (iv) it has boundary control mechanisms. Distinctively from the BSA, the OFA algorithm is based on the Optimal Foraging Theory developed to explain the dietary patterns of animals. In the OFA, the animal foraging is an individual and its position represents a solution. Its time complexity is $O(group\_size \times dimensionality \times iterations)$ and its space complexity is $O(group\_size \times dimensionality \times (iterations + 1))$. The FEO simulates the active and passive electrolocation mechanisms used by sharks and "elephant nose fishes" to find prey. A series of electric waves are generated and reflected back to the fish after hitting the surrounding objects, which creates an electric image that is then analyzed. In the case of the MPA, the different strategies used for finding food and the interaction between predator and prey are modeled in different scenarios through Brownian and Levy strategies. The MPA algorithm complexity is O(iterations × (agent_number × dimensionality + Cost_function_evaluation × agent_number)).

Another aspect specific to all species in their quest to survive is represented by the competition for food, resources or mates. Two metaheuristic optimizers based on competition were identified: Competition over Resources (COR) [230] and the Competitive Optimization Algorithm (COOA) [231]. The COR algorithm mimics the competition for food of wild animals. The groups with the best approach to storing food have improved scores while the worst performance groups are starving and, after a few generations, die and are removed from the population. In the COOA approach, the competition is simulated by the Imperialist Competitive Algorithm [232] and the groups are represented by the populations of various metaheuristics.

Migration behavior is encountered in all major animal groups. Among the first bio-inspired metaheuristics that contain elements specific to migration is the Biogeography-based Optimization (BBO) [233]. However, the BBO imitates a much larger phenomenon—island biogeography—that includes both migration and mutation [234]. Another algorithm that has the migration principle at its core is the Migrating Birds Optimization [50]. As it simulates the features of the "V" flight of birds, the MBO was included in the bird inspired metaheuristic section. The Animals Migration Optimization (AMO) [235] simulates the animal migration model proposed by ecologists and uses two idealized assumptions: (i) the leader animal will survive to the next generation; (ii) the number of animals in the population is fixed. The algorithm has two phases: the migration process (where the individuals respect three rules: move in the same direction as the neighbors, remain close to the neighbors and avoid collision with neighbors) and population update (where some individuals leave the group and others join it).

Table 3 summarizes the algorithms briefly presented in this section and shows a series of examples for improvements and applications. The same structure and idea as in Tables 1 and 2 are applied.

Table 3. Improvements and applications for metaheuristics inspired from general behavior (alphabetically sorted).

Algorithm	Source Code	Modifications and Improvements	Applications
Animals Migration Optimization (AMO) [235]		- inclusion of an interactive learning behavior [236]; - use of Opposition Based Learning [237] - hybridization with Association Rule Mining [238] - population updating step [239]	- bridge reinforcement [240] - multilevel image thresholding [241] - data mining [238] - data clustering analysis [239]
Backtracking Search Algorithm Optimization (BSA) [227]	(MATLAB—author source) https://www.mathworks.com/matlabcentral/fileexchange/44842, accessed on 10 December 2019	- multiple mutation strategies [242] - discrete variant [243] - use of Opposition Based Learning [244] - hybrid mutation and crossover strategy [243] - hybridization with TLBO [245] - constraint handling mechanisms [246]	- casting heat treatment charge plan problem [243] - electricity price forecasting (in combination with adaptive network-based fuzzy inference system) [247] - parameter estimation for frequency-modulated sound waves [227] - engineering design problems [227,246]
Biogeography based Optimization (BBO) [233]		- inclusion of re-sampling [248] - inclusion of mutation strategy [234] - inclusion of chaos maps [249]	- flood susceptibility assessment (in combination with adaptive network-based fuzzy inference system) [157] - soil consolidation (in combination with artificial neural networks) [250] - power fuel cells [234]
Competition over Resources (COR) [230]	(MATLAB—author source) http://freesourcecode.net/matlabprojects/71991/competition-over-resources--a-new-optimization-algorithm-based-on-animals-behavioral-ecology-in-matlab, accessed on 25 June 2020		- building lighting system [251] - magnetic actuators [252]
Hunting Search (HuS) [226]		- hybridization with Harmony Search [253]	- artificial neural network training [253] - steel cellular beams [254]
Marine Predators Algorithm [229]	(MATLAB—author source) au.mathworks.com/matlabcentral/fileexchange/74578, accessed on 08 August 2021	- hybridization with Moth Flame Optimization [255], Teaching-learning based optimization [256] - binary version with V-shaped and S-shaped transfer functions [257]	- parameter extraction of photovoltaic models [258] - multi-level thresholding for image segmentation [255]
Optimal Foraging Algorithm (OFA) [25]	(MATLAB-author source) https://www.mathworks.com/matlabcentral/fileexchange/62593, accessed on 24 April 2020	- chaos [259] - constraint handling mechanisms [259]	- drilling path optimization [260] - SVM Parameter optimization [261] - white blood cell segmentation [259]

3.2. Invertebrates

From the total of algorithms inspired from animals, the ones based on invertebrates represent 38.4%, with the main sub-group indicated by insects (Section 3.2.1). As the number of algorithms inspired from other invertebrate sub-groups were small, the ones not belonging to insects were included in a separate section (Section 3.2.2).

3.2.1. Insects

Although the majority of insects are solitary, several types of insects are organized in colonies or swarms [262]. As insect swarms have several desirable attributes, a high percentage of insect-inspired metaheuristic optimizers belong to the swarm intelligence class.

Swarm intelligence has two main key components: self-organization (global response through interactions among low level components that do not have a central authority) and division of labor (the tasks are performed by specialized individuals) [139,263]. It follows three basic principles: (i) separation (static collision avoidance); (ii) alignment (velocity matching); (iii) cohesion (the tendency of individuals to go towards the center of the mass of the swarm) [264].

While, in the classic swarm approaches, the individuals considered are unisex and perform virtually the same behavior, thus wasting the possibility of adding new operators [265], in the newer bio-inspired metaheuristics researchers began to incorporate different types of individuals in the population(s), and the results obtained show an improvement of several characteristics, such as search ability and population diversity. However, the use of different operators leads to an increase in complexity and, until now, theoretical studies that can explain the influence of these operators and the context in which they are recommended have been very scarce.

Hymenoptera

This order includes some of the best-known social insects: wasps, bees and ants. The main characteristics of these insects are: (i) the presence of a pair of membranous wings; (ii) antennae longer than the head; (iii) complete metamorphosis.

- Bees

The social bees show all the characteristics of eusociality: generation overlapping, separation into fertile and infertile groups, labor division and brood care. In addition, the beehive can be considered as a self-organizing system with multiple agents [266].

In a comprehensive review regarding the algorithms inspired by honey bees, the authors identified five main characteristics that were modeled: (i) mating; (ii) foraging and communication; (iii) swarming; (iv) spatial memory and navigation; (v) division of labor [267]. However, [268] considers that, alongside mating and foraging, the third class is represented by nest-site selection process, and thus proposed the Bee Nest-Site Selection Scheme (BNSS)—a framework for designing optimization algorithms.

In addition to the algorithms presented in [267], other approaches that simulate bee behavior are: Bumblebees (B) [269], Bee Colony Inspired Algorithm (BCiA) [266] and Bumble Bee Mating Optimization (BBMO) [270]. The B algorithm is based on a simplified model of the evolution of bumblebees and can be regarded as a loose implementation of the concepts of the evolutionary model proposed by [271]. On the other hand, the BCiA focuses on the foraging behavior and the BBMO simulates the mating process.

- *Ants*

  Among the first algorithms that simulate ant behavior is the Ant System [272]. However, the best-known approach is the Ant Colony Optimization (ACO), used to find the path of minimum length in a graph. To the authors' knowledge, the only other approach simulating ant behavior, which is not based on the ACO, is Termite-hill [273]. It is a swarm-based algorithm designed for wireless sensor networks, i.e., an on-demand and multipath routing algorithm.

Diptera

- *Flies*

  Due to their short life-span and easiness of breeding and of providing an adequate living environment, the fruit fly is widely studied in laboratory conditions. Consequently, their behavior is known in detail and specific mechanisms for finding food are sources of inspiration for new algorithms. To the authors' knowledge, there are two metaheuristics that simulate the fruit fly: the Fruit Fly Optimization (FOA) and the Drosophila Food Search Optimization (DFO) [274]. Similar to the DFO, the FOA is also based on the Drosophila fly and the literature shows that there are at least two different implementations, proposed by [275] and [276].

- *Mosquitoes*

  The host seeking behavior of female mosquitos is mimicked by the Mosquito host seeking algorithm (MHSA) [277]. The general idea is simple and it is based on the following idealized rules: (i) the mosquito looks for carbon dioxide or some other scent; (ii) if found, the mosquito moves toward the location with the highest concentration; (iii) it descends when the heat radiating from the host is felt. The algorithm was developed specifically to solve the travelling salesman problem and it has several advantages that include: (i) the ability to perform large-scale distributed parallel optimization; (ii) it can describe complex dynamics; (iii) it can perform multi-objective optimization; (iv) it is claimed to be independent of the initial conditions and problem size [278]

Lepidoptera

The insects that belong to this class have wings covered with overlapping small scales. The best-known examples include butterflies and moths.

- *Butterflies*

  The Monarch Butterfly Optimization (MBO) [279] simulates the migration behavior of monarch butterflies through the use of a set of idealized rules: (i) the entire population of butterflies is located in two areas, i.e., Land1 and Land2; (ii) each offspring is generated by the migration operator applied to individuals from Land1 or Land2; (iii) once an offspring is generated, the parent butterfly dies if its fitness is worse than that of the offspring; (iv) the butterflies with the best fitness survive to the next generation. Similar to the MBO, the Monarch Migration Algorithm (MMA) [280] models the migration behavior of monarch butterflies. The main differences between the MBO and the MMA consist in the mechanisms used for movement, for new individual creation and for population size control.

  If the MBO and the MMA focus on migration aspects, the Butterfly Optimizer (BO) simulates the mate-location, behavior–perching and patrolling of male butterflies [281]. The initial BO version is developed for unconstrained optimization and is a dual population algorithm that includes male butterflies and auxiliary butterflies. The Artificial Butterfly Optimization (ABO) [282] is inspired from the same mating strategy as BO. However, the ABO is a single-population optimizer that contains two types of butterflies: sunspot and canopy, and the rules that it follows are different. In the BO, the following rules are considered: (i) the male butterflies are attracted to the highest UV/radiation object; (ii) the best perching position and the flying direction is memorized; (iii) the flying velocity is constant; (iii) the flying direction is changed if necessary [281]. On the other hand, the ABO considers the following generalized rules: (i) all male butterflies attempt to fly towards a better location (sunspot); (ii) in order to occupy a better position, the sunspot butterflies try to fly to the neighbor's sunspot; (iii) the canopy butterflies continually fly towards the sunspot butterflies to contend for the sunspot [282]. In ABO, three flight strategies are considered and their combination leads to two other variants of the algorithm.

The Butterfly Optimization Algorithm (BOA) [283] considers the foraging behavior and focuses on the smell of butterflies as the strategy used for determining the location of food or of a mating partner. In order to model this behavior, a set of idealized rules are used: (i) all butterflies emit fragrances that attract each other; (ii) the movement of the butterfly is random or towards the most fragrant butterfly; (iii) the stimulus intensity is influenced by the landscape of the objective function.

- Moths

The transverse orientation navigation mechanism of moths represents the source of inspiration for the Moth Flame Optimization (MFO) [284]. The population of moths updates their position in accordance with a flame. The group of flames represents the best solutions and serves as guidance for the moths [285]. The complexity of the MFO is O(problem_dimension × iteration × objective_number × population_size × objective_function) [54]. While the MFO contains a population of moths and flames, in the case of the Moth Swarm Algorithm (MSA) [286], which is also inspired by the navigation behavior of moths, the population is formed of three groups of moths: pathfinders (with the ability to discover new areas of the search space), prospectors (that tend to wander in spiral) and onlookers (that drift directly to the solutions obtained by the prospectors). Distinctively from MFO and MSA, the Moth Search (MS) algorithm [287] considers the phototaxis and the Levy flight of the moths as a source of inspiration. In this case, the population is formed of two subpopulations. One follows the Levy movement and the other simulates the straight flight.

Ortoptera

This order includes, among others, insects such as grasshoppers, crickets and locusts. They are insects that move with great agility and have many shapes and characteristics.

The first metaheuristic optimizer that used the locust swarm metaphor is the Locust Swarm [288], a multi-optima search technique developed for non-globally convex search spaces. However, in order to identify the starting points for the search, it uses the PSO as part of its search [288], and one may argue that it is not a new metaheuristic, but a hybridization of the PSO. On the other hand, the Locust Swarm proposed in [289] emulates the interaction of a locust cooperative swarm. Since the two algorithms have the same name, in this work they are referred to as LS1 for the version proposed in [288] and LS2 for the [289] version. LS2 considers both solitary and social behavior, and consists of three parts: initialization, solitary operation and social processes [289]. The solitary phase performs the exploration of the search space, while the social phase is dedicated to exploitation.

The grasshoppers' social interactions represent the basis for the Grasshopper Optimization Algorithm (GOA) [290]. Both larvae, which corresponds to the feeding stage, and the adult form, which corresponds to the exploration stage, are considered. While in nature, the individual evolves from larvae to adult (local, then global), and due to the nature of the search space, the optimization algorithm first needs to find a promising region and, after that, exploit it (global, then local).

Other Insects

- *Hunting*

The mechanisms used by antlions to hunt ants is simulated in the Ant Lion Optimizer (ALO) [291]. Antlions have two phases: larvae, focused on feeding, and adults, focused on mating. The ALO is based on the larvae form and, in order to perform the optimization, a series of conditions are considered: (i) a random walk imitates the movement of an ant; (ii) the random walks are affected by the traps of the antlions; (iii) the pits built by antlions (and the probability of catching ants) are proportional with the antlion fitness; (iv) the random walk range decreases adaptively; (v) the ant is caught when its fitness is worse than that of an antlion; (vi) after catching a prey, the antlion repositions and builds another pit.

- *Mixed behavior*

Inspiration for metaheuristics comes not only from insects that are generally considered useful, e.g., bees, but also from insects that are considered pests, such as cockroaches. Among the first approaches that included the social behavior of cockroaches is the Roach Infestation Optimization (RIO) [292]. It is an adaptation of the PSO that implements three elements: finding the darkness, finding friends and finding food. Other algorithms simulating cockroach behavior are: the Cockroach Swarm Optimization (CSO) [293] and the Cockroach-inspired Swarm Evolution (CSE) [294]. The paper containing the initial CSO version was retracted from the IEEE database due to violations of publication principles, however, this did not stop other researchers from using and improving CSO; Google Scholar indicates that, as of the end of February 2019, there are 32 articles citing the retracted paper. Unlike the RIO, the CSE considers competition, space endurance and migration of cockroaches, beside cooperative behavior [294].

The Dragonfly Algorithm (DA) [264] is inspired from the static (feeding) and dynamic (migratory) swarming behavior of dragonflies. In the feeding stage, the dragonflies are organized into small groups that cover a small area to hunt, through back-and-forth movement with abrupt changes in the flying path. In the dynamic stage, a large group of individuals form a swarm and migrate in one direction over long distances.

The Pity Beetle Algorithm (PBA) [295] is based on the aggregation behavior and search mechanisms used for nesting and food finding of Pityogenesis chalcographus, a beetle also known under the name of "six toothed spruce bark beetle". The PBA follows three stages: searching (where the chemicals emitted by the weakened trees are used to identify a suitable host), aggregation (were multiple individuals feed on the host and attract more individuals—both male and female) and anti-aggregation (that is specific to the situation when the population size surpasses a specific threshold).

Table 4 summarizes the algorithms inspired from inspects and presents a series of examples for improvements and applications. The same structure and idea as in the previous tables are applied.

Table 4. Improvements and applications for insect-inspired metaheuristics (alphabetically sorted).

Algorithm	Source Code	Modifications and Improvements	Applications
Ant Lion Optimizer (ALO) [291]	(MATLAB-author source) http://www.alimirjalili.com/ALO.html, accessed on 6 June 2021	– multi-objective optimization [296] – binary [297] – inclusion of chaos [298]	– Artificial Neural Network training [299] – multi-objective engineering design problems [296] – automatic generation control [300] – feature selection [297]
Bee Colony Inspired Algorithm (BCiA) [266]			– vehicle routing problem with time windows [266]
Bumble Bee Mating Optimization (BBMO) [270]		– inclusion of combinatorial neighborhood topology [301] – parameter adaptation [302]	– multicast routing, traveling salesman problem [302] – feature selection [303] – vehicle routing problem with stochastic demands [301]
Butterfly Optimizer (BO) [281]		– inclusion of constraints [304] – inclusion of covariance matrix [305]	
Butterfly Optimization Algorithm (BOA) [283]	(MATLAB-author source) https://www.mathworks.com/matlabcentral/fileexchange/68209-butterfly-optimization-algorithm-boa/, accessed on 12 December 2019	– inclusion of mutualism principle [306], cross-entropy [307], learning automata [308] – binary approach [309] – the search is modified to use a normal distribution	– maximum power point tracking in photovoltaic systems [310] – feature selection [309]
Pity Beetle Algorithm (PBA) [295]		– new search and population reproduction mechanism, parameter adaptation [311] – inclusion of the opposition-based principle [312]	– wireless multimedia sensors [311] – lung cancer classification [312]
Dragonfly Algorithm (DA) [264]	(MATLAB-author source) http://www.alimirjalili.com/DA.html, accessed on 6 June 2021	– binary [264] – multi-objective [264] – inclusion of memory mechanisms specific to PSO [313], chaos theory [314]	– feature selection [314–316] – proton exchange fuel cells [317] – engineering design [313] – submarine propeller optimization [264]
Drosophila Food Search Optimization (DFO) [274]			– winner takes all circuit [318]
Firefly algorithm (FF) [37]	(MATLAB) http://yarpiz.com/259/ypea112-firefly-algorithm, accessed on 12 December 2019	– chaotic maps [319] – hybridization with Patter Search [320], Harmony Search [321], Group Search Optimizer [322]	– load frequency controller design [323] – ophthalmology [324] – discrete optimization [325]

Table 4. Cont.

Algorithm	Source Code	Modifications and Improvements	Applications
Fruit Fly Optimization (FOA) [276]	(MATLAB—author source) http://www.oitecshop.byethost16.com/FOA.html?i=1, accessed on 15 June 2020	- multi-swarm [326,327] - adaptive cooperative learning [328] - introduction of random perturbation [329] - use of a cloud-based model [330]	- shortest path in mobile ad-hoc networks [331] - image processing [328] - joint replenishment problems [329] - parameter identification of synchronous generator [327]
Grasshopper Optimization Algorithm (GOA) [290]	(MATLAB-author source) http://www.alimirjalili.com/GOA.html, accessed on 6 June 2021	- binary [332] - multi-objective [333] - inclusion of chaos [334] - inclusion of levy flight mechanism [335]	- feature selection [336] - Support Vector Machine optimization [336] - financial stress prediction (in combination with extreme learning machine) [335] - Artificial Neural Network training [332] - decision making for self-driving vehicles [337]
Locust Swarm (LS1) [288]			- joint replenishment problems [338]
Locust Swarm (LS2) [289]	(MATLAB-author source) https://www.mathworks.com/matlabcentral/fileexchange/53271-locust-search-ls-algorithm, accessed on: 20 December 2019		- image segmentation [339,340]
Mayfly optimization algorithm (MA) [341]	(MATLAB-author source) https://in.mathworks.com/matlabcentral/fileexchange/76902-a-mayfly-optimization-algorithm, accessed on 15 August 2021	- hybridization with Harmony Search [342]	- feature selection [342] - optimal design of energy renewable sources (in combination with radial basis neural networks) [343]
Monarch Butterfly Optimization (MBO) [279]	(C++, MATLAB) https://github.com/ggw0122/Monarch-Butterfly-Optimization, accessed on 12 December 2019	- binary adaptation [309] - hybridization with Differential Evolution [344] - inclusion of crossover operator [345] - self-adaptive strategies [346]	- 0–1 knapsack problem [347] - osteoporosis classification (in combination with Artificial Neural Networks) [348] - vehicle routing problem [349]
Mosquito host-seeking algorithm (MHSA) [277]		- inclusion of random walk and game of life [350]	- travelling salesman problem [278]
Moth Flame Optimization (MFO) [284]	(MATLAB-author source) http://www.alimirjalili.com/MFO.html, accessed on 6 June 2021	- inclusion of Gaussian mutation [351] - multi-objective optimization [352] - inclusion of chaos [298,353] - inclusion of levy flight mechanism [354,355]	- non-linear feedback control design [336] - medical diagnosis (in combination with extreme learning machine) [353] - reactor power dispatch [285] - engineering design problems [355]

Table 4. Cont.

Algorithm	Source Code	Modifications and Improvements	Applications
Moth Swarm Algorithm (MSA) [286]	(MATLAB-author source) https://www.mathworks.com/matlabcentral/fileexchange/57822-moth-swarm-algorithm-msa, accessed on 8 February 2020	– inclusion of Opposition Based Learning [357] – hybridization with Gravitational Search Algorithm [358] – inclusion of arithmetic crossover [359] – inclusion of chaos [360]	– power flow [359] – threshold image segmentation [361]
Moth Search (MS) algorithm [287]	(MATLAB-author source) https://in.mathworks.com/matlabcentral/fileexchange/59010-moth-search-ms-algorithm, accessed on 8 February 2020	– inclusion of disruptor operator [362] – binary optimization [363] – alteration at step level [364] – hybridization with Ant Colony Optimization [365]	– photovoltaic parameter identification [362] – knapsack problem [363] – drone placement [366]
Roach Infestation Optimization (RIO) [292]	(C#, VB) https://msdn.microsoft.com/en-us/magazine/mt632275.aspx, accessed on 6 February 2020	– introduction of the center agent concept [367] – addition of cannibalism components [368] – dynamic step size adaptation [369]	– Artificial Neural Networks training [367] – engineering design [367]
Water strider algorithm (WSA) [370]		– inclusion of chaos [371], of quasi-opposition and elite-guide evolution mechanism [372], adaptable parameters [373]	– optimal design of renewable energy systems [372]

3.2.2. Other Invertebrates

This group includes algorithms inspired from different invertebrates that do not belong to the insect class.

- *Arachnids*

The social behavior of spiders, i.e., communication using vibrations throughout the web, represents the source of inspiration for the Social Spider Optimization (SSO) [265]. It emulates a group of spiders that contains both males and females and applies different evolutionary operators to mimic the distinct behaviors typically found in the colony. In addition to the cooperation behavior, a mating operator, applicable only to the strongest individuals, is introduced to increase diversity. A comprehensive review of the SSO, which covers its main variants and applications, is the work of Luque–Chang et al. [374].

Distinctively from the SSO, which models the cooperative behavior and exchange of information through the web, the Social Spider Algorithm (SSA) [375] simulates the foraging behavior of social spiders. The SSA does not distinguish the individuals by sex; all the spiders share the same search operations. Compared to the SSO, the SSA is simpler, it uses a single random move operator and depends on the parameter settings to control the search [375]. A parameter sensitivity analysis (through advanced non-parametric statistical tests) indicated that medium population, small to medium attenuation rate, medium crossover probability and small mutation probability lead to good results for the majority of the problem being tested [376].

- *Crustacea*

The Krill Herd Algorithm (KHA) [377] is inspired from the herding behavior of krill individuals and was developed to solve non-complex optimization problems. It is a population-based approach that uses three main ways to explore the search space: (i) movement, induced by other individuals; (ii) foraging; (iii) random diffusion. A review that covers the main improvements and applications of the KHA is [378]. Newer studies (after 2017) that use the KHA to solve specific problems are presented in Table 5.

- *Annelid worms*

The reproduction mechanisms used by earthworms are the source of inspiration for the Earthwork Optimization Algorithm (EWA) [379]. The idealized rules that the EWA follows are: (i) all the earthworms can produce offspring using only two types of reproduction; (ii) the number of genes of the offspring is the same as the parent's; (iii) the best individuals go directly, without change, to the next generation so as to ensure that the population cannot deteriorate throughout generations.

- *Tunicata*

Tunicates are marine, small bio-luminescent invertebrates with a unique mode of jet propulsion. The movement strategy and the swarming behavior of tunicates was modelled in the Tunicate Swarm Algorithm (TSA) [380], its performance for a set of benchmarking problems being similar with state-of-the-art approaches. The time complexity of the TSA is $O(\text{iterations} \times \text{population\_size} \times \text{dimensionality} \times N)$ where N indicates the jet propulsion and swarm behaviors.

Table 5. Improvements and applications for other invertebrates-based metaheuristics (alphabetically sorted).

Algorithm	Source Code	Modifications and Improvements	Applications
Earthwork Optimization Algorithm (EWA) [379]	(MATLAB—author source) https://in.mathworks.com/matlabcentral/fileexchange/53479-earthworm-optimization-algorithm-ewa?s_tid=FX_rc3_behav, accessed on 15 March 2021	– hybridization with Differential Evolution [381]	– home energy management system [381–383]
Krill Herd Algorithm (KHA) [377]	(MATLAB—author source) https://www.mathworks.com/matlabcentral/fileexchange/55486-krill-herd-algorithm, accessed on 9 February 2020	– hybridization with Ant Colony Optimization [384], Bat Algorithm [385], Clonal Selection [386] – modification at inner lever [387]	– Artificial Neural Network training [388] – planning and scheduling [385] – feature reduction [389] – text clustering [387]
Social Spider Optimization (SSO) [265]	(MATLAB—author source) https://www.mathworks.com/matlabcentral/fileexchange/46942-a-swarm-optimization-algorithm-inspired-in-the-behavior-of-the-social-spider, accessed on 26 April 2020	– modification of the solution generation mechanism [390] – inclusion of rough sets [391] – constraint handling [392]	– reactive power dispatch [390] – minimum number attributes reduction problem [391] – Artificial Neural Network training [393]
Social Spider Algorithm (SSA) [375]	(MATLAB, C++, Python—author source) https://github.com/James-Yu/SocialSpiderAlgorithm, accessed on 18 January 2020	– inclusion of differential mutation [394], chaos [395] – new mutation strategy [396]	– train energy optimization [397] – scheduling [394,395] – load dispatch problem [398]
Tunicate Swarm Algorithm (TSA) [380]	(MATLAB—author source) https://www.mathworks.com/matlabcentral/fileexchange/75182-tunicate-swarm-algorithm-tsa, accessed on 17 July 2021	– inclusion of local escaping operator [399], Levi Flight distribution [400] – hybridization with Salp Swarm Optimizer [401]	– control and operation of automated distribution networks [400] – connectivity and coverage optimization in wireless sensor networks [401]

4. The Exploration–Exploitation Balance

Although based on different ideas, for all metaheuristic optimizers, the mechanisms used to simulate the optimization behaviors are similar. Generally speaking, all the algorithms in this class start with an initial population of potential solutions, usually randomly generated, which is evolved, i.e., modified by a series of mechanisms that can include—among others—selection, crossover and mutation, until a stopping criterion is satisfied. The actual strategies used to perform these steps and the mechanisms used to control the exploration–exploitation balance (EEB) influence the performance behavior and represent the main elements that make the distinction between algorithms.

The research in the area of metaheuristics often mentions the exploration and exploitation aspects of the algorithms; however, these terms have never been formally defined [402]. Informally, exploration is defined as the process of visiting entirely new regions of the search space, also known as the global search ability of the algorithm, while exploration is the process of visiting those regions of the search space within the neighborhood of previously visited points, which represents the local search ability [402]. Pure exploration leads to a decrease in precision but increases the ability of finding new, good solutions, while pure exploitation refines the existing solution and drives the search to a local optimum [403]. Because it indicates how the resources are allocated, knowing the EEB information can be useful to determine the impact of specific aspects of the algorithm [404]. The EEB can be seen from two points of view: (i) exploration and exploitation as opposing forces; (ii) exploration and exploitation as orthogonal forces [404]. However, it was shown that the opposing forces view is a special case of the orthogonal view and, thus, EEB monitoring must involve a metric for the exploration axis and one for the exploitation axis [404].

For evolutionary algorithms, it was shown that different operators, depending on the algorithm, are acting as exploitation or exploration procedures [1]. In population-based algorithms, the EEB is connected to the population diversity: when this is high, the algorithm is explorative and when it is low, the behavior is exploitative [1]. Although a diverse population is a prerequisite rather than a guarantee for the EEB and a good EEB can be reached through other means, e.g., fitness, using diversity is one of the simplest methods for achieving it [402]. Due to the fact that the problems to be solved must be encoded into a binary or real-valued vector, a clear distinction between the genotype (the encoded structure) and the phenotype (the actual problem) must be done. In this context, the diversity can be measured at the genotype level, at the phenotype level or using complex or composite measures that combine the genotype and the phenotype.

In [402], the diversity-based approaches applied to the EEB are classified as: (i) diversity maintenance—in this case it is assumed that the techniques will maintain diversity per se; (ii) diversity control, where feedback from measured individual fitness and or/fitness improvement is used to direct the evolution towards exploration or exploitation; (iii) diversity learning—a long-term history in combination with machine learning approaches is used to learn unexplored search areas; iv) other direct approaches (Figure 2). In the case of diversity maintenance, two categories can be indicated: niching and non-niching. The niching techniques represent extensions of the algorithms to multi-modal domains [405]. One of the most comprehensive definitions for niching is given in [406]: "Niching is a two-step procedure that (a) concurrently or subsequently distributes individuals into distinct basins of attraction and (b) facilitates approximation of the corresponding (local) optimizers".

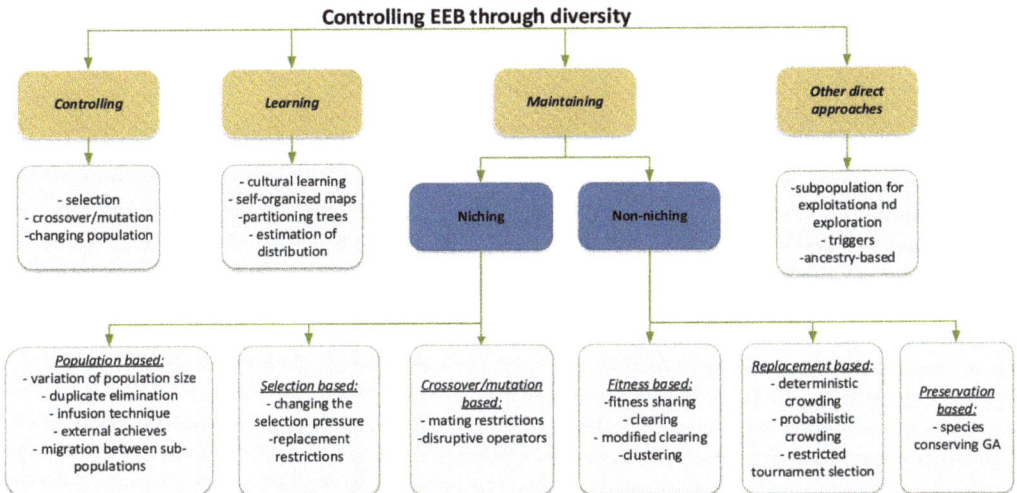

Figure 2. Mechanisms used to control the exploration–exploitation balance through diversity.

Table 6 shows the different mechanisms for the EEB used by the initial versions of the algorithms presented in Section 3. The subsequent modifications performed to the base versions are not considered in this table. The following notations are used: *C* indicates the controlling mechanisms, *L* the learning approaches, *OD* are the other direct approaches and *H.* represents the hybrid techniques. *Pop.* represents the population-based techniques, *Sel.* the selection based, *Crs.* the crossover/mutation based, *Fit.* the fitness based, *Rep.* the replacement based and *Pres.* the preservation-based approaches. In Table 6, in each case a specific approach is encountered in the algorithm, an x is set in the corresponding column.

Table 6. The diversity-based approaches used for EEB by the bio-inspired metaheuristic optimizers (alphabetically sorted).

Algorithm	C	L	Maintaining								OD	Description of the Mechanisms
			Niching					Non-Niching				
			Pop.	Sel.	Crs.	H.	Fit.	Rep.	Pres.	H.		
Backtracking Search Algorithm Optimization (BSA)			x		x						x	– dual-population with a complex crossover approach
Bat Algorithm (BA)			x								x	– a local search is performed to randomly selected best solutions; –new solutions are generated by flying randomly and added to the population if their fitness is good
Bee Colony Inspired Algorithm (BCiA)			x								x	– 2 populations with individuals that migrate between them
Bird Mating Optimizer (BMO)			x		x							– new individuals are randomly inserted after a certain number of generations – the mating behavior of male and female is different
Blind Naked Mole Rats (BNMR)			x									– a new solution is generated and added to the population its fitness is good
Bumblebees (B)			x									– infusion of new individuals in the population and elimination of the worst
Bumble Bee Mating Optimization (BBMO)					x							– there are mating restrictions with division of roles within individuals in the population
Chicken Swarm Optimization (CSO)			x									– different subpopulations are considered
Competition over Resources (COR)			x								x	– 2 subgroups performing a separate search
Cuckoo Search (CS)			x									– worst individuals are removed and new ones are added
Cuckoo Optimization Algorithm (COA)								x				– k-means is used to cluster the cuckoos into groups
Drosophila Food Search Optimization (DFO)											x	– Redundant Search algorithm is used to perform a neighborhood search
Elephant Herding Optimization (EHO)			x									– each iteration, the males- i.e., the worst individuals from each clan are replaced by new individuals (separating operator)
Elephant Search Algorithm (ESA)											x	– the population is formed by two groups: male (that performs exploration) and females (that performs exploitation)
Grey Wolf Optimizer (GWO)												– the balance between exploration and exploitation is performed through control parameters
Hunting Search (HuS)	x											– when the individuals in the population are close together, a reset procedure is applied
Krill Herd Algorithm (KHA)							x					– the fitness function is used to simulate the motion induced by other krill individuals

Table 6. *Cont.*

Algorithm	C	L	Maintaining								OD	Description of the Mechanisms
			Niching				Non-Niching					
			Pop.	Sel.	Crs.	H.	Fit.	Rep.	Pres.	H.		
Locust Swarm (LS2)											x	– the exploration and exploitation steps are one after the other
Lion's Algorithm (LA)			x					x				– k-means is used to perform gender grouping – at pride update, sick/weak cubs are killed/eliminated
Lion Optimization Algorithm (LOA)			x		x							– the behavior of males and females in the pride is different; -there are different populations (nomad and pride); -restricted mating between males and females
Lion Pride Optimization Algorithm (LPOA)			x									– multiple subpopulations (prides) are considered
Monarch Butterfly Optimization (MBO)			x								x	– 2 subpopulations with individuals that migrate between them
Moth Swarm Algorithm (MSA)	x											– uses an adaptive crossover based on diversity
Moth Search (MS) algorithm			x								x	– at each generation the population recombines all the individuals and splits them into 2 groups based on fitness
Pigeon Inspired Optimization (PIO)	x											– the landmark operation implies reducing the number of individuals to half
Satin Bowerbird Optimizer (SBO)								x				– the probability of finding a mate is based on the fitness function
Sperm Whale Algorithm (SWA)					x							– restricted mating (strongest male mates with several females)
Social Spider Optimization (SSO)					x						x	– the individuals are gender oriented, with different behavior; the offspring generation is restricted to the classical male-female mating
Spider Monkey Optimization (SMO)											x	– multiple sub-populations
Spotted Hyena Optimizer (SHO)	x											– the parameters controlling the prey encircling are changed as the search progresses
Squirrel Search Algorithm (SSA)			x									– when seasonal conditions are satisfied, the squirrels are randomly relocated

As it can be observed from Table 6, some strategies are more popular than others; the non-niching techniques based on population are the most used approaches for diversity maintenance. This can be explained by the fact that the bio-inspired metaheuristics are population-based and, therefore, the most intuitive methods consist in modifying the characteristics of the population as a means of improving performance.

5. Algorithm Selection

As it can be observed, the list of algorithms, even when the source of inspiration is restrung to a single category, is extensive. In practice, the most common question is what is the best suited algorithm and how can it can be successfully applied for a specific problem? Unfortunately, answering this question is not an easy task, as by their definition, heuristics can provide sufficiently good solutions to an optimization problem. Thus, depending on the accepted level of precision, there can be more than one heuristic that can generate acceptable solutions in terms of quality. However, performance is not the only aspect that can be taken into account [407]. The issue of algorithm selection was formalized by Rice in [408], and involves: (i) a problem space P; (ii) an algorithm space A; (iii) a mapping PxA onto R (also known as performance model). This implies that there must exist an extensive set of problem instances and features that describe them and the algorithm state at any time [409]. Thus, although advances regarding algorithm selection were made [409,410], the most used strategy in the metaheuristic filed is based on comparison. The work of LaTorre et al. [411] presents a series of methodological guidelines for comparing metaheuristics that involves: (i) selection of benchmarks and refence algorithms; (ii) validation of results (with statistical analysis and visualization); (iii) parameter tunning; (iv) identification of usefulness.

As a demonstration, in this work, the three most used real-world benchmark problems were selected (Table 7) and used to determine (based on already reported results) what the best performing animal-inspired metaheuristics are. The centralized results, organized from the best to the worst solution, are presented in Table 8 for the pressure vessel design, Table 9 for the welded beam design and Table 10 for the tension/compression spring design. All the considered problems are constrained minimization problems and their description can be found in the references from the column "Reported work" of Tables 8–10.

Table 7. Real-world problems characteristics.

Problem	Decision Variables	Inequality Constraints
Tension/Compression spring	3 (diameter, mean coil diameter, number of active coils)	4
Pressure vessel design	4 (thinckness of the shell, thinkness of th head, inner radius, length of the cylindrical section)	4
Welded beam design	4 (thikness of the weld, length of the attached part of the bar, height of the bar and thickess of the bar)	7

Table 8. Solutions for the pressure vessel design problem.

Algorithms	Reported Work	Modified Version	Ts	Th	R	L	Optimal Cost
Sooty Tern Optimization Algorithm (STOA) [54]	[54]	No	0.778095	0.38324	40.31511	200	5879.1253
Emperor Penguin Optimization (EPO) [55]	[55]	No	0.778099	0.383241	40.31512	200	5880.07
Chameleon Swarm Algorithm (ChSA) [224]	[224]	No	12.450698	6.154387	40.31961	200	5885.3327
Memory based Dragonfly algorithm (MHDA) [313]	[313]	Yes	0.778169	0.384649	40.3196	200	5885.3353
COOT [412]	[412]	No	0.77817	0.384651	40.31961	200	5885.3487
Marine Predator Algorithm (MPA) -continuous variant [229]	[229]	No	0.77816876	0.3846497	40.31962	199.99999	5885.3353
Spotted Hyena Optimizer (SHO) [142]	[55]	No	0.77821	0.384889	40.31504	200	5885.5773
Modified Spider Monkey Optimization (SMONM) [203]	[412]	Yes	0.778322	0.384725	40.32759	199.8889	5885.595
African Vulture Optimization Algorithm (AVOA) [57]	[57]	No	0.778954	0.3850374	40.36031	199.43429	5886.67659
Grey Wolf Optimizer (GWO) [136]	[55]	No	0.779035	0.38466	40.32779	199.65029	5889.3689
Dragonfly Algorithm (DA) [264]	[313]	No	0.782825	0.384649	40.3196	200	5923.11
Aquila Optimization (AO) [47]	[47]	No	1.0540	0.182806	59.6219	38.8050	5949.2258
Improved Grasshoper Oprimization (OBLGOA) [413]	[412]	Yes	0.81622	0.4035	42.29113	174.81119	5966.6716
Slime Mould Algorithm (SMA) [414]	[414]	No	0.7931	0.3932	40.6711	196.2178	5994.1857
Harris Hawk Optimization (HHO) [46]	[412]	No	0.81758383	0.4072927	42.09174	176.71963	6000.46259
Improved Artificial bee Colony (I-ABC greedy) [415]	[412]	Yes	0.8125	0.4375	42.0984	176.6369	6059.7124
Firefly Algorithm (FA) [416]	[412]	No	0.8125	0.4375	42.09844	176.63659	6059.7143
Moth-flame Optimization (MFO) [284]	[412]	No	0.8125	0.4375	42.09844	176.63659	6059.7143
Marine Predator Algorithm (MPA) -mixed integer variant [229]	[229]	No	0.8125	0.4375	42.09844	176.63660	6059.7144
Sine-Cosine Grey Wolf Optimizer (SC-GWO) [417]	[412]	Yes	0.8125	0.4375	42.0984	176.6370	6059.7179
Co-evolutionary Differential Evolution (CDE) [418]	[229]	Yes	0.8125	0.4375	42.09841	176.6376	6059.734
Whale Optimization Algorithm (WOA) [134]	[412]	No	0.8125	0.4375	42.09826	176.63899	6059.741
Bacterial foraging Optimization (BFOA) [419]	[229]	No	0.8125	0.4375	42.09639	176.68323	6060.46
Co-evolutionary Particle Swarm Optimization (CPSO) [420]	[313]	Yes	0.8125	0.4375	42.09126	176.7465	6061.077
Artificial Immune System-Genetic Algorithm (HGA-1) [421]	[229]	Yes	0.8125	0.4375	42.0492	177.2522	6065.821
Artificial Immune System-Genetic Algorithm (HGA-2) [421]	[229]	Yes	1.125	0.5625	58.1267	44.5941	6832.583
Harmony Search (HS) [422]	[229]	No	1.125	0.625	58.2789	43.7549	7198.433

Table 9. Solutions for the welded beam design.

Algorithms	Reported Work	Modified Version	τ	σ	Pc	δ	Optimal Cost
Aquila Optimization (AO) [47]	[47]	No	0.1631	3.3652	9.0202	0.2067	1.6566
Butterfly Optimization Algorithm (BOA) [283]	[283]	No	0.1736	2.969	8.7637	0.2188	1.6644
COOT [412]	[412]	No	0.19883	3.33797	9.19199	0.19883	1.6703
Memory based Dragon Fly algorithm(MHDA)) [313]	[313]	Yes	0.20573	3.25312	9.03662	0.20573	1.69525
Slime Mould algorithm (SMA) [414]	[414]	No	0.2054	3.2589	9.0384	0.2058	1.696
Dragonfly Algorithm (DA) [264]	[313]	No	0.19429	3.46681	9.04543	0.2057	1.70808
Tunicate Swarm Algorithm (TSA) [380]	[412]	No	0.20329	3.47114	9.0351	0.20115	1.72102
Seagull optimization algorithm (SOA) [53]	[53]	No	0.205408	3.472316	9.035208	0.20114	1.723485
Emperor Penguin Optimization (EPO) [55]	[55]	No	0.205411	3.472341	9.035215	0.20115	1.723589
Sooty Tern Optimization Algorithm (STOA) [54]	[54]	No	0.205415	3.472346	9.03522	0.20116	1.72359
Improved Artificial bee Colony (I-ABC greedy) [415]	[412]	Yes	0.20573	3.47049	9.03662	0.20573	1.72482
Co-evolutionary Particle Swarm Optimization (CPSO) [420]	[313]	Yes	0.20573	3.47049	9.03662	0.20573	1.72485
Modified Artificial Bee Colony (ABC) [263]	[313]	Yes	0.20573	3.47049	9.03662	0.20573	1.72485
Modified Spider Monkey Optimization (SMONM) [203]	[412]	Yes	0.20573	3.47049	9.03662	0.20573	1.72485
Chameleon Swarm Algorithm (ChSA) [224]	[224]	No	0.205730	3.470489	9.036624	0.20573	1.724852
African Vulture Optimization Algorithm (AVOA) [57]	[57]	No	0.20573	3.470474	9.03662	0.20573	1.724852
Moth-flame Optimization (MFO) [284]	[370]	No	0.20573	3.47049	9.03662	0.20573	1.7249
Water Strider Algorithm (WSA) [370]	[370]	No	0.20573	3.47049	9.03662	0.20573	1.7249
Marine Predator Algorithm (MPA) [229]	[229]	No	0.20573	3.47051	9.03662	0.20573	1.72485
Salp Swarm Algorithm (SSA) [11]	[414]	No	0.2057	3.4714	9.0366	0.2057	1.7249
Derivative free Simulated Annealing (SA) [423]	[313]	Yes	0.20564	3.47258	9.03662	0.20573	1.725
Spotted Hyena Optimizer(SHO) [142]	[412]	No	0.20556	3.47485	9.0358	0.20581	1.72566
Improved Grasshopper Optimization Algorithm (OBLGOA) [413]	[412]	Yes	0.20577	3.47114	9.03273	0.20591	1.7257
Grey Wolf Optimizer (GWO) [136]	[414]	No	0.2057	3.4784	9.0368	0.2058	1.7262
Whale Optimization Algorithm (WOA) [134]	[134]	No	0.205396	3.484293	9.037426	0.20627	1.730499
Harris Hawk Optimization (HHO) [46]	[412]	No	0.20404	3.53106	9.02746	0.20615	1.73199
Sailfish Optimizer (SFO) [222]	[222]	No	0.2038	3.6630	9.0506	0.2064	1.73231
Co-evolutionary Differential Evolution (CDE) [418]	[412]	Yes	0.20314	3.543	9.0335	0.20618	1.73346
Levy Flight Distribution (LFD) [424]	[412]	No	0.1857	3.907	9.1552	0.2051	1.77
Harmony Search and Genetic Algorithm (HSA-GA) [425]	[229]	Yes	0.2231	1.5815	12.8468	0.2245	2.25
Improved harmony Search (HS) [426]	[283]	Yes	0.2442	6.2231	8.2915	0.2443	2.3807
Differential Evolution with stochastic selection (DSS-DE) [427]	[229]	Yes	0.2444	6.1275	8.2915	0.2444	2.381
APPROX [428]	[134]	No	0.2444	6.2189	8.2915	0.2444	2.3815
Ragsdell [428]	[370]	No	0.2455	6.196	8.2915	0.2444	2.38154
David [428]	[134]	No	0.2434	6.2552	8.2915	0.2444	2.3841
Bacterial Foraging Optimization (BFOA) [419]	[229]	No	0.2057	3.4711	9.0367	0.2057	2.3868
Simplex [428]	[370]	No	0.2792	5.6256	7.7512	0.2796	2.5307
Random [428]	[134]	No	0.4575	4.7313	5.0853	0.66	4.1185

Table 10. Solutions for the tension/compression spring.

Algorithms	Reported Work	Modified Version	d	D	N	Optimal Cost
Aquila Optimization (AO) [47]	[47]	No	0.050243	0.35262	10.5425	0.011165
Butterfly Optimization Algorithm (BOA) [283]	[283]	No	0.051343	0.334871	12.9227	0.011965
Emperor Penguin Optimization (EPO) [55]	[55]	No	0.051087	0.342908	12.0898	0.012656
Sooty Tern Optimization Algorithm (STOA) [54]	[54]	No	0.05109	0.34291	12.09	0.012656
FireFly algorithm (BA) [416]	[412]	No	0.05169	0.35673	11.2885	0.012665
Pathfinder algorithm (PFA) [429]	[229]	No	0.051726	0.357629	11.235724	0.012665
Marine Predator Algorithm (MPA) [229]	[229]	No	0.0517244	0.35757003	11.2391955	0.012665
Improved Artificial bee Colony (I-ABC greedy) [415]	[412]	Yes	0.051686	0.356014	11.202765	0.012665
COOT [412]	[412]	No	0.0516527	0.3558442	11.340383	0.012665
African Vulture Optimization Algorithm (AVOA) [57]	[57]	No	0.051669	0.3562553	11.316126	0.0126652
Chameleon Swarm Algorithm (ChSA) [224]	[224]	No	0.051778	0.358851	11.164981	0.0126653
Harris Hawk Optimization (HHO) [46]	[412]	No	0.0517963	0.3593053	11.138859	0.01266
Grey Wolf Optimizer (GWO) [136]	[412]	No	0.05169	0.356737	11.28885	0.012666
Modified Spider Monkey Optimization (SMONM) [203]	[412]	Yes	0.051918	0.362248	10.97194	0.012666
Moth-flame Optimization (MFO) [284]	[412]	No	0.0519944	0.36410932	10.868422	0.012666
Artificial Immune System-Genetic Algorithm (HGA-1) [421]	[229]	Yes	0.051302	0.347475	11.852177	0.012668
Co-evolutionary Differential Evolution (CDE) [418]	[229]	Yes	0.051609	0.354714	11.410831	0.01267
Improved harmony Search (HS) [426]	[283]	Yes	0.051154	0.349871	12.076432	0.012670
Bacterial Foraging Optimization (BFOA) [420]	[229]	No	0.051825	0.359935	11.107103	0.012671
Sine-Cosine Grey Wolf Optimizer (SC-GWO) [417]	[412]	Yes	0.051511	0.352376	11.5526	0.012672
Spotted Hyena Optimizer (SHO) [142]	[412]	No	0.051144	0.343751	12.0955	0.012674
Co-evolutionary Particle Swarm Optimization (CPSO) [420]	[229]	Yes	0.051728	0.357644	11.244543	0.012674
Whale Optimization Algorithm (WOA) [134]	[412]	No	0.051207	0.345215	0.004032	0.012676
Salp Swarm Algorithm (SSA) [11]	[412]	No	0.051207	0.345215	12.004032	0.012676
Improved Grasshopper Optimization Algorithm (OBLGOA) [413]	[412]	Yes	0.0530178	0.38953229	9.6001616	0.012701
Mathematical_optimization [430]	[283]	-	0.053396	0.39918	9.1854	0.012730
Constraint_correction [431]	[283]	-	0.05	0.3159	14.25	0.012833

In Tables 8–10, the column "Reported work" indicates the paper where the specific results were reported and where the control parameters used to obtain those results where indicated. The column "Modified version" indicates if the specific algorithm is a modified version of the base variant. As it can be observed, the majority of the algorithms in the top of the list represent base variants of metaheuristics proposed in the last five years. This indicates that, for this type of constrained problem with a reduced number of parameters, the newer metaheuristics tend to perform better than the older, more known metaheuristics, as well as the classical mathematical approaches.

To determine which algorithm is best suited to all the considered engineering problems, a Condorcet-based approach was applied [432]. It is based on the idea of the voting system, the problem of determining a rank of algorithms becoming an electoral problem. In this context, the considered algorithms represent the candidates and their solutions for each problem indicate the voters. Thus, as a majority-based method, the Condorcet algorithm determines the winner of an election as the candidate who outperforms or is equal to each candidate in a pair-wise comparison. As not all the algorithms considered were tested on all of the problems, the Condorcet algorithm was applied for the metaheuristics tested on all three problems. The obtained results identified the top four metaheuristics as: EPO (45 votes), AO (42 votes), COOT (40 votes) and ChSA (34 votes). In the fifth and sixth pace, at equality with 32 votes, are I-ABC greedy and AVOA. As it can be observed, the difference between the algorithms placed in the first three positions is relatively small (2 votes). Similarly, there is a small difference between the algorithm placed in positions four, five and six. On the other hand, the difference between place three and four is larger (6 votes), indicating that the first three algorithms, when applied for the three engineering problems considered, performed substantially better than the next three. To test if there is a significant difference between the two groups, a t-Test Paired Sample was performed. The results obtained indicate a Pearson correlation of 0.999 and a $P(T <= t)$ two-tail of 0.3168. As it is higher than 0.05, the null hypothesis is accepted, resulting in that there are no statistically significant differences between the results provided by the best three algorithms versus the results provided by the next three best algorithms. Thus, it can be concluded that, although from the 17 algorithms considered EPO is the winner, all of the first six algorithms can provide similar results and can be used successfully for solving the three engineering problems considered.

6. General Issues

Eighty five percent of articles that propose new bio-inspired metaheuristics have a high number of citations, i.e., more than 20/year, in a relatively reduced period in comparison with the norm in the area of artificial intelligence, where, during a year, the average number of citations is around 5. This indicates that the issues of finding good optimizers that can be easily applied to solve different problems is of high interest. However, a high percentage of the research performed and the subsequently published articles is focused on applications. An in-depth analysis of the theoretical aspects that influence the performance of the different operators used and their combination is relatively scarce. However, researchers are trying to correct this aspect and, in the latest years, a series of studies focusing on the analysis of theoretical and practical aspects were published [10,409,411,433–436].

The high number of citations was observed mostly for the algorithms for which the source-code is provided or easy to find. For the majority of these highly cited metaheuristics, the research focused on two main directions: (i) improvement or hybridization and (ii) applications—usually without any analysis or motivation for the selection of a particular algorithm. However, although the rate of publishing new algorithms (and the variants proposed) is high, studies focusing on the aspects that make an algorithm successful or on the mechanisms that lead to improvements in performance are quite rare. Therefore, in order to further advance the knowledge in the area and to establish some comprehensive basis on which newer, faster and efficient approaches are developed and successfully applied to problems from various domains, the mechanisms and the influence of different aspects of the problem/optimizer domain must be analyzed in depth. In the last years, it was observed that the manuscripts publishing new metaheuristics contain a more detailed analysis and comparison with other algorithms. However, these studies are predominantly based on empirical observations gathered from simulations performed on a handful of benchmarks (mathematical functions such as those proposed in the CEC competitions and engineering problems such as welded beam, pressure vessel or tension/compression spring design). The fact that the CEC test problems are considered, until recently, only in C++ and Matlab can be one explanation for the fact that the majority of these metaheuristics are implemented in Matlab.

This paper presented a comprehensive list of metaheuristics, with a focus on animals as a source of inspiration. All the studied works have a similar organization. First, a general description of the domain is presented, followed by the natural mechanisms used as sources of inspiration and a section with the implementation strategies used to simulate the natural mechanisms. In the results section, a set of problems are selected to demonstrate the strengths and weaknesses of the proposed approach. Although this seems straightforward and easy to understand, in the metaheuristics area, the main issues are related to the fact that:

- The biological terminology used is complex and, in many cases, difficult to understand, which conceals that, in the implementation phase, the mechanisms used are simple and well-known and are, in fact, variations on the same theme; the work of [437] tries to shed some light onto the computational mechanisms used by the best-known metaheuristics. Also, in terms of the real-world mechanisms modeled, some of the algorithms are 'weak inspired', in the sense that the so-called modeled behavior is not met in the species that give the name of the algorithm [438];
- After overcoming the terminology barrier, upon a closer analysis, some of the so-called new algorithms not only do not have any novel aspect, but the papers describing them are incomplete or an implementation following the pseudo-code identifies other problems. In this regard, the work of Nguyen Van Thieu is worth mentioning, wherein he strides to implement these in a comprehensive python module with metaheuristics [438], and identified some of these dummy metaheuristics;
- Although some algorithms are inspired from the same source, the mechanisms modeled are different. For example, for Pidgeon inspired approaches, two algorithms were identified: PIO, which focuses on the movement of an individual from point A to point B and POA, which focuses on the movement of pigeons, taking into account the social interactions;
- There are multiple benchmark libraries that can be used and, in the majority of cases, the problems chosen by the authors to test the performance are very varied; thus, a comparison of performance between multiple algorithms based on the published literature is not always possible. The work of [439] presents the winners of some well-known competitions where standard benchmarks are used. In addition, as publishing bad results is sometimes discouraged, only the problems with the best results are chosen. In [2], it was shown that, in the comparison phase, the number of algorithms, the number of problems tested and the statistics used can lead to wrong conclusions if not properly selected;
- In an attempt to create high performance algorithms, the tendency is to include multiple strategies that have proven efficient over the years, e.g., self-adaptation, chaos, local search, etc. However, this has led to over-complicated methods that do not always show a direct correlation between complexity and performance. For example, for two winners of the CEC2016 competition, simpler versions (without operators biased towards 0) proved competitive against a large number of metaheuristics and even performed better for problems with solutions not close to 0 [435].

It can be observed that the source of inspiration follows the main classes identified in the biological taxonomy. Although the inspiration sources are varied and range from the behavior of simple organisms to the mechanisms used for survival of the species by large animals, the simulation of these sources is focused on exploration and exploitation, which translates into mathematical relations that make changes on the individuals. As it can be observed from Table 6, the majority of mechanisms used to control the exploration–exploitation balance in the standard versions belong to the niching class and are population-based. Overall, the manner in which the mechanisms that simulate the real-life behavior of animals are implemented and the combinations used represent the main aspects that differentiate the algorithms and that make them more sensitive or insensitive to the characteristics of the problem being solved, e.g., multi-modality, separability, etc.

Based on the aspects described above, the following potential directions of research can be identified:

- Performance measurement: the issue of performance is a complex aspect, especially taking into account that different metrics can be used. Although the tendency is to see performance as the capability to provide the best solution, other aspects such as complexity, computational resources consumed and stability can be employed. Moreover, how the best solution is identified is usually based on experiments with

mean and standard deviation as validation criteria [440], and a standardization of all these metrics and criteria of evolution can be a further step in the development of a general framework for metaheuristics.
- Performance analysis and improvement: identifying the main mechanisms that make a particular algorithm efficient for a particular class or group of problems. In this context, a better understanding of the exploration–exploitation balance, convergence analysis, diversity and the strategies that focus these aspects to a direction or another would help in providing a better foundation for the improvement of existing variants and creating new ones. In this regard, some studies focusing on these aspects were published (examples include: convergence analysis [441–444], fitness landscape analysis [445–448], exploration–exploitation [449–451]). However, additional research is required to reach field maturity.
- Algorithm selection: procedures and algorithms that can automatically select the best metaheuristic for a specific problem or group of problems. A wide level of applicability is one of the reasons for the popularity of metaheuristics. Thus, better strategies that can allow an easy identification of suitable algorithms are necessary. In this context, in the last few years, various methodologies and strategies to compare and select algorithms were proposed [2,411,433] and recommender systems were developed [409]. However, they are not widely accepted and applied and additional research in simplifying and generalizing these aspects is required.

7. Conclusions

This work is a review that focus on the animal-inspired metaheuristics proposed between 2006–2021. It was observed that, despite the rising number of critiques addressed to the entire metaheuristic community, the trend of proposing algorithms based on novel ideas and sources of inspiration does not seem to slow down considerably. In fact, it maintains the growth rate already observed a few years ago, mainly due to the large area of applicability and popularity of both the older, more established algorithms such as the GA, and newer approaches, for which the tendency is to provide the source code and thus increase the ease of use.

Regarding the animal-based metaheuristics, the most used source of inspiration is represented by the vertebrates, where easily observable behaviors such as food finding and mating are mathematically modeled using various approaches. However, a closer analysis of the inspiration sources indicated that all the main branches of the biological classification are represented in the metaheuristic world. This shows that researchers are actively searching for new ideas in unusual places and are not hindered by the difficulties associated with identifying the mechanisms of the behaviors of hard to analyze sources, such as animals living in remote and difficult to reach areas. In fact, the more exotic the inspiration source and the more uncommon the behavior, the higher the probability of finding new mechanisms that can be translated into truly novel approaches.

The main directions of research that were identified focus on the proposal of new metaheuristics and their application for various types of problems and only a few studies tackle the influence of specific operators or mechanisms on performance. Better performing algorithms are always desired and using nature as a source of inspiration can lead to new advances in this field of metaheuristics. However, attention must be paid not only to the source of inspiration but also to how this inspiration is modelled and put into practice. Similarity with existing variants, performance, complexity, exploration–exploitation balance, proper comparison and use of benchmarks must also be taken into account. An in-depth analysis of all the aspects that influence the performance behavior and the relations with the characteristics of the problems being solved can benefit both the metaheuristic community and the areas where these algorithms are applied.

Author Contributions: E.N.D. design the study and drafted the work. V.D. performed the literature search, revised and completed the manuscript. All authors have read and agreed to the published version of the manuscript.

Funding: This work was supported by project PN-III-P4-ID-PCE no 58/2021 financed by UEFISCDI, Romania.

Acknowledgments: The authors want to thank Florin Leon for his valuable insights and suggestions regarding the use of animal inspired metaheuristics.

Conflicts of Interest: The authors declare no conflict of interest.

References

1. Salcedo-Sanz, S. Modern meta-heuristics based on nonlinear physics processes: A review of models and design procedures. *Phys. Rep.* **2016**, *655*, 1–70. [CrossRef]
2. Hussain, K.; Salleh, M.N.M.; Cheng, S.; Shi, Y. Metaheuristic research: A comprehensive survey. *Artif. Intell. Rev.* **2018**, *52*, 2191–2233. [CrossRef]
3. Nabaei, A.; Hamian, M.; Parsaei, M.R.; Safdari, R.; Samad-Soltani, T.; Zarrabi, H.; Ghassemi, A. Topologies and performance of intelligent algorithms: A comprehensive review. *Artif. Intell. Rev.* **2016**, *49*, 79–103. [CrossRef]
4. Del Ser, J.; Osaba, E.; Molina, D.; Yang, X.S.; Salcedo-Sanz, S.; Camacho, D.; Das, S.; Suganthan, P.N.; Coello, C.A.C.; Herrera, F. Bio-inspired computation: Where we stand and what's next. *Swarm Evol. Comput.* **2019**, *48*, 220–250. [CrossRef]
5. Wolpert, D.H.; Macready, W.G. No free lunch theorems for optimization. *IEEE Trans. Evol. Comput.* **1997**, *1*, 67–82. [CrossRef]
6. Lam, A.; Li, V.O.K. Chemical-Reaction-Inspired Metaheuristic for Optimization. *IEEE Trans. Evol. Comput.* **2009**, *14*, 381–399. [CrossRef]
7. Adam, S.P.; Alexandropoulos, S.A.N.; Pardalos, P.M.; Vrahatis, M.N. No Free Lunch Theorem: A Review. In *Approximation and Optimization: Algorithms, Complexity and Applications*; Demetriou, I.C., Pardalos, P.M., Eds.; Springer International Publishing: Cham, Switzerland, 2019; pp. 57–82.
8. Hosseini, S.; Al Khaled, A. A survey on the Imperialist Competitive Algorithm metaheuristic: Implementation in engineering domain and directions for future research. *Appl. Soft Comput.* **2014**, *24*, 1078–1094. [CrossRef]
9. Doğan, B.; Ölmez, T. A new metaheuristic for numerical function optimization: Vortex Search algorithm. *Inf. Sci.* **2015**, *293*, 125–145. [CrossRef]
10. Stegherr, H.; Heider, M.; Hähner, J. Classifying Metaheuristics: Towards a unified multi-level classification system. *Nat. Comput.* **2020**, 1–17. [CrossRef]
11. Mirjalili, S.; Gandomi, A.H.; Mirjalili, S.Z.; Saremi, S.; Faris, H.; Mirjalili, S.M. Salp Swarm Algorithm: A bio-inspired optimizer for engineering design problems. *Adv. Eng. Softw.* **2017**, *114*, 163–191. [CrossRef]
12. Mirjalili, S.; Mirjalili, S.M.; Hatamlou, A. Multi-verse optimizer: A nature-inspired algorithm for global optimization. *Neural Comput. Appl.* **2016**, *27*, 495–513. [CrossRef]
13. Kirkpatrick, S.; Gelatt, C.D., Jr.; Vecchi, M.P. Optimization by simulated annealing. *Science* **1983**, *220*, 671–680. [CrossRef]
14. Glover, F. Tabu search—Part I. *ORSA J. Comput.* **1989**, *1*, 190–206. [CrossRef]
15. Mladenović, N.; Hansen, P. Variable neighborhood search. *Comput. Oper. Res.* **1997**, *24*, 1097–1100. [CrossRef]
16. Lourenço, H.R.; Martin, O.C.; Stützle, T. Iterated Local Search. In *Handbook of Metaheuristics*; Springer: Boston, MA, USA, 2003; pp. 320–353.
17. Turky, A.M.; Abdullah, S. A multi-population electromagnetic algorithm for dynamic optimisation problems. *Appl. Soft Comput.* **2014**, *22*, 474–482. [CrossRef]
18. Yang, X.S. A New Metaheuristic Bat-Inspired Algorithm. *Nicso 2010 Nat. Inspired Coop. Strateg. Optim.* **2010**, *284*, 65–74.
19. Ahandani, M.A. A diversified shuffled frog leaping: An application for parameter identification. *Appl. Math. Comput.* **2014**, *239*, 1–16. [CrossRef]
20. Zheng, Y.-J. Water wave optimization: A new nature-inspired metaheuristic. *Comput. Oper. Res.* **2015**, *55*, 1–11. [CrossRef]
21. Moghdani, R.; Salimifard, K. Volleyball Premier League Algorithm. *Appl. Soft Comput.* **2018**, *64*, 161–185. [CrossRef]
22. Molina, D.; Poyatos, J.; Del Ser, J.; García, S.; Hussain, A.; Herrera, F. Comprehensive taxonomies of nature-and bio-inspired optimization: Inspiration versus algorithmic behavior, critical analysis recommendations. *Cogn. Comput.* **2020**, *12*, 897–939. [CrossRef]
23. Fausto, F.; Reyna-Orta, A.; Cuevas, E.; Andrade, Á.; Perez-Cisneros, M. From ants to whales: Metaheuristics for all tastes. *Artif. Intell. Rev.* **2019**, *53*, 753–810. [CrossRef]
24. Brabazon, A.; McGarraghy, S. Formal Models of Foraging. In *Foraging-Inspired Optimisation Algorithms*; Brabazon, A., McGarraghy, S., Eds.; Springer International Publishing: Cham, Switzerland, 2018; pp. 23–44.
25. Zhu, G.-Y.; Zhang, W.-B. Optimal foraging algorithm for global optimization. *Appl. Soft Comput.* **2017**, *51*, 294–313. [CrossRef]
26. Brabazon, A.; Cui, W.; O'Neill, M. The raven roosting optimisation algorithm. *Soft Comput.* **2015**, *20*, 525–545. [CrossRef]
27. Brabazon, A.; McGarraghy, S. Introduction to Foraging-Inspired Algorithms. In *Foraging-Inspired Optimisation Algorithms*; Brabazon, A., McGarraghy, S., Eds.; Springer International Publishing: Cham, Switzerland, 2018; pp. 87–101.
28. Askarzadeh, A. Bird mating optimizer: An optimization algorithm inspired by bird mating strategies. *Commun. Nonlinear Sci. Numer. Simul.* **2014**, *19*, 1213–1228. [CrossRef]
29. Yang, X.-S.; Deb, S. Cuckoo search via Lévy flights. In Proceedings of the 2009 World Congress on Nature & Biologically Inspired Computing (NaBIC), IEEE, Coimbatore, India, 9–11 December 2009.

30. Wang, L.; Yin, Y.; Zhong, Y. Cuckoo search with varied scaling factor. *Front. Comput. Sci.* **2015**, *9*, 623–635. [CrossRef]
31. Chawla, M.; Duhan, M. Levy Flights in Metaheuristics Optimization Algorithms—A Review. *Appl. Artif. Intell.* **2018**, *32*, 802–821. [CrossRef]
32. Rakhshani, H.; Rahati, A. Snap-drift cuckoo search: A novel cuckoo search optimization algorithm. *Appl. Soft Comput.* **2017**, *52*, 771–794. [CrossRef]
33. Joshi, A.; Kulkarni, O.; Kakandikar, G.; Nandedkar, V. Cuckoo Search Optimization—A Review. *Mater. Today Proc.* **2017**, *4*, 7262–7269. [CrossRef]
34. Rajabioun, R. Cuckoo Optimization Algorithm. *Appl. Soft Comput.* **2011**, *11*, 5508–5518. [CrossRef]
35. Moosavi, S.H.S.; Bardsiri, V.K. Satin bowerbird optimizer: A new optimization algorithm to optimize ANFIS for software development effort estimation. *Eng. Appl. Artif. Intell.* **2017**, *60*, 1–15. [CrossRef]
36. Yang, X.S.; Deb, S. Eagle strategy using Levy walk and firefly algorithms for stochastic optimization. In *Nature Inspired Cooperative Strategies for Optimization (NICSO 2010)*; Springer: Berlin/Heidelberg, Germany, 2010; pp. 101–111.
37. Yang, X.-S. *Nature-inspired metaheuristic algorithms*; Luniver press: Bristol, UK, 2010.
38. Yang, X.-S.; Deb, S. Two-stage eagle strategy with differential evolution. *Int. J. Bio-Inspired Comput.* **2012**, *4*, 1–5. [CrossRef]
39. Gandomi, A.H.; Yang, X.-S.; Talatahari, S.; Deb, S. Coupled eagle strategy and differential evolution for unconstrained and constrained global optimization. *Comput. Math. Appl.* **2012**, *63*, 191–200. [CrossRef]
40. Talatahari, S.; Gandomi, A.H.; Yang, X.-S.; Deb, S. Optimum design of frame structures using the Eagle Strategy with Differential Evolution. *Eng. Struct.* **2015**, *91*, 16–25. [CrossRef]
41. Storn, R.; Price, K. Differential evolution—A simple and efficient adaptive scheme for global optimization over continuous spaces. *J. Glob. Optim.* **1995**, *11*, 341–359. [CrossRef]
42. Meng, X.; Liu, Y.; Gao, X.; Zhang, H. A New Bio-inspired Algorithm: Chicken Swarm Optimization. In *Advances in Swarm Intelligence, Pt1*; Tan, Y., Shi, Y., Coello, C.A.C., Eds.; Springer: Cham, Switzerland, 2014; pp. 86–94.
43. Askarzadeh, A. A novel metaheuristic method for solving constrained engineering optimization problems: Crow search algorithm. *Comput. Struct.* **2016**, *169*, 1–12. [CrossRef]
44. Jain, M.; Maurya, S.; Rani, A.; Singh, V. Owl search algorithm: A novel nature-inspired heuristic paradigm for global optimization. *J. Intell. Fuzzy Syst.* **2018**, *34*, 1573–1582. [CrossRef]
45. Zhuoran, Z.; Changqiang, H.; Hanqiao, H.; Shangqin, T.; Kangsheng, D. An optimization method: Hummingbirds optimization algorithm. *J. Syst. Eng. Electron.* **2018**, *29*, 386–404.
46. Heidari, A.A.; Mirjalili, S.; Faris, H.; Aljarah, I.; Mafarja, M.; Chen, H. Harris hawks optimization: Algorithm and applications. *Futur. Gener. Comput. Syst.* **2019**, *97*, 849–887. [CrossRef]
47. Abualigah, L.; Yousri, D.; Elaziz, M.A.; Ewees, A.A.; Al-Qaness, M.A.; Gandomi, A.H. Aquila Optimizer: A novel meta-heuristic optimization algorithm. *Comput. Ind. Eng.* **2021**, *157*, 10725. [CrossRef]
48. Mohammadi-Balani, A.; Nayeri, M.D.; Azar, A.; Taghizadeh-Yazdi, M. Golden eagle optimizer: A nature-inspired metaheuristic algorithm. *Comput. Ind. Eng.* **2020**, *152*, 107050. [CrossRef]
49. Sun, J.; Lei, X. Geese-inspired hybrid particle swarm optimization algorithm for traveling salesman problem. In Proceedings of the 2009 International Conference on Artificial Intelligence and Computational Intelligence, IEEE, Shanghai, China, 7–8 November 2009.
50. Duman, E.; Uysal, M.; Alkaya, A.F. Migrating Birds Optimization: A new metaheuristic approach and its performance on quadratic assignment problem. *Inf. Sci.* **2012**, *217*, 65–77. [CrossRef]
51. Goel, S. Pigeon Optimization Algorithm: A Novel Approach for Solving Optimization Problems. In Proceedings of the 2014 International Conference on Data Mining and Intelligent Computing (Icdmic), IEEE, Delhi, India, 5–6 September 2014.
52. Duan, H.; Qiao, P. Pigeon-inspired optimization: A new swarm intelligence optimizer for air robot path planning. *Int. J. Intell. Comput. Cybern.* **2014**, *7*, 24–37. [CrossRef]
53. Dhiman, G.; Kumar, V. Seagull optimization algorithm: Theory and its applications for large-scale industrial engineering problems. *Knowl. Based Syst.* **2018**, *165*, 169–196. [CrossRef]
54. Dhiman, G.; Kaur, A. STOA: A bio-inspired based optimization algorithm for industrial engineering problems. *Eng. Appl. Artif. Intell.* **2019**, *82*, 148–174. [CrossRef]
55. Dhiman, G.; Kumar, V. Emperor penguin optimizer: A bio-inspired algorithm for engineering problems. *Knowl. Based Syst.* **2018**, *159*, 20–50. [CrossRef]
56. Harifi, S.; Khalilian, M.; Mohammadzadeh, J.; Ebrahimnejad, S. Emperor Penguins Colony: A new metaheuristic algorithm for optimization. *Evol. Intell.* **2019**, *12*, 1–16. [CrossRef]
57. Abdollahzadeh, B.; Gharehchopogh, F.S.; Mirjalili, S. African vultures optimization algorithm: A new nature-inspired metaheuristic algorithm for global optimization problems. *Comput. Ind. Eng.* **2021**, *158*, 107408. [CrossRef]
58. Amiri, K.; Niknam, T. Optimal Planning of a Multi-carrier Energy Hub Using the Modified Bird Mating Optimizer. *Iran. J. Sci. Technol. Trans. Electr. Eng.* **2018**, *43*, 517–526. [CrossRef]
59. Ahmadi, M.; Kazemi, K.; Aarabi, A.; Niknam, T.; Helfroush, M.S. Image segmentation using multilevel thresholding based on modified bird mating optimization. *Multimed. Tools Appl.* **2019**, *78*, 23003–23027. [CrossRef]
60. Sadeeq, H.; Abdulazeez, A.; Kako, N.; Abrahim, A. A Novel Hybrid Bird Mating Optimizer with Differential Evolution for Engineering Design Optimization Problems. In Proceedings of the International Conference on Reliable Information and Communication Technology, Johor Bahru, Malaysia, 23–24 April 2017.

61. Zhang, Q.; Yu, G.; Song, H. A hybrid bird mating optimizer algorithm with teaching-learning-based optimization for global numerical optimization. *Stat. Optim. Inf. Comput.* **2015**, *3*, 54–65. [CrossRef]
62. Zhu, J.; Huang, M.; Lu, Z. Bird mating optimizer for structural damage detection using a hybrid objective function. *Swarm Evol. Comput.* **2017**, *35*, 41–52. [CrossRef]
63. Goswami, D.; Chakraborty, S. Multi-objective optimization of electrochemical discharge machining processes: A posteriori approach based on bird mating optimizer. *Opsearch* **2016**, *54*, 306–335. [CrossRef]
64. Skarzadeh, A.; Coelho, L.D.S. Determination of photovoltaic modules parameters at different operating conditions using a novel bird mating optimizer approach. *Energy Convers. Manag.* **2015**, *89*, 608–614. [CrossRef]
65. Zouache, D.; Arby, Y.O.; Nouioua, F.; Ben Abdelaziz, F. Multi-objective chicken swarm optimization: A novel algorithm for solving multi-objective optimization problems. *Comput. Ind. Eng.* **2019**, *129*, 377–391. [CrossRef]
66. Chen, Y.L.; He, P.L.; Zhang, Y.H. Combining Penalty Function with Modified Chicken Swarm Optimization for Constrained Optimization. In Proceedings of the First International Conference on Information Sciences, Machinery, Materials and Energy, Congqing, China, 11–13 April 2015.
67. Wu, D.; Kong, F.; Gao, W.; Shen, Y.; Ji, Z. Improved chicken swarm optimization. In Proceedings of the 2015 IEEE International Conference on Cyber Technology in Automation, Control, and Intelligent Systems (CYBER); IEEE, Shenyang, China, 8–12 June 2015.
68. Khan, A.; Shah, R.; Bukhari, J.; Akhter, N.; Attaullah; Idrees, M.; Ahmad, H. A Novel Chicken Swarm Neural Network Model for Crude Oil Price Prediction. In *Advances on Computational Intelligence in Energy*; Springer: Cham, Switzerland, 2019; pp. 39–58.
69. Liu, D.; Liu, C.; Fu, Q.; Li, T.; Khan, M.I.; Cui, S.; Faiz, M.A. Projection pursuit evaluation model of regional surface water environment based on improved chicken swarm optimization algorithm. *Water Resour. Manag.* **2018**, *32*, 1325–1342. [CrossRef]
70. Banerjee, S.; Chattopadhyay, S. Improved serially concatenated convolution turbo code (SCCTC) using chicken swarm optimization. In Proceedings of the 2015 IEEE Power, Communication and Information Technology Conference (PCITC), IEEE, Bhubaneswar, India, 15–17 October 2015.
71. Javidi, A.; Salajegheh, E.; Salajegheh, J. Enhanced crow search algorithm for optimum design of structures. *Appl. Soft Comput.* **2019**, *77*, 274–289. [CrossRef]
72. Díaz, P.; Pérez-Cisneros, M.; Cuevas, E.; Avalos, O.; Gálvez, J.; Hinojosa, S.; Zaldivar, D. An Improved Crow Search Algorithm Applied to Energy Problems. *Energies* **2018**, *11*, 571. [CrossRef]
73. Hinojosa, S.; Oliva, D.; Cuevas, E.; Pajares, G.; Avalos, O.; Gálvez, J. Improving multi-criterion optimization with chaos: A novel Multi-Objective Chaotic Crow Search Algorithm. *Neural Comput. Appl.* **2018**, *29*, 319–335. [CrossRef]
74. Sayed, G.I.; Hassanien, A.E.; Azar, A.T. Feature selection via a novel chaotic crow search algorithm. *Neural Comput. Appl.* **2017**, *31*, 171–188. [CrossRef]
75. Dos Santos Coelho, L.; Richter, C.; Mariani, V.C.; Askarzadeh, A. Modified crow search approach applied to electromagnetic optimization. In Proceedings of the 2016 IEEE Conference on Electromagnetic Field Computation (CEFC), IEEE, Miami, FL, USA, 11–13 November 2016.
76. Gupta, D.; Sundaram, S.; Khanna, A.; Hassanien, A.E.; De Albuquerque, V.H.C. Improved diagnosis of Parkinson's disease using optimized crow search algorithm. *Comput. Electr. Eng.* **2018**, *68*, 412–424. [CrossRef]
77. Oliva, D.; Hinojosa, S.; Cuevas, E.; Pajares, G.; Avalos, O.; Galvez, J. Cross entropy based thresholding for magnetic resonance brain images using Crow Search Algorithm. *Expert Syst. Appl.* **2017**, *79*, 164–180. [CrossRef]
78. Chi, R.; Su, Y.-X.; Zhang, D.-H.; Chi, X.-X.; Zhang, H.-J. A hybridization of cuckoo search and particle swarm optimization for solving optimization problems. *Neural Comput. Appl.* **2017**, *31*, 653–670. [CrossRef]
79. Feng, Y.; Wang, G.-G.; Feng, Q.; Zhao, X.-J. An Effective Hybrid Cuckoo Search Algorithm with Improved Shuffled Frog Leaping Algorithm for 0-1 Knapsack Problems. *Comput. Intell. Neurosci.* **2014**, *2014*, 857254. [CrossRef]
80. Wang, L.; Zhong, Y. Cuckoo Search Algorithm with Chaotic Maps. *Math. Probl. Eng.* **2015**, *2015*, 1–14. [CrossRef]
81. Khodier, M. Comprehensive study of linear antenna array optimisation using the cuckoo search algorithm. *IET Microw. Antennas Propag.* **2019**, *13*, 1325–1333. [CrossRef]
82. Ikeda, S.; Ooka, R. Metaheuristic optimization methods for a comprehensive operating schedule of battery, thermal energy storage, and heat source in a building energy system. *Appl. Energy* **2015**, *151*, 192–205. [CrossRef]
83. Afzalan, E.; Joorabian, M. An improved cuckoo search algorithm for power economic load dispatch. *Int. Trans. Electr. Energy Syst.* **2014**, *25*, 958–975. [CrossRef]
84. Shokri-Ghaleh, H.; Alfi, A. A comparison between optimization algorithms applied to synchronization of bilateral teleoperation systems against time delay and modeling uncertainties. *Appl. Soft Comput.* **2014**, *24*, 447–456. [CrossRef]
85. Gheisarnejad, M. An effective hybrid harmony search and cuckoo optimization algorithm based fuzzy PID controller for load frequency control. *Appl. Soft Comput.* **2018**, *65*, 121–138. [CrossRef]
86. Mahmoudi, S.; Lotfi, S. Modified cuckoo optimization algorithm (MCOA) to solve graph coloring problem. *Appl. Soft Comput.* **2015**, *33*, 48–64. [CrossRef]
87. Mohammadrezapour, O.; YoosefDoost, I.; Ebrahimi, M. Cuckoo optimization algorithm in optimal water allocation and crop planning under various weather conditions (case study: Qazvin plain, Iran). *Neural Comput. Appl.* **2017**, *31*, 1879–1892. [CrossRef]
88. Bayati, M. Using cuckoo optimization algorithm and imperialist competitive algorithm to solve inverse kinematics problem for numerical control of robotic manipulators. *Proc. Inst. Mech. Eng. Part I J. Syst. Control. Eng.* **2015**, *229*, 375–387. [CrossRef]

89. Dhiman, G.; Oliva, D.; Kaur, A.; Singh, K.K.; Vimal, S.; Sharma, A.; Cengiz, K. BEPO: A novel binary emperor penguin optimizer for automatic feature selection. *Knowl. Based Syst.* **2020**, *211*, 106560. [CrossRef]
90. Kaur, H.; Rai, A.; Bhatia, S.S.; Dhiman, G. MOEPO: A novel Multi-objective Emperor Penguin Optimizer for global optimization: Special application in ranking of cloud service providers. *Eng. Appl. Artif. Intell.* **2020**, *96*, 104008. [CrossRef]
91. Baliarsingh, S.K.; Vipsita, S.; Muhammad, K.; Bakshi, S. Analysis of high-dimensional biomedical data using an evolutionary multi-objective emperor penguin optimizer. *Swarm Evol. Comput.* **2019**, *48*, 262–273. [CrossRef]
92. Dhiman, G. ESA: A hybrid bio-inspired metaheuristic optimization approach for engineering problems. *Eng. Comput.* **2019**, *37*, 323–353. [CrossRef]
93. Baliarsingh, S.K.; Ding, W.; Vipsita, S.; Bakshi, S. A memetic algorithm using emperor penguin and social engineering optimization for medical data classification. *Appl. Soft Comput.* **2019**, *85*, 105773. [CrossRef]
94. Xing, Z. An improved emperor penguin optimization based multilevel thresholding for color image segmentation. *Knowl. Based Syst.* **2020**, *194*, 105570. [CrossRef]
95. Harifi, S.; Mohammadzadeh, J.; Khalilian, M.; Ebrahimnejad, S. Hybrid-EPC: An Emperor Penguins Colony algorithm with crossover and mutation operators and its application in community detection. *Prog. Artif. Intell.* **2021**, *10*, 181–193. [CrossRef]
96. Harifi, S.; Khalilian, M.; Mohammadzadeh, J.; Ebrahimnejad, S. Optimization in solving inventory control problem using nature inspired Emperor Penguins Colony algorithm. *J. Intell. Manuf.* **2020**, *32*, 1361–1375. [CrossRef]
97. Harifi, S.; Khalilian, M.; Mohammadzadeh, J.; Ebrahimnejad, S. Optimizing a Neuro-Fuzzy System Based on Nature-Inspired Emperor Penguins Colony Optimization Algorithm. *IEEE Trans. Fuzzy Syst.* **2020**, *28*, 1110–1124. [CrossRef]
98. Chen, H.; Jiao, S.; Wang, M.; Heidari, A.A.; Zhao, X. Parameters identification of photovoltaic cells and modules using diversification-enriched Harris hawks optimization with chaotic drifts. *J. Clean. Prod.* **2019**, *244*, 118778. [CrossRef]
99. Zhang, Y.; Liu, R.; Wang, X.; Chen, H.; Li, C. Boosted binary Harris hawks optimizer and feature selection. *Eng. Comput.* **2021**, *37*, 3741–3770. [CrossRef]
100. Chen, H.; Heidari, A.A.; Chen, H.; Wang, M.; Pan, Z.; Gandomi, A.H. Multi-population differential evolution-assisted Harris hawks optimization: Framework and case studies. *Future Gener. Comput. Syst.* **2020**, *111*, 175–198. [CrossRef]
101. Essa, F.; Elaziz, M.A.; Elsheikh, A. An enhanced productivity prediction model of active solar still using artificial neural network and Harris Hawks optimizer. *Appl. Therm. Eng.* **2020**, *170*, 115020. [CrossRef]
102. Meng, T.; Pan, Q.-K.; Li, J.-Q.; Sang, H.-Y. An improved migrating birds optimization for an integrated lot-streaming flow shop scheduling problem. *Swarm Evol. Comput.* **2018**, *38*, 64–78. [CrossRef]
103. Segredo, E.; Lalla-Ruiz, E.; Hart, E.; Voß, S. On the performance of the hybridisation between migrating birds optimisation variants and differential evolution for large scale continuous problems. *Expert Syst. Appl.* **2018**, *102*, 126–142. [CrossRef]
104. Sioud, A.; Gagné, C. Enhanced migrating birds optimization algorithm for the permutation flow shop problem with sequence dependent setup times. *Eur. J. Oper. Res.* **2018**, *264*, 66–73. [CrossRef]
105. Zhang, B.; Pan, Q.-K.; Gao, L.; Zhang, X.-L.; Sang, H.-Y.; Li, J.-Q. An effective modified migrating birds optimization for hybrid flowshop scheduling problem with lot streaming. *Appl. Soft Comput.* **2017**, *52*, 14–27. [CrossRef]
106. Gao, L.; Pan, Q.-K. A shuffled multi-swarm micro-migrating birds optimizer for a multi-resource-constrained flexible job shop scheduling problem. *Inf. Sci.* **2016**, *372*, 655–676. [CrossRef]
107. Niroomand, S.; Hadi-Vencheh, A.; Şahin, R.; Vizvári, B. Modified migrating birds optimization algorithm for closed loop layout with exact distances in flexible manufacturing systems. *Expert Syst. Appl.* **2015**, *42*, 6586–6597. [CrossRef]
108. Pan, Q.-K.; Dong, Y. An improved migrating birds optimisation for a hybrid flowshop scheduling with total flowtime minimisation. *Inf. Sci.* **2014**, *277*, 643–655. [CrossRef]
109. Jain, M.; Singh, V.; Rani, A. A novel nature-inspired algorithm for optimization: Squirrel search algorithm. *Swarm Evol. Comput.* **2019**, *44*, 148–175. [CrossRef]
110. Andrea, H.; Aranguren, I.; Oliva, D.; Abd Elaziz, M.; Cuevas, E. Efficient image segmentation through 2D histograms and an improved owl search algorithm. *Int. J. Mach. Learn. Cybern.* **2021**, *12*, 131–150. [CrossRef]
111. El-Ashmawi, W.H.; Elminaam, D.S.A.; Nabil, A.M.; Eldesouky, E. A chaotic owl search algorithm based bilateral negotiation model. *Ain Shams Eng. J.* **2020**, *11*, 1163–1178. [CrossRef]
112. Mandal, A.K.; Sen, R.; Chakraborty, B. Binary owl search algorithm for feature subset selection. In Proceedings of the 2019 IEEE 10th International Conference on Awareness Science and Technology (iCAST), IEEE, Morioka, Japan, 23–25 October 2019.
113. Zhong, Y.; Wang, L.; Lin, M.; Zhang, H. Discrete pigeon-inspired optimization algorithm with Metropolis acceptance criterion for large-scale traveling salesman problem. *Swarm Evol. Comput.* **2019**, *48*, 134–144. [CrossRef]
114. Wang, H.; Zhang, Z.; Dai, Z.; Chen, J.; Zhu, X.; Du, W.; Cao, X. Heterogeneous pigeon-inspired optimization. *Sci. China Inf. Sci.* **2019**, *62*, 70205. [CrossRef]
115. Yang, Z.; Duan, H.; Fan, Y.; Deng, Y. Automatic Carrier Landing System multilayer parameter design based on Cauchy Mutation Pigeon-Inspired Optimization. *Aerosp. Sci. Technol.* **2018**, *79*, 518–530. [CrossRef]
116. Deng, Y.; Duan, H. Control parameter design for automatic carrier landing system via pigeon-inspired optimization. *Nonlinear Dyn.* **2016**, *85*, 97–106. [CrossRef]
117. Qiu, H.; Duan, H. Multi-objective pigeon-inspired optimization for brushless direct current motor parameter design. *Sci. China Ser. E Technol. Sci.* **2015**, *58*, 1915–1923. [CrossRef]

118. Zhang, B.; Duan, H. Predator-Prey Pigeon-Inspired Optimization for UAV Three-Dimensional Path Planning. In *Advances in Swarm Intelligence, Icsi 2014, Pt Ii*; Tan, Y., Shi, Y., Coello, C.A.C., Eds.; Springer: Cham, Switzerland, 2014; pp. 96–105.
119. Jiang, F.; He, J.; Zeng, Z. Pigeon-inspired optimization and extreme learning machine via wavelet packet analysis for predicting bulk commodity futures prices. *Sci. China Inf. Sci.* **2019**, *62*, 70204. [CrossRef]
120. Torabi, S.; Safi-Esfahani, F. Improved Raven Roosting Optimization algorithm (IRRO). *Swarm Evol. Comput.* **2018**, *40*, 144–154. [CrossRef]
121. Torabi, S.; Safi-Esfahani, F. A dynamic task scheduling framework based on chicken swarm and improved raven roosting optimization methods in cloud computing. *J. Supercomput.* **2018**, *74*, 2581–2626. [CrossRef]
122. Zhang, S.; Zhou, Y.; Luo, Q. A Complex-Valued Encoding Satin Bowerbird Optimization Algorithm for Global Optimization. *Evolving Systems* **2021**, *12*, 191–205. [CrossRef]
123. El-Hay, E.; El-Hameed, M.; El-Fergany, A. Steady-state and dynamic models of solid oxide fuel cells based on Satin Bowerbird Optimizer. *Int. J. Hydrogen Energy* **2018**, *43*, 14751–14761. [CrossRef]
124. Dhiman, G.; Singh, K.K.; Soni, M.; Nagar, A.; Dehghani, M.; Slowik, A.; Kaur, A.; Sharma, A.; Houssein, E.H.; Cengiz, K. MOSOA: A new multi-objective seagull optimization algorithm. *Expert Syst. Appl.* **2020**, *167*, 114150. [CrossRef]
125. Che, Y.; He, D. A Hybrid Whale Optimization with Seagull Algorithm for Global Optimization Problems. *Math. Probl. Eng.* **2021**, *2021*, 1–31.
126. Das, G.; Panda, R. Seagull-Cuckoo Search Algorithm for Function Optimization. In Proceedings of the 2021 6th International Conference for Convergence in Technology (I2CT), IEEE, Maharashtra, India, 2–4 April 2021.
127. Jia, H.; Xing, Z.; Song, W. A New Hybrid Seagull Optimization Algorithm for Feature Selection. *IEEE Access* **2019**, *7*, 49614–49631. [CrossRef]
128. Ali, H.H.; Fathy, A.; Kassem, A.M. Optimal model predictive control for LFC of multi-interconnected plants comprising renewable energy sources based on recent sooty terns approach. *Sustain. Energy Technol. Assess.* **2020**, *42*, 100844. [CrossRef]
129. Addi, N.S.; Abdullah, S.; Hamdan, A.R. Multi-population cooperative bat algorithm-based optimization of artificial neural network model. *Inf. Sci.* **2015**, *294*, 628–644.
130. Rekaby, A. Directed Artificial Bat Algorithm (DABA)-A new bio-inspired algorithm. In Proceedings of the 2013 International Conference on Advances in Computing, Communications and Informatics (ICACCI), IEEE, Mysore, India, 22–25 August 2013.
131. Topal, A.O.; Altun, O. A novel meta-heuristic algorithm: Dynamic Virtual Bats Algorithm. *Inf. Sci.* **2016**, *354*, 222–235. [CrossRef]
132. Kaveh, A.; Farhoudi, N. A new optimization method: Dolphin echolocation. *Adv. Eng. Softw.* **2013**, *59*, 53–70. [CrossRef]
133. Ebrahimi, A.; Khamehchi, E. Sperm whale algorithm: An effective metaheuristic algorithm for production optimization problems. *J. Nat. Gas Sci. Eng.* **2016**, *29*, 211–222. [CrossRef]
134. Mirjalili, S.; Lewis, A. The Whale Optimization Algorithm. *Adv. Eng. Softw.* **2016**, *95*, 51–67. [CrossRef]
135. Gharehchopogh, F.S.; Gholizadeh, H. A comprehensive survey: Whale Optimization Algorithm and its applications. *Swarm Evol. Comput.* **2019**, *48*, 1–24. [CrossRef]
136. Mirjalili, S.; Mirjalili, S.M.; Lewis, A. Grey Wolf Optimizer. *Adv. Eng. Softw.* **2014**, *69*, 46–61. [CrossRef]
137. KKumar, V.; Kumar, D. An astrophysics-inspired Grey wolf algorithm for numerical optimization and its application to engineering design problems. *Adv. Eng. Softw.* **2017**, *112*, 231–254. [CrossRef]
138. Fong, S.; Deb, S.; Yang, X.-S. A heuristic optimization method inspired by wolf preying behavior. *Neural Comput. Appl.* **2015**, *26*, 1725–1738. [CrossRef]
139. Bansal, J.C.; Sharma, H.; Jadon, S.S.; Clerc, M. Spider Monkey Optimization algorithm for numerical optimization. *Memetic Comput.* **2014**, *6*, 31–47. [CrossRef]
140. Khishe, M.; Mosavi, M.R. Chimp optimization algorithm. *Expert Syst. Appl.* **2020**, *149*, 113338. [CrossRef]
141. Abdollahzadeh, B.; Gharehchopogh, F.S.; Mirjalili, S. Artificial gorilla troops optimizer: A new nature-inspired metaheuristic algorithm for global optimization problems. *Int. J. Intell. Syst.* **2021**, *36*, 5887–5958. [CrossRef]
142. Dhiman, G.; Kumar, V. Spotted hyena optimizer: A novel bio-inspired based metaheuristic technique for engineering applications. *Adv. Eng. Softw.* **2017**, *114*, 48–70. [CrossRef]
143. Rajakumar, B.R. The Lion's Algorithm: A New Nature-Inspired Search Algorithm. *Procedia Technol.* **2012**, *6*, 126–135. [CrossRef]
144. Yazdani, M.; Jolai, F. Lion Optimization Algorithm (LOA): A nature-inspired metaheuristic algorithm. *J. Comput. Des. Eng.* **2015**, *3*, 24–36. [CrossRef]
145. Kaveh, A.; Mahjoubi, S. Lion Pride Optimization Algorithm: A meta-heuristic method for global optimization problems. *Sci. Iran.* **2018**, *25*, 3113–3132. [CrossRef]
146. I Mohammad, T.M.H.; Mohammad, H.B.; Shirzadi, M.T.M.H.; Bagheri, M.H. A novel meta-heuristic algorithm for numerical function optimization: Blind, naked mole-rats (BNMR) algorithm. *Sci. Res. Essays* **2012**, *7*, 3566–3583. [CrossRef]
147. Deb, S.; Fong, S.; Tian, Z. Elephant search algorithm for optimization problems. In Proceedings of the 2015 Tenth International Conference on Digital Information Management (ICDIM), IEEE, Jeju, Korea, 21–23 October 2015.
148. Wang, G.G.; Deb, S.; Coelho, L.D.S. Elephant Herding Optimization. In Proceedings of the 2015 3rd International Symposium on Computational and Business Intelligence, IEEE, Bali, Indonesia, 7–9 December 2015.
149. Wang, G.G.; Deb, S.; Gao, X.Z.; Coelho, L.D.S. A new metaheuristic optimisation algorithm motivated by elephant herding behaviour. *Int. J. Bio-Inspired Comput.* **2016**, *8*, 394. [CrossRef]

150. Osaba, E.; Yang, X.-S.; Fister, I.; Del Ser, J.; Lopez-Garcia, P.; Vazquez-Pardavila, A.J. A Discrete and Improved Bat Algorithm for solving a medical goods distribution problem with pharmacological waste collection. *Swarm Evol. Comput.* **2019**, *44*, 273–286. [CrossRef]
151. Chakri, A.; Khelif, R.; Benouaret, M.; Yang, X.-S. New directional bat algorithm for continuous optimization problems. *Expert Syst. Appl.* **2017**, *69*, 159–175. [CrossRef]
152. Meng, X.-B.; Gao, X.; Liu, Y.; Zhang, H. A novel bat algorithm with habitat selection and Doppler effect in echoes for optimization. *Expert Syst. Appl.* **2015**, *42*, 6350–6364. [CrossRef]
153. Yılmaz, S.; Kucuksille, E.U. A new modification approach on bat algorithm for solving optimization problems. *Appl. Soft Comput.* **2015**, *28*, 259–275. [CrossRef]
154. Fister, I., Jr.; Fister, D.; Yang, X.S. A hybrid bat algorithm. *arXiv* **2013**, arXiv:1303.6310.
155. Mirjalili, S.; Mirjalili, S.M.; Yang, X.S. Binary bat algorithm. *Neural Comput. Appl.* **2014**, *25*, 663–681. [CrossRef]
156. Hong, W.-C.; Li, M.-W.; Geng, J.; Zhang, Y. Novel chaotic bat algorithm for forecasting complex motion of floating platforms. *Appl. Math. Model.* **2019**, *72*, 425–443. [CrossRef]
157. Ahmadlou, M.; Karimi, M.; Alizadeh, S.; Shirzadi, A.; Parvinnejhad, D.; Shahabi, H.; Panahi, M. Flood susceptibility assessment using integration of adaptive network-based fuzzy inference system (ANFIS) and biogeography-based optimization (BBO) and BAT algorithms (BA). *Geocarto Int.* **2018**, *34*, 1252–1272. [CrossRef]
158. Dao, T.-K.; Pan, T.-S.; Nguyen, T.-T.; Pan, J.-S. Parallel bat algorithm for optimizing makespan in job shop scheduling problems. *J. Intell. Manuf.* **2015**, *29*, 451–462. [CrossRef]
159. Osaba, E.; Yang, X.-S.; Diaz, F.; Lopez-Garcia, P.; Carballedo, R. An improved discrete bat algorithm for symmetric and asymmetric Traveling Salesman Problems. *Eng. Appl. Artif. Intell.* **2016**, *48*, 59–71. [CrossRef]
160. Bahmani-Firouzi, B.; Azizipanah-Abarghooee, R. Optimal sizing of battery energy storage for micro-grid operation management using a new improved bat algorithm. *Int. J. Electr. Power Energy Syst.* **2014**, *56*, 42–54. [CrossRef]
161. Gandomi, A.H.; Yang, X.-S.; Alavi, A.H.; Talatahari, S. Bat algorithm for constrained optimization tasks. *Neural Comput. Appl.* **2012**, *22*, 1239–1255. [CrossRef]
162. Hasançebi, O.; Teke, T.; Pekcan, O. A bat-inspired algorithm for structural optimization. *Comput. Struct.* **2013**, *128*, 77–90. [CrossRef]
163. Taherdangkoo, M.; Shirzadi, M.H.; Yazdi, M.; Bagheri, M.H. A robust clustering method based on blind, naked mole-rats (BNMR) algorithm. *Swarm Evol. Comput.* **2013**, *10*, 1–11. [CrossRef]
164. Kaur, M.; Kaur, R.; Singh, N.; Dhiman, G. SChoA: A newly fusion of sine and cosine with chimp optimization algorithm for HLS of datapaths in digital filters and engineering applications. *Eng. Comput.* **2021**, 1–29. [CrossRef]
165. Khishe, M.; Mosavi, M. Classification of underwater acoustical dataset using neural network trained by Chimp Optimization Algorithm. *Appl. Acoust.* **2019**, *157*, 107005. [CrossRef]
166. Kaveh, A.; Hosseini, P. A simplified dolphin echolocation optimization method for optimum design of trusses. *Iran Univ. Sci. Technol.* **2014**, *4*, 381–397.
167. Daryan, A.S.; Palizi, S.; Farhoudi, N. Optimization of plastic analysis of moment frames using modified dolphin echolocation algorithm. *Adv. Struct. Eng.* **2019**, *22*, 2504–2516. [CrossRef]
168. Gholizadeh, S.; Poorhoseini, H. Optimum design of steel frame structures by a modified dolphin echolocation algorithm. *Struct. Eng. Mech.* **2015**, *55*, 535–554. [CrossRef]
169. Lenin, K.; Reddy, B.R.; Kalavathi, M.S. Dolphin echolocation algorithm for solving optimal reactive power dispatch problem. *Int. J. Comput.* **2014**, *12*, 1–15.
170. Topal, A.O.; Yildiz, Y.E.; Ozkul, M. Improved Dynamic Virtual Bats Algorithm for Global Numerical Optimization. In Proceedings of the World Congress on Engineering and Computer Science, San Francisco, CA, USA, 25–27 October 2017.
171. Elhosseini, M.A.; El Sehiemy, R.A.; Rashwan, Y.I.; Gao, X. On the performance improvement of elephant herding optimization algorithm. *Knowl. Based Syst.* **2019**, *166*, 58–70. [CrossRef]
172. Jafari, M.; Salajegheh, E.; Salajegheh, J. An efficient hybrid of elephant herding optimization and cultural algorithm for optimal design of trusses. *Eng. Comput.* **2018**, *35*, 781–801. [CrossRef]
173. Sadouki, S.C.; Tari, A. Multi-objective and discrete Elephants Herding Optimization algorithm for QoS aware web service composition. *RAIRO Oper. Res.* **2019**, *53*, 445–459. [CrossRef]
174. Tuba, E.; Capor-Hrosik, R.; Alihodzic, A.; Jovanovic, R.; Tuba, M. Chaotic elephant herding optimization algorithm. In Proceedings of the 2018 IEEE 16th World Symposium on Applied Machine Intelligence and Informatics (SAMI); IEEE, Kosice and Herlany, Slovakia, 7–10 February 2018.
175. Xu, H.; Cao, Q.; Fu, H.; Fu, C.; Chen, H.; Su, J. Application of Support Vector Machine Model Based on an Improved Elephant Herding Optimization Algorithm in Network Intrusion Detection. In *International CCF Conference on Artificial Intelligence, Xuzhou, China, 22–23 August 2019*; Springer: Singapore, 2019.
176. Tuba, E.; Alihodzic, A.; Tuba, M. Multilevel image thresholding using elephant herding optimization algorithm. In Proceedings of the 2017 14th International Conference on Engineering of Modern Electric Systems (EMES), IEEE, Oradea, Romania,, 1–2 June 2017.
177. Tuba, E.; Ribic, I.; Capor-Hrosik, R.; Tuba, M. Support Vector Machine Optimized by Elephant Herding Algorithm for Erythemato-Squamous Diseases Detection. *Procedia Comput. Sci.* **2017**, *122*, 916–923. [CrossRef]

178. Pichpibul, T. Modified Elephant Search Algorithm for Distribution of Snack Food in Thailand. In Proceedings of the 2nd International Conference on Intelligent Systems, Metaheuristics & Swarm Intelligence, ACM, Phuket, Thailand, 24–25 March 2018.
179. Tian, Z.; Fong, S.; Wong, R.; Millham, R. Elephant search algorithm on data clustering. In Proceedings of the 2016 12th International Conference on Natural Computation, Fuzzy Systems and Knowledge Discovery (ICNC-FSKD), IEEE, Changsha, China, 13–15 August 2016.
180. Deb, S.; Tian, Z.; Fong, S.; Wong, R.; Millham, R.; Wong, K.K.L. Elephant search algorithm applied to data clustering. *Soft Comput.* **2018**, *22*, 6035–6046. [CrossRef]
181. Deb, S.; Fong, S.; Tian, Z.; Wong, R.K.; Mohammed, S.; Fiaidhi, J. Finding approximate solutions of NP-hard optimization and TSP problems using elephant search algorithm. *J. Supercomput.* **2016**, *72*, 3960–3992. [CrossRef]
182. Abdel-Basset, M.; El-Shahat, D.; El-Henawy, I.; de Albuquerque, V.H.C.; Mirjalili, S. A new fusion of grey wolf optimizer algorithm with a two-phase mutation for feature selection. *Expert Syst. Appl.* **2020**, *139*, 112824. [CrossRef]
183. Gupta, S.; Deep, K. A novel Random Walk Grey Wolf Optimizer. *Swarm Evol. Comput.* **2019**, *44*, 101–112. [CrossRef]
184. Lu, C.; Gao, L.; Yi, J. Grey wolf optimizer with cellular topological structure. *Expert Syst. Appl.* **2018**, *107*, 89–114. [CrossRef]
185. Qais, M.H.; Hasanien, H.M.; Alghuwainem, S. Augmented grey wolf optimizer for grid-connected PMSG-based wind energy conversion systems. *Appl. Soft Comput.* **2018**, *69*, 504–515. [CrossRef]
186. Emary, E.; Zawbaa, H.M.; Hassanien, A.E. Binary grey wolf optimization approaches for feature selection. *Neurocomputing* **2016**, *172*, 371–381. [CrossRef]
187. Mirjalili, S.; Saremi, S.; Mirjalili, S.M.; Coelho, L.D.S. Multi-objective grey wolf optimizer: A novel algorithm for multi-criterion optimization. *Expert Syst. Appl.* **2016**, *47*, 106–119. [CrossRef]
188. Mirjalili, S.; Aljarah, I.; Mafarja, M.; Heidari, A.A.; Faris, H. Grey Wolf optimizer: Theory, literature review, and application in computational fluid dynamics problems. In *Nature-Inspired Optimizers*; Springer: Cham, Switzerland, 2020; pp. 87–105.
189. Nahak, N.; Sahoo, S.R.; Mallick, R.K. Design of dual optimal UPFC based PI controller to damp low frequency oscillation in power system. In Proceedings of the Technologies for Smart-City Energy Security and Power (ICSESP), IEEE, Bhubaneswar, India, 28–30 March 2018.
190. Emary, E.; Zawbaa, H.M.; Grosan, C. Experienced Gray Wolf Optimization Through Reinforcement Learning and Neural Networks. *IEEE Trans. Neural Networks Learn. Syst.* **2017**, *29*, 681–694. [CrossRef]
191. Mohanty, S.; Subudhi, B.; Ray, P.K. A New MPPT Design Using Grey Wolf Optimization Technique for Photovoltaic System Under Partial Shading Conditions. *IEEE Trans. Sustain. Energy* **2015**, *7*, 181–188. [CrossRef]
192. Rajakumar, B. Lion algorithm for standard and large scale bilinear system identification: A global optimization based on Lion's social behavior. In Proceedings of the 2014 IEEE Congress on Evolutionary Computation (CEC); IEEE, Beijing, China, 6–11 July 2014.
193. Marichelvam, M.; Manimaran, P.; Geetha, M. Solving flexible job shop scheduling problems using a hybrid lion optimisation algorithm. *Int. J. Adv. Oper. Manag.* **2018**, *10*, 91–108. [CrossRef]
194. Paraskar, S.; Singh, D.K.; Tapre, P.C. Lion algorithm for generation rescheduling based congestion management in deregulated power system. In Proceedings of the 2017 International Conference on Energy, Communication, Data Analytics and Soft Computing (ICECDS), IEEE, Chennai, India, 1–2 August 2017.
195. Sowmiyasree, S.; Sumitra, P. Lion Optimization Algorithm Using Data Mining Classification and Clustering Models. *GSJ* **2018**, *6*, 219–226.
196. Kaveh, A.; Mahjoubi, S. Optimum Design of Double-layer Barrel Vaults by Lion Pride Optimization Algorithm and a Comparative Study. *Structures* **2018**, *13*, 213–229. [CrossRef]
197. Engy, E.; Ali, E.; Sally, E.-G. An optimized artificial neural network approach based on sperm whale optimization algorithm for predicting fertility quality. *Stud. Inform. Control.* **2018**, *27*, 349–358.
198. Sharma, N.; Kaur, A.; Sharma, H.; Sharma, A.; Bansal, J.C. Chaotic Spider Monkey Optimization Algorithm with Enhanced Learning. In *Soft Computing for Problem Solving*; Springer: Singapore, 2018; pp. 149–161.
199. Sharma, A.; Sharma, H.; Bhargava, A.; Sharma, N.; Bansal, J.C. Optimal power flow analysis using lévy flight spider monkey optimisation algorithm. *Int. J. Artif. Intell. Soft Comput.* **2017**, *5*, 320–352. [CrossRef]
200. Gupta, K.; Deep, K.; Bansal, J.C. Improving the Local Search Ability of Spider Monkey Optimization Algorithm Using Quadratic Approximation for Unconstrained Optimization. *Comput. Intell.* **2016**, *33*, 210–240. [CrossRef]
201. Sharma, A.; Sharma, A.; Panigrahi, B.K.; Kiran, D.; Kumar, R. Ageist Spider Monkey Optimization algorithm. *Swarm Evol. Comput.* **2016**, *28*, 58–77. [CrossRef]
202. Sharma, A.; Sharma, H.; Bhargava, A.; Sharma, N.; Bansal, J.C. Optimal placement and sizing of capacitor using Limaçon inspired spider monkey optimization algorithm. *Memetic Comput.* **2016**, *9*, 311–331. [CrossRef]
203. Singh, P.R.; Elaziz, M.A.; Xiong, S. Modified Spider Monkey Optimization based on Nelder–Mead method for global optimization. *Expert Syst. Appl.* **2018**, *110*, 264–289. [CrossRef]
204. Singh, U.; Salgotra, R.; Rattan, M. A Novel Binary Spider Monkey Optimization Algorithm for Thinning of Concentric Circular Antenna Arrays. *IETE J. Res.* **2016**, *62*, 736–744. [CrossRef]
205. Tripathy, D.; Sahu, B.K.; Patnaik, B.; Choudhury, N.D. Spider monkey optimization based fuzzy-2D-PID controller for load frequency control in two-area multi source interconnected power system. In Proceedings of the 2018 Technologies for Smart-City Energy Security and Power (ICSESP), IEEE, Bhubaneswar, India, 29–30 March 2018.

206. Ehteram, M.; Karami, H.; Farzin, S. Reducing Irrigation Deficiencies Based Optimizing Model for Multi-Reservoir Systems Utilizing Spider Monkey Algorithm. *Water Resour. Manag.* **2018**, *32*, 2315–2334. [CrossRef]
207. Cheruku, R.; Edla, D.R.; Kuppili, V. SM-RuleMiner: Spider monkey based rule miner using novel fitness function for diabetes classification. *Comput. Biol. Med.* **2017**, *81*, 79–92. [CrossRef] [PubMed]
208. Dhiman, G.; Kumar, V. Multi-objective spotted hyena optimizer: A Multi-objective optimization algorithm for engineering problems. *Knowl. Based Syst.* **2018**, *150*, 175–197. [CrossRef]
209. Dhiman, G.; Kaur, A. Spotted hyena optimizer for solving engineering design problems. In Proceedings of the 2017 International Conference on Machine Learning and Data Science (MLDS), IEEE, Noida, India, 14–15 December 2017.
210. Luo, Q.; Li, J.; Zhou, Y.; Liao, L. Using Spotted Hyena Optimizer for Training Feedforward Neural Networks. In Proceedings of the International Conference on Intelligent Computing, Wuhan, China, 15–18 August 2018.
211. Dhiman, G.; Kaur, A. Optimizing the Design of Airfoil and Optical Buffer Problems Using Spotted Hyena Optimizer. *Designs* **2018**, *2*, 28. [CrossRef]
212. Abdel-Basset, M.; Manogaran, G.; El-Shahat, D.; Mirjalili, S. A hybrid whale optimization algorithm based on local search strategy for the permutation flow shop scheduling problem. *Futur. Gener. Comput. Syst.* **2018**, *85*, 129–145. [CrossRef]
213. Mafarja, M.M.; Mirjalili, S. Hybrid Whale Optimization Algorithm with simulated annealing for feature selection. *Neurocomputing* **2017**, *260*, 302–312. [CrossRef]
214. Kumar, N.; Hussain, I.; Singh, B.; Panigrahi, B.K. MPPT in Dynamic Condition of Partially Shaded PV System by Using WODE Technique. *IEEE Trans. Sustain. Energy* **2017**, *8*, 1204–1214. [CrossRef]
215. Kaveh, A.; Ghazaan, M.I. Enhanced whale optimization algorithm for sizing optimization of skeletal structures. *Mech. Based Des. Struct. Mach.* **2016**, *45*, 345–362. [CrossRef]
216. Sun, W.Z.; Wang, J.S. Elman Neural Network Soft-Sensor Model of Conversion Velocity in Polymerization Process Optimized by Chaos Whale Optimization Algorithm. *IEEE Access* **2017**, *5*, 13062–13076. [CrossRef]
217. Medani, K.B.O.; Sayah, S.; Bekrar, A. Whale optimization algorithm based optimal reactive power dispatch: A case study of the Algerian power system. *Electr. Power Syst. Res.* **2018**, *163*, 696–705. [CrossRef]
218. Mehne, H.H.; Mirjalili, S. A parallel numerical method for solving optimal control problems based on whale optimization algorithm. *Knowl. Based Syst.* **2018**, *151*, 114–123. [CrossRef]
219. Tharwat, A.; Moemen, Y.S.; Hassanien, A.E. Classification of toxicity effects of biotransformed hepatic drugs using whale optimized support vector machines. *J. Biomed. Inform.* **2017**, *68*, 132–149. [CrossRef] [PubMed]
220. Dao, T.-K.; Pan, T.-S.; Pan, J.-S. A multi-objective optimal mobile robot path planning based on whale optimization algorithm. In Proceedings of the 2016 IEEE 13th International Conference on Signal Processing (ICSP); IEEE, Chengdu, China, 6–10 November 2016.
221. Hassanien, A.E.; Abd Elfattah, M.; Aboulenin, S.; Schaefer, G.; Zhu, S.Y.; Korovin, I. Historic handwritten manuscript binarisation using whale optimisation. In Proceedings of the 2016 IEEE International Conference on Systems, Man, and Cybernetics (SMC), IEEE, Budapest, Hungary, 9–12 October 2016.
222. Shadravan, S.; Naji, H.; Bardsiri, V. The Sailfish Optimizer: A novel nature-inspired metaheuristic algorithm for solving constrained engineering optimization problems. *Eng. Appl. Artif. Intell.* **2019**, *80*, 20–34. [CrossRef]
223. Kumar, N.; Singh, N.; Vidyarthi, D.P. Artificial lizard search optimization (ALSO): A novel nature-inspired meta-heuristic algorithm. *Soft Comput.* **2021**, *25*, 6179–6201. [CrossRef]
224. Braik, M.S. Chameleon Swarm Algorithm: A bio-inspired optimizer for solving engineering design problems. *Expert Syst. Appl.* **2021**, *174*, 114685. [CrossRef]
225. Boettcher, S.; Percus, A. Nature's way of optimizing. *Artif. Intell.* **2000**, *119*, 275–286. [CrossRef]
226. Oftadeh, R.; Mahjoob, M.; Shariatpanahi, M. A novel meta-heuristic optimization algorithm inspired by group hunting of animals: Hunting search. *Comput. Math. Appl.* **2010**, *60*, 2087–2098. [CrossRef]
227. Civicioglu, P. Backtracking Search Optimization Algorithm for numerical optimization problems. *Appl. Math. Comput.* **2013**, *219*, 8121–8144. [CrossRef]
228. Haldar, V.; Chakraborty, N. A novel evolutionary technique based on electrolocation principle of elephant nose fish and shark: Fish electrolocation optimization. *Soft Comput.* **2016**, *21*, 3827–3848. [CrossRef]
229. Faramarzi, A.; Heidarinejad, M.; Mirjalili, S.; Gandomi, A.H. Marine Predators Algorithm: A nature-inspired metaheuristic. *Expert Syst. Appl.* **2020**, *152*, 113377. [CrossRef]
230. Mohseni, S.; Gholami, R.; Zarei, N.; Zadeh, A.R. Competition over resources: A new optimization algorithm based on animals behavioral ecology. In Proceedings of the 2014 International Conference on Intelligent Networking and Collaborative Systems (INCoS), Salerno, Italy, 10–12 September 2014.
231. Sharafi, Y.; Khanesar, M.A.; Teshnehlab, M. COOA: Competitive optimization algorithm. *Swarm Evol. Comput.* **2016**, *30*, 39–63. [CrossRef]
232. Atashpaz-Gargari, E.; Lucas, C. Imperialist competitive algorithm: An algorithm for optimization inspired by imperialistic competition. In Proceedings of the IEEE Congress on Evolutionary Computation, Singapore, 25–28 September 2007.
233. Simon, D. Biogeography-based optimization. *IEEE Trans. Evol. Comput.* **2008**, *12*, 702–713. [CrossRef]
234. Niu, Q.; Zhang, L.; Li, K. A biogeography-based optimization algorithm with mutation strategies for model parameter estimation of solar and fuel cells. *Energy Convers. Manag.* **2014**, *86*, 1173–1185. [CrossRef]

235. Li, X.; Zhang, J.; Yin, M. Animal migration optimization: An optimization algorithm inspired by animal migration behavior. *Neural Comput. Appl.* **2013**, *24*, 1867–1877. [CrossRef]
236. Lai, Z.; Feng, X.; Yu, H. An Improved Animal Migration Optimization Algorithm Based on Interactive Learning Behavior for High Dimensional Optimization Problem. In Proceedings of the 2019 International Conference on High Performance Big Data and Intelligent Systems (HPBD&IS), IEEE, Shenzhen, China, 9–11 May 2019.
237. Cao, Y.; Li, X.; Wang, J. Opposition-Based Animal Migration Optimization. *Math. Probl. Eng.* **2013**, *2013*, 1–7. [CrossRef]
238. Son, L.H.; Chiclana, F.; Kumar, R.; Mittal, M.; Khari, M.; Chatterjee, J.M.; Baik, S.W. ARM–AMO: An efficient association rule mining algorithm based on animal migration optimization. *Knowl. Based Syst.* **2018**, *154*, 68–80. [CrossRef]
239. Ma, M.; Luo, Q.; Zhou, Y.; Chen, X.; Li, L. An Improved Animal Migration Optimization Algorithm for Clustering Analysis. *Discret. Dyn. Nat. Soc.* **2015**, *2015*, 1–12. [CrossRef]
240. Morales, A.; Crawford, B.; Soto, R.; Lemus-Romani, J.; Astorga, G.; Salas-Fernández, A.; Rubio, J.M. Optimization of Bridges Reinforcement by Conversion to Tied Arch Using an Animal Migration Algorithm. In Proceedings of the International Conference on Industrial, Engineering and Other Applications of Applied Intelligent Systems, Graz, Austria, 9–11 July 2019.
241. Farshi, T.R. A multilevel image thresholding using the animal migration optimization algorithm. *Iran J. Comput. Sci.* **2018**, *2*, 9–22. [CrossRef]
242. Tsai, H.-C. Improving backtracking search algorithm with variable search strategies for continuous optimization. *Appl. Soft Comput.* **2019**, *80*, 567–578. [CrossRef]
243. Zhou, J.; Ye, H.; Ji, X.; Deng, W. An improved backtracking search algorithm for casting heat treatment charge plan problem. *J. Intell. Manuf.* **2017**, *30*, 1335–1350. [CrossRef]
244. Lin, J. Oppositional backtracking search optimization algorithm for parameter identification of hyperchaotic systems. *Nonlinear Dyn.* **2014**, *80*, 209–219. [CrossRef]
245. Chen, D.; Zou, F.; Lu, R.; Wang, P. Learning backtracking search optimisation algorithm and its application. *Inf. Sci.* **2017**, *376*, 71–94. [CrossRef]
246. Zhang, C.; Lin, Q.; Gao, L.; Li, X. Backtracking Search Algorithm with three constraint handling methods for constrained optimization problems. *Expert Syst. Appl.* **2015**, *42*, 7831–7845. [CrossRef]
247. Pourdaryaei, A.; Mokhlis, H.; Illias, H.A.; Kaboli, S.H.A.; Ahmad, S. Short-Term Electricity Price Forecasting via Hybrid Backtracking Search Algorithm and ANFIS Approach. *IEEE Access* **2019**, *7*, 77674–77691. [CrossRef]
248. Ma, H.; Fei, M.; Simon, D.; Chen, Z. Biogeography-based optimization in noisy environments. *Trans. Inst. Meas. Control.* **2014**, *37*, 190–204. [CrossRef]
249. Saremi, S.; Mirjalili, S.; Lewis, A. Biogeography-based optimisation with chaos. *Neural Comput. Appl.* **2014**, *25*, 1077–1097. [CrossRef]
250. Pham, B.T.; Nguyen, M.D.; Bui, K.-T.T.; Prakash, I.; Chapi, K.; Bui, D.T. A novel artificial intelligence approach based on Multi-layer Perceptron Neural Network and Biogeography-based Optimization for predicting coefficient of consolidation of soil. *Catena* **2018**, *173*, 302–311. [CrossRef]
251. Mendes, L.A.; Freire, R.Z.; Coelho, L.D.S.; Moraes, A.S. Minimizing computational cost and energy demand of building lighting systems: A real time experiment using a modified competition over resources algorithm. *Energy Build.* **2017**, *139*, 108–123. [CrossRef]
252. Bouchekara, H.R.; Nahas, M. Optimization of magnetic actuators using competition over resources algorithm. In *Progress in Electromagnetics Research Symposium-Fall (PIERS-FALL), Singapore, 19–22 November 2017*; IEEE: Singapore, 2017.
253. Kulluk, S. A novel hybrid algorithm combining hunting search with harmony search algorithm for training neural networks. *J. Oper. Res. Soc.* **2013**, *64*, 748–761. [CrossRef]
254. Doğan, E.; Erdal, F. Hunting search algorithm based design optimization of steel cellular beams. In Proceedings of the 15th Annual Conference Companion on Genetic and Evolutionary Computation, New York, NY, USA, 6–10 July 2013.
255. Elaziz, M.A.; Ewees, A.A.; Yousri, D.; Alwerfali, H.S.N.; Awad, Q.A.; Lu, S.; Al-Qaness, M.A.A. An Improved Marine Predators Algorithm With Fuzzy Entropy for Multi-Level Thresholding: Real World Example of COVID-19 CT Image Segmentation. *IEEE Access* **2020**, *8*, 125306–125330. [CrossRef] [PubMed]
256. Zhong, K.; Luo, Q.; Zhou, Y.; Jiang, M. TLMPA: Teaching-learning-based Marine Predators algorithm. *AIMS Math.* **2021**, *6*, 1395–1442. [CrossRef]
257. Abdel-Basset, M.; Mohamed, R.; Chakrabortty, R.K.; Ryan, M.; Mirjalili, S. New binary marine predators optimization algorithms for 0–1 knapsack problems. *Comput. Ind. Eng.* **2020**, *151*, 106949. [CrossRef]
258. Ridha, H.M. Parameters extraction of single and double diodes photovoltaic models using Marine Predators Algorithm and Lambert W function. *Sol. Energy* **2020**, *209*, 674–693. [CrossRef]
259. Sayed, G.I.; Solyman, M.; Hassanien, A.E. A novel chaotic optimal foraging algorithm for unconstrained and constrained problems and its application in white blood cell segmentation. *Neural Comput. Appl.* **2018**, *31*, 7633–7664. [CrossRef]
260. Zhang, W.-B.; Zhu, G.-Y. Drilling Path Optimization by Optimal Foraging Algorithm. *IEEE Trans. Ind. Informatics* **2017**, *14*, 2847–2856. [CrossRef]
261. Sayed, G.I.; Soliman, M.; Hassanien, A.E. Modified optimal foraging algorithm for parameters optimization of support vector machine. In Proceedings of the International Conference on Advanced Machine Learning Technologies and Applications, Cairo, Egypt, 22–24 February 2018.

262. Srivastava, S.; Sahana, S.K. *The Insects of Innovative Computational Intelligence*; Springer: Singapore, 2017.
263. Akay, B.; Karaboga, D. A modified Artificial Bee Colony algorithm for real-parameter optimization. *Inf. Sci.* **2012**, *192*, 120–142. [CrossRef]
264. Mirjalili, S. Dragonfly algorithm: A new meta-heuristic optimization technique for solving single-objective, discrete, and multi-objective problems. *Neural Comput. Appl.* **2015**, *27*, 1053–1073. [CrossRef]
265. Cuevas, E.; Cienfuegos, M.; Zaldivar-Navarro, D.; Perez-Cisneros, M.A. A swarm optimization algorithm inspired in the behavior of the social-spider. *Expert Syst. Appl.* **2013**, *40*, 6374–6384. [CrossRef]
266. Häckel, S.; Dippold, P. The Bee Colony-inspired Algorithm (BCiA): A two-stage approach for solving the vehicle routing problem with time windows. In Proceedings of the 11th Annual Genetic and Evolutionary Computation Conference, ACM, Montreal, Canada, 8–12 July 2009.
267. Rajasekhar, A.; Lynn, N.; Das, S.; Suganthan, P. Computing with the collective intelligence of honey bees—A survey. *Swarm Evol. Comput.* **2017**, *32*, 25–48. [CrossRef]
268. Diwold, K.; Beekman, M.; Middendorf, M. Honeybee optimisation–an overview and a new bee inspired optimisation scheme. In *Handbook of Swarm Intelligence, In Handbook of Swarm Intelligence*; Springer: Berlin/Heidelberg, Germany, 2011; pp. 295–327.
269. Comellas, F.; Martínez-Navarro, J. Bumblebees: A multiagent combinatorial optimization algorithm inspired by social insect behaviour. In Proceedings of the 1st ACM/SIGEVO Summit on Genetic and Evolutionary Computation, GEC'09; ACM, Shanghai, China, 12–14 June 2009.
270. Marinakis, Y.; Marinaki, M.; Matsatsinis, N. A Bumble Bees Mating Optimization Algorithm for Global Unconstrained Optimization Problems. In *Nature Inspired Cooperative Strategies for Optimization (NICSO 2010), Granada, Spain, 12–15 May 2010*; Springer: Berlin/Heidelberg, Germany, 2010; pp. 305–318.
271. Shnerb, N.M.; Louzoun, Y.; Bettelheim, E.; Solomon, S. The importance of being discrete: Life always wins on the surface. *Proc. Natl. Acad. Sci. USA* **2000**, *97*, 10322–10324. [CrossRef] [PubMed]
272. Dorigo, M.; Maniezzo, V.; Colorni, A. *The Ant System: An Autocatalytic Optimizing Process*; Politecnico di Milano: Milan, Italy, 1991; pp. 1–21.
273. Zungeru, A.M.; Ang, L.-M.; Seng, K.P. Termite-hill: Performance optimized swarm intelligence based routing algorithm for wireless sensor networks. *J. Netw. Comput. Appl.* **2012**, *35*, 1901–1917. [CrossRef]
274. Das, K.N.; Singh, T.K. Drosophila Food-Search Optimization. *Appl. Math. Comput.* **2014**, *231*, 566–580. [CrossRef]
275. Abidin, Z.Z.; Arshad, M.R.; Ngah, U.K. A Simulation Based Fly Optimization Algorithm for Swarms of Mini Autonomous Surface Vehicles Application. Available online: http://nopr.niscair.res.in/handle/123456789/11731 (accessed on 3 January 2021).
276. Pan, W.-T. A new Fruit Fly Optimization Algorithm: Taking the financial distress model as an example. *Knowl. Based Syst.* **2012**, *26*, 69–74. [CrossRef]
277. Feng, X.; Lau, F.C.M.; Gao, D. *A New Bio-Inspired Approach to the Traveling Salesman Problem*; Springer: Berlin/Heidelberg, Germany, 2009.
278. Feng, X.; Lau, F.C.; Yu, H. A novel bio-inspired approach based on the behavior of mosquitoes. *Inf. Sci.* **2013**, *233*, 87–108. [CrossRef]
279. Wang, G.-G.; Deb, S.; Cui, Z. Monarch butterfly optimization. *Neural Computing and Applications* **2015**, *31*, 1995–2014. [CrossRef]
280. Bhattacharjee, K.K.; Sarmah, S.P. Monarch Migration Algorithm for optimization problems. In Proceedings of the IEEE International Conference on Industrial Engineering and Engineering Management;IEEE, Bali, Indonesia, 4–7 December 2016.
281. Kumar, A.; Misra, R.K.; Singh, D. Butterfly optimizer. In Proceedings of the 2015 IEEE Workshop on Computational Intelligence: Theories, Applications and Future Directions, WCI 2015; IEEE, Kanpur, India, 14–17 December 2015.
282. Qi, X.; Zhu, Y.; Zhang, H. A new meta-heuristic butterfly-inspired algorithm. *J. Comput. Sci.* **2017**, *23*, 226–239. [CrossRef]
283. Arora, S.; Singh, S. Butterfly optimization algorithm: A novel approach for global optimization. *Soft Comput.* **2018**, *23*, 715–734. [CrossRef]
284. Mirjalili, S. Moth-flame optimization algorithm: A novel nature-inspired heuristic paradigm. *Knowl. Based Syst.* **2015**, *89*, 228–249. [CrossRef]
285. Mei, R.N.S.; Sulaiman, M.H.; Mustaffa, Z.; Daniyal, H. Optimal reactive power dispatch solution by loss minimization using moth-flame optimization technique. *Appl. Soft Comput.* **2017**, *59*, 210–222.
286. Mohamed, A.-A.A.; Mohamed, Y.S.; El-Gaafary, A.A.; Hemeida, A.M. Optimal power flow using moth swarm algorithm. *Electr. Power Syst. Res.* **2017**, *142*, 190–206. [CrossRef]
287. Wang, G.-G. Moth search algorithm: A bio-inspired metaheuristic algorithm for global optimization problems. *Memetic Comput.* **2016**, *10*, 151–164. [CrossRef]
288. Chen, S. Locust Swarms-A new multi-optima search technique. In Proceedings of the 2009 IEEE Congress on Evolutionary Computation. IEEE, Trondheim, Norway, 18–21 May 2009.
289. Cuevas, E.; Gonzalez, A.; Zaldívar, D.; Perez-Cisneros, M.A. An optimisation algorithm based on the behaviour of locust swarms. *Int. J. Bio-Inspired Comput.* **2015**, *7*, 402. [CrossRef]
290. Saremi, S.; Mirjalili, S.; Lewis, A. Grasshopper Optimisation Algorithm: Theory and application. *Adv. Eng. Softw.* **2017**, *105*, 30–47. [CrossRef]
291. Mirjalili, S. The Ant Lion Optimizer. *Adv. Eng. Softw.* **2015**, *83*, 80–98. [CrossRef]

292. Havens, T.C.; Spain, C.J.; Salmon, N.G.; Keller, J.M. Roach infestation optimization. In Proceedings of the 2008 IEEE Swarm Intelligence Symposium, St. Louis, MO, USA, 21–23 September 2008.
293. ZhaoHui, C.; HaiYan, T. Cockroach swarm optimization. In Proceedings of the 2010 2nd International Conference on Computer Engineering and Technology, Chengdu, China, 16–19 April 2010.
294. Wu, S.-J.; Wu, C.-T. A bio-inspired optimization for inferring interactive networks: Cockroach swarm evolution. *Expert Syst. Appl.* **2015**, *42*, 3253–3267. [CrossRef]
295. Kallioras, N.A.; Lagaros, N.D.; Avtzis, D.N. Pity beetle algorithm – A new metaheuristic inspired by the behavior of bark beetles. *Adv. Eng. Softw.* **2018**, *121*, 147–166. [CrossRef]
296. Mirjalili, S.; Jangir, P.; Saremi, S. Multi-objective ant lion optimizer: A multi-objective optimization algorithm for solving engineering problems. *Appl. Intell.* **2016**, *46*, 79–95. [CrossRef]
297. Emary, E.; Zawbaa, H.M.; Hassanien, A.E. Binary ant lion approaches for feature selection. *Neurocomputing* **2016**, *213*, 54–65. [CrossRef]
298. Emary, E.; Zawbaa, H.M. Impact of Chaos Functions on Modern Swarm Optimizers. *PLoS ONE* **2016**, *11*, e0158738. [CrossRef] [PubMed]
299. Heidari, A.A.; Faris, H.; Mirjalili, S.; Aljarah, I.; Mafarja, M. Ant Lion optimizer: Theory, literature review, and application in multi-layer perceptron neural networks. In *Nature-Inspired Optimizers*; Springer: Cham, Switzerland, 2020; pp. 23–46.
300. Raju, M.; Saikia, L.C.; Sinha, N. Automatic generation control of a multi-area system using ant lion optimizer algorithm based PID plus second order derivative controller. *Int. J. Electr. Power Energy Syst.* **2016**, *80*, 52–63. [CrossRef]
301. Marinakis, Y.; Marinaki, M. Combinatorial neighborhood topology bumble bees mating optimization for the vehicle routing problem with stochastic demands. *Soft Comput.* **2014**, *19*, 353–373. [CrossRef]
302. Marinakis, Y.; Marinaki, M.; Migdalas, A. An Adaptive Bumble Bees Mating Optimization algorithm. *Appl. Soft Comput.* **2017**, *55*, 13–30. [CrossRef]
303. Marinaki, M.; Marinakis, Y. A bumble bees mating optimization algorithm for the feature selection problem. *Int. J. Mach. Learn. Cybern.* **2014**, *7*, 519–538. [CrossRef]
304. Kumar, A.; Maini, T.; Misra, R.K.; Singh, D. *Butterfly Constrained Optimizer for Constrained Optimization Problems*; Springer: Singapore, 2019.
305. Kumar, A.; Misra, R.K.; Singh, D. Improving the local search capability of effective butterfly optimizer using covariance matrix adapted retreat phase. In Proceedings of the 2017 IEEE Congress on Evolutionary Computation (CEC), IEEE, Donostia, Spain, 5–8 June 2017.
306. Sharma, S.; Saha, A.K. m-MBOA: A novel butterfly optimization algorithm enhanced with mutualism scheme. *Soft Comput.* **2019**, *24*, 4809–4827. [CrossRef]
307. Li, G.; Shuang, F.; Zhao, P.; Le, C. An Improved Butterfly Optimization Algorithm for Engineering Design Problems Using the Cross-Entropy Method. *Symmetry* **2019**, *11*, 1049. [CrossRef]
308. Arora, S.; Anand, P. Learning automata-based butterfly optimization algorithm for engineering design problems. *Int. J. Comput. Mater. Sci. Eng.* **2018**, *7*, 1850021. [CrossRef]
309. Arora, S.; Anand, P. Binary butterfly optimization approaches for feature selection. *Expert Syst. Appl.* **2018**, *116*, 147–160. [CrossRef]
310. Aygül, K.; Cikan, M.; Demirdelen, T.; Tumay, M. Butterfly optimization algorithm based maximum power point tracking of photovoltaic systems under partial shading condition. *Energy Sources Part A Recovery Util. Environ. Eff.* **2019**, 1–19. [CrossRef]
311. Wang, Y.-J.; Sun, P. One-Way Pioneer Guide Pity Beetle Algorithm: A New Evolutionary Algorithm for Solving Global Optimization Problems. *IEEE Access* **2020**, *8*, 203270–203293. [CrossRef]
312. Priya, M.M.M.A.; Jawhar, D.S.J.; Geisa, D.J.M. Optimal Deep Belief Network with Opposition based Pity Beetle Algorithm for Lung Cancer Classification: A DBNOPBA Approach. *Comput. Methods Programs Biomed.* **2021**, *199*, 105902. [CrossRef]
313. KS, S.R.; Murugan, S. Memory based Hybrid Dragonfly Algorithm for numerical optimization problems. *Expert Syst. Appl.* **2017**, *83*, 63–78.
314. Sayed, G.I.; Tharwat, A.; Hassanien, A.E. Chaotic dragonfly algorithm: An improved metaheuristic algorithm for feature selection. *Appl. Intell.* **2018**, *49*, 188–205. [CrossRef]
315. Hariharan, M.; Sindhu, R.; Vijean, V.; Yazid, H.; Nadarajaw, T.; Yaacob, S.; Polat, K. Improved binary dragonfly optimization algorithm and wavelet packet based non-linear features for infant cry classification. *Comput. Methods Programs Biomed.* **2018**, *155*, 39–51. [CrossRef] [PubMed]
316. Mafarja, M.; Heidari, A.A.; Faris, H.; Mirjalili, S.; Aljarah, I. Dragonfly algorithm: Theory, literature review, and application in feature selection. In *Nature-Inspired Optimizers*; Springer: Cham, Switzerland, 2020; pp. 47–67.
317. El-Hay, E.A.; El-Hameed, M.A.; El-Fergany, A.A. Improved performance of PEM fuel cells stack feeding switched reluctance motor using multi-objective dragonfly optimizer. *Neural Comput. Appl.* **2018**, *31*, 6909–6924. [CrossRef]
318. Das, K.N.; Singh, T.K.; Baishnab, K.L. Parameter Optimization of Winner-Take-All Circuit for Attention Shift Using Drosophila Food-Search Optimization Algorithm. In *Proceedings of Fourth International Conference on Soft Computing for Problem Solving*; Springer: New Delhi, India, 2015. [CrossRef]
319. Fister, I.J.; Perc, M.; Kamal, S.M. A review of chaos-based firefly algorithms: Perspectives and research challenges. *Appl. Math. Comput.* **2015**, *252*, 155–165. [CrossRef]

320. Sahu, R.K.; Panda, S.; Pradhan, P.C. Design and analysis of hybrid firefly algorithm-pattern search based fuzzy PID controller for LFC of multi area power systems. *Int. J. Electr. Power Energy Syst.* **2015**, *69*, 200–212. [CrossRef]
321. Tahershamsi, A.; Kaveh, A.; Sheikholeslami, R.; Kazemzadeh Azad, S. An improved firefly algorithm with harmony search scheme for optimization of water distribution systems. *Sci. Iran.* **2014**, *21*, 1591–1607.
322. George, G.; Parthiban, L. Multi objective hybridized firefly algorithm with group search optimization for data clustering. In Proceedings of the 2015 IEEE International Conference on Research in Computational Intelligence and Communication Networks, Kolkata, India, 20–22 November 2015.
323. Abd-Elazim, S.; Ali, E.S. Load frequency controller design of a two-area system composing of PV grid and thermal generator via firefly algorithm. *Neural Comput. Appl.* **2016**, *30*, 607–616. [CrossRef]
324. Dey, N.; Samanta, S.; Chakraborty, S.; Das, A.; Chaudhuri, S.S.; Suri, J.S. Firefly Algorithm for Optimization of Scaling Factors During Embedding of Manifold Medical Information: An Application in Ophthalmology Imaging. *J. Med Imaging Heal. Inform.* **2014**, *4*, 384–394. [CrossRef]
325. Sayadi, M.K.; Hafezalkotob, A.; Naini, S.G.J. Firefly-inspired algorithm for discrete optimization problems: An application to manufacturing cell formation. *J. Manuf. Syst.* **2013**, *32*, 78–84. [CrossRef]
326. Chen, H.; Li, S.; Heidari, A.A.; Wang, P.; Li, J.; Yang, Y.; Wang, M.; Huang, C. Efficient multi-population outpost fruit fly-driven optimizers: Framework and advances in support vector machines. *Expert Syst. Appl.* **2019**, *142*, 112999. [CrossRef]
327. Yuan, X.; Dai, X.; Zhao, J.; He, Q. On a novel multi-swarm fruit fly optimization algorithm and its application. *Appl. Math. Comput.* **2014**, *233*, 260–271. [CrossRef]
328. Ding, G.; Dong, F.; Zou, H. Fruit fly optimization algorithm based on a hybrid adaptive-cooperative learning and its application in multilevel image thresholding. *Appl. Soft Comput.* **2019**, *84*, 105704. [CrossRef]
329. Wang, L.; Shi, Y.; Liu, S. An improved fruit fly optimization algorithm and its application to joint replenishment problems. *Expert Syst. Appl.* **2015**, *42*, 4310–4323. [CrossRef]
330. Wu, L.; Zuo, C.; Zhang, H. A cloud model based fruit fly optimization algorithm. *Knowl. Based Syst.* **2015**, *89*, 603–617. [CrossRef]
331. Darwish, S.M.; Elmasry, A.; Ibrahim, S.H. Optimal Shortest Path in Mobile Ad-Hoc Network Based on Fruit Fly Optimization Algorithm. In Proceedings of the International Conference on Advanced Machine Learning Technologies and Applications, Cairo, Egypt, 28–30 March 2019.
332. Mafarja, M.; Aljarah, I.; Faris, H.; Hammouri, A.I.; Al-Zoubi, A.M.; Mirjalili, S. Binary grasshopper optimisation algorithm approaches for feature selection problems. *Expert Syst. Appl.* **2018**, *117*, 267–286. [CrossRef]
333. Mirjalili, S.Z.; Mirjalili, S.; Saremi, S.; Faris, H.; Aljarah, I. Grasshopper optimization algorithm for multi-objective optimization problems. *Appl. Intell.* **2018**, *48*, 805–820. [CrossRef]
334. Arora, S.; Anand, P. Chaotic grasshopper optimization algorithm for global optimization. *Neural Comput. Appl.* **2018**, *31*, 4385–4405. [CrossRef]
335. Luo, J.; Chen, H.; Zhang, Q.; Xu, Y.; Huang, H.; Zhao, X. An improved grasshopper optimization algorithm with application to financial stress prediction. *Appl. Math. Model.* **2018**, *64*, 654–668. [CrossRef]
336. Aljarah, I.; Al-Zoubi, A.M.; Faris, H.; Hassonah, M.A.; Mirjalili, S.; Saadeh, H. Simultaneous Feature Selection and Support Vector Machine Optimization Using the Grasshopper Optimization Algorithm. *Cogn. Comput.* **2018**, *10*, 478–495. [CrossRef]
337. Shi, Y.; Li, Y.; Fan, J.; Wang, T.; Yin, T. A Novel Network Architecture of Decision-Making for Self-Driving Vehicles Based on Long Short-Term Memory and Grasshopper Optimization Algorithm. *IEEE Access* **2020**, *8*, 155429–155440. [CrossRef]
338. Cui, L.; Deng, J.; Wang, L.; Xu, M.; Zhang, Y. A novel locust swarm algorithm for the joint replenishment problem considering multiple discounts simultaneously. *Knowl. Based Syst.* **2016**, *111*, 51–62. [CrossRef]
339. Cuevas, E.; Zaldívar, D.; Perez-Cisneros, M. Automatic Segmentation by Using an Algorithm Based on the Behavior of Locust Swarms. In *Applications of Evolutionary Computation in Image Processing and Pattern Recognition*; Springer International Publishing: Cham, Switzerland, 2016; pp. 229–269.
340. Cuevas, E.; Gonzalez, A.; Fausto, F.; Zaldívar, D.; Perez-Cisneros, M.A. Multithreshold Segmentation by Using an Algorithm Based on the Behavior of Locust Swarms. *Math. Probl. Eng.* **2015**, *2015*, 1–25. [CrossRef]
341. Zervoudakis, K.; Tsafarakis, S. A mayfly optimization algorithm. *Comput. Ind. Eng.* **2020**, *145*, 106559. [CrossRef]
342. Bhattacharyya, T.; Chatterjee, B.; Singh, P.K.; Yoon, J.H.; Geem, Z.W.; Sarkar, R. Mayfly in Harmony: A New Hybrid Meta-Heuristic Feature Selection Algorithm. *IEEE Access* **2020**, *8*, 195929–195945. [CrossRef]
343. Ramasamy, K.; Ravichandran, C.S. Optimal design of renewable sources of PV/wind/ FC generation for power system reliability and cost using MA-RBFNN approach. *Int. J. Energy Res.* **2021**, *45*, 10946–10962. [CrossRef]
344. Yazdani, S.; Hadavandi, E. LMBO-DE: A linearized monarch butterfly optimization algorithm improved with differential evolution. *Soft Comput.* **2018**, *23*, 8029–8043. [CrossRef]
345. Wang, G.-G.; Deb, S.; Zhao, X.; Cui, Z. A new monarch butterfly optimization with an improved crossover operator. *Oper. Res.* **2016**, *18*, 731–755. [CrossRef]
346. Wang, G.-G.; Zhao, X.; Deb, S. A Novel Monarch Butterfly Optimization with Greedy Strategy and Self-Adaptive. *2015 Second. Int. Conf. Soft Comput. Mach. Intell.* **2015**, *2015*, 45–50.
347. Feng, Y.; Yang, J.; Wu, C.; Lu, M.; Zhao, X.-J. Solving 0–1 knapsack problems by chaotic monarch butterfly optimization algorithm with Gaussian mutation. *Memetic Comput.* **2016**, *10*, 135–150. [CrossRef]

348. Devikanniga, D.; Raj, R.J.S. Classification of osteoporosis by artificial neural network based on monarch butterfly optimisation algorithm. *Heal. Technol. Lett.* **2018**, *5*, 70–75. [CrossRef]
349. Chen, S.; Chen, R.; Gao, J. A Monarch Butterfly Optimization for the Dynamic Vehicle Routing Problem. *Algorithms* **2017**, *10*, 107. [CrossRef]
350. Zhu, Y.; Feng, X.; Yu, H. *Mosquito Host-Seeking Algorithm Based on Random Walk and Game of Life*; Springer International Publishing: Cham, Switzerland, 2018.
351. Xu, Y.; Chen, H.; Heidari, A.A.; Luo, J.; Zhang, Q.; Zhao, X.; Li, C. An efficient chaotic mutative moth-flame-inspired optimizer for global optimization tasks. *Expert Syst. Appl.* **2019**, *129*, 135–155. [CrossRef]
352. Savsani, V.; Tawhid, M.A. Non-dominated sorting moth flame optimization (NS-MFO) for multi-objective problems. *Eng. Appl. Artif. Intell.* **2017**, *63*, 20–32. [CrossRef]
353. Wang, M.; Chen, H.; Yang, B.; Zhao, X.; Hu, L.; Cai, Z.; Huang, H.; Tong, C. Toward an optimal kernel extreme learning machine using a chaotic moth-flame optimization strategy with applications in medical diagnoses. *Neurocomputing* **2017**, *267*, 69–84. [CrossRef]
354. Wu, Z.; Shen, D.; Shang, M.; Qi, S. Parameter Identification of Single-Phase Inverter Based on Improved Moth Flame Optimization Algorithm. *Electr. Power Components Syst.* **2019**, *47*, 456–469. [CrossRef]
355. Li, Z.; Zhou, Y.; Zhang, S.; Song, J. Lévy-Flight Moth-Flame Algorithm for Function Optimization and Engineering Design Problems. *Math. Probl. Eng.* **2016**, *2016*, 1–22. [CrossRef]
356. Mehne, S.H.H.; Mirjalili, S. Moth-Flame Optimization Algorithm: Theory, Literature Review, and Application in Optimal Nonlinear. *Nat. Inspired Optim. Theor. Lit. Rev. Appl.* **2020**, *810*, 143.
357. Luo, Q.; Yang, X.; Zhou, Y. Nature-inspired approach: An enhanced moth swarm algorithm for global optimization. *Math. Comput. Simul.* **2018**, *159*, 57–92. [CrossRef]
358. Shilaja, C.; Arunprasath, T. Optimal power flow using Moth Swarm Algorithm with Gravitational Search Algorithm considering wind power. *Future Gener. Comput. Syst.* **2019**, *98*, 708–715.
359. Duman, S. A Modified Moth Swarm Algorithm Based on an Arithmetic Crossover for Constrained Optimization and Optimal Power Flow Problems. *IEEE Access* **2018**, *6*, 45394–45416. [CrossRef]
360. Guvenc, U.; Duman, S.; Hınıslıoglu, Y. Chaotic moth swarm algorithm. In Proceedings of the 2017 IEEE International Conference on INnovations in Intelligent SysTems and Applications (INISTA), IEEE, Gdynia, Poland, 3–5 July 2017.
361. Zhou, Y.; Yang, X.; Ling, Y.; Zhang, J. Meta-heuristic moth swarm algorithm for multilevel thresholding image segmentation. *Multimedia Tools Appl.* **2018**, *77*, 23699–23727. [CrossRef]
362. Fathy, A.; Elaziz, M.A.; Sayed, E.; Olabi, A.; Rezk, H. Optimal parameter identification of triple-junction photovoltaic panel based on enhanced moth search algorithm. *Energy* **2019**, *188*, 116025. [CrossRef]
363. Feng, Y.-H.; Wang, G.-G. Binary moth search algorithm for discounted {0-1} knapsack problem. *IEEE Access* **2018**, *6*, 10708–10719. [CrossRef]
364. Strumberger, I.; Bacanin, N. Modified Moth Search Algorithm for Global Optimization Problems. *Int. J. Comput.* **2018**, *3*, 44–48.
365. Strumberger, I.; Tuba, E.; Bacanin, N.; Beko, M.; Tuba, M. Hybridized moth search algorithm for constrained optimization problems. In Proceedings of the 2018 International Young Engineers Forum (YEF-ECE), IEEE, Costa da Caparica, Portugal, 4 May 2018.
366. Strumberger, I.; Sarac, M.; Markovic, D.; Bacanin, N. Moth Search Algorithm for Drone Placement Problem. *Int. J. Comput.* **2018**, *3*, 75–80.
367. Tsai, H.-C. Roach infestation optimization with friendship centers. *Eng. Appl. Artif. Intell.* **2015**, *39*, 109–119. [CrossRef]
368. Obagbuwa, I.C.; Adewumi, A.O. A modified roach infestation optimization. In Proceedings of the 2014 IEEE Conference on Computational Intelligence in Bioinformatics and Computational Biology, Honolulu, HI, USA, 21–24 May 2014.
369. Obagbuwa, I.C.; Adewumi, A.O.; Adebiyi, A.A. A dynamic step-size adaptation roach infestation optimization. In Proceedings of the 2014 IEEE International Advance Computing Conference, Gurgaon, India, 21–22 February 2014.
370. Kaveh, A.; Eslamlou, A.D. Water strider algorithm: A new metaheuristic and applications. *Structures* **2020**, *25*, 520–541. [CrossRef]
371. Kaveh, A.; Amirsoleimani, P.; Eslamlou, A.D.; Rahmani, P. Frequency-constrained optimization of large-scale dome-shaped trusses using chaotic water strider algorithm. *Struct.* **2021**, *32*, 1604–1618. [CrossRef]
372. Xu, Y.-P.; Ouyang, P.; Xing, S.-M.; Qi, L.-Y.; Khayatnezhad, M.; Jafari, H. Optimal structure design of a PV/FC HRES using amended Water Strider Algorithm. *Energy Rep.* **2021**, *7*, 2057–2067. [CrossRef]
373. Kaveh, A. Water Strider Optimization Algorithm and Its Enhancement. In *Advances in Metaheuristic Algorithms for Optimal Design of Structures*; Kaveh, A., Ed.; Springer International Publishing: Cham, Switzerland, 2021; pp. 783–848.
374. Luque-Chang, A.; Cuevas, E.; Fausto, F.; Zaldívar, D.; Pérez, M. Social Spider Optimization Algorithm: Modifications, Applications, and Perspectives. *Math. Probl. Eng.* **2018**, *2018*, 1–29. [CrossRef]
375. Yu, J.J.; Li, V.O. A social spider algorithm for global optimization. *Appl. Soft Comput.* **2015**, *30*, 614–627. [CrossRef]
376. James, J.; Li, V.O. Parameter sensitivity analysis of social spider algorithm. In Proceedings of the 2015 IEEE Congress on Evolutionary Computation (CEC), IEEE, Sendai, Japan, 25–28 May 2015.
377. Gandomi, A.H.; Alavi, A.H. Krill herd: A new bio-inspired optimization algorithm. *Commun. Nonlinear Sci. Numer. Simul.* **2012**, *17*, 4831–4845. [CrossRef]
378. Wang, G.-G.; Gandomi, A.H.; Alavi, A.H.; Gong, D. A comprehensive review of krill herd algorithm: Variants, hybrids and applications. *Artif. Intell. Rev.* **2017**, *51*, 119–148. [CrossRef]

379. Wang, G.G.; Deb, S.; Coelho, L.D.S. Earthworm optimization algorithm: A bio-inspired metaheuristic algorithm for global optimization problems. *Int. J. Bio-Inspired Comput.* **2015**, *1*, 1. [CrossRef]
380. Kaur, S.; Awasthi, L.K.; Sangal, A.; Dhiman, G. Tunicate Swarm Algorithm: A new bio-inspired based metaheuristic paradigm for global optimization. *Eng. Appl. Artif. Intell.* **2020**, *90*, 103541. [CrossRef]
381. Javaid, N.; Ullah, I.; Zarin, S.S.; Kamal, M.; Omoniwa, B.; Mateen, A. *Differential-Evolution-Earthworm Hybrid Meta-heuristic Optimization Technique for Home Energy Management System in Smart Grid*; Springer International Publishing: Cham, Switzerland, 2019.
382. Faraz, S.H.; Ur Rehman, S.; Sarwar, M.A.; Ali, I.; Farooqi, M.; Javaid, N. *Comparison of BFA and EWA in Home Energy Management System Using RTP*; Springer International Publishing: Cham, Switzerland, 2018.
383. Ali, M.; Abid, S.; Ghafar, A.; Ayub, N.; Arshad, H.; Khan, S.; Javaid, N. *Earth Worm Optimization for Home Energy Management System in Smart Grid*; Springer International Publishing: Cham, Switzerland, 2018.
384. Wang, H.; Yi, J.-H. An improved optimization method based on krill herd and artificial bee colony with information exchange. *Memetic Comput.* **2017**, *10*, 177–198. [CrossRef]
385. Chansombat, S.; Musikapun, P.; Pongcharoen, P.; Hicks, C. A Hybrid Discrete Bat Algorithm with Krill Herd-based advanced planning and scheduling tool for the capital goods industry. *Int. J. Prod. Res.* **2018**, *57*, 6705–6726. [CrossRef]
386. Abdel-Basset, M.; Wang, G.-G.; Sangaiah, A.K.; Rushdy, E. Krill herd algorithm based on cuckoo search for solving engineering optimization problems. *Multimedia Tools Appl.* **2017**, *78*, 3861–3884. [CrossRef]
387. Abualigah, L.M.; Khader, A.T.; Hanandeh, E.S. Hybrid clustering analysis using improved krill herd algorithm. *Appl. Intell.* **2018**, *48*, 4047–4071. [CrossRef]
388. Asteris, P.G.; Nozhati, S.; Nikoo, M.; Cavaleri, L.; Nikoo, M. Krill herd algorithm-based neural network in structural seismic reliability evaluation. *Mech. Adv. Mater. Struct.* **2018**, *26*, 1146–1153. [CrossRef]
389. Das, S.R.; Kuhoo; Mishra, D.; Rout, M. An optimized feature reduction based currency forecasting model exploring the online sequential extreme learning machine and krill herd strategies. *Phys. A Stat. Mech. its Appl.* **2018**, *513*, 339–370. [CrossRef]
390. Nguyen, T.T.; Vo, D.N. Improved social spider optimization algorithm for optimal reactive power dispatch problem with different objectives. *Neural Comput. Appl.* **2019**, *32*, 5919–5950. [CrossRef]
391. El Aziz, M.A.; Hassanien, A.E. An improved social spider optimization algorithm based on rough sets for solving minimum number attribute reduction problem. *Neural Comput. Appl.* **2017**, *30*, 2441–2452. [CrossRef]
392. Cuevas, E.; Cienfuegos, M. A new algorithm inspired in the behavior of the social-spider for constrained optimization. *Expert Syst. Appl.* **2014**, *41*, 412–425. [CrossRef]
393. Mirjalili, S.Z.; Saremi, S.; Mirjalili, S.M. Designing evolutionary feedforward neural networks using social spider optimization algorithm. *Neural Comput. Appl.* **2015**, *26*, 1919–1928. [CrossRef]
394. Zhou, G.; Zhou, Y.; Zhao, R. Hybrid social spider optimization algorithm with differential mutation operator for the job-shop scheduling problem. *J. Ind. Manag. Optim.* **2021**, *17*, 533–548. [CrossRef]
395. Xavier, V.M.A.; Annadurai, S. Chaotic social spider algorithm for load balance aware task scheduling in cloud computing. *Clust. Comput.* **2018**, *22*, 287–297.
396. Elsayed, W.; Hegazy, Y.; Bendary, F.; El-Bages, M. Modified social spider algorithm for solving the economic dispatch problem. *Eng. Sci. Technol. Int. J.* **2016**, *19*, 1672–1681. [CrossRef]
397. Sung, H.-K.; Jung, N.-G.; Huang, S.-R.; Kim, J.-M. Application of Social Spider Algorithm to Optimize Train Energy. *J. Electr. Eng. Technol.* **2019**, *14*, 519–526. [CrossRef]
398. Yu, J.J.; Li, V.O. A social spider algorithm for solving the non-convex economic load dispatch problem. *Neurocomputing* **2016**, *171*, 955–965. [CrossRef]
399. Houssein, E.H.; Helmy, B.E.-D.; Elngar, A.A.; Abdelminaam, D.S.; Shaban, H. An Improved Tunicate Swarm Algorithm for Global Optimization and Image Segmentation. *IEEE Access* **2021**, *9*, 56066–56092. [CrossRef]
400. Fetouh, T.; Elsayed, A.M. Optimal Control and Operation of Fully Automated Distribution Networks Using Improved Tunicate Swarm Intelligent Algorithm. *IEEE Access* **2020**, *8*, 129689–129708. [CrossRef]
401. Chelliah, J.; Kader, N. Optimization for connectivity and coverage issue in target-based wireless sensor networks using an effective multiobjective hybrid tunicate and salp swarm optimizer. *Int. J. Commun. Syst.* **2020**, *34*, e4679.
402. Črepinšek, M.; Liu, S.H.; Mernik, M. Exploration and exploitation in evolutionary algorithms: A survey. *ACM Comput. Surv.* **2013**, *43*, 1–33. [CrossRef]
403. Cuevas, E.; Echavarría, A.; Zaldívar, D.; Pérez-Cisneros, M. A novel evolutionary algorithm inspired by the states of matter for template matching. *Expert Syst. Appl.* **2013**, *40*, 6359–6373. [CrossRef]
404. Corriveau, G.; Guilbault, R.; Tahan, A.; Sabourin, R. Review of phenotypic diversity formulations for diagnostic tool. *Appl. Soft Comput.* **2013**, *13*, 9–26. [CrossRef]
405. Shir, O.M. Niching in Evolutionary Algorithms. In *Handbook of Natural Computing*; Rozenberg, G., Bäck, T., Kok, J.N., Eds.; Springer: Berlin/Heidelberg, Germany, 2012; pp. 1035–1069.
406. Preuss, M. Niching prospects. In *International Conference on Bioinspired Optimization Methods and Their Applications*; Filipic, B., Silic, J., Eds.; Josef Stefan Institute: Lublijana, Slovenia, 2006.
407. Silberholz, J.; Golden, B. Comparison of metaheuristics. In *Handbook of Metaheuristics*; Springer: Boston, MA, USA, 2010; pp. 625–640.
408. Rice, J.R. The Algorithm Selection Problem. *Adv. Comput.* **1976**, *15*, 65–118.

409. Misir, M.; Sebag, M. Alors: An algorithm recommender system. *Artif. Intell.* **2017**, *244*, 291–314. [CrossRef]
410. Bischl, B.; Kerschke, P.; Kotthoff, L.; Lindauer, M.; Malitsky, Y.; Fréchette, A.; Hoos, H.; Hutter, F.; Leyton-Brown, K.; Tierney, K.; et al. ASlib: A benchmark library for algorithm selection. *Artif. Intell.* **2016**, *237*, 41–58. [CrossRef]
411. LaTorre, A.; Molina, D.; Osaba, E.; Del Ser, J.; Herrera, F. Fairness in bio-inspired optimization research: A prescription of methodological guidelines for comparing meta-heuristics. *arXiv* **2020**, arXiv:2004.09969.
412. Naruei, I.; Keynia, F. A new optimization method based on COOT bird natural life model. *Expert Syst. Appl.* **2021**, *183*, 115352. [CrossRef]
413. Ewees, A.A.; Elaziz, M.A.; Houssein, E.H. Improved grasshopper optimization algorithm using opposition-based learning. *Expert Syst. Appl.* **2018**, *112*, 156–172. [CrossRef]
414. Li, S.; Chen, H.; Wang, M.; Heidari, A.A.; Mirjalili, S. Slime mould algorithm: A new method for stochastic optimization. *Futur. Gener. Comput. Syst.* **2020**, *111*, 300–323. [CrossRef]
415. Sharma, T.K.; Abraham, A. Artificial bee colony with enhanced food locations for solving mechanical engineering design problems. *J. Ambient. Intell. Humaniz. Comput.* **2019**, *11*, 267–290. [CrossRef]
416. Yang, X.-S.; Hosseini, S.S.S.; Gandomi, A. Firefly Algorithm for solving non-convex economic dispatch problems with valve loading effect. *Appl. Soft Comput.* **2012**, *12*, 1180–1186. [CrossRef]
417. Gupta, S.; Deep, K.; Moayedi, H.; Foong, L.K.; Assad, A. Sine cosine grey wolf optimizer to solve engineering design problems. *Eng. Comput.* **2021**, *37*, 3123–3149. [CrossRef]
418. Huang, F.-Z.; Wang, L.; He, Q. An effective co-evolutionary differential evolution for constrained optimization. *Appl. Math. Comput.* **2007**, *186*, 340–356. [CrossRef]
419. Mezura-Montes, E.; Hernández-Ocana, B. Bacterial foraging for engineering design problems: Preliminary results. In *Memorias del 4o Congreso Nacional de Computación Evolutiva (COMCEV'2008)*; Centro de Investigación en Matemáticas: Guanajuato, México, 2008.
420. He, Q.; Wang, L. An effective co-evolutionary particle swarm optimization for constrained engineering design problems. *Eng. Appl. Artif. Intell.* **2007**, *20*, 89–99. [CrossRef]
421. Bernardino, H.S.; Barbosa, H.J.; Lemonge, A.C.; Fonseca, L.G. A new hybrid AIS-GA for constrained optimization problems in mechanical engineering. In *2008 IEEE Congress on Evolutionary Computation (IEEE World Congress on Computational Intelligence)*; IEEE: Hong Kong, China, 2008.
422. Lee, K.S.; Geem, Z.W. A new meta-heuristic algorithm for continuous engineering optimization: Harmony search theory and practice. *Comput. Methods Appl. Mech. Eng.* **2005**, *194*, 3902–3933. [CrossRef]
423. Hedar, A.-R.; Fukushima, M. Derivative-Free Filter Simulated Annealing Method for Constrained Continuous Global Optimization. *J. Glob. Optim.* **2006**, *35*, 521–549. [CrossRef]
424. Houssein, E.H.; Saad, M.R.; Hashim, F.; Shaban, H.; Hassaballah, M. Lévy flight distribution: A new metaheuristic algorithm for solving engineering optimization problems. *Eng. Appl. Artif. Intell.* **2020**, *94*, 103731. [CrossRef]
425. Hwang, S.-F.; He, R.-S. A hybrid real-parameter genetic algorithm for function optimization. *Adv. Eng. Informatics* **2006**, *20*, 7–21. [CrossRef]
426. Mahdavi, M.; Fesanghary, M.; Damangir, E. An improved harmony search algorithm for solving optimization problems. *Appl. Math. Comput.* **2007**, *188*, 1567–1579. [CrossRef]
427. Zhang, M.; Luo, W.; Wang, X. Differential evolution with dynamic stochastic selection for constrained optimization. *Inf. Sci.* **2008**, *178*, 3043–3074. [CrossRef]
428. Ragsdell, K.M.; Phillips, D.T. Optimal Design of a Class of Welded Structures Using Geometric Programming. *J. Eng. Ind.* **1976**, *98*, 1021–1025. [CrossRef]
429. Yapici, H.; Cetinkaya, N. A new meta-heuristic optimizer: Pathfinder algorithm. *Appl. Soft Comput.* **2019**, *78*, 545–568. [CrossRef]
430. Belegundu, A.D.; Arora, J.S. A study of mathematical programming methods for structural optimization. Part I: Theory. *Int. J. Numer. Methods Eng.* **1985**, *21*, 1583–1599. [CrossRef]
431. Arora, J.S. *Introduction to Optimum Design*; Elsevier: Cambridge, MA, USA, 2004.
432. Montague, M.; Aslam, J.A. Condorcet fusion for improved retrieval. In Proceedings of the Eleventh International Conference on Information and Knowledge Management, ACM, McLean, VA, USA, 4–9 December 2002.
433. Osaba, E.; Carballedo, R.; Diaz, F.; Onieva, E.; Masegosa, A.D.; Perallos, A. Good practice proposal for the implementation, presentation, and comparison of metaheuristics for solving routing problems. *Neurocomputing* **2018**, *271*, 2–8. [CrossRef]
434. Piotrowski, A.P.; Napiorkowski, M.J.; Napiorkowski, J.J.; Osuch, M.; Kundzewicz, Z.W. Are modern metaheuristics successful in calibrating simple conceptual rainfall–runoff models? *Hydrol. Sci. J.* **2017**, *62*, 606–625. [CrossRef]
435. Piotrowski, A.P.; Napiorkowski, J.J. Some metaheuristics should be simplified. *Inf. Sci.* **2018**, *427*, 32–62. [CrossRef]
436. Tzanetos, A.; Dounias, G. Nature inspired optimization algorithms or simply variations of metaheuristics? *Artif. Intell. Rev.* **2021**, *54*, 1841–1862. [CrossRef]
437. Lones, M.A. Mitigating Metaphors: A Comprehensible Guide to Recent Nature-Inspired Algorithms. *arXiv* **2019**, arXiv:1902.08001. [CrossRef]
438. Van Thieu, N. The State-of-the-art MEta-Heuristics Algorithms in PYthon (MEALPY). 2021. Available online: https://pypi.org/project/mealpy/ (accessed on 2 August 2021).
439. Molina, D.; Latorre, A.; Herrera, F. An Insight into Bio-inspired and Evolutionary Algorithms for Global Optimization: Review, Analysis, and Lessons Learnt over a Decade of Competitions. *Cogn. Comput.* **2018**, *10*, 517–544. [CrossRef]

440. Veček, N.; Črepinšek, M.; Mernik, M. On the influence of the number of algorithms, problems, and independent runs in the comparison of evolutionary algorithms. *Appl. Soft Comput.* **2017**, *54*, 23–45. [CrossRef]
441. Squillero, G.; Tonda, A. Divergence of character and premature convergence: A survey of methodologies for promoting diversity in evolutionary optimization. *Inf. Sci.* **2016**, *329*, 782–799. [CrossRef]
442. Liu, H.-L.; Chen, L.; Deb, K.; Goodman, E. Investigating the Effect of Imbalance Between Convergence and Diversity in Evolutionary Multi-objective Algorithms. *IEEE Trans. Evol. Comput.* **2016**, *21*, 408–425. [CrossRef]
443. Wright, J.; Jordanov, I. Convergence properties of quantum evolutionary algorithms on high dimension problems. *Neurocomputing* **2019**, *326–327*, 82–99. [CrossRef]
444. Chen, Y.; He, J. Average Convergence Rate of Evolutionary Algorithms II: Continuous Optimization. *arXiv* **2018**, arXiv:1810.11672.
445. Shirakawa, S.; Nagao, T. Bag of local landscape features for fitness landscape analysis. *Soft Comput.* **2016**, *20*, 3787–3802. [CrossRef]
446. Yang, S.; Li, K.; Li, W.; Chen, W.; Chen, Y. Dynamic Fitness Landscape Analysis on Differential Evolution Algorithm. In *Bio-inspired Computing–Theories and Applications: 11th International Conference, BIC-TA 2016, Xi'an, China, 28–30 October 2016, Revised Selected Papers, Part II*; Gong, M., Pan, L., Song, T., Zhang, G., Eds.; Springer: Singapore, 2016; pp. 179–184.
447. Aleti, A.; Moser, I.; Grunske, L. Analysing the fitness landscape of search-based software testing problems. *Autom. Softw. Eng.* **2016**, *24*, 603–621. [CrossRef]
448. Liang, J.; Li, Y.; Qu, B.; Yu, K.; Hu, Y. *Mutation Strategy Selection Based on Fitness Landscape Analysis: A Preliminary Study*; Springer: Singapore, 2020.
449. Hussain, K.; Salleh, M.N.M.; Cheng, S.; Shi, Y. On the exploration and exploitation in popular swarm-based metaheuristic algorithms. *Neural Comput. Appl.* **2018**, *31*, 7665–7683. [CrossRef]
450. Chen, Y.; He, J. Exploitation and Exploration Analysis of Elitist Evolutionary Algorithms: A Case Study. *arXiv* **2020**, arXiv:2001.10932.
451. Morales-Castañeda, B.; Zaldívar, D.; Cuevas, E.; Fausto, F.; Rodríguez, A. A better balance in metaheuristic algorithms: Does it exist? *Swarm Evol. Comput.* **2020**, *54*, 100671. [CrossRef]

Article

A ResNet50-Based Method for Classifying Surface Defects in Hot-Rolled Strip Steel

Xinglong Feng [1,†], Xianwen Gao [1,*] and Ling Luo [2,†]

1 College of Information Science and Engineering, Northeastern University, Shenyang 110819, China; 1610238@stu.neu.edu.cn
2 Moviebook Technology Co., Ltd., Beijing 100027, China; ling_luo@moviebook.cn
* Correspondence: gaoxianwen@mail.neu.edu.cn
† These authors contributed equally to this work.

Citation: Feng, X.; Gao, X.; Luo, L. A ResNet50-Based Method for Classifying Surface Defects in Hot-Rolled Strip Steel. *Mathematics* 2021, 9, 2359. https://doi.org/10.3390/math9192359

Academic Editors: Florin Leon, Mircea Hulea and Marius Gavrilescu

Received: 13 August 2021
Accepted: 18 September 2021
Published: 23 September 2021

Publisher's Note: MDPI stays neutral with regard to jurisdictional claims in published maps and institutional affiliations.

Copyright: © 2021 by the authors. Licensee MDPI, Basel, Switzerland. This article is an open access article distributed under the terms and conditions of the Creative Commons Attribution (CC BY) license (https://creativecommons.org/licenses/by/4.0/).

Abstract: Hot-rolled strip steel is widely used in automotive manufacturing, chemical and home appliance industries, and its surface quality has a great impact on the quality of the final product. In the manufacturing process of strip steel, due to the rolling process and many other reasons, the surface of hot rolled strip steel will inevitably produce slag, scratches and other surface defects. These defects not only affect the quality of the product, but may even lead to broken strips in the subsequent process, seriously affecting the continuation of production. Therefore, it is important to study the surface defects of strip steel and identify the types of defects in strip steel. In this paper, a scheme based on ResNet50 with the addition of FcaNet and Convolutional Block Attention Module (CBAM) is proposed for strip defect classification and validated on the X-SDD strip defect dataset. Our solution achieves a classification accuracy of 94.11%, higher than more than a dozen other compared deep learning models. Moreover, to adress the problem of low accuracy of the algorithm in classifying individual defects, we use ensemble learning to optimize. By integrating the original solution with VGG16 and SqueezeNet, the recognition rate of oxide scale of plate system defects improved by 21.05 percentage points, and the overall defect classification accuracy improved to 94.85%.

Keywords: hot rolled strip steel; deep learning; surface defects; defect classification

1. Introduction

Hot-rolled strip steel is produced by rolling the billet at a temperature higher than the recrystallization temperature and then going through a series of processes such as phosphorus removal, finishing, polishing, edge cutting and straightening. Hot-rolled strip steel has good processing performance and strong coverage ability, which is widely used in automobile manufacturing, home appliance manufacturing, shipbuilding and chemical industry, etc. In the manufacturing process of strip steel, for various reasons [1–3], surface defects will inevitably arise, and these defects cannot be completely overcome by improving the process. Therefore, the detection of surface defects in hot rolled strip is an important part of hot rolled strip production and is closely related to the surface quality of the strip. Figure 1 shows the quality inspection process of surface defects in the actual production of a steel mill.

As shown in Figure 1, the hot rolled strip is first inspected by the hot rolled strip quality inspection system. The system takes high speed images of the top and bottom surfaces of the strip steel and determines the images that may have surface defects and passes them to the quality inspector. Since hot rolled strip passes through the quality inspection system very quickly, often in less than two minutes for a roll of strip to pass through the system, the quality inspector must judge the pictures coming from the system quickly. Strip steels judged to be normal by the quality inspectors will go directly to the next process, while coils judged to require further treatment will be given more specific treatment by the next batch of quality inspectors. Since the previous steps would have blocked the problematic steel coils, this batch of quality inspectors have more time to analyze the steel coils with surface defects and thus give the next instructions. After a series of processing, the finished strip coil is finally obtained as shown in Figure 2.

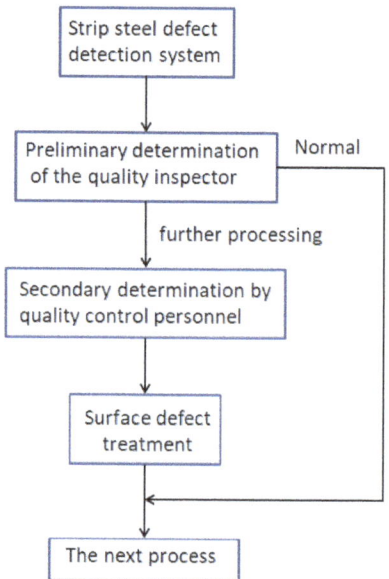

Figure 1. Flow chart of strip defect detection.

Figure 2. The finished steel coils.

Although the above solution can meet the steel mill's requirements for strip surface quality, this solution has the following shortcomings: Firstly, strip production often takes place throughout the day, which requires quality inspectors who make preliminary judgments to work at night, and long hours of night work are detrimental to their health [4]. Secondly, for quality inspectors, the work of observing defective pictures for a long time is not only easy to produce visual fatigue but also very boring, and therefore easy to produce errors [5]. Last but not least, the work of quality inspection increases the cost of the steel mill because of the large amount of manpower required.

The main reason for the current use of manual further testing on the basis of the strip surface defect detection system is that the accuracy of the existing system is not yet as good as that of the quality inspectors. So the key question is how to improve the accuracy of this system to reach the average level of quality control workers. The strip surface defect detection system commonly used in steel mills today is shown in Figure 3.

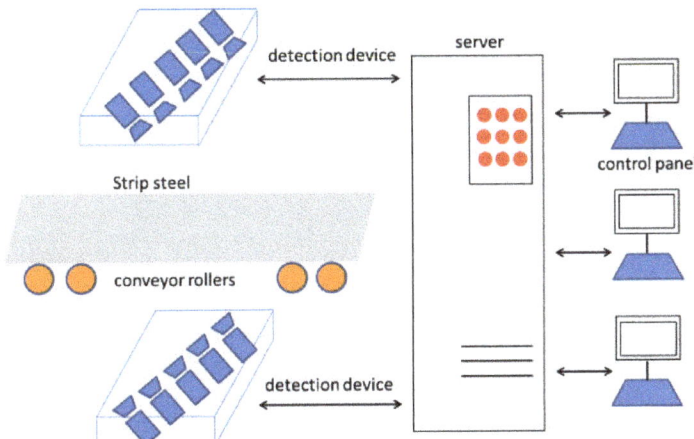

Figure 3. The strip surface defect detection system.

As shown in Figure 3, a schematic diagram of strip surface defect detection [6] is shown. The conveyor rollers rotate and drive the strip through the inspection device at high speed, and the inspection device takes high-speed images of the strip surface. Inspection devices generally include industrial cameras, industrial light sources, protection devices, etc.; images of the strip surface taken by the inspection devices are transmitted to the server, which processes them by the algorithm in the server. The server extracts samples that may be defective and sends them to the quality control personnel at the console for determination while storing them for later inspection. The hardware of the current strip surface defect detection system is sufficient to meet the use of detection, while the algorithm in the server determines the final accuracy of the strip defect classification. Therefore, to address the shortcomings of the existing system mentioned in the previous section, we try to improve the classification accuracy of the interserver algorithm. the contributions of this paper are shown below:

- We combine ResNet50, FcaNet and CBAM to propose a fused network for the classification of surface defects in hot-rolled steel strips.
- We validate the proposed algorithm on the X-SDD dataset [7], compare it with several deep learning models, and design ablation experiments to verify the effectiveness of the algorithm.

2. Related Work
2.1. Machine Learning Based Methods

There are many ways to classify surface defects in strip steel, and scholars have conducted many studies and proposed many schemes in this field. Xiao et al. [8] proposed an evolutionary classifier with Bayesian kernel (BYEC), which can be tuned with a small sample set to better fit the model of a new production line. Firstly, the classifier is designed by introducing rich features to cover the details of the defects. A series of support vector machines (SVMs) are then constructed from a random subspace of features. Finally, the Bayesian classifier is trained as an evolutionary kernel that is fused with the results of the sub-SVMs to form a comprehensive classifier. Gong et al. [9] proposed a novel multiclass classifier, i.e., support vector hyper-spheres with insensitivity to noise (INSVHs), in order to improve the classification accuracy and efficiency of steel plate surface defects. On the one hand, the INSVHs classifier introduces the bouncing sphere loss to reduce its sensitivity to the noise around the decision boundary. On the other hand, the INSVHs classifier reduces the detrimental effect of label noise and enhances the beneficial effect of important samples by increasing the local intra-class sample density weights. Chu et al. [10] proposed a novel support vector machine with adjustable hyper-sphere (AHSVM) focusing on the classification of strip surface defects. Meanwhile, a new multi-class classification method is proposed. AHSVM originates from the support vector data description and employs hyperspheres to solve the classification problem. AHSVM can follow two principles: marginal maximization and intra-class dispersion minimization. In addition, the hypersphere of AHSVM is tunable, which makes the final classification hypersphere optimal for the training dataset. Luo et al. [11] proposed a generalized completed local binary patterns (GCLBP) framework. Two variants of the improved completion local binary pattern (ICLBP) and the improved completion noise-invariant local structure pattern (ICNLP) are developed under the GCLBP framework for steel surface defect classification. Unlike the traditional local binary pattern variants, descriptive information hidden in non-uniform patterns is innovatively mined for better defect representation. After binarizing the strip surface defect images, Hu et al. [12] combined the defect target images and their corresponding binarized images to extract three types of image features, including geometric features, grayscale features and shape features. For the support vector machine-based classification model, they use Gaussian radial basis as the kernel function, determine the model parameters by cross-validation, and use a one-versus-one approach for multi-class classifiers. Zhang et al. [13] proposed a feature selection method based on a filtering approach combined with an implicit Bayesian classifier to improve the efficiency of defect identification and reduce the complexity of computation. The details of the method are: a large set of image features is initially obtained based on the discrete wavelet transform feature extraction method. Then three feature selection methods (including correlation-based feature selection, consistency subset evaluator [CSE], and information gain) are used to optimize the feature space.

2.2. Deep Learning Based Methods

Although the above traditional machine learning-based schemes are effective to some extent for the classification of strip defects, their effectiveness often relies on feature extraction. The feature extraction-based schemes often require manual operations and expert knowledge, which limits the generality of the algorithms. In recent years, convolutional neural networks (CNN) have gradually received more and more attention from scholars due to their advantages of automatic feature extraction. Fu [14] proposed a compact and effective CNN model that emphasizes the training of low-level features and combines multiple receptive fields for fast and accurate classification of steel surface defects. The solution uses a pre-trained SqueezeNet as the backbone architecture. It requires only a small number of defect-specific training samples to achieve high accuracy recognition on a diversity-enhanced test dataset containing steel surface defects with severe non-uniform illumination, camera noise and motion blur. Liu et al. [15] used GoogLeNet as the base model and added identity mapping to it, which was improved to some extent. The network achieved a measured speed of 125 FPS (Frames Per Second), which fully meets the real-time requirements of the actual steel strip production line. Zhou et al. [16] designed a CNN containing seven layers, including two convolutional layers, two subsampling layers, and two fully connected layers. The experimental results confirm that their proposed method is quite simple, effective and robust for the classification of surface defects in hot rolled steel sheets. Konovalenko I et al. [17] used a deep learning model based on ResNet50 as the base classifier to perform classification experiments on planar images with three types of damage, and the results showed that the model has excellent recognition ability, high speed and accuracy at the same time. Yi et al. [18] proposed an end-to-end surface defect recognition system for steel strip surface inspection. The system is based on a symmetric wrap-around salinity map for surface defect detection and a deep CNN that uses the defect images directly as input and the defect class as output for defect classification. CNNs are trained purely on the original defect images and learn the defect features from the network training, which avoids the separation between feature extraction and image classification, resulting in an end-to-end defect recognition pipeline. Deep learning-based strip defect classification schemes have shown relatively better performance than traditional machine learning schemes, however, the current research has the following shortcomings: Firstly, most of the current studies are based on the NEU surface defect dataset [19], which is balanced among the six categories. However, in the actual field of strip production, the frequency of various types of defects is not the same. Therefore, on the one hand, it is necessary to study on a dataset with unbalanced samples. On the other hand, the attention mechanism has been shown to improve the accuracy of CNN [20–22] for it can make the algorithm focus more attention on the valuable information in the image; while current research rarely introduces the attention mechanism to improve the classification accuracy of strip surface defects.

3. Method

3.1. Introduction of ResNet

As the deep learning-based network evolves, its structure is deepening; while this helps the network to perform more complex feature pattern extraction, it may also introduce the problem of gradient disappearance or gradient explosion. "Gradient disappearance" and "gradient explosion" can lead to the following shortcomings: (1) Long training time but network convergence becomes very difficult or even non-convergent. (2) The network performance will gradually saturate and even begin to degrade, known as the degradation problem of deep networks. To solve such problems, He et al. [23] proposed the ResNet network, which makes it possible to obtain a good performance and efficiency of the network even when the number of network layers is very deep (even over 1000 layers). The deep residual learning framework of ResNet is shown in Figure 4.

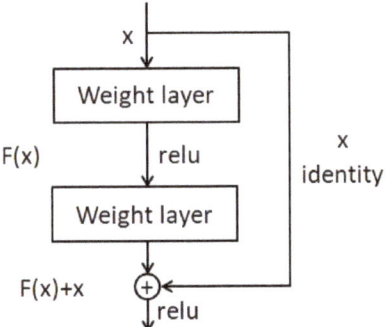

Figure 4. Residual learning: a building block.

As shown in Figure 4, there is an identity mapping in the residual module of ResNet that causes the output of the network to change from F (x) to F (x) + x. The training error of a deep network is generally higher than that of a shallow network. However, adding multiple layers of constant mapping (y = x) to a shallow network turns it into a deep network, and such a deep network can get the same training error as a shallow network. This shows that the layers of constant mapping are better trained. For the residual network, when the residual is 0, the stacking layer only does constant mapping at this time, and according to the above conclusion, theoretically the network performance will not degrade at least.

3.2. Introduction of CBAM

Woo et al. [24] proposed the convolutional block attention module (CBAM) in 2018, a simple and effective attention module for feed-forward convolutional neural networks. The significance of attention has been extensively studied in the previous literature [25–28]. Attention not only tells people where to focus their attention, it also improves representation of interest. Representation can be improved by using attentional mechanisms: focusing on important features and suppressing unnecessary ones. The structure of CBAM is shown in Figure 5. The CBAM module has two sequential sub-modules: channel attention model and spatial attention model.

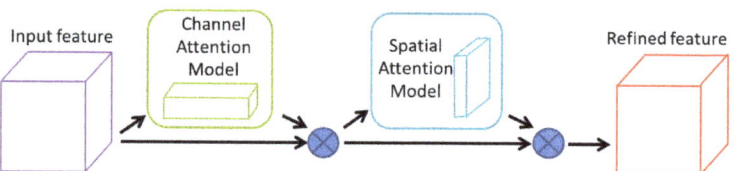

Figure 5. The Convolutional Block Attention Module.

Given an intermediate layer feature map named **F** with dimension $C \times H \times W$ as input, CBAM sequentially generates a 1-dimensional channel attention map (with dimension $\mathbf{Mc} \in C \times 1 \times 1$) and a 2-dimensional spatial attention map (with dimension $\mathbf{Ms} \in 1 \times H \times W$). The overall CBAM attention process can be summarized by the following equation:

$$\mathbf{F}' = \mathbf{Mc}(\mathbf{F}) \otimes \mathbf{F}, \tag{1}$$

$$\mathbf{F}'' = \mathbf{Ms}(\mathbf{F}) \otimes \mathbf{F}', \tag{2}$$

where $\otimes$ represents the one-to-one multiplication of the corresponding elements, and during the multiplication, the attention values are broadcasted (copied) accordingly: the channel attention values are broadcasted along the spatial dimension and vice versa. F'' is the output of the final attention weights. The schematic diagrams of the channel attention mechanism and the spatial attention mechanism are shown in Figure 6.

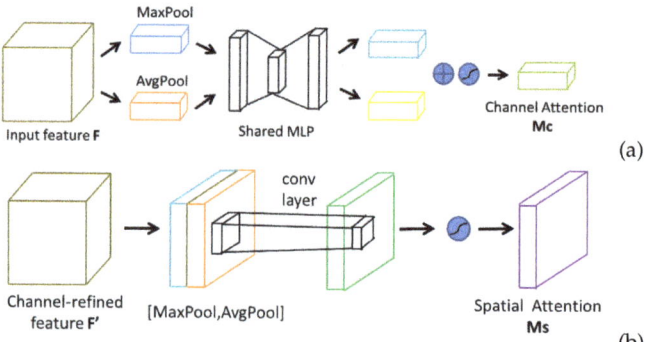

Figure 6. The Convolutional Block Attention Moudle: (**a**) Channel Attention Moudle. (**b**) Spatial Attention Moudle.

3.3. Introduction of FcaNet

In general, when calculating the channel attention, each channel will need a learnable scalar value to calculate the attention weight behind the scalar calculation function is generally used Global Average Pooling (GAP). However, GAP is not so perfect, and the simple de-averaging method discards a lot of information and does not fully capture the diversity of each channel. In order to obtain sufficient information about the diversity of each channel, Qin et al. [29] proved that GAP is a special form of discrete cosine transform (DCT), and based on this proof, generalized channel attention to the frequency domain and proposed FcaNet, a channel attention network using multiple frequencies. Assuming that X is the input feature map, the channel attention mechanism can be written as Equation (3) [30,31]:

$$att = sigmoid(fc(gap(X))), \qquad (3)$$

where att reprents the attention vector, $sigmoid$ reprents the sigmoid function, fc is the maping functions and gap is GAP. Once this attention vector is obtained, each channel can be scaled by the corresponding elements of this attention vector to obtain the output of the channel attention mechanism:

$$X^*_:,i,:,: := att_i X_:,i,:,:, \qquad s.t. \quad i \in 0,1,\ldots,C-1 \qquad (4)$$

where X^* reprents the out of attention mechanism, att_i is the i-th element of attention vector, and $X_:,i,:,:$ is the i-th channel of input. The DCT is defined as Equation (5) [32]

$$f_k = \sum_{i=0}^{L-1} x_i cos(\pi k/L(i+1/2)), \qquad s.t. \quad k \in 0,1,\ldots,L-1 \qquad (5)$$

Here, f is the spectrum of the DCT, x is the input, and L is the length of x. The 2-dimensional DCT can be written as:

$$f^{2d}_{h,w} = \sum_{i=0}^{H-1}\sum_{j=0}^{W-1} x^{2d}_{i,j} cos(\pi h/H(i+1/2))cos(\pi w/W(j+1/2)), \qquad (6)$$

$$s.t. \quad i \in 0,1,\ldots,H-1, j \in 0,1,\ldots,W-1$$

where f is the 2D DCT frequency spectrum, x is the input, H is the heght of x, and W reprents the width of x. The inverse transformation of 2D DCT can be written as:

$$x^{2d}_{i,j} = \sum_{h=0}^{H-1}\sum_{w=0}^{W-1} f^{2d}_{h,w} cos(\pi h/H(i+1/2))cos(\pi w/W(j+1/2)), \qquad (7)$$

$$s.t. \quad i \in 0,1,\ldots,H-1, j \in 0,1,\ldots,W-1$$

With the definition of channel attention and DCT, we can summarize two points: (1) Existing methods use GAP as preprocessing when doing channel attention. (2) DCT can be viewed as a weighted sum of inputs, and the weights are the cosine part of Equations (6) and (7). For more details, please refer to the reference [29].

3.4. Our Method

In terms of model selection, we choose CNN as the backbone network because the CNN model has the following advantages: The CNN learns local patterns and captures promising semantic information. Moreover, it is also known to be efficient compared to other model types for it has less number of parameters [33,34]. Considering the excellent performance achieved by ResNet50 in the field of strip classification defects, we decided to use it as the backbone network of our method. On this basis, since CBAM, FcaNet attention mechanism can weight the relevant parameters, making the algorithm focus on more and more valuable information; therefore, we add CBAM and FcaNet to improve the performance of the original model. The overall structure diagram of our proposed method is shown in Figure 7.

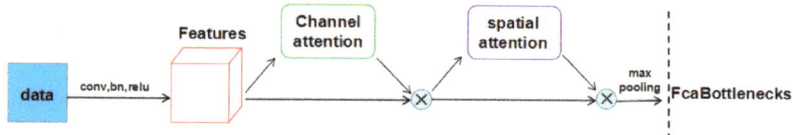

Figure 7. The overall structure of the method in this paper.

We adopt the FcaBottleneck instead of the Bottleneck structure in the original ResNet50 and place the spatial attention mechanism and the channel attention mechanism before the FcaBottleneck. In other words, we adopt CBAM outside the Bottleneck of ResNet for improvement and FcaNet inside the Bottleneck for improvement, so that the original Bottleneck, is converted to FcaBottleneck. The difference between the original Bottleneck in ResNet50 and the FcaBottleneck after the addition of FcaNet is shown in Figure 8.

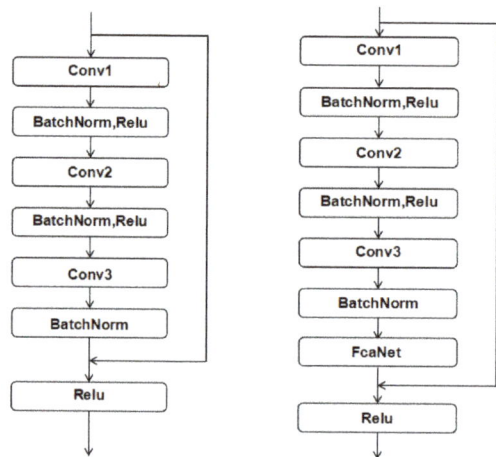

Figure 8. The Bottleneck and FcaBottleneck.

As shown in Figure 8, The flow chart on the left is Bottleneck and the flow chart on the right is FcaBottleneck. We can see the main difference between Bottleneck and FcaBottleneck: FcaBottleneck has an additional layer of FcaNet than Bottleneck. The code details can be found at: https://github.com/Fighter20092392/ResNet50-CBAM-FcaNet (accessed on 5 July 2021).

In contrast to other studies that added attentional mechanisms, we paired two different attentional mechanisms instead of adding only a single one. Moreover, we place CBAM and FcaNet inside and outside of the block, so that the attention mechanism can be fully functional. Whether such an improved scheme will improve the classification accuracy of strip surface defects will be verified by experiments next.

4. Experiments

4.1. Introduction of the Dataset

We choose the newly proposed X-SDD [7] strip surface defect dataset to validate the proposed method in this paper. The X-SDD dataset contains 7 types of 1360 surface defects in hot rolled strip: 238 slag inclusions, 397 red iron sheet, 122 iron sheet ash, 134 surface scratches, 63 oxide scale of plate system, 203 finishing roll printing and 203 oxide scale of temperature system. the size of original images is 128×128 pixels with 3 channel JPG format. The defect pattern in this dataset is shown in Figure 9.

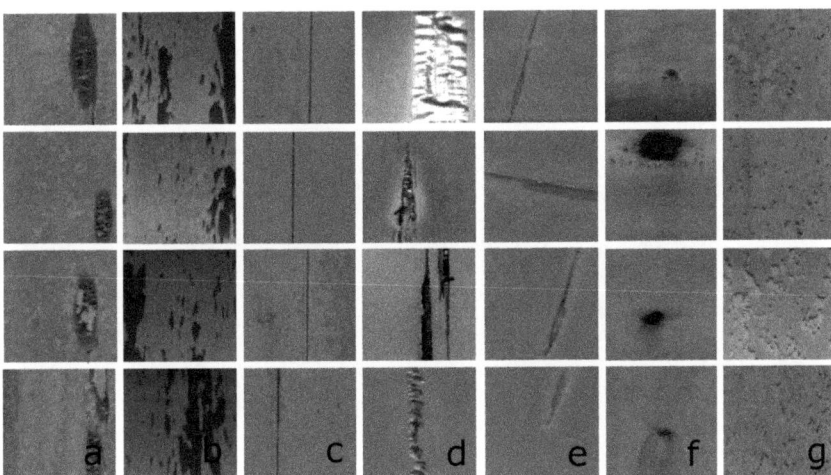

Figure 9. Samples of seven kinds of typical surface on X-SDD. (**a**) oxide scale of plate system. (**b**) red iron sheet. (**c**) surface scratches. (**d**) slag inclusions. (**e**) finishing roll printing. (**f**) iron sheet ash. (**g**) oxide scale of temperature system.

4.2. Experimental Settings

The experiments were conducted under the Win10 operating system and the PyTorch deep learning framework. The hardware configuration for the experiments was a single card NVIDIA RTX3060 GPU, an Intel Core i7-9700 CPU and a 64GB of RAM. In the experiment, the input size is set to 224 × 224 pixels, the batch size is set to 16 (Generally speaking, the batch size value should be set as large as possible within the allowed range of video memory), the learning rate is set to 0.0001 based on experience, the Adam optimizer is used for optimization, and the number of training epochs is 100 (Note that as the batch size increases, the epoch must be increased to force the model to maintain the same accuracy.). We use 70% of the defective images in the X-SDD dataset for the trainset and the remaining 30% of the images for the testset.

4.3. Experimental Results

In order to make the experimental results more convincing, we chose several indicators for comparison, including: Accuary, Macro-Recall, Macro-Precision and Macro-F1. The above indicators are derived as shown in Equations (8)–(12).

$$n\_correct = TP_0 + TP_1 + \ldots + TP_{N-1} \tag{8}$$

$$Accuary = \frac{n\_correct}{n\_total} \tag{9}$$

$$Macro - Recall = (\frac{TP_0}{TP_0 + FN_0} + \frac{TP_1}{TP_1 + FN_1} + \ldots + \frac{TP_{N-1}}{TP_{N-1} + FN_{N-1}}) \times \frac{1}{N} \tag{10}$$

$$Macro - Precision = (\frac{TP_0}{TP_0 + FP_0} + \frac{TP_1}{TP_1 + FP_1} + \ldots + \frac{TP_{N-1}}{TP_{N-1} + FP_{N-1}}) \times \frac{1}{N} \tag{11}$$

$$Macro - F1 = (\frac{2P_0 R_0}{P_0 + R_0} + \frac{2P_1 R_1}{P_1 + R_1} + \ldots + \frac{2P_{N-1} R_{N-1}}{P_{N-1} + R_{N-1}}) \times \frac{1}{N} \tag{12}$$

where n_total represents the total number of samples in the testset; N is the total number of defect types and in this paper the value of N is 7; $TP_0, TP_1, \ldots, TP_{N-1}$ represents the number of true cases in each category, i.e., the number that classifies the positive cases correctly. We have chosen several deep learning models for comparison: AlexNet [35],

MobileNet v3 [36], Xception [37], ShuffleNet [38], EspNet v2 [39], GhostNet [40], VGG16, VGG19 [41], ResNet101 and ResNet152 [23]. The experimental results are shown in Table 1.

Table 1. The experimental results.

Model	Accuary	Macro-Recall	Macro-Precision	Macro-F1
AlexNet	90.69%	82.79%	88.95%	84.21%
MobileNet v3	91.67%	87.95%	91.83%	88.59%
Xception	91.18%	84.30%	90.28%	85.37%
ShuffleNet	89.71%	84.76%	89.44%	84.87%
EspNet v2	86.52%	82.46%	84.10%	81.88%
GhostNet	89.22%	82.99%	87.16%	83.91%
ResNet101	92.40%	86.29%	93.30%	88.02%
ResNet152	89.22%	**89.10%**	87.26%	87.54%
VGG16	89.71%	86.64%	88.68%	87.47%
VGG19	86.52%	86.06%	88.98%	86.86%
RegVGG B1g2	88.48%	80.33%	92.54%	81.34%
Our Method	**93.87%**	87.33%	**94.35%**	**88.71%**

As shown in Table 1, our method achieves better than other compared models in terms of Accuracy, Macro-Precision and Macro-F1. Among them, our method achieved 93.87% in Accuracy, which is 1.47 percentage points igher than the second place ResNet101. Our method achieved the third place in the Macro-Recall metric by 1.77 percentage points lower than ResNet152 and 0.62 percentage points lower than MobileNet v3. One possible reason for the low Recall metric of our method is that Accuracy and Recall tend to affect each other, and our method focuses on improving Accuracy at the expense of Recall to some extent. Nevertheless, considering that our method is better than ResNet152 and MobileNet v3 in other metrics; therefore, our method has an advantage over ResNet152 as well as MobileNet v3.

4.4. Ablation Experiments

In order to verify the improvement of our method over the original ResNet50, the following ablation experiment is designed to analyze the effect. We compare the scheme proposed in this paper with ResNet50 and ResNet50+CBAM to analyze the effectiveness of our improved scheme. The results of the ablation experiments are shown in Table 2.

Table 2. The results of the ablation experiments.

Model	Accuary	Macro-Recall	Macro-Precision	Macro-F1
ResNet50	92.40%	86.45%	94.08%	88.32%
ResNet50+CBAM	92.65%	**88.62%**	91.40%	**89.71%**
ResNet50+CBAM+FcaNet	**93.87%**	87.33%	**94.35%**	88.71%

As can be seen in Table 2, the improved scheme of ResNet50+CBAM compared to ResNet50 has some improvement in Accuary, Macro-Recall and Macro-F1. This shows that the CBAM attention mechanism makes the algorithm pay more attention to the valuable information of images in spatial and channels, which in turn improves the classification ability of the algorithm. The only shortcoming is that the ResNet50+CBAM model is 2.68 percentage points lower than the ResNet50 model in the Macro-Precision metric. In contrast, our proposed ResNet50+CBAM+FcaNet scheme achieves higher scores than the ResNet50 model in all four metrics, which indicates that our proposed approach is more effective in improving the results. From a practical point of view, the most important of the four metrics is the Accuracy metric, and the scheme in this paper achieves the highest Accuracy. This indicates that our proposed ResNet50+CBAM+FcaNet method has more practical application value.

The confusion matrix of our proposed method is shown in Figure 10. The horizontal and vertical coordinates of 0–6 in the figure represent the oxide scale of plate system, red iron sheet, surface scratches, slag inclusions, finishing roll printing, iron sheet ash and oxide scale of temperature system, respectively. As can be seen from the confusion matrix, our method can classify most defect categories very accurately, with less accuracy only in the case of oxide scale of temperature system. Our model has 7 correct classifications and 12 incorrect classifications for oxide scale of plate system, with a correct classification rate of only 36.84% for this type of defect. The reason for this result may be that the amount of data on oxide scale of temperature system is small and the algorithm fails to learn effectively for this type of defect. A possible solution to this problem is to perform more data augmentation for this class of defects, using multiple models for cascading or ensemble learning. In the next part of this paper, we will try to solve the problem by using an ensemble learning approach.

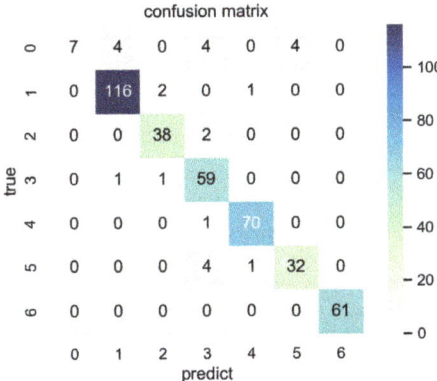

Figure 10. The confusion matrix.

4.5. Comparison of Model Complexity

The comparison results of the number of parameters and computation of the model are shown in Table 3. As can be seen from Table 3, the proposed method in this paper is basically the same in terms of number of parameters and computational effort compared with the original ResNet50. This shows that the improvement in the effect of our proposed method does not come from an increase in the number of participants but from a more rational structure. Compared with heavyweight deep learning models such as ResNet101, ResNet152 and VGG16, our method has the advantage of smaller number of parameters and computational complexity. The computational and parametric quantities of our method are only 35.60% and 44.77% of those of ResNet152, respectively. As can be seen from Table 1, ResNet152 has some advantages over our method in terms of recall, but our method is much better than ResNet152 in terms of the number of parameters and computational complexity. Compared to lightweight deep learning models such as EspNet v2 with 0.092 G of computation and 0.638 M of parameters, our method requires more hardware resources. In the future, model pruning, quantization, and knowledge distillation can be used to reduce the computational effort and number of parameters of the model, making it easier to deploy.

Table 3. The comparision of model complexity.

Model	Flops (G)	Params (M)
AlexNet	0.309	14.596
MobileNet v3	0.300	4.317
Xception	4.617	20.822
ShuffleNet	0.132	0.860
EspNet v2	0.092	0.638
GhostNet	0.213	3.127
ResNet50	4.109	23.522
ResNet101	7.832	42.515
ResNet152	11.557	58.158
VGG16	15.484	138.358
VGG19	19.647	143.667
RepVGG B1g2	9.815	43.748
ResNet50+CBAM	4.111	23.523
Our Method	4.114	26.038

4.6. The Ensemble Model

We use three models for integration, the sub-models are ResNet50+CBAM+FcaNet, VGG16 and SqueezeNet. The three sub-models were chosen because they differ in principle and meet the need for diversity in ensemble learning. We set the weights of ResNet50+CBAM+FcaNet, VGG16 and SqueezeNet to 1.2, 0.9, 0.9 respectively. The weights of the models are not all set equal; this is to facilitate the final choice of the integrated model when all three sub-models have different output values. The output of the ensemble model is shown in Figure 11.

As can be seen in Figure 11, the number of correctly classified oxide scale of plate system defects is 11, and the number of incorrectly classified defects is 8. The classification accuracy of this category of defects is 57.89%, which is 21.05% higher than the 36.84% of the ResNet50+CBAM+FcaNet model. The results of the ensemble model on each metric are shown in Table 4.

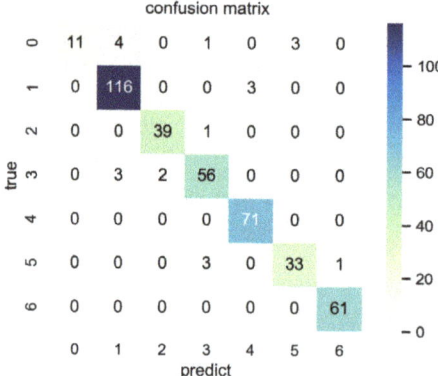

Figure 11. The output of the ensemble model.

Table 4. The effect of ensemble model.

Model	Accuary	Macro-Recall	Macro-Precsion	Macro-F1
The ensemble model	94.85%	90.71%	95.04%	92.06%

Comparing the results in Table 4 with those in Tables 1 and 2 shows that the ensemble model outperforms all comparison models in all four metrics. The ensemble model achieves a good score of over 90% on all indicators, which indicates that the improved model is more balanced on all indicators. In summary, it is effective to improve the model using ensemble learning.

5. Discussion and Conclusions

In this paper, we propose a ResNet50+CBAM+FcaNet model for the problem of classifying surface defects in hot-rolled strip steel. After validation on the newly proposed X-SDD dataset, our proposed algorithm achieves 93.87% accuracy on the testset, which is better than more than ten other comparative algorithms. In addition, our method still achieves better results relative to other comparison models on Macro-Precision, Macro-F1, and third place on Macro-Recall. The above results show the effectiveness of the algorithm proposed in this paper. Combining CBAM with FcaNet helps to improve the accuracy of ResNet50, and we argue that this improved approach will also be applicable to other models. In the next work, we may verify through more experiments which combination of this scheme and which model will achieve optimal results. Although our previous paper [7] showed that a RepVGG based scheme with an added attention mechanism may be superior in terms of effectiveness; according to Table 3, the ResNet50 based approach has an advantage over RepVGG in terms of number of parameters and computational complexity, i.e., it is easier to deploy in practice.

In order to further confirm the effectiveness of the attention mechanism, an ablation experiment is designed to verify. The ablation experiment verifies that adding the attention mechanism can effectively improve the classification accuracy of the algorithm, but while the overall accuracy is improved, the classification accuracy of individual categories may be reduced, which in turn affects the overall Recall and F1 metrics. Since the number of categories in the X-SDD dataset we use is unbalanced among categories, our approach will favor improving the accuracy of the categories with larger sample sizes at the expense of the accuracy of the categories with smaller sample sizes. In the case of a category with a smaller sample size, the accuracy of the category may be significantly reduced, as well as the overall Recall being significantly affected, simply because a few more samples are misclassified than before. The analysis of Figure 10 shows that the main factor affecting the Macro-Recall of our method is the low accuracy of the classification of oxide scale of plate system.

To solve the low classification accuracy of ResNet50+CBAM+FcaNet model on the oxide scale of plate system, we improve the original scheme. We introduce the concept of ensemble learning by integrating the original ResNet50+CBAM+FcaNet with VGG16 and SqueezeNet. We believe that the integration of multiple models can alleviate the problem of low classification level of a single model on a particular category to some extent, because the focus of different models may be different. In the selection of the ensemble sub-models, we fully consider the diversity of sub-models; the final selection of sub-models covers three models with different characteristics, such as with and without attention mechanism, heavy weight network and lightweight network. The final experimental results show that the ensemble model is optimal in all four indicators, and the classification accuracy of oxide scale of plate system has been improved substantially.

Although our proposed ResNet50+CBAM+FcaNet model and the improved ensemble model both achieve good results, there are still some areas that can be improved. Firstly, there is still some room for further improvement in the effectiveness of the model for the category imbalance problem. In this paper, the model integration is carried out in a weighting way, while other ensemble methods such as probabilistic summation can also be considered. In addition, modifying the loss function may also improve the classification accuracy for classes with small sample sizes. Secondly, we combine two attention mechanisms-CBAM and FcaNet with ResNet50, while more attention mechanisms can be considered for combination. Modification of the existing attention mechanism or proposing a new attention mechanism based on the characteristics of the steel strip surface defects may also yield good results.

After classifying the surface defects of hot rolled strip, different treatments are often required depending on the severity of the defects. Therefore, in the future, we will compile a dataset of the degree of surface defects of hot rolled steel strip and design an algorithm to classify the degree of surface defects of hot rolled steel strip. We may introduce the newly proposed MLP-mixer [42] algorithm into the field of strip defects and improve the original algorithm to make it more suitable for the context of strip defect classification.

Author Contributions: Conceptualization, X.F. and L.L.; methodology, X.F. and L.L.; software, L.L.; validation, X.F., L.L. and X.G.; formal analysis, L.L.; investigation, X.F.; resources, X.G.; data curation, X.F.; writing—original draft preparation, X.F., L.L. and X.G.; writing—review and editing, X.F. and L.L.; visualization, X.F.; supervision, X.F. and L.L.; project administration, X.G.; funding acquisition, X.G. All authors have read and agreed to the published version of the manuscript.

Funding: This research was funded by National Science Foundation under Grant 61573087, 61573088, 62173072 and 62173073.

Institutional Review Board Statement: Not applicable.

Informed Consent Statement: Not applicable.

Data Availability Statement: Not applicable.

Acknowledgments: This work is done when Xinglong Feng was an intern at xBang Inc., Shenyang, China. Thanks their support for this work.

Conflicts of Interest: The authors declare no conflict of interest.

References

1. Kumar, A.; Das, A.K. Evolution of microstructure and mechanical properties of Co-SiC tungsten inert gas cladded coating on 304 stainless steel. *Eng. Sci. Technol. Int. J.* **2020**, *24*, 591–604. [CrossRef]
2. Afanasieva, L.E.; Ratkevich, G.V.; Ivanova, A.I.; Novoselova, M.V.; Zorenko, D.A. On the Surface Micromorphology and Structure of Stainless Steel Obtained via Selective Laser Melting. *J. Surf. Investig. X-ray Synchrotron Neutron Tech.* **2018**, *12*, 1082–1087. [CrossRef]
3. Gromov, V.E.; Gorbunov, S.V.; Ivanov, Y.F.; Vorobiev, S.V.; Konovalov, S.V. Formation of surface gradient structural-phase states under electron-beam treatment of stainless steel. *J. Surf. Investig. X-ray Synchrotron Neutron Tech.* **2011**, *5*, 974–978. [CrossRef]
4. Youkachen, S.; Ruchanurucks, M.; Phatrapomnant, T.; Kaneko, H. Defect Segmentation of Hot-rolled Steel Strip Surface by using Convolutional Auto-Encoder and Conventional Image processing. In Proceedings of the 2019 10th International Conference of Information and Communication Technology for Embedded Systems (IC-ICTES), Bangkok, Thailand, 25–27 March 2019; pp. 1–5. [CrossRef]
5. Ashour, M.W.; Khalid, F.; Halin, A.A.; Abdullah, L.N.; Darwish, S.H. Surface defects classification of hot-rolled steel strips using multi-directional shearlet features. *Arab. J. Sci. Eng.* **2019**, *44*, 2925–2932. [CrossRef]
6. Luo, Q.; Fang, X.; Sun, Y.; Liu, L.; Ai, J.; Yang, C.; Simpson, O. Surface Defect Classification for Hot-Rolled Steel Strips by Selectively Dominant Local Binary Patterns. *IEEE Access* **2019**, *7*, 23488–23499. [CrossRef]
7. Feng, X.; Gao, X.; Luo, L. X-SDD: A New Benchmark for Hot Rolled Steel Strip Surface Defects Detection. *Symmetry* **2021**, *13*, 706. [CrossRef]
8. Xiao, M.; Jiang, M.; Li, G.; Xie, L.; Yi, L. An evolutionary classifier for steel surface defects with small sample set. *EURASIP J. Image Video Process.* **2017**, *2017*, 1–13. [CrossRef]
9. Gong, R.; Chu, M.; Yang, Y.; Feng, Y. A multi-class classifier based on support vector hyper-spheres for steel plate surface defects. *Chemom. Intell. Lab. Syst.* **2019**, *188*, 70–78. [CrossRef]
10. Chu, M.; Liu, X.; Gong, R.; Zhao, J. Multi-class classification method for strip steel surface defects based on support vector machine with adjustable hyper-sphere. *J. Iron Steel Res. Int.* **2018**, *25*, 706–716. [CrossRef]
11. Luo, Q.; Sun, Y.; Li, P.; Simpson, O.; Tian, L.; He, Y. Generalized completed local binary patterns for time-efficient steel surface defect classification. *IEEE Trans. Instrum. Meas.* **2018**, *68*, 667–679. [CrossRef]
12. Hu, H.; Li, Y.; Liu, M.; Liang, W. Classification of defects in steel strip surface based on multiclass support vector machine. *Multimed. Tools Appl.* **2014**, *69*, 199–216. [CrossRef]
13. Zhang, Z.F.; Liu, W.; Ostrosi, E.; Tian, Y.; Yi, J. Steel strip surface inspection through the combination of feature selection and multiclass classifiers. *Eng. Comput.* **2020**, *38*, 1831–1850. [CrossRef]
14. Fu, G.; Sun, P.; Zhu, W.; Yang, J.; Cao, Y.; Yang, M.Y.; Cao, Y. A deep-learning-based approach for fast and robust steel surface defects classification. *Opt. Lasers Eng.* **2019**, *121*, 397–405. [CrossRef]

15. Liu, Y.; Geng, J.; Su, Z.; Yin, Y. Real-time classification of steel strip surface defects based on deep CNNs. In *Proceedings of 2018 Chinese Intelligent Systems Conference*; Springer: Singapore, 2019; pp. 257–266.
16. Zhou, S.; Chen, Y.; Zhang, D.; Xie, J.; Zhou, Y. Classification of surface defects on steel sheet using convolutional neural networks. *Mater. Technol.* **2017**, *51*, 123–131.
17. Konovalenko, I.; Maruschak, P.; Brezinová, J.; Viňáš, J.; Brezina, J. Steel Surface Defect Classification Using Deep Residual Neural Network. *Metals* **2020**, *10*, 846. [CrossRef]
18. Yi, L.; Li, G.; Jiang, M. An end to end steel strip surface defects recognition system based on convolutional neural networks. *Steel Res. Int.* **2017**, *88*, 1600068. [CrossRef]
19. Song, K.; Yan, Y. Micro Surface defect detection method for silicon steel strip based on saliency convex active contour model. *Math. Probl. Eng.* **2013**, *2013*, 429094. [CrossRef]
20. Wang, F.; Jiang, M.; Qian, C.; Yang, S.; Li, C.; Zhang, H.; Wang, X.; Tang, X. Residual attention network for image classification. In Proceedings of the 2017 IEEE Conference on Computer Vision and Pattern Recognition (CVPR), Honolulu, HI, USA, 21–26 July 2017; pp. 6450–6458.
21. Wang, X.; Girshick, R.; Gupta, A.; He, K. Non-local neural networks. In Proceedings of the IEEE Conference on Computer Vision and Pattern Recognition, Salt Lake City, UT, USA, 18–23 June 2018; pp. 7794–7803.
22. Cao, Y.; Xu, J.; Lin, S.; Wei, F.; Hu, H. Gcnet: Non-local networks meet squeeze-excitation networks and beyond. In Proceedings of the IEEE/CVF International Conference on Computer Vision Workshops, Seoul, Korea, 27–28 October 2019.
23. He, K.; Zhang, X.; Ren, S.; Sun, J. Deep residual learning for image recognition. In Proceedings of the IEEE Conference on Computer Vision and Pattern Recognition, Las Vegas, NV, USA, 27–30 June 2016; pp. 770–778.
24. Woo, S.; Park, J.; Lee, J.Y.; Kweon, I.S. Cbam: Convolutional block attention module. In Proceedings of the European Conference on Computer Vision (ECCV), Munich, Germany, 8–14 September 2018; pp. 3–19.
25. Bahdanau, D.; Cho, K.; Bengio, Y. Neural machine translation by jointly learning to align and translate. *arXiv* **2014**, arXiv:1409.0473.
26. Xu, K.; Ba, J.; Kiros, R.; Cho, K.; Courville, A.; Salakhudinov, R.; Zemel, R.; Bengio, Y. Show, attend and tell: Neural image caption generation with visual attention. In Proceedings of the International Conference on Machine Learning (PMLR), Lille, France, 7–9 July 2015; pp. 2048–2057.
27. Gregor, K.; Danihelka, I.; Graves, A.; Rezende, D.; Wierstra, D. Draw: A recurrent neural network for image generation. In Proceedings of the International Conference on Machine Learning (PMLR), Lille, France, 7–9 July 2015; pp. 1462–1471.
28. Jaderberg, M.; Simonyan, K.; Zisserman, A. Spatial transformer networks. *arXiv* **2015**, arXiv:1506.02025.
29. Qin, Z.; Zhang, P.; Wu, F.; Li, X. FcaNet: Frequency Channel Attention Networks. *arXiv* **2020**, arXiv:2012.11879.
30. Hu, J.; Shen, L.; Sun, G. Squeeze-and-excitation networks. In Proceedings of the IEEE Conference on Computer Vision and Pattern Recognition, Salt Lake City, UT, USA, 18–23 June 2018; pp. 7132–7141.
31. Qilong, W.; Banggu, W.; Pengfei, Z.; Li, P.; Zuo, W.; Hu, Q. ECA-Net: Efficient Channel Attention for Deep Convolutional Neural Networks. *arXiv* **2020**, arXiv:1910.03151.
32. Ahmed, N.; Natarajan, T.; Rao, K.R. Discrete cosine transform. *IEEE Trans. Comput.* **1974**, *100*, 90–93. [CrossRef]
33. Jeon, M.; Jeong, Y.S. Compact and accurate scene text detector. *Appl. Sci.* **2020**, *10*, 2096. [CrossRef]
34. Vu, T.; Van Nguyen, C.; Pham, T.X.; Luu, T.M.; Yoo, C.D. Fast and efficient image quality enhancement via desubpixel convolutional neural networks. In Proceedings of the European Conference on Computer Vision (ECCV) Workshops, Glasgow, UK, 23–28 August 2018; pp. 243–259.
35. Krizhevsky, A.; Sutskever, I.; Hinton, G.E. Imagenet classification with deep convolutional neural networks. *Adv. Neural Inf. Process. Syst.* **2012**, *25*, 1097–1105. [CrossRef]
36. Howard, A.; Sandler, M.; Chu, G.; Chen, L.; Chen, B.; Tan, M.; Wang, W.; Zhu, Y.; Pang, R.; Vasudevan, V.; et al. Searching for mobilenetv3. In Proceedings of the IEEE/CVF International Conference on Computer Vision, Seoul, Korea, 27–28 October 2019; pp. 1314–1324.
37. Chollet, F. Xception: Deep learning with depthwise separable convolutions. In Proceedings of the IEEE Conference on Computer Vision and Pattern Recognition, Honolulu, HI, USA, 21–26 July 2017; pp. 1251–1258.
38. Zhang, X.; Zhou, X.; Lin, M.; Sun, J. Shufflenet: An extremely efficient convolutional neural network for mobile devices. In Proceedings of the IEEE Conference on Computer Vision and Pattern Recognition, Salt Lake City, UT, USA, 18–23 June 2018; pp. 6848–6856.
39. Mehta, S.; Rastegari, M.; Shapiro, L.; Hajishirzi, H. Espnetv2: A light-weight, power efficient, and general purpose convolutional neural network. In Proceedings of the IEEE/CVF Conference on Computer Vision and Pattern Recognition, Seoul, Korea, 27–28 October 2019; pp. 9190–9200.
40. Han, K.; Wang, Y.; Tian, Q.; Guo, J.; Xu, C.; Xu, C. Ghostnet: More features from cheap operations. In Proceedings of the IEEE/CVF Conference on Computer Vision and Pattern Recognition, Seattle, WA, USA, 14–19 June 2020; pp. 1580–1589.
41. Simonyan, K.; Zisserman, A. Very deep convolutional networks for large-scale image recognition. *arXiv* **2014**, arXiv:1409.1556.
42. Tolstikhin, I.; Houlsby, N.; Kolesnikov, A.; Beyer, L.; Zhai, X.; Unterthiner, T.; Yung, J.; Steiner, A.; Keysers, D.; Uszkoreit, J.; et al. Mlp-mixer: An all-mlp architecture for vision. *arXiv* **2021**, arXiv:2105.01601.

Article

Dynamic Programming Algorithms for Computing Optimal Knockout Tournaments

Amelia Bădică [1], Costin Bădică [2,*], Ion Buligiu [1], Liviu Ion Ciora [1] and Doina Logofătu [3]

[1] Department of Statistics and Business Informatics, University of Craiova, 200585 Craiova, Romania; amelia.badica@edu.ucv.ro (A.B.); ion.buligiu@edu.ucv.ro (I.B.); liviu.ciora@edu.ucv.ro (L.I.C.)
[2] Department of Computers and Information Technology, University of Craiova, 200585 Craiova, Romania
[3] Faculty of Computer Science and Engineering, Frankfurt University of Applied Sciences, Nibelungenplatz 1, 60318 Frankfurt am Main, Germany; logofatu@fb2.fra-uas.de
* Correspondence: costin.badica@edu.ucv.ro

Abstract: We study competitions structured as hierarchically shaped single-elimination tournaments. We define optimal tournaments by maximizing attractiveness such that the topmost players will have the chance to meet in higher stages of the tournament. We propose a dynamic programming algorithm for computing optimal tournaments and we provide its sound complexity analysis. Based on the idea of the dynamic programming approach, we also develop more efficient deterministic and stochastic sub-optimal algorithms. We present experimental results obtained with the Python implementation of all the proposed algorithms regarding the optimality of solutions and the efficiency of the running time.

Keywords: optimization; knockout tournament; dynamic programming algorithm; computational complexity; combinatorics

Citation: Bădică, A.; Bădică, C.; Buligiu, I.; Ciora, L.I.; Logofătu, D. Dynamic Programming Algorithms for Computing Optimal Knockout Tournaments. *Mathematics* **2021**, *9*, 2480. https://doi.org/10.3390/math9192480

Academic Editor: Fabio Caraffini

Received: 7 September 2021
Accepted: 27 September 2021
Published: 4 October 2021

Publisher's Note: MDPI stays neutral with regard to jurisdictional claims in published maps and institutional affiliations.

Copyright: © 2021 by the authors. Licensee MDPI, Basel, Switzerland. This article is an open access article distributed under the terms and conditions of the Creative Commons Attribution (CC BY) license (https://creativecommons.org/licenses/by/4.0/).

1. Introduction

Tournament design is a combinatorial problem with many theoretical implications, as well as with a lot of practical applications. There are many types of tournaments that have been theoretically analyzed and practically used in various contexts. Basically, there are two main principles used in tournament design: "round-robin" principle and "knockout" principle. They can be used in isolation or combined for obtaining different tournaments designs, depending on various factors, such as the number of players, time available to carry out the tournament, and application domain.

In this paper we propose a formal definition of competitions that have the shape of single-elimination tournaments, also known as knockout tournaments. We introduce methods to quantitatively evaluate the attractiveness and competitiveness of a given tournament. We consider that a tournament is more attractive if competition is encouraged in higher stages, i.e., higher-ranked players will have the chance to meet in higher stages of the tournament, thus increasing the stake of their matches.

In knockout tournaments, the result of each match is always a win of one of the two players, i.e., draws are not possible. A knockout tournament is hierarchically structured as a binary tree such that each leaf represents one player or team that is enrolled in the tournament, while each internal node represents a game of the tournament.

The tournament is carried out in a series of rounds. If there is a number N of players equal to a power of 2, for example $N = 8 = 2^3$ then the tournament tree is a complete binary tree with all the players entering the tournament in the first round; however, in the general case, the number of players might not be a power of 2, for example $N = 9$. In this case some of the players will receive waivers thus entering the tournament directly in the second round, while the rest of the players will enter the tournament in the first round.

In this paper we significantly extend our preliminary results reported in [1] for fully balanced tournaments (the number of players is $N = 2^k$) to general tournaments where the number of players can be an arbitrary natural number, not necessarily a power of 2. Our new results are summarized as follows:

1. An exact formula for counting the total number of knockout tournaments in the general case, showing that the number of tournaments grows very large with the number of players.
2. A tournament cost function based on players' quota that assigns a higher cost to those tournaments where highly ranked players tend to meet in higher stages, thus making the tournament more attractive and competitive.
3. An exact dynamic programming algorithm for computing optimal tournaments in the general case.
4. A more efficient generic sub-optimal algorithm derived from the idea of the dynamic programming approach.
5. Deterministic and stochastic versions of the generic sub-optimal algorithm.
6. The complexity analysis of all the proposed algorithms.
7. The implementation issues of the proposed algorithms using Python, as well as the experimental results obtained with our implementation.

2. Related Works

Tournament design attracted research in operations research, combinatorics, and statistics. The problem is also related to intelligent planning and activity scheduling, broadly covered also by artificial intelligence.

There are two main principles used in tournament design, namely the "round-robin" principle and "knockout" principle, and they can be used in isolation or combined for obtaining different tournaments designs. The "round-robin" principle states that in a tournament, each two players should meet at least once, sometimes exactly once. "Knockout", also known as the "elimination" principle states that players are eliminated after a certain number of games, sometimes exactly after one game.

For example, in a round-robin tournament in which each two players should meet exactly once, an important aspect is the scheduling of the tournament, a problem also known as league scheduling [2]. This is an important component of the tournament design. Note that the "round robin" principle can also be applied with restrictions. Consider for example a two-team tournament, in which each team has the same number of players. Each game involves two players from different teams and any two players from different teams must play exactly once. In this case, we can still apply the round-robin principle, but players of the same team are not allowed to play. This type of tournament is called a bipartite tournament. A good coverage of the combinatorial aspects of round-robin tournaments can be found in monograph [2].

On the other hand, in a knockout tournament where two players will play at most one game and after each game exactly one player advances in the tournament, while the other is kicked off, the tournament schedule results directly from the tournament design, i.e., no separate scheduling stage is needed.

Note also that the round-robin and knock-out principles can be combined into a single tournament design. Consider for example the UEFA Champions League football tournament. In the groups' phase, round-robin is used inside each group, while after the group phase, knockout is used to determine the tournament winner.

In this paper we consider knockout tournaments in which two players meet at most once. These are specialized tournaments with possible applications in sports (e.g., football and tennis tournaments), online games (e.g., online poker), and election processes. While in the former case there is a relatively low number of players, thus not raising special computational challenges, the number of players in massive multiplayer online games can grow such that computing the optimal tournament becomes a more difficult problem.

A comprehensive analysis of knockout tournaments is proposed in [3,4]. Traditionally, the method for designing a tournament involves two stages: (i) tournament structure design and (ii) seeding. In the first stage, the structure of the tournament tree is proposed.

In the second stage, players are assigned to each leaf of the tournament tree; however, as we can argue that this method of tournament design has some limitations, we proposed different "integrated" approach. One limitation is for example the fact that once fixed, the tournament structure cannot be changed. On one hand this will result in a smaller search space during the seeding process, on the other hand it limits the total number of tournament designs. So our model of tournament trees includes both the tree structure, as well as the players' seeding; a separate seeding process is not necessary.

Research in combinatorics of knockout tournaments also produced interesting mathematical results. The structure of a knockout tournament can be modeled as a special kind of binary tree, called an Otter tree [5]. The number of knockout tournament structures (i.e., prior to seeding) for N players is given by the Wedderburn—Etherington number [6] of order N that is known to have an exponential growth approximately equal to $0.3188 \times \frac{2.4832^N}{N^{1.5}}$.

An important aspect concerns the factors that can be used to evaluate a tournament design. This problem has been also considered in previous works [3,4]. An interesting discussion of economic aspects of tournament attractiveness, such as spectator interest, is provided by [7].

A method for augmenting a tournament with probabilistic information based on tournament results was proposed in [8]. The effectiveness of tournament plans based on dominance graphs is studied in [6]. The tournament problem was also a source of inspiration for programming competitions [9]. A more recent work addressing competitiveness development and ranking precision of tournaments is [10].

For example, the more recent work [3,4] proposes a probabilistic approach to define the tournament cost, by including in their model the win–loss probabilities of each game between two players i and j. On one hand, we can question the robustness of such values. On the other hand, we recognize that some approximations of such values might be empirically obtained based on several factors, e.g., global player rankings (when available) or on the history of games between the players (if a nonempty history exists). In our work we do not use this information, i.e., we assume by default that there are equal winning chances for the players of each game. While this simplification clearly has drawbacks, it has the advantage of enabling a clean algorithm design based on dynamic programming principles. Our approach can be extended by adding probabilistic information to the cost function, but then it will require further analysis of algorithmic solutions within our "integrated" approach.

A theoretical investigation of knockout tournaments is provided by [11]. Their analysis is focused only on tournaments with a power of 2 number of players, i.e., similar with [1], but definitely less general than in the current work, where an arbitrary number of players is considered. Interesting results of this work concern the discussion of new seeding approaches named "equal gap" and "increasing competitive", as well as the investigation of their theoretical properties.

There has been also theoretical interest in analyzing the possible outcomes of knockout tournaments. Upper and lower bounds of winning probabilities of players of a random knockout tournament are provided in [12]. Note that the analysis is focused on the random knockout tournament where the definition of matches to be carried out in each round is defined randomly. Moreover, this work assumes as [3,4], that the win–loss probabilities of each match between two players are known.

There is also interest in the literature in designing new formats of knockout tournaments. For example, a new format based on actively involving the teams in defining the tournament format, was recently proposed in [13] for the specific competition of UEFA Champions League. The proposed format was coined "Choose your opponent" with the claimed benefit to make group stages more exciting. The authors also show how this model can be used for the objective of maximizing the number of home games during the knockout stage.

Knockout tournament structures are sometimes called tournament brackets. According to [14], two types of tournament brackets are possible: fixed and adaptive. In fixed brackets, the tournament structure is fixed, while in adaptive brackets pairings in stage $i+1$ are defined based on winners of stage i. Our approach is clearly fixed, with the difference that we use an integrated approach to define both the structure as well as the seeding. What is different in [14] is the fact that authors look for optimizing a fixed bracket by using utility functions and Bayesian optimal design. They propose a simulated annealing algorithm to optimize the expected value of a given utility function on a fixed tournament bracket. While interesting, this endeavor is clearly different from our approach. We plan however to investigate in the future the suitability of extending integrated approach and proposed algorithms by incorporating probabilistic information.

Clearly tournaments have a lot of practical applications, for example in the sports' domain. In this context, the recent work [15] provides an interesting discussion on the economics of sports from operations research, as well as practical applicability perspectives. The discussion is centered around several paradoxes of tournament rankings, with clear examples from the practice of tournament design.

3. Knockout Tournaments

We consider hierarchically structured knockout tournaments such that the result of each match is always a win of one of the two players, i.e., draws are not possible. A tournament is modeled as a binary tree such that each leaf node represents a unique player and each internal node represents a game between two players and its winner.

Definition 1 (Tournaments). *Let Σ be a finite nonempty set of players. We define the set of trees $\mathcal{T}(\Sigma)$ with leaves Σ as follows:*

1. *If $\Sigma = \{i\}$ is a singleton set then $\mathcal{T}(\Sigma) = \{i\}$, i.e., there is a single tree containing a single node i.*
2. *If Σ_1 and Σ_2 are two disjoint sets of players then let $\Sigma = \Sigma_1 \cup \Sigma_2$. Then:*

$$\mathcal{T}(\Sigma) = \{t | t = \{t_1, t_2\}, t_1 \in \Sigma_1, t_2 \in \Sigma_2\} \quad (1)$$

Note that the set notation in Equation (1) implies that the trees are not ordered, i.e., the order of the left and right branches does not matter.

Example 1. *We consider examples of tournaments for sets of players with $1, 2, 3$ and 4 elements:*

1. *If $\Sigma = \{1\}$ then $\mathcal{T}(\Sigma) = \{1\}$.*
2. *If $\Sigma = \{1,2\}$ then $\mathcal{T}(\Sigma) = \{\{1,2\}\}$.*
3. *If $\Sigma = \{1,2,3\}$ then $\mathcal{T}(\Sigma) = \{\{\{1,2\},3\},\{\{1,3\},2\},\{\{3,2\},1\}\}$.*
4. *If $\Sigma = \{1,2,3,4\}$ then $\mathcal{T}(\Sigma) = \{\{\{1,2\},\{3,4\}\},\{\{1,3\},\{2,4\}\},\{\{1,4\},\{2,3\}\}, \{\{\{1,2\},3\},4\},\ldots\}$. It is not difficult to see that in this case there are 15 trees.*

Some of the tournaments introduced in Example 1 are depicted graphically in Figure 1. Observe that the tournaments on the first row (labeled "a" and "b") involve a number of elements that is a power of two ($2 = 2^1$ and $4 = 2^2$, respectively) and are fully balanced. However, the tournaments on the second row are not fully balanced, although the lower rightmost tournament involves $4 = 2^2$ players. However, intuitively, the tournament with three players (labeled "c") should be accepted, as player 3 will enter the tournament only 1 round after players 1 and 2, i.e., it has a sense of "balancing". However, the lower rightmost tournament with four players (labeled "d") is not acceptable, as player 4 received an exemption from playing in the first two rounds, and this is considered unfair.

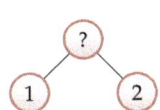
a. Balanced tournament with two players.

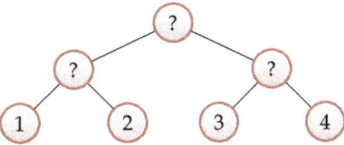
b. Balanced tournament with four players.

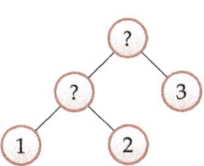
c. Balanced tournament with three players.

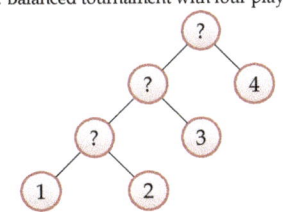
d. Unbalanced tournament with four players.

Figure 1. Tournaments of two and three players (first column) and four players (second column).

Proposition 1 (Counting tournaments). *The set $\mathcal{T}(\Sigma)$ with $|\Sigma| = N$ players contains:*

$$\frac{(2N-2)!}{(N-1)! \times 2^{N-1}} \qquad (2)$$

elements.

Proof. The number of full binary tree structures with N leaves is equal to C_{N-1} where C_N is Catalan's number [16] defined by:

$$C_N = \frac{1}{N+1} \binom{2N}{N} \qquad (3)$$

Now, each permutation of the N players can be attached to the leaves of a binary tree, thus obtaining $N! \cdot C_{N-1}$ trees. However, the branches of each internal node can be exchanged, resulting in the same tree. There are $N-1$ internal nodes and therefore a total number of 2^{N-1} exchanges, resulting a number of trees given by:

$$\frac{N! \cdot C_{N-1}}{2^{N-1}} = \frac{(2N-2)!}{(N-1)! \cdot 2^{N-1}} \qquad (4)$$

q.e.d. □

Example 2. *For example, if $N = 3$ we obtain $\frac{4!}{2! \times 2^2} = 3$ trees, while if $N = 4$ we obtain $\frac{6!}{3! \times 2^3} = 15$ trees. These results are consistent with Example 1.*

A valid tournament should be balanced, i.e., each player should play (almost) the same number of games to win the tournament.

Analyzing the tournaments from Example 1 and Figure 1 we can observe that if $|\Sigma| \leq 3$ then each element of $\mathcal{T}(\Sigma)$ represents a valid tournament. However, if $|\Sigma| = 4$ then only 3 trees of $\mathcal{T}(\Sigma)$ represent valid tournaments. For example, $\{\{1,2\},\{3,4\}\}$ is a valid tournament as each player should play exactly two games to win the tournament. In this case we have a fully balanced tournament consisting of $N = 2^2$ players. Moreover, $\{\{1,2\},3\}$ is also considered a valid tournament, as players 1 and 2 must play two games to win, while player 3 must play one game to win, i.e., has an exemption for the first round (the difference between the number of games played by each player is at most 1). However, $\{\{\{1,2\},3\},4\}$ is not a valid tournament, as players 1 and 2 must play three games to win the tournament, while player 4 must play a single match to win the tournament (the difference between the number of games played by each player is above 1, i.e., more than one exemption for a player is considered unfair).

Observe that a tree representing a valid tournament has the property that all its leaves are of height n or $n+1$ for a suitable value of n. Actually, the value of n can be determined from the given number of players N of the tournament and it represents the number of rounds of the tournament.

Let us consider a tournament with n rounds. It is not difficult to see that the maximum number of players is $N_{max} = 2^n$ and it is obtained when in the first round we have a maximum number of 2^{n-1} games; therefore, for a tournament with n rounds we have:

$$2^{n-1} < N \leq 2^n \quad (5)$$

Observe that from Equation (5) it follows that:

$$n = \lceil \log_2 N \rceil \quad (6)$$

Definition 2 (Balanced (valid) tournaments). *Let $n \in \mathbb{N}$ be the number of rounds. Let Σ be a nonempty set of N players such that conditions (5) and (6) are fulfilled. Then the set $\mathcal{T}_n(\Sigma)$ of balanced trees with n layers representing the set of balanced (valid) tournaments with n rounds is defined as follows:*

1. *If $n = 0$ then $N = 1$ so we have a singleton set $\Sigma = \{i\}$. In this case $\mathcal{T}_0(\Sigma) = \{i\}$.*
2. *If $n \geq 1$, $t_1 \in \mathcal{T}_{n-1}(\Sigma_1)$, $t_2 \in \mathcal{T}_{n-1}(\Sigma_2)$, $\Sigma_1 \cap \Sigma_2 = \emptyset$ and $\Sigma_1 \cup \Sigma_2 = \Sigma$ then $t = \{t_1, t_2\} \in \mathcal{T}_n(\Sigma)$.*
3. *If $n \geq 2$, $t_1 \in \mathcal{T}_{n-1}(\Sigma_1)$, $t_2 \in \mathcal{T}_{n-2}(\Sigma_2)$ is a fully balanced tree (i.e., $|\Sigma_2| = 2^{n-2}$), $\Sigma_1 \cap \Sigma_2 = \emptyset$ and $\Sigma_1 \cup \Sigma_2 = \Sigma$ then $t = \{t_1, t_2\} \in \mathcal{T}_n(\Sigma)$.*

If $n \geq 1$ then there are $N \in 2^{n-1} + 1 \ldots 2^n$ players. Then a tree $t \in \mathcal{T}_n(\Sigma)$ can be obtained either (i) by joining two balanced trees with $n-1$ layers or (ii) by joining one balanced tree with $n-1$ layers and one fully balanced tree with $n-2$ layers (all its leaves are on layer $n-2$), so in both cases the balancing condition of t is properly preserved.

Proposition 2 (Structure of a balanced tournament). *Let $t \in \mathcal{T}_n(\Sigma)$ be a tournament of N players such that n is defined by Equation (6). Then the number of players starting in the first round is $\beta = 2N - 2^n$ and the number of players starting in the second round (waivers) is $\gamma = 2^n - N$. Moreover, if $n \geq 1$ then the number of internal nodes of level 2 in the tree is equal to $\alpha = N - 2^{n-1}$, i.e., $\beta = 2\alpha$ and $\gamma = 2^{n-1} - \alpha$.*

Proof. First observe that if the number of players is a power of 2, i.e., $N = 2^n$, then $\alpha = 2^{n-1} = N/2$, $\beta = N$, and $\gamma = 0$. This is trivially true, as in this case the tournament is fully balanced and all the players start in the first round (there are no exemptions).

The proof for the general case can be shown by induction on $n \in \mathbb{N}$.

For $n = 0$ the tournament has $N = 1$ players. In this case there is a single balanced tournament with $\gamma = 0$ and $\beta = 1$, so the property trivially holds.

For $n = 1$ the tournament has $N = 2$ players. In this case there is a single balanced tournament with $\alpha = 1$, $\beta = 2$ and $\gamma = 0$, so the property trivially holds.

Let us now assume that the property holds for $k = 0, 1, \ldots, n$ and let us prove it for $k = n + 1$. There are two cases.

Case 1. If $n \geq 1$, $t_1 \in \mathcal{T}_n(\Sigma_1)$, $t_2 \in \mathcal{T}_n(\Sigma_2)$, $\Sigma_1 \cap \Sigma_2 = \emptyset$ and $\Sigma_1 \cup \Sigma_2 = \Sigma$, let us consider $t = \{t_1, t_2\} \in \mathcal{T}_{n+1}(\Sigma)$ such that the second condition of Definition 2 is fulfilled. According to the induction hypothesis we have $\gamma_i = 2^n - N_i$, $\beta_i = 2N_i - 2^n$, $\alpha_i = N_i - 2^{n-1}$ for $i = 1, 2$ and $N = N_1 + N_2$. Then $\gamma = \gamma_1 + \gamma_2 = 2^{n+1} - (N_1 + N_2) = 2^{n+1} - N$. Similarly $\beta = \beta_1 + \beta_2 = 2N - 2^{n+1}$ and $\alpha = \alpha_1 + \alpha_2 = N - 2^n$ q.e.d.

Case 2. If $n \geq 2$, $t_1 \in \mathcal{T}_n(\Sigma_1)$, $t_2 \in \mathcal{T}_{n-1}(\Sigma_2)$ is a fully balanced tree (i.e., $|\Sigma_2| = 2^{n-1}$), $\Sigma_1 \cap \Sigma_2 = \emptyset$ and $\Sigma_1 \cup \Sigma_2 = \Sigma$, let us consider $t = \{t_1, t_2\} \in \mathcal{T}_{n+1}(\Sigma)$ such that the third condition of Definition 2 is fulfilled. According to the induction hypothesis we have $\gamma_1 = 2^n - N_1$, $\beta_1 = 2N_1 - 2^n$, $\alpha_1 = N_1 - 2^{n-1}$, $\gamma_2 = 0$, $\beta_2 = 2^{n-1}$ and $\alpha_2 = 2^{n-2}$, $N_2 = 2(n-1)$, and $N = N_1 + 2^{n-1}$. Then $\beta = \beta_1 = 2(N_1 + 2^{n-1}) - 2^n - 2^n = 2N - 2^{n+1}$. Similarly $\gamma = \gamma_1 + \beta_2 = 2^n - N_1 + 2^{n-1} = 2^n + 2^{n-1} + 2^{n-1} - N = 2^{n+1} - N$ and similarly for $\alpha = \alpha_1 = N_1 - 2^{n-1} = N - 2^{n-1} - 2^{n-1} = N - 2^n$ q.e.d.

The relations $\beta = 2\alpha$ and $\gamma = 2^{n-1} - \alpha$ can be now easily checked. □

Example 3 (Tournament structure design). *Let us consider a tournament with $N = 5$ players. In this case $n = 3$, $\gamma = 2^3 - 5 = 3$, $\alpha = 2^2 - 3 = 1$, $\beta = 2 \times 5 - 2^3 = 2$. A tree representing a tournament with five players will have three layers such that the first layer consists of $\beta = 2$ leaves (players) and the second layer consists of $2^{n-1} = 2^2 = 4$ nodes among which there is $\alpha = 1$ internal node and $\gamma = 3$ leaves (players). One such a balanced tournament is depicted in Figure 2.*

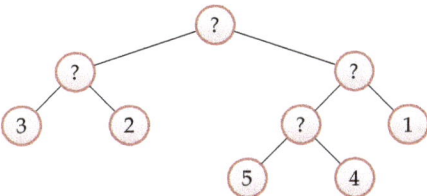

Figure 2. A balanced tournament of five players.

Proposition 3 (Counting balanced tournaments). *The set $\mathcal{T}_n(\Sigma)$ with $|\Sigma| = N$ players contains:*

$$\frac{N! \cdot \binom{2^{n-1}}{\gamma}}{2^{N-1}} \quad (7)$$

elements.

Proof. There are $\binom{2^{n-1}}{\gamma}$ ways of choosing how the γ players will enter second round. Their ordering matters so we multiply with $\gamma!$. Moreover those γ players are arbitrarily chosen from the set of N players, so we multiply with $\binom{N}{\gamma}$. Finally, the ordering of those β remaining players that enter first round matters, so we also multiply with $\beta!$. For each internal node of the tree, exchanging its left and right sub-tree is a tournament invariant. There are 2^{N-1} independent ways of exchanging left and right sub-tree of the tree, so we must divide by 2^{N-1}. We obtain:

$$\frac{\binom{N}{\gamma} \cdot \binom{2^{n-1}}{\gamma} \cdot \gamma! \cdot \beta!}{2^{N-1}} = \frac{N! \cdot \binom{2^{n-1}}{\gamma}}{2^{N-1}} \quad (8)$$

A simpler proof is obtained by thinking about structures (i.e., "shapes") of tournament trees. The selection of the "locations" of those γ players entering second round can be achieved in $\binom{2^{n-1}}{\gamma}$ ways. For each tree structure defined in this way there are $N!$ permutations of the leaves (players), thus defining a total number of $N! \cdot \binom{2^{n-1}}{\gamma}$ balanced tournament trees. Finally we divide by 2^{N-1} and we obtain Equation (7). □

Example 4. *Let us check the number of balanced tournaments for several cases.*

1. *The number of balanced tournaments with $N = 5$ players can be obtained as follows:*

$$\frac{5! \times \binom{2^2}{3}}{2^4} = 30 \quad (9)$$

It is not difficult to verify that this result is correct. In this case $\beta = 2$. There are $\binom{5}{2} = 10$ ways of selecting those two players that will enter first round. We have one separate tournament by letting the winner of the game of these two players playing against each of the remaining three players in the second round. So in total there are $10 \times 3 = 30$ balanced tournaments with five players.

2. For $n = 3$ players we obtain:

$$\frac{3! \times \binom{2^1}{1}}{2^2} = 3 \tag{10}$$

3. If $N = 2^n$ then $\gamma = 0$, thus we obtain our result for fully balanced tournaments from [1] stating that the total number of fully balanced tournaments is given by:

$$\frac{(2^n)!}{2^{2^n-1}}. \tag{11}$$

The number of fully balanced tournaments with two rounds is $\frac{(2^2)!}{2^{2^2-1}} = \frac{4!}{8} = 3$. Observe that applying the formula, we obtain 315 fully balanced tournaments with three rounds. Let us obtain this result using a different reasoning. Let us count the number of a set with eight elements consisting of two subsets of four elements each. There are $\binom{8}{4}/2 = 35$ possibilities, as we consider the four combinations of eight elements, and we divide by two as the order of the subsets of a partition does not matter; however, for each set of each partition there are three fully balanced tournaments of three rounds, so multiplying we obtain a total of nine possibilities. So the number of three-stage tournaments is $9 \times 35 = 315$.

4. Optimal Tournaments

Each round of a tournament with n rounds defines possible games between players. Note that in a given tournament any two players can play in a game at one and only one of its rounds. This follows from the fact that for any two leaves of a binary tree there is a unique closest common ancestor. It follows that the tournament round $s_{i,j}$ where players i, j can meet is a unique value in $1, 2, \ldots, n$ and it is well defined. For example, referring to the tournament shown in Figure 2, $s_{2,4} = 3$, $s_{1,5} = 2$, and $s_{4,5} = 1$.

Intuitively, the higher the quotations of players i and j, the better it is to let them meet in a higher stage of the tournament in order to increase the stakes of their games.

We assume in what follows that a quotation $q_i \in (0, +\infty)$ is available for each player $i \in \Sigma$. Quotations can be obtained from the players' current ranking (as for example in international tennis tournaments ATP and WTA) or by other means.

Definition 3 (Tournament cost). *Let $t \in \mathcal{T}_n(\Sigma)$ be a tournament with n rounds and let $s_{i,j}^t \in \{1, 2, \ldots, n\}$ be the stage of t where players i, j can meet. Let $q_i > 0$ be the quotations of players for all $i \in \Sigma$. The cost of t is defined as:*

$$Cost(t) = \sum_{i,j \in \Sigma, i<j} q_i q_j s_{ij}^t \tag{12}$$

Definition 4 (Optimal tournament). *A tournament such that its cost computed with Equation (12) is maximal is called an* optimal tournament *and it is defined by:*

$$\begin{aligned} OptC(\Sigma) &= \max_{t \in \mathcal{T}_n(\Sigma)} Cost(t) \\ t^* &= \operatorname{argmax}_{t \in \mathcal{T}_n(\Sigma)} Cost(t) \end{aligned} \tag{13}$$

Obviously, better ranked players have a higher quotation. We assume that if player i has rank r_i then its quota is q_i such that whenever $r_i < r_j$ we have $q_i > q_j$. For example, if there are $N = 2^n$ players then we can choose $q_i = N + 1 - r_i$ for all $i = 1, \ldots, N$.

Example 5. *Let us consider the tournaments t_1, t_2, and t_3 with three players shown in Figure 3. Let us introduce:*

$$\begin{aligned} A &= q_1 q_2 + q_1 q_3 + q_2 q_3 \\ S &= q_1 + q_2 + q_3 \end{aligned} \tag{14}$$

We obtain:

$$\begin{aligned}
\text{Cost}(t_1) &= q_2 q_3 1 + q_1 q_2 2 + q_1 q_3 2 = A + q_1(q_2 + q_3) = A + q_1(S - q_1) \\
\text{Cost}(t_2) &= A + q_2(S - q_2) \\
\text{Cost}(t_3) &= A + q_3(S - q_3)
\end{aligned} \quad (15)$$

The ordering of the costs depends on the ordering of the values of the function $q(S - q)$ for $q = q_1, q_2, q_3 \in [0, S]$. This function is monotonically increasing on $[0, S/2]$ and monotonically decreasing on $[S/2, S]$. Observe that if $q_i \leq S/2$, i.e., if neither player gets more than a half of the total quotation stake, then the ordering of the costs is given by the ordering of the quotations q_i.

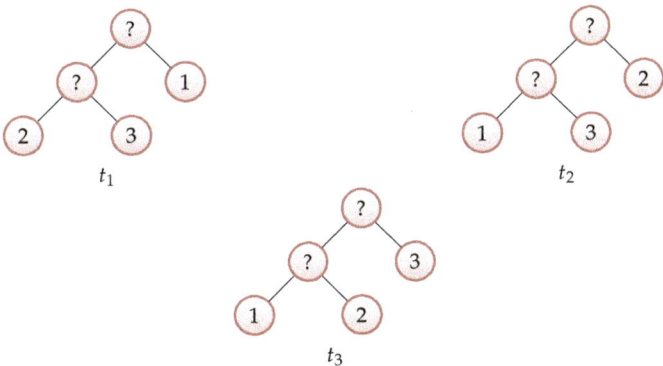

Figure 3. Balanced tournaments with three players.

Example 6. Let us consider four players (see Table 1). We assume that each player has a unique rank from 1 to 4. Now, if we choose $q_i = 5 - r_i$ then, using this approach for defining players' quota, player 2 with rank 4 is assigned quotation $q_2 = 1$. We consider the three tournaments $t_1, t_2, t_3 \in T_2(\{1, 2, 3, 4\})$ from Figure 4. According to Equation (12), the cost of a tournament $t \in T_2(\{1, 2, 3, 4\})$ is:

$$\text{Cost}(t) = \sum_{1 \leq i < j \leq 4}^{4} s_{ij}^t q_i q_j \quad (16)$$

$$\text{Cost}(t) = s_{12}^t \cdot 4 \times 1 + s_{13}^t \cdot 4 \times 2 + s_{14}^t \cdot 4 \times 3 + s_{23}^t \cdot 1 \times 2 + s_{24}^t \cdot 1 \times 3 + s_{34}^t \cdot 2 \times 3$$

Substituting stage values s_{ij}^t for each tournament from Table 2 into Equation (16) we obtain the tournaments' cost values from Table 2. We observe that in this case the best tournament is t_3. Actually it can be easily checked that the best tournament is t_3 for whatever values of the quota that are decreasingly ordered according to the ranks.

Table 1. Players' ranking and quotation for $n = 4$.

Player i	Rank r_i	Quota $q_i = n + 1 - r_i$
1	1	4
4	2	3
3	3	2
2	4	1

Table 2. Games playing stages for each tournament of Figure 4 and their costs.

	s^{t_1}	1	2	3	4	s^{t_2}	1	2	3	4	s^{t_3}	1	2	3	4
	1		2	1	2	1		2	2	1	1		1	2	2
	2			2	1	2			1	2	2			2	2
	3				2	3				2	3				1
	Cost		59					56					60		
			Cost of t_1					Cost of t_2					Cost of t_3		

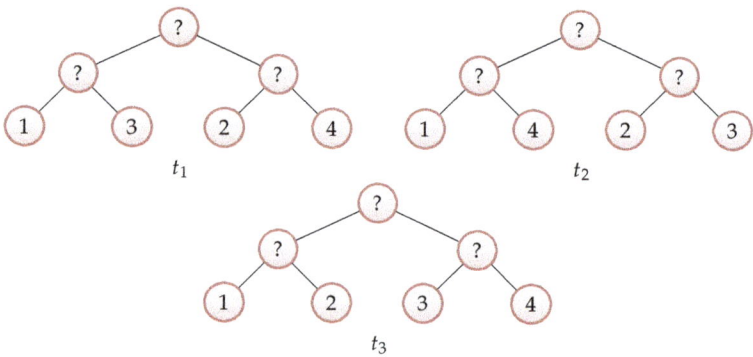

Figure 4. Balanced tournaments with four players.

5. Dynamic Programming Algorithm for Computing Optimal Tournaments

For any set of players Σ we denote by q_Σ the sum of quotations of the players in Σ.

$$q_\Sigma = \sum_{i \in \Sigma} q_i \tag{17}$$

Proposition 4 (Recurrence for tournament cost). *Let $t \in \mathcal{T}_n(\Sigma)$ be an n-stage tournament such that Σ is a finite nonempty set with $N \in 2^{n-1} + 1 \ldots 2^n$ elements. Then:*

$$\text{Cost}(t) = \begin{cases} 0 & n = 0 \\ q_i \cdot q_j & n = 1, N = 2, \Sigma = \{i, j\} \\ \text{Cost}(t_1) + \text{Cost}(t_2) + nq_{\Sigma_1}q_{\Sigma_2} & n \geq 2, t = \{t_1, t_2\}, t_1 \in \mathcal{T}(\Sigma_1), \\ & t_2 \in \mathcal{T}(\Sigma_2), \Sigma_1 \cup \Sigma_2 = \Sigma, \Sigma_1 \cap \Sigma_2 = \emptyset, \\ & t_1 \text{ and } t_2 \text{ are balanced, at most one has } n-2 \\ & \text{levels and in this case it is fully balanced} \end{cases} \tag{18}$$

Proof. If $n = 0$ then Σ has a single player so the result is obvious, as no games are played to determine the winner of the tournament.

If $n = 1$ then Σ has two players i and j so the result is obvious, as a single game is played to determine the winner of the tournament, between player i and player j.

If $n \geq 2$ then $t = \{t_1, t_2\}$. If $i \in \Sigma_1$ and $j \in \Sigma_2$ then $s_{ij}^t = n$. So Equation (12) gives:

$$\begin{aligned}\text{Cost}(t) = \sum_{i,j \in \Sigma_1, i<j} q_i q_j s_{ij}^{t_1} + \sum_{i,j \in \Sigma_2, i<j} q_i q_j s_{ij}^{t_2} + \sum_{i \in \Sigma_1, j \in \Sigma_2} q_i q_j s_{ij}^{t} = \\ \text{Cost}(t_1) + \text{Cost}(t_2) + n \sum_{i \in \Sigma_1, j \in \Sigma_2} q_i q_j = \text{Cost}(t_1) + \text{Cost}(t_2) + nq_{\Sigma_1}q_{\Sigma_2}\end{aligned} \tag{19}$$

The conditions from the Equation (18) follow directly from the recursive definition of balanced tournaments (Definition 2). □

Proposition 5 (Recurrences for optimal tournaments).

1. *The optimal tournament cost OptC introduced by Equation (13) can be defined recursively as follows:*

$$OptC(\Sigma) = \begin{cases} 0 & n = 0, |\Sigma| = 1 \\ q_i \cdot q_j & n = 1, \Sigma = \{i, j\} \\ \max_{\substack{\Sigma_1 \cup \Sigma_2 = \Sigma \\ \Sigma_1 \cap \Sigma_2 = \emptyset}} OptC(\Sigma_1) + OptC(\Sigma_2) + n q_{\Sigma_1} q_{\Sigma_2} & n \geq 2, 2^{n-1} < |\Sigma| \leq 2^n \\ & 2^{n-2} \leq |\Sigma_1| \leq \\ & |\Sigma_2| \leq 2^{n-1} \end{cases} \quad (20)$$

2. *The optimal tournament can be determined by recording the pairs of subsets $Opt(\Sigma) = (\Sigma_1, \Sigma_2 = \Sigma \setminus \Sigma_1)$ that maximize OptC in Equation (20) for $n \geq 2$, $2^{n-1} < |\Sigma| \leq 2^n$ as follows:*

$$OptS(\Sigma) = \underset{\Sigma_1, \Sigma_2 \subseteq \Sigma}{\arg\max} \, OptC(\Sigma_1) + OptC(\Sigma_2) + n q_{\Sigma_1} q_{\Sigma_2} \quad (21)$$

Note that the limits of argmax in Equation (21) must satisfy the conditions from the third branch of Equation (20). Note also that it is enough to record $Opt(\Sigma) = \Sigma_1$ as $\Sigma_2 = \Sigma \setminus \Sigma_1$.

Proof. The proof follows by applying the maximization operation in Equation (18) and observing that the term $n q_{\Sigma_1} q_{\Sigma_2}$ does not depend on $t = \{t_1, t_2\}$. The condition $|\Sigma_1| \leq |\Sigma_2|$ ensures that a pair $\{\Sigma_1, \Sigma_2\}$ is uniquely considered (otherwise each pair will be considered twice as $\{\Sigma_1, \Sigma_2\}$ and $\{\Sigma_2, \Sigma_1\}$).

Moreover, the sets $OptS(\Sigma)$ can be used to construct an optimal tournament. Let $\Sigma^n = \Sigma$. We define: $\Sigma^{n-1} = OptS(\Sigma^n), \ldots, \Sigma^0 = OptS(\Sigma^1)$. Then the optimal tournament t^* can be defined recursively as follows:

$$t^* = t^n(\Sigma^n)$$
$$t^i(\Sigma^i) = \begin{cases} \{t_1^{i-1}(\Sigma^{i-1}), t_2^{i-1}(\Sigma^i \setminus \Sigma^{i-1})\} & i \geq 1 \\ j & i = 0, \Sigma^0 = \{j\} \end{cases} \quad (22)$$

□

Proposition 5 (Equation (20) in particular) can be used to design a dynamic programming algorithm for computing the optimal tournament and its cost. The dynamic programming algorithm can be implemented either with a bottom-up approach or using a top-down approach with memoization [17]. We will explore these possibilities in what follows by deriving a bottom-up dynamic programming algorithm for fully balanced tournaments as well as a top-down dynamic programming algorithm with memoization for the general case.

5.1. Top-Down Dynamic Programming Algorithm with Memoization

Proposition 6 (Top-down recursive application of Equation (20)). *Let us assume that we want to compute $OptC(\Sigma)$ for $|\Sigma| = N$, $N \geq 2$, $2^{n-1} + 1 \leq N \leq 2^n$. According to Proposition 5, we must recursively explore all tournaments of shape $t = \{t_1, t_2\}$ such that $t_i \in T(\Sigma_i)$, $|\Sigma_i| = N_i$, $i = 1, 2$, $N = N_1 + N_2$. Then we should recursively apply Equation (20) only for the following values of N_1:*

$$\max\{N - 2^{n-1}, 2^{n-2}\} \leq N_1 \leq \lfloor N/2 \rfloor \quad (23)$$

Proof. As $N = N_1 + N_2$ and $N_1 \leq N_2$ we obtain:

$$N_1 \leq \lfloor N/2 \rfloor \quad (24)$$

As $N = N_1 + N_2$ and $2^{n-2} \leq N_1 \leq 2^{n-1}$ we obtain:

$$2^{n-2} \leq N_1 \leq 2^{n-1}$$
$$N - 2^{n-1} \leq N_1 \leq N - 2^{n-2} \tag{25}$$

Combining (24) and (25) we obtain:

$$\max\{N - 2^{n-1}, 2^{n-2}\} \leq N_1 \leq \min\{2^{n-1}, N - 2^{n-2}, \lfloor N/2 \rfloor\} \tag{26}$$

It is not difficult to see that $\lfloor N/2 \rfloor \leq N/2 \leq 2^{n-1}$ and $\lfloor N/2 \rfloor \leq N/2 < N - 2^{n-2}$ so the value of right-hand side of Equation (26) is $\lfloor N/2 \rfloor$, q.e.d. □

Observe that for fully balanced tournaments $N = 2^n$, so this top-down recursive process will generate subsets of sizes $2^{n-1}, 2^{n-2} \ldots 2, 1$.

Example 7. *Let us illustrate the application Equation (20) for $N = |\Sigma| = 25$ players. The results are summarized in Table 3. It follows that solving the problem for a set of 25 players requires the solving of all subproblems corresponding to its subsets of $1, 2, \ldots, 16$ players; however, solving the problem for a set of 15 players requires the solving of all subproblems corresponding to its subsets of $1, 2, 3, 4, 7, 8$ players. Moreover, solving the problem for a set of 14 players requires the solving of all subproblems corresponding to its subsets of $1, 2, 3, 4, 6, 7, 8$ players, while solving the problem for a set of 12 or 13 players requires the solving of all subproblems corresponding to its subsets of $1, 2, 3, 4, 5, 6, 7, 8$ players.*

Table 3. Games playing stages for each tournament and tournament costs.

N	$n = \lceil \log_2 n \rceil$	$\lfloor \frac{N}{2} \rfloor$	$N - 2^{n-1}$	2^{n-2}	N_1^{min}, N_1^{max}	N_2^{min}, N_2^{max}
25	5	12	9	8	9, 12	13, 16
16	4	8	8	4	8, 8	8, 8
15	4	7	7	4	7, 7	8, 8
14	4	7	6	4	6, 7	7, 8
13	4	6	5	4	5, 6	7, 8
12	4	6	4	4	4, 6	6, 8
11	4	5	3	4	4, 5	6, 7
10	4	5	2	4	4, 5	5, 6
9	4	4	1	4	4, 4	5, 5
8	3	4	4	2	4, 4	4, 4
7	3	3	3	2	3, 3	4, 4
6	3	3	2	2	2, 3	3, 4
5	3	2	1	2	2, 3	2, 3
4	2	2	2	1	2, 2	2, 2
3	2	1	1	1	1, 1	2, 2
2	1		solved directly			
1	0		solved directly			

Proposition 6 sets the iteration bounds for exploring the subsets of players in the top-down approach. Combining the results of Propositions 5 and 6, we obtain the top-down approach for computing optimal balanced tournaments; see Algorithm 1.

Algorithm 1 $OptTourCostTD(\Sigma, N, q)$ top-down dynamic programming algorithm with memoization for computing the cost of the optimal tournament.

Global: $OptC$, initially $\emptyset$, maps subsets of players to costs of optimal tournaments.
$OptS$, initially $\emptyset$, maps subsets to sub-subsets for building optimal tournaments.

Input: $N \in \mathbb{N}^*$ represents the number of players.
q. Vector of size N representing the players' quota.
Σ such that $|\Sigma| = N$. Σ represents the set of players.

Output: C_{max}. Cost of the optimal tournament for set Σ of players.

$n \leftarrow \lceil \log_2 N \rceil$
if $n = 0$ **then**
$\quad C_{max} \leftarrow 0$
$\quad S_{max} \leftarrow \emptyset$
else if $n = 1$ (i.e., $\Sigma = \{i, j\}$) **then**
$\quad C_{max} \leftarrow q_i * q_j$
$\quad S_{max} \leftarrow \{i\}$
else
$\quad P1 \leftarrow 2^{n-2}$
$\quad P2 \leftarrow 2 * P1$
$\quad k_{max} \leftarrow \lfloor N/2 \rfloor$
$\quad$**if** $N \leq P1 + P2$ **then**
$\quad\quad k_{min} \leftarrow P1$
$\quad$**else**
$\quad\quad k_{min} \leftarrow N - P2$
$\quad$**end if**
$\quad C_{max} \leftarrow -\infty$
$\quad$**for** $k = k_{min}, k_{max}$ **do**
$\quad\quad$**for** $\Sigma_1 \subseteq \Sigma$ s.t. $|\Sigma_1| = k$ **do**
$\quad\quad\quad$**if** $\Sigma_1 \notin OptC$ **then**
$\quad\quad\quad\quad C_1 \leftarrow OptTourCostTD(\Sigma_1, k, q)$
$\quad\quad\quad$**else**
$\quad\quad\quad\quad C_1 \leftarrow OptC[\Sigma_1]$
$\quad\quad\quad$**end if**
$\quad\quad\quad \Sigma_2 \leftarrow \Sigma \setminus \Sigma_1$
$\quad\quad\quad$**if** $\Sigma_2 \notin OptC$ **then**
$\quad\quad\quad\quad C_2 \leftarrow OptTourCostTD(\Sigma_2, k, q)$
$\quad\quad\quad$**else**
$\quad\quad\quad\quad C_2 \leftarrow OptC[\Sigma_2]$
$\quad\quad\quad$**end if**
$\quad\quad\quad C \leftarrow C_1 + C_2$
$\quad\quad\quad ql \leftarrow 0$
$\quad\quad\quad$**for** $i \in \Sigma_1$ **do**
$\quad\quad\quad\quad ql \leftarrow ql + q_i$
$\quad\quad\quad$**end for**
$\quad\quad\quad qr \leftarrow 0$
$\quad\quad\quad$**for** $i \in \Sigma_2$ **do**
$\quad\quad\quad\quad qr \leftarrow qr + q_i$
$\quad\quad\quad$**end for**
$\quad\quad\quad C \leftarrow C + k * ql * qr$
$\quad\quad\quad$**if** $C > Cmax$ **then**
$\quad\quad\quad\quad Cmax \leftarrow C$
$\quad\quad\quad\quad Smax \leftarrow \Sigma_1$
$\quad\quad\quad$**end if**
$\quad\quad$**end for**
$\quad$**end for**
end if
$OptC[\Sigma] \leftarrow C_{max}$
$OptS[\Sigma] \leftarrow S_{max}$

5.2. Bottom-Up Dynamic Programming Algorithm for Fully Balanced Tournaments

The cost of the optimal tournament is computed with the help of $OptC$ vector that is indexed by all the subsets of Σ generated recursively by Equation (20), starting from the topmost set Σ. Note that for a fully balanced tournament we have $|\Sigma| = 2^n$ and the process will generate exactly all the subsets of Σ of cardinal: $2^0, 2^1, \ldots 2^n$. Note that in this case the size of $OptC$ can be determined as:

$$S_n = \sum_{i=0}^{2^n} \binom{2^n}{2^i} \tag{27}$$

Additionally we must save in vector $OptS$ of size S_n the subsets Σ' determined using Equation (21), such that we can reuse them to construct the optimal tournament using Equation (22). Our proposed algorithm is presented as Algorithm 2.

Algorithm 2 $OptTourCostBU(\Sigma, N = 2^n, q)$ bottom up dynamic programming algorithm for computing the cost of optimal fully balanced tournaments.

Input: $N = 2^n, n \in \mathbb{N}$. N represents the number of players.
$\quad\quad\quad$ q. Vector of size N representing the players' quota.
$\quad\quad\quad$ Σ such that $|\Sigma| = 2^n$. Σ represents the set of players.
Output: $OptC$. Vector of costs of the optimal sub-tournaments.
$\quad\quad\quad\quad$ $OptS$. Vector of sets to construct the optimal tournament.
1: **for** $i = 1, N$ **do**
2: $\quad$ $OptC[\{i\}] \leftarrow 0$
3: **end for**
4: **for** $k = 1, n$ **do**
5: $\quad$ **for** $\Sigma_1 \subseteq \Sigma$ s.t. $|\Sigma_1| = 2^k$ **do**
6: $\quad\quad$ $Cmax \leftarrow -\infty$
7: $\quad\quad$ **for** $\Sigma' \subseteq \Sigma_1$ s.t. $|\Sigma'| = 2^{k-1}$ **do**
8: $\quad\quad\quad$ $C \leftarrow OptC(\Sigma') + OptC(\Sigma_1 \setminus \Sigma')$
9: $\quad\quad\quad$ $ql \leftarrow 0$
10: $\quad\quad\quad$ **for** $i \in \Sigma'$ **do**
11: $\quad\quad\quad\quad$ $ql \leftarrow ql + q_i$
12: $\quad\quad\quad$ **end for**
13: $\quad\quad\quad$ $qr \leftarrow 0$
14: $\quad\quad\quad$ **for** $i \in \Sigma_1 \setminus \Sigma'$ **do**
15: $\quad\quad\quad\quad$ $qr \leftarrow qr + q_i$
16: $\quad\quad\quad$ **end for**
17: $\quad\quad\quad$ $C \leftarrow C + k * ql * qr$
18: $\quad\quad\quad$ **if** $C > Cmax$ **then**
19: $\quad\quad\quad\quad$ $Cmax \leftarrow C$
20: $\quad\quad\quad\quad$ $Smax \leftarrow \Sigma'$
21: $\quad\quad\quad$ **end if**
22: $\quad\quad$ **end for**
23: $\quad\quad$ $OptC[\Sigma_1] \leftarrow Cmax$
24: $\quad\quad$ $OptS[\Sigma_1] \leftarrow Smax$
25: $\quad$ **end for**
26: **end for**

5.3. Computing an Optimal Tournament

Note that both Algorithms 1 and 2 determine the $OptS$ structure that records the split points for each subset of players according to Equation (20). The $OptS$ structure can be used to actually build an optimal tournament according to Algorithm 3 using Equation (22).

Algorithm 3 $OptTour(\Sigma, N, OptS)$ algorithm for computing the optimal tournament.

Input: Σ representing the set of players.
$N \in \mathbb{N}^*$ representing the number of players.
$OptS$. Structure determined either by Algorithm 1 or by Algorithm 2.
Output: Returns the optimal tournament.
1: **if** $N = 1$ (i.e., $\Sigma = \{j\}$) **then**
2:     **return** j
3: **end if**
4: $\Sigma_1 \leftarrow OptS(\Sigma)$
5: $N_1 \leftarrow |\Sigma_1|$
6: $\Sigma_2 \leftarrow \Sigma \setminus \Sigma_1$
7: $t_1 \leftarrow OptTour(\Sigma_1, N_1, OptS)$
8: $t_2 \leftarrow OptTour(\Sigma_2, N - N_1, OptS)$
9: **return** $\{t_1, t_2\}$

5.4. Correctness and Complexity Results

Proposition 7 (Correctness of Algorithms 1–3).

a. *The value $C_{max} = OptC[\Sigma]$ computed by Algorithms 1 and 2 represents the cost of the optimal tournament in both cases.*
b. *The tournament determined by Algorithm 3 is the optimal tournament.*

Proof.
Proof of a. Algorithms 1 and 2 compute the values of $OptC$ and $OptS$ either in top-down or bottom-up fashion for all the subsets that are required to determine the optimal tournament for the set Σ of players. The computation follows Equations (20) and (21); therefore the correctness of this point follows from Propositions 5 and 6.

Proof of b. Algorithm 3 computes the optimal tournament using Equations (22). As values of $OptS$ are correctly determined according to point "a", it follows that the tournament computed by Algorithm 3 is the optimal tournament. □

Proposition 8 (Complexity of Algorithms 2 and 3). *Let us consider tournaments with N players.*

a. *Space complexity of Algorithm 2 is $\Theta\left(\frac{2^N}{\sqrt{N}}\right)$.*
b. *Time complexity of Algorithm 2 is $\Theta((2\sqrt{2})^N)$.*
c. *Space complexity of Algorithm 1 is $\Theta\left(\frac{2^N}{\sqrt{N}}\right)$ for fully balanced tournaments and $O(2^N)$ in the general case.*
d. *Time complexity of Algorithm 1 is $O(3^N \cdot N\sqrt{N})$.*
e. *Time complexity of Algorithm 3 is $\Theta(N)$.*

Proof. The proof is using the Stirling approximation of the factorial, written in inequality form ([18]), in fact showing that $p! = \Theta\left(\sqrt{p}\left(\frac{p}{e}\right)^p\right)$:

$$\sqrt{2\pi} \leq \frac{p!}{\sqrt{p}\left(\frac{p}{e}\right)^p} \leq e \tag{28}$$

Using this observation it is not difficult to prove that:

$$\binom{2p}{p} = \Theta\left(\frac{2^{2p}}{\sqrt{p}}\right) \tag{29}$$

Proof of a. The space complexity of Algorithms 2 and 3 is given by the size of structures $OptC$ and $OptS$ (see Equation (27)); however, the asymptotically dominant term of this summation is $\binom{2^n}{2^{n-1}}$. Then the result follows using Equation (29) for $p = 2^{n-1}$.

Proof of b. Algorithm 2 contains one "for" loop (lines 4–26) including other three nested "for" loops. The first inner "for" loop (lines 5–25) is executed $\binom{N}{2^k}$ times. The second inner "for" loop (lines 7–22) is executed $\binom{2^k}{2^{k-1}}$ times. The third inner "for" loop (lines 10–12 and 14–16) is executed 2^{k-1} times. The total number of steps of Algorithm 2 is given by:

$$\sum_{k=1}^{n} \binom{N}{2^k}\binom{2^k}{2^{k-1}} 2^{k-1} \tag{30}$$

Observe that the asymptotically dominant term of this summation is obtained for $k = n - 1$ and it can be transformed using Equation (29), thus concluding the proof:

$$\binom{N}{2^{n-1}}\binom{2^{n-1}}{2^{n-2}} 2^{n-2} = \Theta\left(\frac{2^{2^n}}{\sqrt{2^{n-1}}} \frac{2^{2^{n-1}}}{\sqrt{2^{n-2}}} 2^{n-2}\right) = \Theta\left(\frac{1}{\sqrt{2}} 2^N 2^{N/2}\right) = \Theta((2\sqrt{2})^N) \tag{31}$$

Proof of c. If N is a power of two, i.e., we have a fully balanced tournament, the memory consumption is exactly as in case a, so the first result follows trivially. Otherwise, the space consumption of tables $OptC$ and $OptS$ has an upper bound given by the size of the power set of Σ, and the result follows immediately.

Proof of d. Let us first observe that in order to determine the cost of an optimal tournament with N players we need to know the costs of optimal tournaments for N_1 and N_2 players such that $N = N_1 + N_2$ and condition of Equation (23) holds. It is not difficult to observe that:

$$\frac{N}{4} < N_1 \leq N_2 < \frac{3N}{4} \tag{32}$$

First note that the upper bound of N_2 follows from the lower bound of N_1, so it is enough to show the lower bound of N_1. Let us assume by contradiction that:

$$\frac{N}{4} \geq \max\{N - 2^{n-1}, 2^{n-2}\} \tag{33}$$

It follows that:

$$\frac{N}{2} = \frac{N}{4} + \frac{N}{4} \geq N - 2^{n-1} + 2^{n-2} = N - 2^{n-2} \tag{34}$$

so:

$$N \leq 2^{n-1} \tag{35}$$

and thus contradicting (5).

It is easier to analyze the complexity of Algorithm 1 by thinking "bottom-up" rather than "top-down". The complexity will be the same, as the role of the memorization technique is just to evaluate exactly once the cost of a tournament for each subset of players. So we must determine the cost for a subset of $i = \frac{N}{4}, \frac{N}{4} + 1, \ldots, \frac{3N}{4}$ players ($\lfloor \cdot \rfloor$ can be omitted without losing generality); therefore the total running time has the following upper bound:

$$\sum_{p=\frac{N}{4}}^{p=\frac{3N}{4}} \binom{N}{p} \sum_{k=\frac{p}{4}}^{\frac{3p}{4}} \binom{p}{k} k \tag{36}$$

Although $\binom{p}{k} = \binom{p}{p-k}$, so grouping terms with complementary binomial coefficients of inner sum of (36), noticing that $\binom{p}{k} \leq \binom{p}{\frac{p}{2}}$ and using (29), the inner sum has an upper bound of:

$$p \cdot \sum_{k=\frac{p}{4}}^{\frac{p}{2}} \binom{p}{k} \leq p \cdot \frac{p}{4} \cdot \Theta\left(\frac{2^p}{\sqrt{p}}\right) \leq N\sqrt{N} \cdot \Theta(2^p) \tag{37}$$

Now, substituting (37) in (36) we obtain the following upper bound of the running time:

$$N\sqrt{N} \cdot \sum_{p=\frac{N}{4}}^{p=\frac{3N}{4}} \binom{N}{p} \cdot \Theta(2^p) \leq N\sqrt{N} \cdot \sum_{p=0}^{p=N} \binom{N}{p} \cdot \Theta(2^p) = \Theta(3^N \cdot N\sqrt{N}) \quad (38)$$

As this is in fact only an upper bound of our running time, the result of point e follows (i.e., with O rather than Θ).

Proof of e. Observe that the time complexity of Algorithm 3 satisfies the recursive equation $T(|\Sigma|) = T(|\Sigma_1|) + T(|\Sigma_2|)$. Unfolding this equation with the substitution method yields an asymptotic execution time $\Theta(|\Sigma|) = \Theta(N)$. □

5.5. Sub-Optimal Algorithms

The dynamic programming approach for construing optimal tournaments has the disadvantage that the full exploration of the search space becomes prohibitive for larger tournaments. Our experiments (see Section 6) clearly show that this approach is unfeasible for tournaments of more than $n = 16$ players; however, the dynamic programming algorithms can be easily adapted to explore a smaller size of the search space, leading to sub-optimal solutions. The exploration strategy can be used to tune the trade-off between the complexity of the computation and the "gap" between the provided sub-optimal solution and the actual optimal solution.

The resulting sub-optimal algorithms follow a strictly top-down approach that can be the best described as divide-and-conquer. At each decision point, rather than exploring all pairs of subsets (Σ_1, Σ_2) satisfying conditions of Equation (20), only few such pairs (ideally only 1) are selected for exploration. This selection strategy can be deterministic, based on heuristic principles, or stochastic based on stochastic sampling subsets of Σ satisfying the conditions of Equation (20). The strictly top-down approach has the advantage that it avoids the use of temporary structures *OptS* and *OptC* and of the additional algorithm *OptTour* to build the solution. Rather, the top-down approach will build the solution directly, using the recursive divide-and-conquer approach.

The general approach of a sub-optimal algorithm following a top-down divide and conquer approach is presented as Algorithm 4. Note that this algorithm is using a specific strategy to explore only a few subsets of Σ defined by $Strategy(\Sigma, k_{min}, k_{max}) \subseteq 2^\Sigma$ and satisfying the conditions of Equation (20).

Proposition 9 (Complexity of Algorithm 4). *Let us consider tournaments for N players and let n be the number of stages of a balanced tournament. Then the time complexity of Algorithm 4 is $O(N^{1+\log_2 s})$ where s is the average number of subsets of Σ explored by the strategy of the algorithm.*

Proof. It is not difficult to observe that the time complexity of Algorithm 4 satisfies the following recurrence:

$$T(|\Sigma|) = s \cdot (T(|\Sigma_1|) + T(|\Sigma \setminus \Sigma_1|)) \quad (39)$$

Applying the substitution method for Equation (39) we obtain:

$$T(|\Sigma|) = \sum_{i \in \Sigma} s^{h_i} \cdot O(1) \quad (40)$$

For each $i \in \Sigma$, h_i from Equation (40) denotes the height of leaf i in the tournament tree, so $h_i \leq n \leq \log_2 N$. So:

$$T(|\Sigma|) = N \cdot O(s^{\log_2 N}) = O(N^{1+\log_2 s}) \quad (41)$$

q.e.d. □

Algorithm 4 $OptTourCostSubOpt(\Sigma, N, q, S)$ top-down divide-and-conquer algorithm for computing a sub-optimal tournament.

Input: $N \in \mathbb{N}^*$ represents the number of players.
       q. Vector of size N representing the players' quota.
       Σ such that $|\Sigma| = N$. Σ represents the set of players.
       $S = \sum_{i \in \Sigma} q_i$. Sum of players' quotations.
Output: C_{max}. Cost of the sub-optimal tournament for set Σ of players.
         t_{max}. Sub-optimal tournament tree.

1: $n \leftarrow \lceil \log_2 N \rceil$
2: **if** $n = 0$ (i.e., $\Sigma = \{i\}$) **then**
3:     $C_{max} \leftarrow 0$
4:     $t_{max} \leftarrow i$
5: **else if** $n = 1$ (i.e., $\Sigma = \{i, j\}$) **then**
6:     $C_{max} \leftarrow q_i * q_j$
7:     $t_{max} \leftarrow \{i, j\}$
8: **else**
9:     $P1 \leftarrow 2^{n-2}$
10:    $P2 \leftarrow 2 * P1$
11:    $k_{max} \leftarrow \lfloor N/2 \rfloor$
12:    **if** $N \leq P1 + P2$ **then**
13:      $k_{min} \leftarrow P1$
14:    **else**
15:      $k_{min} \leftarrow N - P2$
16:    **end if**
17:    $C_{max} \leftarrow -\infty$
18:    **for** $\Sigma_1 \in Strategy(\Sigma, k_{min}, k_{max})$ **do**
19:      $S_1 \leftarrow 0$
20:      **for** $i \in \Sigma_1$ **do**
21:         $S_1 \leftarrow S_1 + q_i$
22:      **end for**
23:      $k \leftarrow |\Sigma_1|$
24:      $(C_1, t_1) \leftarrow OptTourCostSubOpt(\Sigma_1, k, q, S_1)$
25:      $(C_2, t_2) \leftarrow OptTourCostSubOpt(\Sigma \setminus \Sigma_1, N - k, q, S - S_1)$
26:      $C \leftarrow C_1 + C_2 + k * S_1 * (S - S_1)$
27:      **if** $C > C_{max}$ **then**
28:         $C_{max} \leftarrow C$
29:         $t_{max} \leftarrow (t_1, t_2)$
30:      **end if**
31:    **end for**
32: **end if**
33: **return** (C_{max}, t_{max})

Observe that if $s = 1$ then the time complexity of Algorithm 4 is linear in the number N of players. Moreover, if $s > 1$ then the time complexity of the algorithm is polynomial in N and the degree of the polynomial grows logarithmically with s.

5.5.1. Deterministic Sub-Optimal Algorithms

We define a deterministic sub-optimal algorithm by letting Σ_1 consist of the smallest set of players $1, 2, \ldots, k$ such that $k \geq k_{min}$ and $q_{\Sigma_1} > q_{\Sigma}/2$.

The rationale of this choice is to try to make the product $q_{\Sigma_1} q_{\Sigma_2}$ from Equation (20) as high as possible. As $q_{\Sigma_1} + q_{\Sigma_2} = q_{\Sigma}$ is constant, we try to make the values q_{Σ_1} and q_{Σ_2} as close as possible, while maintaining the constraints on the size of subset Σ_1.

We can define three variants of the deterministic sub-optimal algorithm by considering the sequence of players' quotations to be: (i) unchanged, i.e., as it was provided as input; (ii) increasingly sorted; (iii) decreasingly sorted.

5.5.2. Stochastic Sub-Optimal Algorithms

We define a stochastic sub-optimal algorithm by letting Σ_1 consist of a family of randomly chosen subsets of players of $\Sigma_1 \subseteq \Sigma$ such that $k_{max} \geq |\Sigma_1| \geq k_{min}$. This is easily achieved by randomly choosing the number of players k uniformly distributed in $[k_{min}, k_{max}]$ and then randomly choosing a subset Σ_1 of k elements and uniformly distributed in Σ.

The number of chosen subsets Σ_1 explored by the algorithm is a parameter denoted by N_{sample} and it usually has a low number, as it directly influences the complexity of the algorithm according to Proposition 9, $s = N_{sampl}$. For example, if $N_{sampl} = 1$ the complexity of the algorithm is $O(N)$, if $N_{sampl} = 2$ the complexity of the algorithm is $O(N^2)$ and if $N_{sampl} = 4$ the complexity of the algorithm is $O(N^3)$.

6. Implementation and Experiments

6.1. Implementation Issues

There were several issues that we had to address by our experimental implementation of Algorithms 1–3.

Firstly, we have chosen to represent sets as arrays of bits, as well as using the integer value that is equivalent to the binary representation as an array of bits.

Secondly, for generating subsets of given size (i.e., combinations) we have used Algorithm 7.2.1.2L from [19] for generating permutations with repetitions of binary arrays. Basically the subsets representing combinations of k elements of a set with $n \geq k$ elements are all the permutations with repetitions of a binary vector of n elements containing exactly k elements equal to 1.

Thirdly, we had to choose an efficient representation of *OptC* and *OptS* structures. Their operation is crucial for the efficient implementation of some of our algorithms. As for our implementation we have chosen Python platform, we decided to implement *OptC* and *OptS* using subset-indexed dictionaries that map subsets of Σ to costs and to subsets necessary for building optimal tournaments, respectively. The subsets representing the dictionary keys are defined as integer values of their characteristic vector in binary format. As Python dictionaries are efficiently implemented using hash tables, an average $O(1)$ time complexity is expected for lookup operations.

Finally, for the implementation of the random selection of subsets we have used the array of bits representation of sets and we have applied the *random.permute* function from *NumPy* package to return a randomly permuted array representing a random subset.

6.2. Experimental Results

Our experiments were developed in Python 3.7.3 using Jupyter Notebook on an x64-based PC with a 2 cores / 4 threads Intel© Core™i7-5500U CPU at 2.40 GHz running Windows 10 (The experimental code is available at http://software.ucv.ro/~cbadica/tour.zip) (accessed on 2 September 2021).

According to our findings, there are no algorithms directly available to be compared with our own proposals. There are two main causes for this. First, we consider an integrated approach of tournament design, rather than a process involving two separated stages for structure design and seeding. Secondly, we do not use probabilistic information in our cost function, thus hindering the direct comparison of tournament cost values.

We took a different path for evaluating our proposals. We have implemented optimal algorithms, as well as several versions of sub-optimal algorithms and then compared their outcomes in terms of running time and optimality. So finally we have implemented and experimentally evaluated eight algorithms, as presented in Table 4.

Table 4. Table presenting the list of implemented optimal and sub-optimal algorithms for balanced tournaments.

Algorithm	Description	Optimality	Players' Number
$OptTD$	Dynamic programming top-down with memoization approach—Algorithm 1	Optimal	Arbitrary natural number
$OptBU$	Dynamic programming bottom-up approach—Algorithm 2	Optimal	Exact power of 2
$SubOptD_1$	Deterministic sub-optimal approach—Algorithm 4 with deterministic strategy and unchanged quotation sequence	Sub-optimal	Arbitrary natural number
$SubOptD_2$	Deterministic sub-optimal approach—Algorithm 4 with deterministic strategy and increasingly sorted quotation sequence	Sub-optimal	Arbitrary natural number
$SubOptD_3$	Deterministic sub-optimal approach—Algorithm 4 with deterministic strategy and decreasingly sorted quotation sequence	Sub-optimal	Arbitrary natural number
$SubOptS_1$	Stochastic sub-optimal approach—Algorithm 4 with stochastic strategy and $N_{sampl} = 1$	Sub-optimal	Arbitrary natural number
$SubOptS_2$	Stochastic sub-optimal approach—Algorithm 4 with stochastic strategy and $N_{sampl} = 2$	Sub-optimal	Arbitrary natural number
$SubOptS_3$	Stochastic sub-optimal approach Algorithm 4 with stochastic strategy and $N_{sampl} = 3$	Sub-optimal	Arbitrary natural number

Note that for the optimal algorithms there are at least two restrictions that hinder a complete experimental comparison with the rest of the algorithms. Firstly, their high computational complexity limits their applicability only to small number of players. Secondly, the dynamic programming bottom-up algorithm works only for a number of players that is an exact power of 2. We have only checked it for $N = 4, 8, 16$.

Our data set includes multiple sequences of players' quotations. For each $N = 3, 4, \ldots 50$ we generated a sequence of quotations $q_1, q_2, \ldots, q_N$ with integer values q_i randomly chosen with a uniform distribution in the interval $[q_{min} = 1, q_{max} = 9]$. This data set was used to experimentally evaluate the algorithms from Table 4, as follows:

1. All the sequences of the data set were used for testing algorithms $SubOptD_i$ and $SubOptS_i$ for $i = 1, 2, 3$.
2. The optimal algorithm $OptTD$ was evaluated only for sequences corresponding to $N = 3, \ldots, 16$ players. The reason is that the algorithm has a too high computational complexity and we limited the running time of each problem instance to 5 min.
3. The optimal algorithm $OptBU$ was evaluated only for sequences corresponding to $N = 4, 8, 16$ players. The reason is both the high computational complexity of the algorithm and the fact that this algorithm was designed to work only with a number of players that is an exact power of 2.

For each algorithm, we recorded the (sub-)optimal cost of the output tournament, as well as the running time. Stochastic algorithms $SubOptS_i$ for $i = 1, 2, 3$ were evaluated by repeating their execution 10 times for each input sequence of quotations from the data set and recording the minimum, maximum, and average costs, as well as the average running times.

Figure 5 presents the sub-optimal costs produced by $SubOptD_i$ and $SubOptS_i$ algorithms for $i = 1, 2, 3$. The figure plots costs C_i produced by algorithms $SubOptD_i$ for $i = 1, 2, 3$, as well as average costs CSa_i produced by algorithms $SubOptS_i$ for $i = 1, 2, 3$ and the maximum cost CSM_3 produced by the 10-times repeated execution of algorithm $SubOptS_3$. Note that we included maximum cost only for this case, as it should be obvious that it is expected that stochastic algorithm $SubOptS_3$ will produce the best results among $SubOptS_i$ for $i = 1, 2, 3$ because it uses the highest number of samplings $N_{sampl} = 3$.

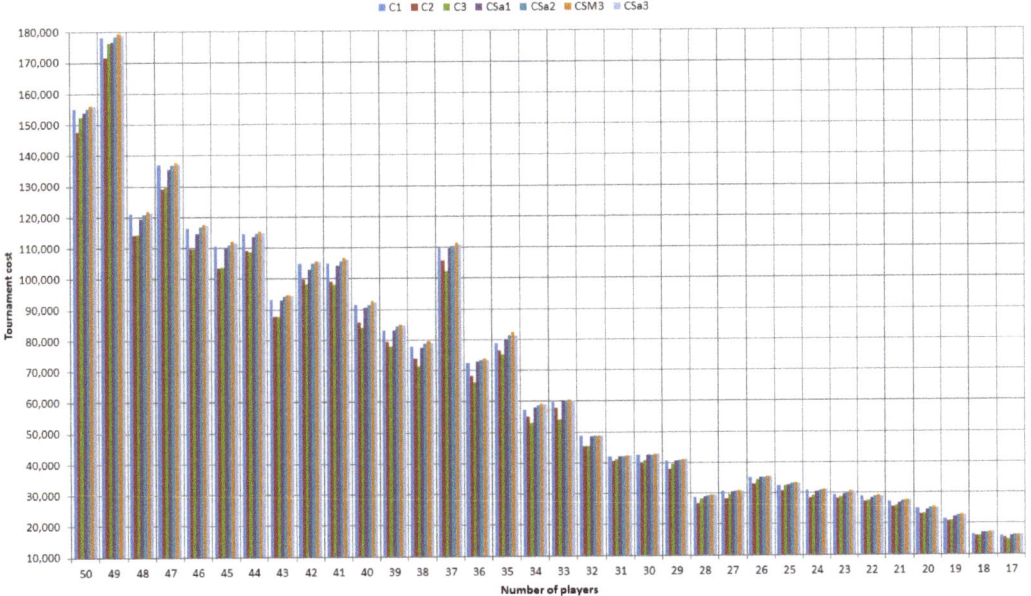

Figure 5. Sub-optimal tournament costs determined by sub-optimal algorithms on different quotations' sequences for various number of players.

Analyzing Figure 5, we first observe that the relative difference of the costs produced by the various algorithms on the same input sequence is rather low. This is expected, as quotations q_i were generated as integer values from a small interval $1 \leq q_i \leq 9$ while the cost tends to reach significantly higher values. For example, analyzing in detail the results obtained for the sample with $N = 50$ players we observe that the relative difference between the smallest and the highest cost obtained (147,555 and 156,203) is of only 5.53%. We can also notice that the best results among sub-optimal algorithms were obtained by algorithm $SubOptS_3$, while the worst results were obtained by algorithms $SubOptD_2$ and $SubOptD_3$. It might look a bit surprising that algorithm $SubOptD_1$ appears to be superior to $SubOptD_2$ and $SubOptD_3$; however, taking into account how the data set was generated, this could be explained by the fact that algorithm $SubOptD_1$ is actually using a random permutation of the quotations' sequence that provides a better balance of the total quotation distribution between the two subsets Σ_1 and $\Sigma \setminus \Sigma_1$ (see Algorithm 4) than algorithms $SubOptD_2$ and $SubOptD_3$. One final remark, also observed experimentally, is that algorithms $SubOptD_2$ and $SubOptD_3$ produce the same sub-optimal costs if the number of players is an exact power of 2 (i.e., 16 and 32 on Figure 5). This is an immediate consequence on the logic behind their strategy definition.

Figure 6 presents the running times TD_i and TS_i of algorithms $SubOptD_i$ and $SubOptS_i$ for $i = 1, 2, 3$. The time figures are given in milliseconds and presented on a logarithmic scale and they were computed by taking the average for 10 executions of the algorithm on the same input data. First observe that deterministic versions $SubOptD_i$ are the fastest and they have virtually almost the same running times. This can be easily explained by the low computational complexity of the implementation of their underlying strategies. Basically, their strategies use the same mechanism, while the additional sorting of the sequence of quotations adds a negligible cost as it is performed before the actual core processing of the algorithms. Second, the highest execution time is achieved, as expected, by $SubOptS_3$. This algorithm has the highest computational complexity among sub-optimal algorithms, as it is using three subset samples during the top-down search. From Figure 6 it also follows that the highest average execution time was obtained for the sequence of quotations with $N = 46$ players and its value was 2.49 s.

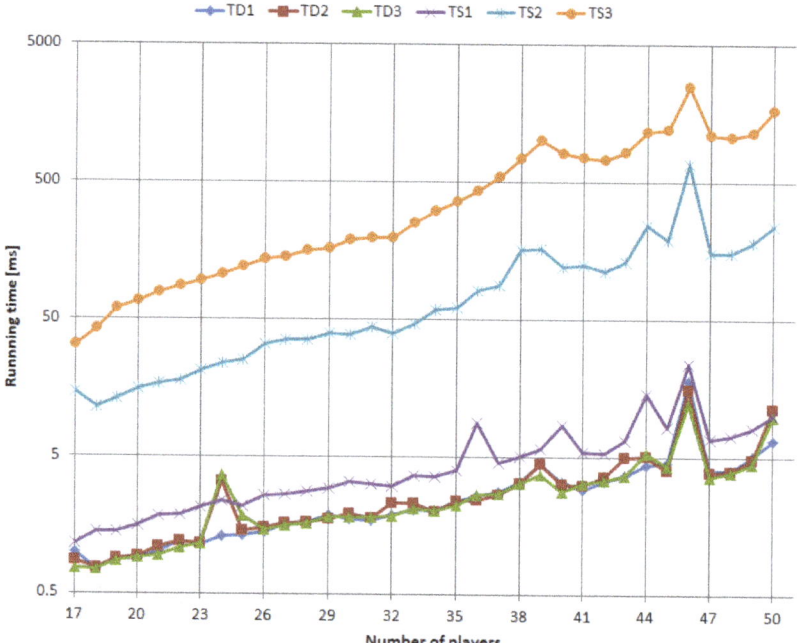

Figure 6. Running times on logarithmic scale of sub-optimal algorithms on different quotations' sequences for a variable number of players.

Figure 7 presents results obtained with optimal algorithms $OptTD$ and $OptBU$, as well as their comparison with results obtained by sub-optimal algorithm $SubOptS_3$ for $N = 3, 4, \ldots, 16$ players, on our input data set.

In Figure 7a we show the comparison of relative maximum and average costs obtained by algorithm $SubOptS_3$ ($CSM3$ and $CSa3$) with the actual optimal cost obtained by algorithm $OptTD$. The relative sub-optimal cost is a measure $C_r \in (0, 1]$ computed with Equation (42) using the absolute values of sub-optimal cost C_0 and optimal cost $C_1 \geq C_0 > 0$. Observe that $C_r = 1$ if and only if $C_1 = C_0$, i.e., if the algorithm providing sub-optimal cost C_0 is in fact optimal. Note that the computation of the relative sub-optimal cost assumes the exact value C_1 of the optimal cost is known. In our case, this value is known, as it was determined using the *OptTD* optimal algorithm for $N = 3, 4, \ldots, 16$ players.

$$C_r = \frac{C_0}{C_1} \qquad (42)$$

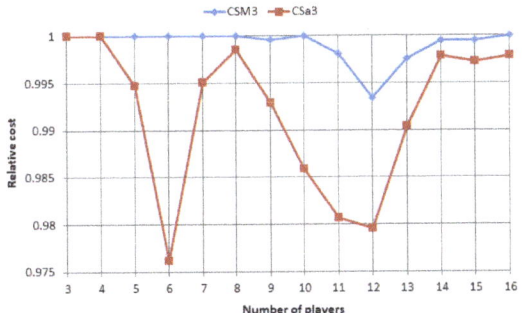 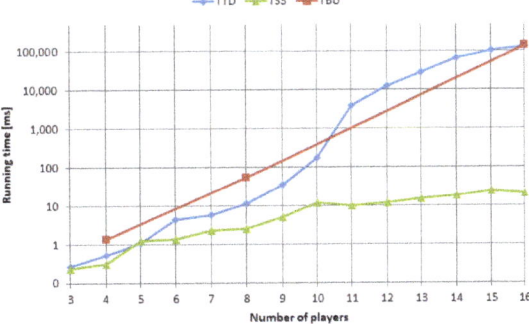

(**a**). Comparison of relative maximum and average costs obtained by algorithm $SubOptS_3$ ($CSM3$ and $CSa3$) with optimal cost obtained by algorithm $SubOptD_1$ for $N = 3, 4, \ldots, 16$ players, on our input data set.

(**b**). Comparison of running times of algorithms *OptTD*, *OptBU*, and $SubOptS_3$ for $N = 3, 4, \ldots, 16$ players, on our input data set. Time values are plotted on a logarithmic scale.

Figure 7. Comparing costs and running times of our implemented algorithms.

In Figure 7b we show the comparison of running times TTD, TBU, and $TS3$ of algorithms *OptTD*, *OptBU*, and $SubOptS_3$ for $N = 3, 4, \ldots, 16$ players. The running times were evaluated by repeating the algorithm execution 10 times for the same input data. They are plotted on a logarithmic scale. Observe that by far the most efficient among them is algorithm $SubOpt_3$. The linear increasing trend of TTD and TBU on the logarithmic scale is consistent to our findings that the complexity of algorithms *OptTD* and *OptBU* is exponential with the number of tournament players. Note that this tendency is also observed on the plot of TBU, for which the values were recorded only for an exact power of two of the number of players, i.e., $N = 4, 8, 16$. Moreover, the sub-linear increasing trend of $TS3$ on the logarithmic scale is consistent with the fact that algorithm $SubOpt_3$ has a polynomial time complexity.

7. Conclusions

In this paper we defined optimal competitions structured as hierarchically shaped single-elimination tournaments. The optimality criterion aimed to maximize tournament attractiveness by letting the topmost players meet in higher stages of the tournament. We proposed a dynamic programming algorithm for computing optimal tournaments and we provided a thorough analysis of its correctness and computational complexity. Based on the idea of the dynamic programming approach, we also developed deterministic and stochastic sub-optimal algorithms. We realized an experimental evaluation of the proposed algorithms by providing experimental results that we obtained with their Python implementation. The results addressed the optimality of solutions and the efficiency of the running time.

Author Contributions: Conceptualization, A.B. and C.B.; methodology, A.B. and C.B.; software, C.B.; formal analysis, C.B. and A.B.; investigation, I.B., L.I.C. and D.L.; writing, C.B. and A.B. All authors have read and agreed to the published version of the manuscript.

Funding: This research received no external funding.

Institutional Review Board Statement: Not applicable.

Informed Consent Statement: Not applicable.

Conflicts of Interest: The authors declare no conflict of interest.

References

1. Bădică, A.; Bădică, C.; Buligiu, I.; Ciora, L.I.; Logofătu, D. Optimal Knockout Tournaments: Definition and Computation. In Proceedings of the Large Scale Scientific Computing—LSSC'2021, LNCS, Sozopol, Bulgaria, 7–11 June 2021, in press.
2. Anderson, I. *Combinatorial Designs and Tournaments*; Oxford University Press Inc.: New York, NY, USA, 1997.
3. Vu, T.; Shoham, Y. Fair Seeding in Knockout Tournaments. *ACM Trans. Intell. Syst. Technol.* **2011**, *3*, 9:1–9:17. [CrossRef]
4. Vu, T.D. Knockout Tournament Design: A Computational Approach. Ph.D. Thesis, Stanford University, Stanford, CA, USA, 2010.
5. Bóna, M.; Flajolet, P. Isomorphism and symmetries in random phylogenetic trees. *J. Appl. Probab.* **2009**, *46*, 1005–1019. [CrossRef]
6. Maurer, W. On Most Effective Tournament Plans with Fewer Games than Competitors. *Ann. Statist.* **1975**, *3*, 717–727. [CrossRef]
7. Dagaev, D.; Suzdaltsev, A. Tournament design allows for spectator interest increase. *Front. Econ. Res.* **2015**. Available online: https://voxeu.org/article/tournament-design-allows-spectator-interest-increase (accessed on 23 August 2021).
8. Hartigan, J.A. Probabilistic Completion of a Knockout Tournament. *Ann. Math. Statist.* **1966**, *37*, 495–503. [CrossRef]
9. CodeChef. Tennis Tournament. Available online: https://www.codechef.com/COOK27/problems/TOURNAM (accessed on 9 August 2021).
10. Bao, N.P.H.; Xiong, S.; Iida, H. Reaper Tournament System. In *Intelligent Technologies for Interactive Entertainment. INTETAIN 2017. Lecture Notes of the Institute for Computer Sciences, Social Informatics and Telecommunications Engineering*; Chisik, Y., Holopainen, J., Khaled, R., Luis Silva, J., Alexandra Silva, P., Eds.; Springer: Cham, Switzerland, 2018; Volume 215, pp. 16–33.
11. Karpov, A. *A Theory of Knockout Tournament Seedings*; Discussion Paper Series; University of Heidelberg, Department of Economics: Heidelberg, Germany, 2015; Volume 600.
12. Adler, I.; Cao, Y.; Karp, R.; Peköz, E.A.; Ross, S.M. Random Knockout Tournaments. *Oper. Res.* **2017**, *65*, 1589–1596. [CrossRef]
13. Guyon, J. "Choose Your Opponent": A New Knockout Design for Hybrid Tournaments. *J. Sport. Anal.* **2021**, 1–21, pre-press. [CrossRef]
14. Hennessy, J.; Glickman, M. Bayesian optimal design of fixed knockout tournament brackets. *J. Quant. Anal. Sports* **2016**, *12*, 1–15. [CrossRef]
15. Csató, L. *Tournament Design. How Operations Research Can Improve Sports Rules*; Palgrave Macmillan: London, UK, 2021.
16. Stojadinović, T. On Catalan numbers. *Teach. Math.* **2015**, *XVIII*, 16–24.
17. Cormen, T.H.; Leiserson, C.E.; Rivest, R.L.; Stein, C. *Introduction to Algorithms*, 3rd ed.; The MIT Press: Cambridge, MA, USA; London, UK, 2009; pp. 365–367.
18. Dutka, J. The early history of the factorial function. *Arch. Hist. Exact Sci.* **1991**, *43*, 225–249. [CrossRef]
19. Knuth, D.E. *The Art of Computer Programming, Volume 4A: Combinatorial Algorithms, Part 1*; Addison-Wesley Professional: Boston, MA, USA, 2011; pp. 319–320.

Article

Spatial-Temporal Traffic Flow Control on Motorways Using Distributed Multi-Agent Reinforcement Learning [†]

Krešimir Kušić [1,*,‡], Edouard Ivanjko [1,‡], Filip Vrbanić [1], Martin Gregurić [1] and Ivana Dusparic [2]

1 Faculty of Transport and Traffic Sciences, University of Zagreb, Vukelićeva Street 4, HR-10 000 Zagreb, Croatia; edouard.ivanjko@fpz.unizg.hr (E.I.); filip.vrbanic@fpz.unizg.hr (F.V.); martin.greguric@fpz.unizg.hr (M.G.)
2 School of Computer Science and Statistics, Trinity College Dublin, Dublin 2, Ireland; ivana.dusparic@scss.tcd.ie
* Correspondence: kresimir.kusic@fpz.unizg.hr
† This paper is an extended version of our paper published in the Proceedings of the 2021 IEEE Intelligent Transportation Systems Conference (ITSC).
‡ These authors contributed equally to this work.

Abstract: The prevailing variable speed limit (VSL) systems as an effective strategy for traffic control on motorways have the disadvantage that they only work with static VSL zones. Under changing traffic conditions, VSL systems with static VSL zones may perform suboptimally. Therefore, the adaptive design of VSL zones is required in traffic scenarios where congestion characteristics vary widely over space and time. To address this problem, we propose a novel distributed spatial-temporal multi-agent VSL (DWL-ST-VSL) approach capable of dynamically adjusting the length and position of VSL zones to complement the adjustment of speed limits in current VSL control systems. To model DWL-ST-VSL, distributed W-learning (DWL), a reinforcement learning (RL)-based algorithm for collaborative agent-based self-optimization toward multiple policies, is used. Each agent uses RL to learn local policies, thereby maximizing travel speed and eliminating congestion. In addition to local policies, through the concept of remote policies, agents learn how their actions affect their immediate neighbours and which policy or action is preferred in a given situation. To assess the impact of deploying additional agents in the control loop and the different cooperation levels on the control process, DWL-ST-VSL is evaluated in a four-agent configuration (DWL4-ST-VSL). This evaluation is done via SUMO microscopic simulations using collaborative agents controlling four segments upstream of the congestion in traffic scenarios with medium and high traffic loads. DWL also allows for heterogeneity in agents' policies; cooperating agents in DWL4-ST-VSL implement two speed limit sets with different granularity. DWL4-ST-VSL outperforms all baselines (W-learning-based VSL and simple proportional speed control), which use static VSL zones. Finally, our experiments yield insights into the new concept of VSL control. This may trigger further research on using advanced learning-based technology to design a new generation of adaptive traffic control systems to meet the requirements of operating in a nonstationary environment and at the leading edge of emerging connected and autonomous vehicles in general.

Keywords: intelligent transport systems; traffic control; spatial-temporal variable speed limit; multi-agent systems; reinforcement learning; distributed W-learning; urban motorways

1. Introduction

Everyday commuting in densely populated urban areas is accompanied by repetitive traffic jams, representing an evident violation of urban life quality. Urban motorways, as an integrated part of the urban road network, are consequently affected by congestion. Variable speed limit (VSL) is an efficient traffic control strategy to improve motorways' Level of service. VSL controls the speed limit in real time by displaying a specific speed limit on variable message signs (VMS). The speed limit value adapts to different traffic situations depending on weather conditions, accidents, traffic jams, etc [1]. The main objective of VSL is to improve traffic safety and throughput on motorways due to the concept of speed homogenization [2] and mainstream traffic flow control (MTFC) [3], respectively. VSL aims to ensure stable traffic flow in motorway areas affected by recurrent bottlenecks. VSL thus has a dual effect: it prevents and alleviates congestion. Typically, problems occur on urban motorways near on-ramps. A higher volume of traffic at the on-ramp can disrupt the main traffic flow and cause a bottleneck activation.

Several VSL control strategies have been suggested in the literature based on different VSL measures and methodologies, such as rule-based VSL activated by predefined threshold values (e.g., flow, speed or density) [4,5], the usage of metaheuristics to optimize VSL [6], optimal control [7], and model-predictive control [8]. The most prominent VSL design (among classical controllers) uses feedback control [3,9], where the speed limit is calculated based on current measurements of traffic conditions, such as traffic density.

However, in recent years, there has been an increasing interest in improving VSL optimization by taking advantage of machine learning techniques with a focus on reinforcement learning (RL). An overview of the existing literature can be found in [10]. RL has a proven track record of solving various complex control problems, including transportation and related control optimization problems, and achieving considerable improvements in transportation management efficiency [11–14]. In particular, RL provides the ability to solve complex Markov decision processes (MDPs) and find a near-optimal solution for discrete-event stochastic systems while not requiring an analytical model of the system to be controlled [15]. In addition, RL-based control systems can continuously improve their performance over time by adapting control policies to newly recognized states of the environment (adaptive control).

The majority of studies in RL-VSL are based on a single objective [16,17] or multiple objectives implemented as a single control policy (strategy) [18–20]. However, large-scale control systems might have various, often conflicting objectives with heterogeneous time and space scales (simultaneous optimization of ramp metering and VSL [21]) or different levels of priorities (safety contrary to throughput [22] or throughput contrary to higher traveling speeds [23]). In practice, VSL is usually applied on several consecutive motorway sections. Thus, the VSL application area should be split into several shorter VSL sections upstream of the bottleneck area to ensure smooth (gradual) spatial adjustment of the speed limits. This can be modelled and solved by multi-agent RL-based control approaches where each agent (VSL controller) sets speed limits on its controlled motorway section [22–24].

Although VSL has been extensively studied and some VSL approaches are being used in practice, there are some open questions in the design of the VSL system itself, on which there is very little research. The critical detail for efficient VSL is the design and placement of the VSL zones. In particular, two practical questions arise: how long should the VSL application zone be, and where should it be placed (in other words, how far should the end of the VSL zone be from the bottleneck) to achieve optimal VSL performance. In general, it can be concluded from [25,26] that different lengths and positions of the VSL application area for different speed limits and different traffic congestion intensities (the spatial variation of the congestion characteristic) significantly affect VSL performance.

To address this problem, in our previous work [23], we proposed a distributed spatio-temporal multi-agent VSL control based on RL (DWL-ST-VSL) with dynamic VSL zone allocation. The DWL-ST-VSL controller dynamically adjusts the configuration of VSL zones and speed limits.

In addition to the results and conclusions in [23], in this study we seek to confirm the extended applicability of DWL-ST-VSL to control longer dynamic VSL application areas with more agents. Therefore, the present study makes the following contributions:

- Extension of the applicability and behaviour analysis of DWL4-ST-VSL by increasing the number of learning agents from the original two to four;
- Evaluation of the performance of DWL4-ST-VSL in controlling speed limits on a longer motorway segment using collaborative agents;

- Assessment of the impact of dynamic VSL zone allocations on traffic flow optimization and comparison to the VSL controllers with static zones in traffic conditions with spatially varying congestion characteristics.

An experimental approach is used to verify suggested solutions using simulation experiments. Thus, the present experiment will give data-based evidence about the potential usefulness of extended DWL4-ST-VSL control with adaptive VSL zones when deployed on longer motorway segments. Results and analysis will provide insights into the modelling of DWL4-ST-VSL and the impact of agents' collaboration on system performance when used to control traffic flow on a longer motorway segment. This is a crucial aspect for the development of adaptive controllers in particular, but also for research investigating reliable and more efficient RL-based VSL.

We hypothesize that the extension of DWL-ST-VSL will contribute to the ability to dynamically configure VSL zones. The fact that agents can collaborate using remote policies results in a better response to moving congestion because they can collectively assemble a larger number of feasible VSL application areas. A certain number of configurations can more appropriately respond to the current downstream congestion. We anticipate that a DWL-ST-VSL system with more agents will use its additional adaptive feature to adjust VSL zones to resolve congestion as much as possible without suppressing the upstream traffic itself. As a result, we expect a further reduction in the overall travel time of the system, and that a smoother speed transition can be achieved by spatially deploying multiple VSL agents. This is more in line with what the VSL implementation should fulfill to achieve a smooth "harmonized" speed transition. Using an adjustable VSL application area supported by multiple dynamically configurable VSL zones reduces the need for severe speed reduction. Agents in upstream zones can prepare vehicles for conditions in downstream VSL zones by slightly decreasing speed limits. This is necessary since speed limits in downstream zones may be lower due to the proximity of the bottleneck. Therefore, this can help to harmonize traffic flow better in order to avoid undesirable effects, such as shockwaves.

Thus, in this paper, we propose an extended version of the DWL-ST-VSL strategy that allows dynamic spatiotemporal VSL zone allocation on a wider motorway section with four VSL learning-based agents (DWL4-ST-VSL). To provide smoother speed limit control, DWL4-ST-VSL implements two speed limit sets with different granularities on the observed motorway section. DWL4-ST-VSL enables automatic, systematic learning in setting up the sufficiently accurate VSL zone configuration (selection is learnt rather than manually designed) for efficient VSL operation under a fluctuating traffic load. From a technical perspective, the physical VMS could soon be replaced (or enhanced) by advanced technologies (vehicle-to-infrastructure communication, e.g.,an intelligent speed assistance (ISA) system [27]). Thus, static placement of physical VMSs would no longer be an obstacle to the dynamic adaptation of VSL zone configurations in real motorway applications.

To set up DWL4-ST-VSL, the distributed W-learning (DWL) algorithm is used. DWL is an RL-based multi-agent algorithm for collaborative agent-based self-optimization with respect to multiple policies. It relies only on local learning and interactions. Therefore, no information about the joint state-action space is shared between agents, which means that the complexity of the model does not increase exponentially with the number of agents. DWL was originally proposed in [28] and successfully applied for controlling traffic signals at multiple urban intersections with different priorities as objectives. It has also been successfully applied for speed limit control on a small urban motorway segment using two agents (DWL2-ST-VSL), as introduced in our previous paper [23].

Thus, in this study, we investigate the applicability of extended DWL4-ST-VSL in terms of the number of learning agents, their behavior, and their impact on traffic flow control, emphasizing an application on a longer motorway segment.

The proposed DWL4-ST-VSL is evaluated using the microscopic simulator, simulation of urban mobility (SUMO) [29], in two scenarios with medium and high traffic loads. Its performance is compared with three baselines: no control (NO-VSL), simple proportional speed controller (SPSC) [30], and W-learning VSL (WL-VSL). The experimental results confirm the feasibility of the proposed extended DWL4-ST-VSL approach with the observed improvement in traffic parameters in the bottleneck area and system travel time of the motorway as a whole. Finally, DWL4-ST-VSL is envisioned as a new approach to dynamically adjust speed limits in space and time, anticipating the practical aspect of vehicle speed control that may be found in the leading-edge of connected and autonomous Vehicles or ISA in general.

The structure of this article is organized as follows: Section 2 discusses related work in the area of RL application in VSL control. Section 3 introduces the DWL algorithm. Section 4 provides insight into the modeling of VSL as a multi-agent DWL problem. Section 5 describes the simulation set-up, and Section 6 delivers the results and analysis of our experiments. The discussion can be found in Section 7. Section 8 summarizes our results and conclusions.

2. Related Work

VSL increases the level of service of motorways by adjusting the speed limit on sections according to the prevailing traffic conditions. The speed limit is posted on the VMS located on a certain section of the motorway, through which drivers are informed of the permitted speed on that section. Usually, warnings about the cause of a speed limit's setting (congestion, slippery pavement, etc.) are also presented.

2.1. Concept of VSL

VSL is used to increase motorway efficiency in areas with frequent recurring bottlenecks [25]. Bottlenecks emerge in motorway sections that present a change in geometry, including on- and off-ramps, lane drops, uphill grade sections, tunnels, accidents etc. At such locations, the upstream traffic volume q_{in} of the motorway periodically may exceed the bottleneck capacity q_{cap}. Once the demand exceeds the bottleneck capacity, congestion starts to form [26]. Even if the downstream motorway section gets released, the accumulated queue shifting upstream of the bottleneck will further reduce the capacity of the upstream part of the motorway. This is known as the capacity drop phenomenon [31], wherein reduced outflow is measured once the bottleneck is active. To eliminate or prevent the activation of a bottleneck, the inflow into the bottleneck must be less than the outflow from the bottleneck q_{out} (see Figure 1). By applying an appropriate speed limit upstream of the bottleneck, VSL can effectively reduce the inflow $q_{VSL} \approx q_{cap}$ into the bottleneck while the outflow capacity is restored. Therefore, VSL seeks to keep bottleneck capacity stable in conditions of increased traffic demand to prevent the capacity drop in a bottleneck area. Otherwise, queues will form at the bottleneck.

The effects of VSL on traffic flow were studied in [32–34]. VSL control measures were first used to improve traffic safety on motorways by harmonizing traffic [2,35,36]. These strategies provide speed limits around the critical speed at which capacity is reached. They are based on the assumption that lower speed limits reduce spatial variations in speed (thus increasing the homogenization of speed), flow, and density on motorways. Thus, the suggested scheme can smooth out the incoming traffic towards the congestion point to avoid undesirable effects, such as shockwaves. As shown in [37], the speed limit is one among multiple dependent factors that impact the level of crash risk on motorways. Mainly, reduced speed variance is considered to solve both the road safety operation level and the risk of capacity drop [2]. However, the available studies do not provide clear evidence of increased capacity at the expense of harmonization (when reported, increased throughput is within an interval of 5–10%).

The second type of VSL control regulates the incoming traffic towards the bottleneck area by restricting mainstream flow and is often referred to as MTFC [3]. Thus, the goal of MTFC is to eliminate or prevent bottleneck activation and capacity drop.

2.2. VSL Control Strategies

Over the years, various VSL control approaches have been suggested based on different system configurations and methodologies, e.g., optimal control, model predictive control, feedback control and shock wave theory [38]. Feedback-based VSL controllers base their speed limit changes as reactive responses (corrective behavior) to the deviation of a controlled process variable (e.g., traffic density) from the reference (e.g., predefined desired density value in the bottleneck) [39]. Feedback-based VSL can be extended by model predictive control and work in a coordinated fashion to address the shortage of delayed responses. However, model predictive control generally does not guarantee the stability of the control loop and is much more computationally intensive [9]. Although feedback-based VSL is much more efficient and robust to current traffic data, such controllers are tuned to a specific range of traffic load and are not adaptive. If traffic patterns and traffic load change significantly, the controller may not be able to achieve the desired state in a timely manner and may, therefore, operate suboptimally [17].

Over the last few years, there has been a renewed interest in improving VSL optimization through control concepts based on RL [16–18,21,40]. In [17], it is shown that RL-VSL can yield better results when applied to system travel time optimization in the case of recurrent motorway congestion as compared to a two-loop feedback cascade VSL control structure. The results reported that the feedback-based VSL controller could lead to a delayed response to the fluctuating traffic load when controlling the bottleneck density. On the contrary, the RL-VSL can learn traffic patterns that trigger the activation of a bottleneck through the learning process. Hence, in some cases, RL-VSL can anticipate bottleneck activation and respond proactively.

In [18], the control policy of RL-VSL was further improved by enriching the agent's state variables with predictive information about the expected traffic state by forecasting the speed and density of the controlled motorway segment by running parallel simulations. RL can be integrated with function approximation techniques (linear or nonlinear). Approximations address the large dimensionality problem of storing state-action values in the computer's memory [15] and enable the computer to work with continuous state/action variables, which, in the end, plagues many real systems with a large solution space, such as RL-VSL [21,40]. Nonlinear function approximation techniques may improve control if an underlying controlled process is nonlinear and nonstationary, as is the case with motorway traffic flow control [18,41].

In [22], a multi-agent VSL with two objectives was tested. Flow control aims to increase throughput in the bottleneck, while traffic safety policy aims to reduce the speed difference between adjacent controlled motorway segments. Each policy was learned and evaluated separately. According to the defined objective, VSL agents have to learn an optimal joint strategy (policy) using distributed Q-learning. The results indicated an improvement in vehicle stops and total travel time compared to the no control case. Similarly, in [19], a Q-learning-based coordinated hard shoulder control strategy and VSL was introduced. In [42], a dynamic control cycle was suggested to compute the optimal duration of control cycles in VSL. Dynamic control cycles were proven to perform better than those which were fixed. The suggested strategy enables adjustable time lengths of each control cycle regarding current traffic states and speed limits, allowing VSL to respond appropriately to time-varying traffic conditions.

In [24], we extended RL-VSL [40] in a multi-agent structure. Using the W-learning (WL) algorithm [43], two RL-VSL agents learn to jointly control two motorway segments in front of a congestion area. WL gave better results in tested traffic scenarios, including dynamic and static traffic loads, and proved suitable as a multi-policy optimization technique in VSL when used for noncooperating agents.

We also analyzed several manually configured WL-VSL configurations, including different VSL zone lengths and their distances relative to the bottleneck area. The results confirmed that changes in VSL zone configurations affect the traffic flow control process differently. These results are consistent with the findings in [25,26] regarding the optimal location and length of VSL application area.

Thereby, we hypothesized whether VSL performance under such conditions could be improved by having the VSL controller dynamically adjust the length and location of the VSL zone (adjustable VSL application area, "similar to the concept of dynamic control cycle suggested in [42]") in response to changing congestion rather than using static VSL zones. In [23], we confirmed our hypothesis experimentally for a two-agent system. In particular, for spatially and temporally varying traffic congestion, dynamic VSL zone allocation proved to be advantageous over static VSL zones (fixed length and location). The appropriate adaptive VSL zone configurations were learned using DWL-ST-VSL without the need for manual setup. In this paper, we experimentally show the need for more complex multi-agent VSL (e.g., a four-agent system) to control a longer motorway segment.

2.3. Spatial Based VSL

The value of the speed limit and the proper placement of the VMS prior to the occurrence of congestion (see Figure 1) are essential factors for an efficient VSL system. Pioneers in defining important theoretical assumptions with evidence for optimal VSL application areas are the following works [25,26]. In [25], a simulation approach is used to determine the optimal location and length of the VSL application area concerning its distance from the bottleneck. Stepwise variation of the length of the VSL application area and acceleration area is used to show the dependence between the lengths and the system travel time, measured in total time spent (TTS) [veh·h].

The recent results of [26] provide new insights into the optimal placement of the VSL application area compared to previous findings, and the given results are confirmed analytically. It is shown that the general assumption that the lower the speed limits, the larger the distance between the VSL application area and the bottleneck should be (to enable vehicles to reach the critical speed before entering a bottleneck) is not always the case. Instead, the results indicate that at a higher value of the speed limit, the distance between the VSL zone and the bottleneck should be larger. In [44], the authors address the same problem, but in the context of the optimal distance between the merging area and the traffic light on the mainstream to achieve the most efficient merging of vehicles in combination with the real-time traffic control strategy used (MTFC with traffic lights instead of VSL). Additionally, in [45], the authors point out the problem of the optimal VSL zone design for the optimization of the bottleneck. Therefore, they propose three VSL zones: the critical VSL zone for regulating the discharge section flow to match the bottleneck's capacity, the VSL zone for the potentially congested area (mainstream storage), and the VSL zone upstream of the congestion tail. The analysis performed in [46] suggested a VSL control model that is able to determine whether the section is congested or not based on predefined thresholds (density, speed, and acceleration), and this information is used to determine the VSL start station. In [47], the bilevel programming model is used to find the most appropriate speed limits and corresponding locations of VMSs in VSL control. The first objective of the bilevel programming model was to optimize the number and speed limits of VMSs by creating a model for a minimum comprehensive accident rate. The second objective was modelled to optimize the locations of VMSs by solving the improved maximum information benefit model. The results presented confirm that appropriate speed limits and proper placement of VMSs can reduce the average queue length, total delay, and total stop frequency of vehicles in motorway work zones.

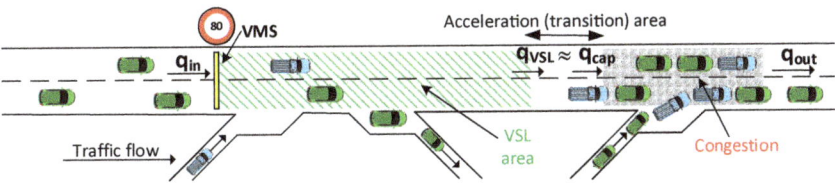

Figure 1. Application of VSL for bottleneck control [10].

Although the results of the above-mentioned analyses point to a possible feasible direction for addressing optimal VSL zone placement, in general, the results and findings indicate that there is no absolute guideline for where the VSL zone should be placed for optimal performance. Instead, it appears that the near-optimal placement of VSL zones depends on the location and intensity of congestion and the speed limit values used in that context.

Given that the congestion characteristic varies in time and space due to stochastic traffic behavior, we have experimentally confirmed the usefulness of the DWL-ST-VSL concept of dynamic VSL zone allocation for speed limit control in [23]. We also demonstrated that DWL-WL-VSL agents and the motorway system could benefit from collaborating to select appropriate actions, not only for their own policies, but also for the policies of the other agents they affect.

Therefore, this paper aims to provide simulation proof of the extended concept of DWL4-ST-VSL and its applicability to speed limit control on a longer motorway segment, which is more in line with what is required in the real world to achieve harmonized traffic flow control. The analysis gives detailed insight into the steps of modeling DWL4-ST-VSL and provides some interesting details on the pros and cons of the proposed algorithm. These are our primary research motivations for implementing an enhanced version of the DWL4-ST-VSL strategy that learns appropriate speed limits and spatiotemporal VSL zone configurations in an automated manner using the DWL algorithm on a longer motorway segment. Four cooperative agents operating upstream of the bottleneck area will be tested in the suggested configuration.

3. Multi-Agent Based Reinforcement Learning

This section presents the essential elements needed to understand RL-based techniques and the DWL algorithm.

3.1. Reinforcement Learning

RL is a simulation-based technique that is useful in large-scale and complex MDPs [48]. It combines the principle of the Monte Carlo method with the principle of dynamic programming, which in RL is called the temporal difference method. In RL, simulation can be used to generate samples of the value function of a complex system (rather than finding an explicit model), which are then averaged to obtain the expected value of the value function. Therefore, transition probabilities are not required in RL (model-free technique). This avoids the curse of dimensionality (a potentially large number of states which leads to the well-known curses of dynamic programming: the curse of modeling and the curse of dimensionality) [15].

3.2. Q-Learning

Q-Learning is an off-policy RL algorithm that perceives and interacts with the environment at each control time step by performing actions and receiving feedback (rewards). Thus, the Q-Learning function $Q(x_t, a_t)$ learns to associate an action a_t with the expected long-term payoff (reward) for performing that action in a given state x_t [49]. How good action is in a given state is expressed as a Q-value. Q-function is learned using the following iterative update rule:

$$Q_i(x_t, a_t) := (1 - \alpha_Q)Q_i(x_t, a_t) + \alpha_Q(r_{t+1} + \gamma \max_{a' \in A} Q_i(x_{t+1}, a')). \tag{1}$$

The performed action a_t in state x_t stimulates a state transition to the new state x_{t+1}, from which an optimal action is a'. Depending on this transition, the agent receives a reward r_{t+1}. The parameter α_Q is the learning rate that controls how fast the Q-values are adjusted. The discount factor γ controls the importance of future rewards. Various exploration/exploitation strategies (e.g., ϵ-greedy) are used to search the solution space, i.e., to ensure that the agent sufficiently explores its environment and learns the appropriate action in a given state.

3.3. W-Learning

The WL algorithm proposed in [43] was designed to manage competition between multiple tasks. In particular, an individual policy is implemented as a separate Q-learning process designed by its own state space. The goal is to learn Q-values for state-action pairs for each policy, where a single policy can be viewed as an agent. At each control time step, each policy nominates an action based on Q-values. Applying WL for each state x of each of their policies, the agent learns what happens concerning the reward received if the nominated action is not performed (rated using a W-value for a given state $W(x)$). Thus, an agent only needs local knowledge—what state x_t it was in, whether the nominated action was obeyed or not, the state transition x_{t+1}, and the received reward r_{t+1}.

Hence, all policies recommend new actions. Nevertheless, only one action is executed (suggested by the "winner policy") based on the highest W-value (if not, this policy will suffer the highest deviation). Each policy updates its own Q_i function using the winning action a_k and its own received reward r_i. W_i values are updated only for policies that were not obeyed ($i \neq k$) using the following update rule:

$$W_i(x_t) := (1 - \alpha_W)W_i(x_t)$$
$$+ \alpha_W(1 - \alpha_Q)^\omega (Q_i(x_t, a_i) - (r_{i,t+1} + \gamma \max_{a' \in A} Q_i(x_{t+1}, a'))), \tag{2}$$

where learning rate α_W and delaying rate ω ($\omega > 0$) control the convergence of W_i.

Thus, WL can be seen as a fair resolution of competition. Competition results in fragmentation of the state-space between the different agents, thus allowing any collection of agents. Eventually, they will divide up state-space among them based on the deviations they cause to each other. The winner of a state (determined by highest $W(x)$) is the agent that is most likely to suffer the highest deviation if it does not win. Eventually, agents are aware of their competition indirectly by the interference they cause.

3.4. Distributed W-Learning

The DWL algorithm proposed in [28] enables an agent $Ai \in A = \{A_1, \ldots, A_n\}$ to learn to select actions that match its local policies while learning how its actions affect its neighbours $Aj \in A$, and to give different weights to the preferences of its neighbours when selecting an action. To prompt an agent Ai to consider the action preference of its neighbours (i.e., to cooperate), each agent implements, in addition to its own local policy $LP_i = \{LP_{i1}, \ldots, LP_{il}\}$, a "remote" policy $RP_i = \{RP_{ij1}, \ldots, RP_{ijr}\}$ for each of the local policies LP_{jl} used on each of its neighbours. To help neighbour Aj implement its local policy, remote policy RP_i receives a reward r_{ijr} every time a neighbour's local policy LP_{jl} receives a reward r_{jl} ($r_{ijr} = r_{jl}$).

RP_i enables heterogeneous agents to collaborate, implement different policies, and have different actions and state spaces. Thus, the DWL scheme lets an agent adapt to the other agents, since their dynamics are generally changeable. Each agent implements its policy as a combination of a Q- and a WL process. Q-values are associated with each of its state-action pairs, while W-values are associated with states. In the learning process, an agent Ai learns Q-values for remote-state/local-action pairs and W-values for local/remote states, through which it learns the influence of its local actions on the states of its neighbours Aj. Thus, DWL does not need a global knowledge or central component. It relies on local learning and interactions with its neighbours, local rewards from the environment, and local actions.

To learn how its actions affect its neighbors, at each control time step, the agent receives information about the current states of its neighbours and the rewards they have received. All local and remote policies nominate an action with an associated W-value. Nominations for LP_i actions are treated with full W-values. In contrast, RP_i nominations are scaled by a cooperation coefficient C ($0 \leq C \leq 1$) to enable an agent to weigh the action preferences of its neighbours. $C = 0$ indicates a non-cooperative local agent, i.e., it does not consider the performance of its neighbours when picking an action. For $C = 1$, the local agent is entirely cooperative, implying that it cares about its neighbours' performance as much as its own.

The action performed at the given control time step (one that wins the competition between policies) is selected based on the highest W-value (W_{win}) after scaling the remote W-values by C:

$$W_{win} = max(W_{il}, C \times W_{ijr}), \qquad (3)$$

where W_{il} and W_{ijr} are W-values nominated by LP_i and RP_i policies of agent Ai, respectively.

4. Modeling Spatial VSL as a DWL-ST-VSL Problem

So far, DWL has been successfully applied to the problem of controlling urban intersections on a larger scale network with a larger number of agents [28]. DWL has also proven successful in the VSL control optimization problem [23] on a smaller motorway segment. Nevertheless, it has never been tested for its extended applicability to motorway traffic control with a higher number of deployed VSL agents. Thus, in our extended DWL4-ST-VSL framework, four neighbouring agents ($Ai, i = 1, 2, 3, 4$) control the speed limit and VSL zone configuration (length and position) on their own motorway section. Each agent in DWL4-ST-VSL perceives its local environment through agent states and rewards (see Figure 2). Thus, in the proposed multi-agent control optimization problem, the agent states x_t, actions a_t, and reward functions r_{t+1} are modelled as follows.

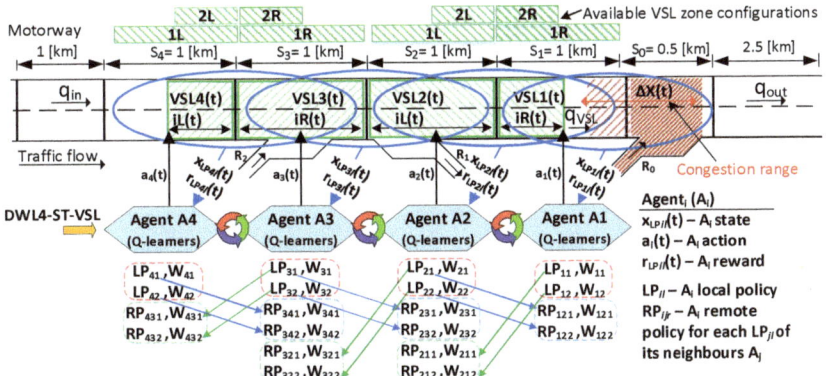

Figure 2. DWL4-ST-VSL configuration scheme.

4.1. State Description

As stated in [18], defining a compact Markovian state representation for motorways is difficult because many external factors influence traffic flow: e.g., weather conditions, motorway geometry (curvature, slope), etc., which are hard to model precisely. Augmenting the state by additional information, such as observing more sections (e.g., the density measured on the motorway section further upstream from the congestion location and the on-ramp queue length, primarily to provide a predictive component in terms of motorway demand [21]) or including information from the past in states, may improve the algorithm's performance. Though this increases solution space, it can be overcome by the function approximation technique [18,40]. However, in DWL modelling, the observation of the agent's neighborhood is available through remote policies. Nevertheless, the observability of the state must be assured. An example of a partially observable state is the usage of flow rate for states. From traffic flow theory, macroscopic variables describe traffic conditions (speed, density, flow). As a result of the nonlinearity of the fundamental diagram (flow-density relationship) [39], the same traffic flow rate can be observed for a density value below critical density with high speed (stable flow) and a density value above critical density with low speed (unstable flow). Thus, the traffic condition is uniquely determined by using the information of traffic density. Therefore, we use speed and density measurements to omit the agents' confusion, thus uniquely determining traffic conditions. As a result, the negative effect of imperfect and incomplete perception of agents' partially observable states in our MDP modeling is reduced.

The inclusion of the speed measurement of the neighboring segments into the state can enhance the learning process, particularly at the beginning of the learning process, when agents cause interference by randomly performing actions (exploration). Besides, low speed indicates traffic flow disruption provoked by congestion. Speeds are encoded in the variable V_n, which corresponds to the measured average vehicle speed $\bar{v}_{n,t}$ at time t in motorway section S_n ($n = 0, 1, 2, 3, 4$), as shown in Figure 2. Each speed measurement can fall into one of four intervals defined with boundary points (50, 76, 101 [km/h]).

Current traffic density $\bar{\rho}_{n,t}$ measured in the motorway section S_n, is stored in the variable P_n. Each measurement can fall into one of twelve intervals defined by the boundary points (15, 20, 23, 26, 29, 32, 35, 38, 45, 55, 65 [veh/km/lane]). Additionally, the state space contains information about the agent's action from the previous control time step, thereby enabling modelling restrictions on the action space by making it state dependent, which is explained in more detail in the following subsection.

Therefore, $A1$'s local policy LP_{11} at time t senses state $x = (a_{1,t-1}, V_0, P_0, P_1)$, while LP_{12} $x = (a_{1,t-1}, V_1, P_0, P_1)$. $A2$'s LP_{21} senses state $x = (a_{2,t-1}, V_1, P_1, P_2,)$, while LP_{22} $x = (a_{2,t-1}, V_2, P_1, P_2)$. Similarly, $A3$'s LP_{31} senses state $x = (a_{3,t-1}, V_2, P_2, P_3,)$, while LP_{32} $x = (a_{3,t-1}, V_3, P_2, P_3)$. Finally, $A4$'s LP_{41} senses state $x = (a_{4,t-1}, V_3, P_3, P_4,)$, while LP_{42} $x = (a_{4,t-1}, V_4, P_3, P_4)$ (see Figure 2).

4.2. Action Space

Each element in the action sets (4) and (5) consists of two variables. The upper one represents the speed limit [km/h] in section S_n, while the lower one represents an active VSL zone (indexes for the left (iL)/right (iR) configuration; see Figure 2). Agent $A1$ controls the speed limit and the length of the VSL zone in section S_1, while $A2$ controls section S_2, and so on. In this way, the agent's winning policy (either LP_i or RP_i) will define the speed limit and the VSL zone configuration for a given motorway section.

$$\mathcal{A}_{1,2,DWL} = \left\{ \begin{Bmatrix} 60 \\ 1 \end{Bmatrix}, \begin{Bmatrix} 60 \\ 2 \end{Bmatrix}, \begin{Bmatrix} 80 \\ 1 \end{Bmatrix}, \begin{Bmatrix} 80 \\ 2 \end{Bmatrix}, \begin{Bmatrix} 100 \\ 1 \end{Bmatrix}, \begin{Bmatrix} 100 \\ 2 \end{Bmatrix}, \begin{Bmatrix} 120 \\ 1 \end{Bmatrix}, \begin{Bmatrix} 120 \\ 2 \end{Bmatrix} \right\} \quad (4)$$

$$\mathcal{A}_{3,4,DWL} = \left\{ \begin{Bmatrix} 90 \\ 1 \end{Bmatrix}, \begin{Bmatrix} 90 \\ 2 \end{Bmatrix}, \begin{Bmatrix} 100 \\ 1 \end{Bmatrix}, \begin{Bmatrix} 100 \\ 2 \end{Bmatrix}, \begin{Bmatrix} 110 \\ 1 \end{Bmatrix}, \begin{Bmatrix} 110 \\ 2 \end{Bmatrix}, \begin{Bmatrix} 120 \\ 1 \end{Bmatrix}, \begin{Bmatrix} 120 \\ 2 \end{Bmatrix} \right\} \quad (5)$$

Q-values in (DWL2 and DWL4)-ST-VSL are stored in a $Q_{|X| \times |\mathcal{A}_{DWL}|}$ matrix, where X is a finite set containing the indices of the coded states of the Cartesian product of the input traffic variables ($|X| = 4608$ and $|\mathcal{A}_{DWL}| = |\mathcal{A}_{1,2,DWL}| = |\mathcal{A}_{3,4,DWL}| = 8$). This seems to be a large solution space for learning optimal Q-values using (1). Nevertheless, the feasible solution space was reduced by constraining the action selection in the nomination process explained in the continuation. Thus, Q-matrix can be considered a sparse matrix, and there is no need to search the whole space.

The consecutive speed limit change within a section (n) must satisfy constraint $|a_{t-1,n} - a_{t,n}| \leq 20$ in the case of agents $A1$ and $A2$, which use action set $A_{1,2,DWL}$. In the case of $A3$ and $A4$ ($A_{3,4,DWL}$), the constraint is $|a_{t-1,n} - a_{t,n}| \leq 10$. This ensures a smooth and safe speed transition between the upstream free-flow and the congested downstream flow characterized by lower vehicle speeds due to the bottleneck. Thus, the final set of actions allowed for the agent Ai at time t depends on the previously executed actions of the agent. This constraint also implies that the next possible action (a') in the update process of the W- and Q-values (see update rules (1) and (2)) must be bounded based on a_t. Thus, each time the Q-value is updated, a possible subset of the allowed actions is considered. E.g., if $a_{k,t-1} = A_{1,2,DWL}(7)$, then the available action subset at time t is $A^*_{1,2,DWL} = \{A_{1,2,DWL}(5), A_{1,2,DWL}(6), A_{1,2,DWL}(7), A_{1,2,DWL}(8)\}$. Therefore, the previous action in the state space is used to uniquely distinguish between states' transitions given the constrained subset of actions between control time steps. This constraint is implicitly modelled in the update rule (1). It addresses a unique row in Q-matrix ($Q(x,a)$) and the reachable entries in that row, corresponding to a given action index. Feasible entries in the particular row correspond to original indexes of elements from the original action set). Thus, only such entries in $Q(x,a)$ are reachable in updating Q- and W-values and in the action nomination process while using "argmax" in Q-learning. Otherwise, the oscillation in the values of elements in a particular state (row) will be present. Thus, Q-values will not converge to a stationary policy, and action nomination in a particular state will constantly switch no matter how long the learning period is. Eventually, a stable agent diminishes the nonstationarity effect in the learning problem of the other agents.

In this way, it is not necessary to model constraints directly in the rewards. It is still ensured that DWL4-ST-VSL operates according to the advised safety rules on maximum allowable speed changes.

It is important to note that the constraints on the spatial difference of speed limit values between two adjacent VSL zones on the motorway are not explicitly considered in this setup. It is assumed that agents communicate information about congestion intensity and locations via remote policies. Thus, the difference in spatial speed limits should be reasonable in terms of optimal traffic flow control. This is also aided by DWL's ability to implement two sets of speed limits with different granularity simultaneously. Action set (4) is for agents $A1$ and $A2$, which are closer to the bottleneck. The finer action set (5) is for upstream agents $A3$ and $A4$. The finer actions aim to slightly adjust the speeds of the arriving vehicles before they enter the VSL application areas controlled by downstream agents. In this way, agents smooth out the incoming traffic towards the congestion point, thus avoiding the undesirable sudden deceleration of vehicles and effects such as shockwaves.

4.3. Reward Function

In [18], the minimization of the total time spent (TTS) of vehicles on the observed motorway segment over a given time interval was successfully used as an objective in RL-VSL control. Therefore, we also use the TTS measure for reward. The variable $TTS_{n,t+1}$ measures TTS between two control time steps t and $t+1$ on the motorway section n. In this way, an agent receives feedback about how good its action was. Each agent must learn to strike a balance between two conflicting policies. In the case of an inactive bottleneck, the penalty will be lower for a higher speed limit. Contrary, when congestion occurs, it is required to gradually reduce the speed limit in upstream sections to control the incoming traffic towards the congestion point so as to maintain the traffic volume near the operational capacity of the active bottleneck. Thus, each policy seeks to optimize its objective as follows.

4.3.1. Local Policy for Stable-Flow Control

The local policy LP_{i1} of an agent Ai aims to learn the speed limit to ensure a reduction of TTS by promoting, when possible, higher traveling speeds in stable-flow conditions. To achieve this goal, the LP_{i1} reward is:

$$r_{LPi1,t+1} = \begin{cases} 0, \text{ if } min\{\bar{v}_{n,t+1} \mid n = i, i-1\} \geq 102 \\ -TTS_{n,t+1}, \ n = i \text{ otherwise} \end{cases}, \quad (6)$$

thereby favoring average vehicle speeds above 102 [km/h].

In a certain percentage, LP_{i1} is activated in saturated flow during the transition from free-flow to congested flow and vice versa. Therefore, it prepares traffic for the second policy (LP_{i2}), which dominates in oversaturated (congested) conditions. After congestion has started to resolve by deploying LP_{i2}, and the congestion intensity reduces to a certain level, LP_{i1} helps restore traffic to free flow (higher traveling speeds) as soon as possible by gradually increasing the speed limit. Thus, LP_{i1} seeks to reduce traffic recovery time. Finally, the states perceived by LP_{i1} satisfy the minimum requirements to determine whether the flow in the agent's neighborhood is a stable flow or deviating from it. Thus, the agent can recognize when the higher speed limits for free flow can be implemented or not.

4.3.2. Local Policy for Unstable-Flow (Congested) Traffic Control

Local policy LP_{i2} aims to reduce TTS in the downstream motorway section in the case of an active bottleneck. Thus, an agent must learn and apply appropriate speed limits to restrict the inflow into the bottleneck until the discharge capacity is restored. If not, congestion will grow, and consequently, it will increase its penalty in proportion to:

$$r_{LPi2,t+1} = -\beta TTS_{n,t+1}, \ n = i-1, \quad (7)$$

where coefficient β controls the agent's sensitivity to congestion. Instead of using only downstream congestion information, LP_{i2} uses information about the upcoming traffic flow (current speed and density) from the section $S_n, n = i$. This can be considered a prediction of the forthcoming traffic flow (how fast and with what volume it will arrive) into the downstream congested section $S_n, n = i-1$. In this way, the description of traffic conditions (states) is extended to include more unique traffic characteristics for more efficient congestion control.

4.3.3. Remote Policies

Cooperation between agents is based on remote policies. Thus, an agent Ai learns additional remote policies ($RP_{ij1}, \ldots, RP_{ijr}$) that complement its neighbouring agent's local policies. In order to know how Ai's local actions a_t affect the neighbours' states, the agent updates the remote policies by the information it receives about its neighbours' current states and the rewards that neighbour agents have received (Figure 2). Our experiments consider that agents' communication is perfect (no loss of information and no breakdown of agents is assumed).

4.4. Winner Action

In DWL4-ST-VSL, an agent Ai's experience (Q-values for local-state/action pairs and Q-values for remote-state/local-action pairs) for each policy are respectively stored in Q_{ik} matrices. In the case of agents $A1$ and $A4$ ($k = 1, \ldots, 4$), while for $A2$ and $A3$ ($k = 1, \ldots, 6$). At the same time, for each of the states of each of its policies, an agent learns W-values of what happens in terms of the reward received if the action nominated by that policy is not performed [43]. This is expressed as a W-value ($W(x_{i,t})$) and stored in W_{ik} matrices in each case. With the knowledge gained from these matrices, all policies (local and remote) propose new actions. The action $a_{k,t}$ that wins the competition between policies at this time step is the one with the highest W-value (W_{max}) (computed using (3)) [28]. After the state transition

$x_t \mapsto x_{t+1}$, each agent's local policy receives its unique reward ($r_{LPi1,t+1}, r_{LPi2,t+1}$) and state ($x_{LPi1,t+1}, x_{LPi2,t+1}$) depending on the consequences of the executed action $a_{k,t}$. The remote policies RP_{ijr} obtain rewards and state information from their neighbour agent by querying the neighbour's local policies states/rewards ($x_{LPj1,t+1}, r_{LPj1,t+1}, x_{LPj2,t+1}$, and $r_{LPj2,t+1}$). Then, all policies update their Q-values (for the winning action $a_{k,t}$), while only the policies that were not obeyed update their W-value. The above process is repeated for all agents.

5. Simulation Set-Up

To evaluate whether the dynamic assignment of VSL zones and cooperation between agents with DWL have an advantage over static VSL zones with non-cooperative agents, we compare DWL4-ST-VSL with our previous work on WL-VSL [24]. To verify the advantages of learning approaches over classical VSL control, we also compare DWL4-ST-VSL with SPSC [30]. It is important to note that the calibration procedure of the simulated motorway section is not included because a synthetic model with different traffic loads was used for this analysis. The objective of this study is to evaluate the impact of dynamically adjusting the VSL zone configurations and the different number of agents in DWL-ST-VSL on the optimization of traffic flow within an active bottleneck and the motorway as a whole.

5.1. Simulation Model

The simulation framework used consists of the microscopic simulator SUMO (version 1.8.0) and the Python programming environment. We referred to the software version because the simulation output in the new version may differ slightly from the simulation in the previous version, as the simulator source code is constantly being improved and updated.

The motorway model is based on the model used in [23]. It is divided into 5 main sections, $S_n, n = 0, 1, 2, 3, 4$. To ensure all combinations of VSL zones used in these experiments (see Figure 2) and to measure spatio-temporal characteristics of the traffic flow, the entire simulation model is divided into smaller links (each 50 m long). The speed limit is simulated along with the computed configuration of the VSL zones for the chosen control time by directly assigning the allowed speeds to the corresponding links. The new speed limit and the configuration of the VSL zones are, thus, calculated by agents for each control time step $T_c = 150$ [s]. In our previous work [23], this T_c value was chosen from multiple tests. The used value is in the range of the foremost values found by the sensitivity analysis of control cycle lengths performed in [42]. The bottleneck is generated on the motorway section S_0. Each simulation lasts 1.5 h, and all learning-based VSL approaches were trained in 14,000 simulations.

5.2. Traffic Scenarios

To evaluate the DWL4-ST-VSL control solution's feasibility and behavior and determine whether agent cooperation and dynamic VSL zone assignment with DWL has advantages over VSL control approaches with static VSL zone configuration (WL-VSL and SPSC), we tested it under medium and high traffic loads. The input traffic data used were synthetic data, and the calibration process of the simulated model is not within the scope of this analysis. Therefore, the driver behavior and vehicle characteristics were modelled using the *Krauss* car-following model with the default settings in SUMO [29].

5.2.1. Medium Traffic Load

In the downstream section S_0 (Figure 2), a bottleneck is induced by an increase in traffic demand at the on-ramp R_0. The generated bottleneck is the primary test for DWL4-ST-VSL with dynamic VSL zone allocation. In this traffic scenario, the demand at on-ramp R_0 changes over time (see Figure 3). For the highest demand at on-ramp R_0, 1315 [veh/h], slower vehicles entering the motorway interact with the mainstream traffic in the merge area. Consequently, this causes disturbances, which triggers the activation of the bottleneck, and congestion appears. Traffic flow at ramps R_1 and R_2 remain constant for both traffic scenarios, with the flow of 385 and 230 [veh/h], respectively. The mainstream flow entering the bottleneck area has a constant rate of 1385 [veh/h/lane]. The traffic flow consists of 94% cars, 3% buses and 3% trucks.

5.2.2. High Traffic Load

The induced congestion is much more significant in this traffic scenario than in the medium scenario. In particular, an increase is generated by a 7.22% higher traffic mainstream demand entering the bottleneck area relative to the medium traffic scenario. This is the test for DWL4-ST-VSL emphasizing the dynamic adjustment of VSL zones. Since the congestion tail propagates much more upstream through the motorway, it can be expected that different VSL zone configurations will be used compared to the medium traffic scenario.

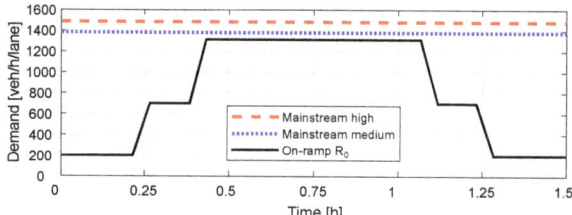

Figure 3. Tested traffic scenarios.

5.3. Baselines of SPSC and WL-VSL

In the case of baselines, the best static VSL zone configuration $S_{2,(2L)} + S_{1,(1R)}$ (see Figure 2) and parameters were selected from several tests conducted for the medium traffic load.

In the case of SPSC [30], the gain ($Kv = 4.5$) and activation threshold (traffic density of 23 [veh/km/lane]) were selected from several tests.

The same best static VSL configuration is also used for the WL-VSL case. In WL-VSL, two local policies were used. Local policy LP_1 aims to maintain a higher speed on controlled motorway sections, while LP_2 aims to reduce congestion in the presence of an active bottleneck. The observed state variables for LP_1 are densities within sections S_1 and $S_{2,(2L)}$, and for LP_2 densities within sections S_0 and S_1, the "bottleneck region". Each element of the action set contains two variables (the section S_1 and $S_{2,(2L)}$ speed limits). In this way, the winning policy will set speed limits for both sections [24]. The two rewards associated with the mentioned policies were modelled as follows:

$$r_{LP1,t+1} = \begin{cases} 0, & \text{if } min\{\bar{v}_{n,t+1} \mid n = 0,1,2\} \geq 100 \\ -0.4(2TTS_{2,(2L),t+1} + TTS_{1,(1R),t+1}) & \text{otherwise} \end{cases}, \quad (8)$$

$$r_{LP2,t+1} = -TTS_{0,t+1}. \quad (9)$$

5.4. DWL-ST-VSL Parameters

For both (DWL2 and DWL4)-ST-VSL and WL-VSL we use the "learning Q (somewhat) before learning W" scheme [43], controlled by $\alpha_W(1 - \alpha_Q)^\omega$ part in (2), where $\alpha_Q = \frac{1}{n(x,a)}$ and $\alpha_W = \frac{1}{n(x)}$ depend on the number of visits to $Q_i(x,a)$. Thus, the weight is larger when an agent is sure of what it is doing in a given state. This is indicated by a higher frequency of nominating a particular action based on the highest Q-value. The parameter $\omega = 1.5$ controls how fast W converges and was selected from multiple tests. The author of WL [43], in his demonstrated example, used $\omega = 3$. The parameter $\gamma = 0.8$ was chosen from [24]. The exploration probability is decreased by the parameter $\epsilon = \exp \frac{-log(20)N}{6000}$, which decreases with the number of simulation runs N [23]. In the DWL-ST-VSL nomination process (3), the cooperation between agents is controlled by remote policies (RP_i) via a cooperation coefficient C. The cooperation levels we test are $C \in \{0, 0.25, 0.5, 0.75, 1\}$.

5.4.1. DWL2-ST-VSL Parameters

To keep the W-values of the local policies comparable to the W-values of the remote policies, we scale the reward function (7) by the factor $\beta = 0.75$ in the case of agents $A2$ and for agent $A1$ $\beta = 1.25$. This is necessary because sections $(S_n, n = 1, 2)$ are longer than S_0, which affects the final comparison in choosing the winning action since the W-values are bounded by Q_{max} and Q_{min}. The bounds on the Q-values depend on the reward values r_{min} and r_{max} [43].

5.4.2. DWL4-ST-VSL Parameters

Similarly, to keep the W-values of local policies comparable to the W-values of remote policies in the case of DWL4-ST-VSL, we scale the reward function (7) by a factor $\beta = 0.75$ for the case of agents $(Ai, i = 2, 3, 4)$ and for agent $A1$ $\beta = 1.25$.

6. Simulation Results

The VSL strategies are evaluated using the overall TTS and measured on the entire simulated motorway segment (including ramps). Traffic parameters, average speed and density are measured in the bottleneck area (section S_0). The results presented in Figures 4–7 are from the exploitation phase. We analyzed the specific response behavior of the allocation of dynamic VSL zones compared to the case of static zones. The space-time congestion analysis is used to analyze the spatiotemporal behaviour of dynamic VSL zone allocation and its impact on traffic flow control. To assess the benefits of cooperation between agents using DWL's remote policies, we also evaluate the impact of the cooperation coefficient on agent performance. As a measure of the learning rate of proposed agent-based learning VSL approaches, the convergence curves of overall motorway TTS during the training (learning) process are shown in Figure 8.

It is important to note that the purpose of this study is not to show the extent to which DWL4-ST-VSL can improve traffic, but to investigate how the dynamic (spatiotemporal) adaptation of VSL zone configurations and the increased number of learning agents affect the traffic control optimization problem. Thus, an improvement over baseline should be considered primarily as a comparative measure between two different VSL approaches, the commonly used static VSL zones and the new paradigm with dynamic VSL zone allocation, rather than as an absolute measure of performance.

6.1. Comparison of Dynamic VSL Zone Allocation and Static VSL Zones

Note that the baselines use the best static VSL zone configuration found for a medium traffic load. Using a medium and a high load in our experimental setup, we simulate significant differences in the spatial displacement of the congestion tail. In this way, we illustrate the benefits and necessity of adaptive spatiotemporal VSL control. Different VSL zone configurations per traffic scenario are learned (without requiring manual setup) and dynamically assigned using DWL4-ST-VSL to better respond to spatially propagating traffic congestion. At the same time, the experiment highlighted the weaknesses of the static VSL zone configuration, which performs suboptimally under high traffic load. Therefore, the VSL zones in VSL with the static VSL zone configuration must be manually set up each time the traffic pattern changes, which is not practical.

6.1.1. Medium Traffic Load

The simulations performed show that the best combination for establishing static VSL zones is $S_{2,(2L)} + S_{1,(1R)}$. In this case, VSL is able to control congestion in the case of medium traffic load. In DWL2-ST-VSL, by additionally activating VSL zones within the S_2 section during the highest congestion peak (around $t = 1$ [h]), the agent $A2$ helps its downstream neighbour $A1$, which contributes to an even more effective congestion resolution than the baselines (SPSC and WL-VSL) with static VSL zones. In DWL4-ST-VSL, the agents closest to the congestion ($A1$ and $A2$) are assisted by upstream agents ($A3$ and $A4$) that activate additional VSL zones within S_3 and S_4 just before the highest congestion peak

($t = 1$ [h]) (for a shorter period than DWL2-ST-VSL). In this way, agents $A3$ and $A4$ help their downstream neighbours. Similar results to those found for DWL2-ST-VSL were observed.

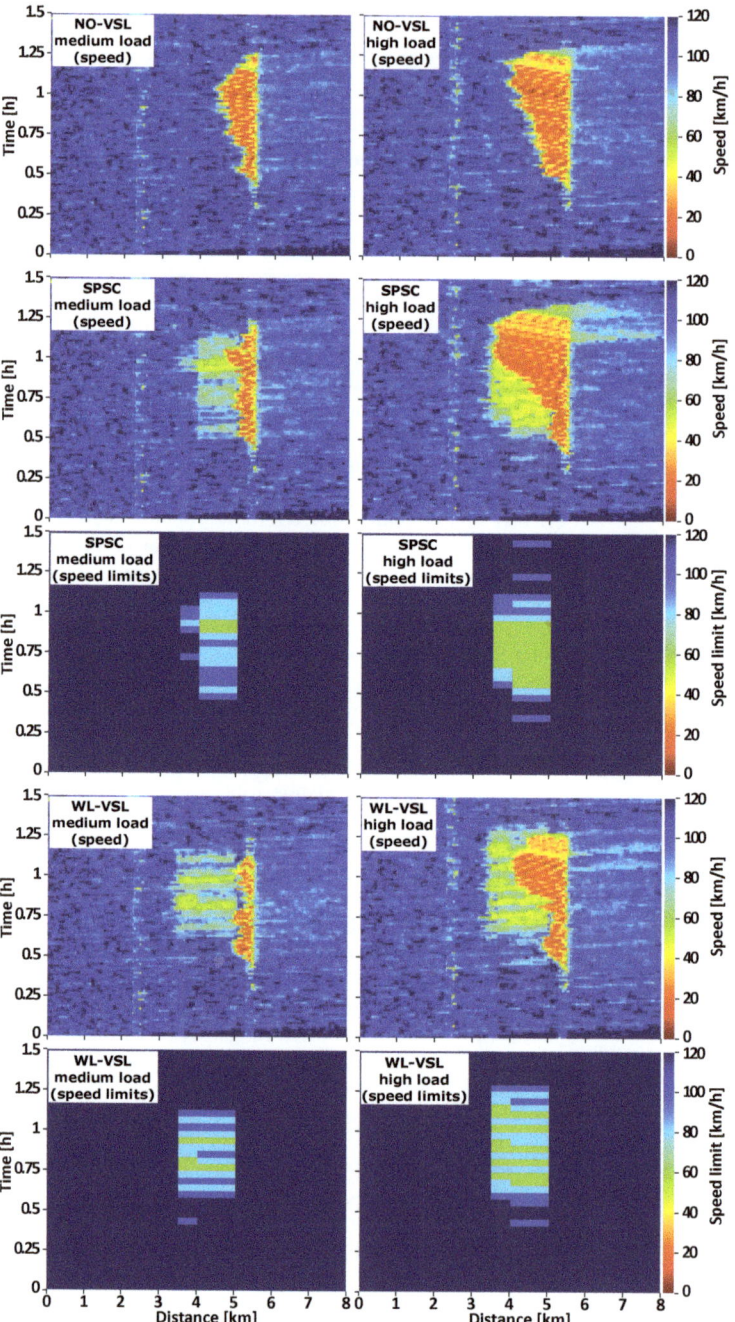

Figure 4. Space-time diagrams for simulated scenarios with static VSL zones.

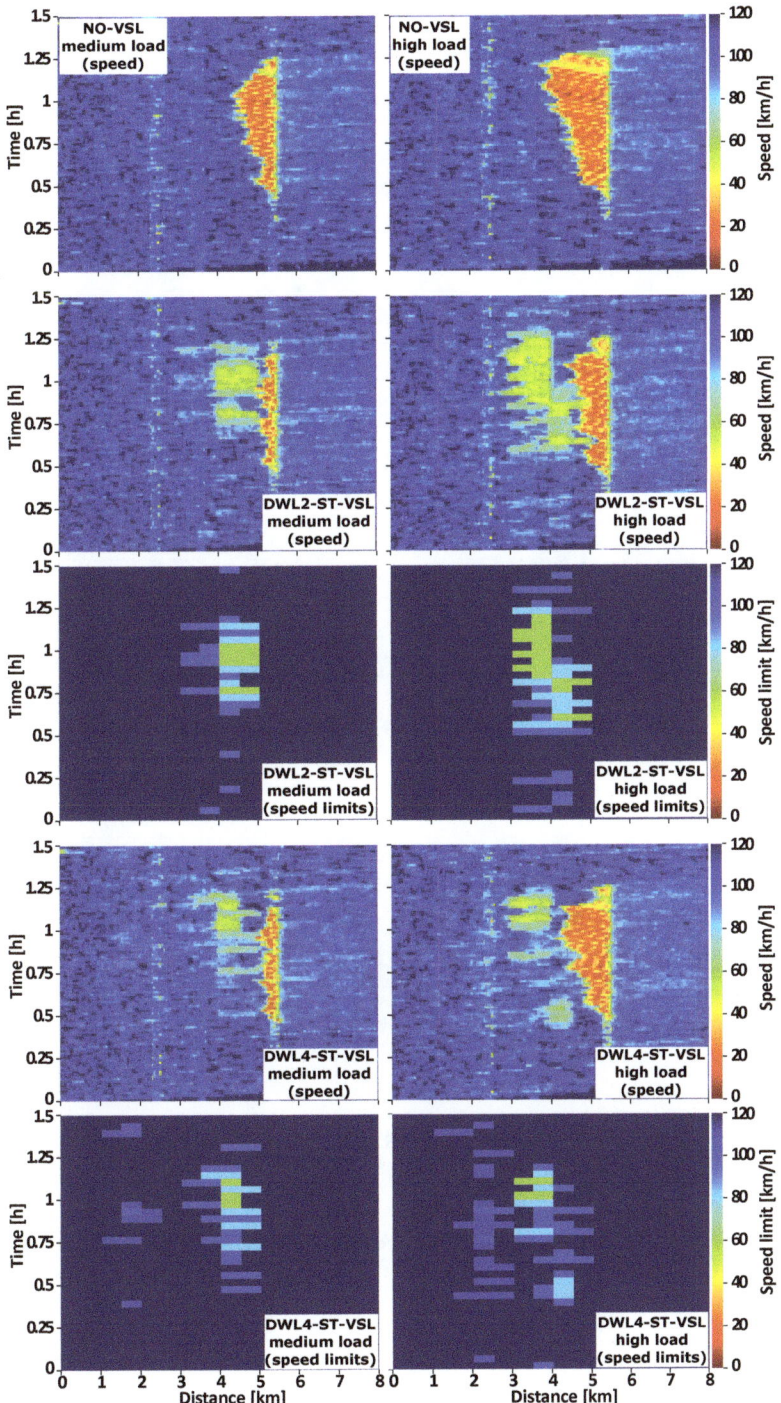

Figure 5. Space-time diagrams for simulated scenarios with multi-agent dynamic VSL zones.

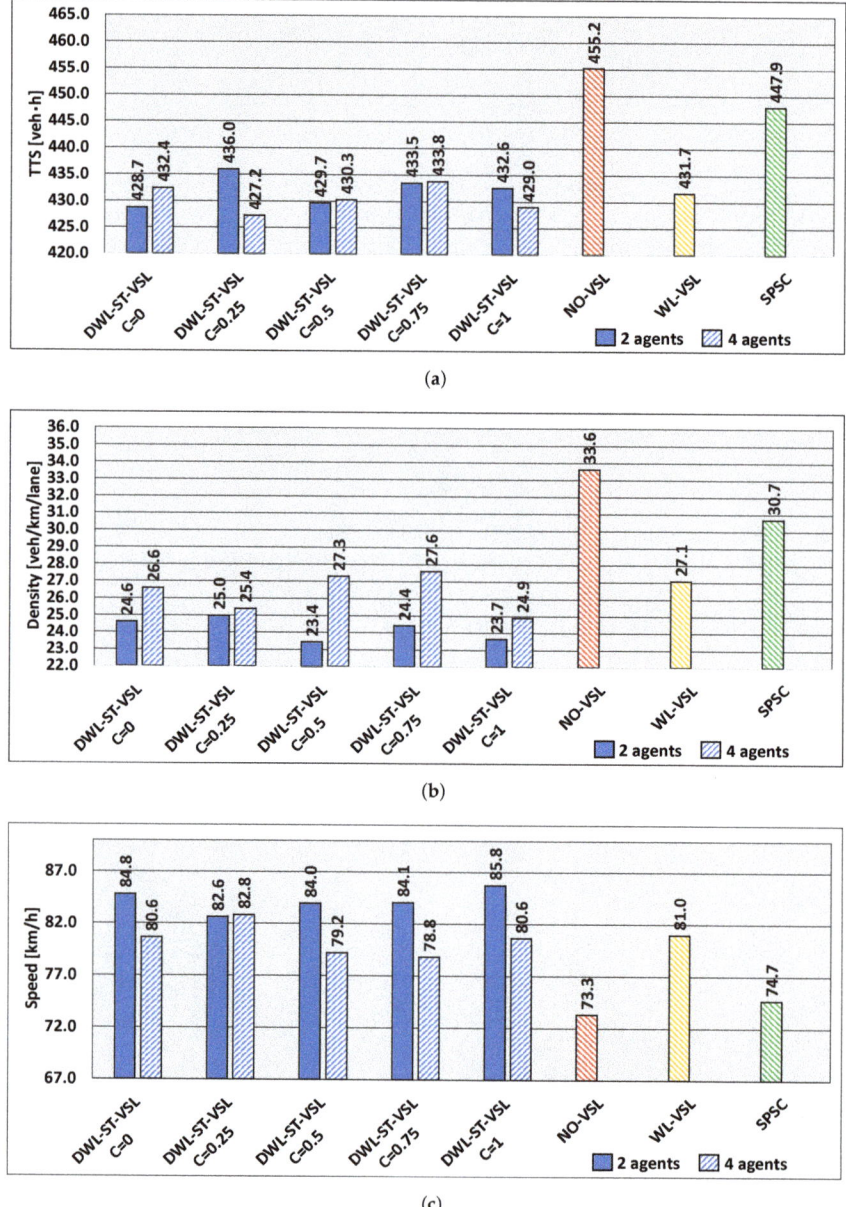

Figure 6. Traffic parameters for different levels of cooperation for the medium traffic load scenario. (**a**) TTS in the overall network. (**b**) Average traffic density in section S_0. (**c**) Average vehicle speed in section S_0.

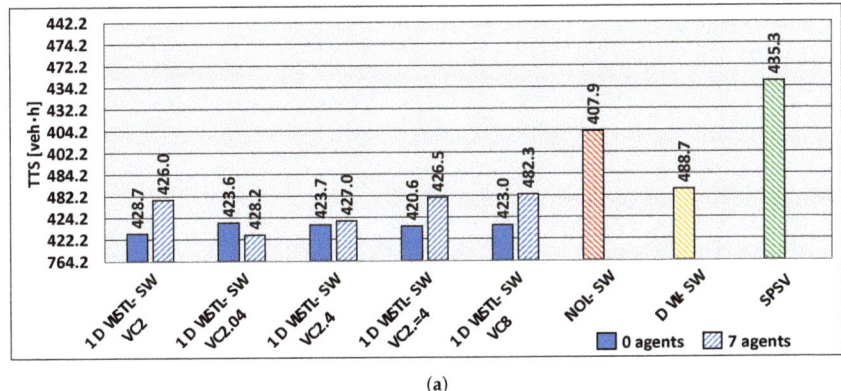

(a)

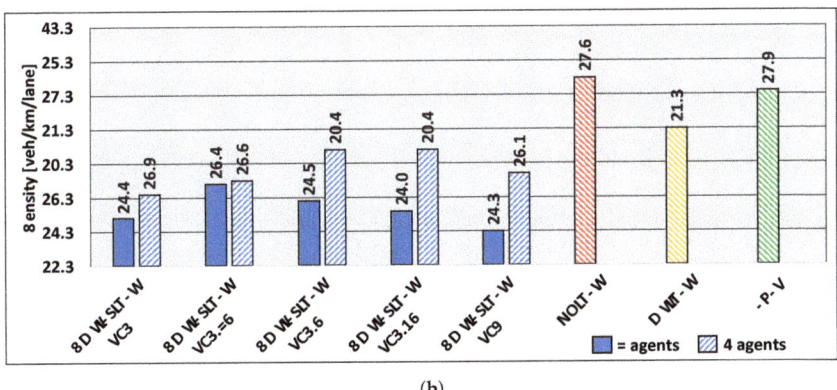

(b)

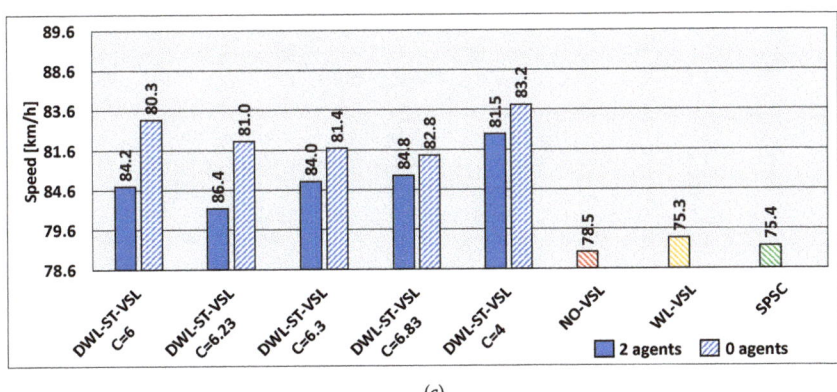

(c)

Figure 7. Traffic parameters for different levels of cooperation for the high traffic load scenario. (**a**) TTS in the overall network. (**b**) Average traffic density in section S_0. (**c**) Average vehicle speed in section S_0.

6.1.2. High Traffic Load

The performed simulations indicate that the static VSL zones perform suboptimally in a high traffic scenario. By applying different VSL zone configurations during the simulation within $S_n, n = 1, 2$ by DWL2-ST-VSL, and within sections $S_n, n = 1, 2, 3, 4$ in the case of DWL4-ST-VSL, they contribute more notably to congestion clearing than baselines, which results from the gradual adjustment of the VSL application area. In the DWL2-ST-VSL case, agents started with stronger activation of the speed limits and VSL zones in section S_1 at the beginning of the congestion. Over time, the congestion starts to propagate upstream through the motorway. The agents begin to use the VSL zones principally in sections S_1 and S_2, while finally, for the highest congestion peak, the VSL zones are primarily activated in section S_2.

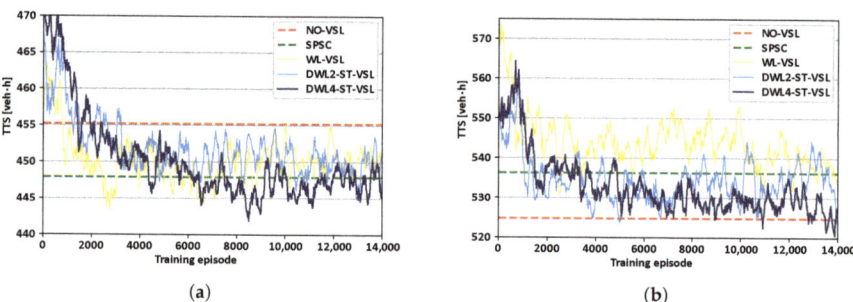

Figure 8. The convergence of TTS during the training process. (**a**) Medium traffic load scenario. (**b**) High traffic load scenario.

In the case of DWL4-ST-VSL, VSL zones are activated mainly in all VSL sections at the onset of congestion (somewhat more sparsely for agent $A3$, while agent $A4$ was almost not activated at all). Agents $A1$ and $A2$ preferred a shorter VSL zone configuration, while $A3$ preferred a longer one. The application of shorter VSL zones in the downstream sections S_1 and S_2 could be due to the additional support provided by the upstream agents, particularly the speed limits applied by agent $A3$, which reduced the need for longer VSL zones and sudden decreases of the speed limit. As congestion increases, it can be seen in Figure 5 that the area of inactive VSL zones increases between upstream and downstream sections, primarily due to the use of shorter VSL zones by agent $A3$ and sparsely activated VSL zones by $A2$. After $t = 0.75$ [h], agent $A2$ starts applying speed limits again in response to the sudden increase of the queue ahead of the bottleneck (faster propagation of the congestion upstream through the motorway). As congestion intensity approaches its peak, agent $A2$ promotes a longer VSL zone, including lower speed limits. Agent $A1$ is mostly inactive during this time period, thus forming an additional valuable transition zone [26] between the active VSL application area and the congestion tail. A somewhat unexpected behavior during the highest congestion peak is observed for agent $A4$, which did not apply speed limits below 120 [km/h] while $A3$ was not active for 3 control steps (Figure 5). In the next section, we will make some arguments that we believe can help explain this unexpected agent behavior.

Nevertheless, both DWL2-ST-VSL and DWL4-ST-VSL adjusted the VSL zones to the spatially moving tail of the resulting congestion. This control strategy is more pronounced in the case of the high congestion scenario, in which agents attempt to create an additional artificial moving bottleneck to reduce the outflow from it and, thus, relieve the congested area. From Figure 5, it can be seen that the agents aim to create such a VSL configuration that ensures the additional space (without speed limit) between the VSL zones and the congested tail. This can be viewed as an acceleration zone after the VSL zone, allowing vehicles to accelerate to the critical speed (at which capacity is reached) before entering the congested tail, as indicated in [25]. This feature of DWL-ST-VSL is very useful compared to the static VSL zone (fixed configuration) and confirms the findings that the higher the speed limit, the farther the VSL application zone should be from the bottleneck, which has been recently proven analytically in [26].

6.2. Space-Time Congestion Analysis

Space-time diagrams are interesting for visualizing how traffic conditions evolve along the observed motorway segment. The on-ramp R_0 in S_0 is located at $x = 5.3$ [km].

DWL2-ST-VSL ranges from $x = 3$ to $x = 5$ [km], while DWL4-ST-VSL ranges from $x = 1$ to $x = 5$ [km]. The best configuration of the static VSL zones (WL-VSL, SPSC) ranges from $x = 3.5$ to $x = 5$ [km]. The initial transition area [26] after the VSL zone starts at $x = 5$ [km] to the on-ramp R_0 and can be changed if the configuration of the VSL zones changes during agents' operations in DWL-ST-VSL (in particular $A1$ and $A2$).

6.2.1. Medium Traffic Load

In Figures 4 and 5, the mixed shades of red and orange correspond to congestion where vehicles are traveling at low speeds. The patterns of red stripes represent the propagation of the shock wave upstream through the motorway. Congestion begins at about $t = 0.4$ [h] in the bottleneck area and propagates upstream. After the demand on the on-ramp R_0 decreases, the congestion decreases and finally dissipates at $t = 1.25$ [h].

In both DWL-ST-VSL control strategies, the congestion (red) area is much smaller than in the baseline cases. The mixed shades of yellow-green-light blue in front of the congestion area correspond to the speed of vehicles obeying the speed limits (60–100 [km/h]) within active VSL zones. Such an artificially generated moving bottleneck (adaptive VSL area) with a significantly higher average travelling speed than the one measured in the congestion area still reduces inflow into the congestion area, which helps to resolve congestion more efficiently than baselines. In response to spatially varying congestion, both DWL-ST-VSL produce more stable downstream flow than the best baselines with static VSL zones. In the medium load scenario, congestion propagates upstream from the bottleneck to location $x = 4.4$ [km]. In the case of DWL2-ST-VSL and DWL4-ST-VSL, the propagation is reduced to $x = 5$ [km], which is an improvement of 66.7% compared to NO-VSL. Finally, the average density in the congested area (bottleneck S_0 and directly affected upstream section S_1) is reduced from 26.0 in NO-VSL to 20.0 [veh/km/lane] in the case of DWL2-ST-VSL, an improvement of 23.1%. The improvement for simulated DWL4-ST-VSL is 20.8%.

6.2.2. High Traffic Load

Again, both DWL-ST-VSL versions win the competition. For DWL2-ST-VSL and DWL4-ST-VSL, the congestion area is smaller than for baselines. During the simulated scenario, different combinations of VSL zones were applied to respond to the changing congestion intensities and moving congestion tail. In this way, DWL2-ST-VSL and DWL4-ST-VSL are able to reduce the congestion area much more effectively than the baselines with static VSL zones. In the case of NO-VSL for the high-load scenario, the congestion spreads upstream from the bottleneck to the location $x = 3.8$ [km]. When DWL2-ST-VSL is applied, the propagation is reduced to near $x = 4.4$ [km], an improvement of 40%. Using the extended version with four agents (DWL4-ST-VSL), propagation is reduced to about $x = 4.2$ [km], an improvement of 26.7%. Finally, the average traffic density in the congested area (S_0 and S_1) is reduced from the original 34.1 to 28.7 [veh/km/lane] by using DWL2-ST-VSL, an improvement of 15.8%. In the case of DWL4-ST-VSL, the improvement achieved is 14.7%. Just for comparison, in the case of WL-VSL with static zones, the congestion propagates near $x = 4$ [km], resulting in negligible improvement. A similar behavior is observed in the case of SPSC, eventually degrading the system performance.

6.3. Level of Cooperation Analysis

To evaluate the benefits of cooperation between agents using the DWL's concept of remote policies, we also assess the effects of the cooperation coefficient on agent performance. The effects of different levels of agent collaboration on system performance are presented in Figures 6 and 7. The analysis was performed for medium and high traffic loads (Figure 3).

6.3.1. Medium Traffic Load

It can be seen that all DWL-based approaches outperform the baselines used in our experiment. The lowest TTS value is obtained with DWL4-ST-VSL and is 427.2 [veh·h] for $C = 0.25$. Compared to the NO-VSL case, ($TTS = 455.2$ [veh·h]), a reduction of 6.2% (Figure 6a). The best density is 23.4 [veh/km/lane] for $C = 0.5$ in the case of DWL2-ST-VSL, while it is 33.6 for the case of NO-VSL, an improvement of 30.4% (Figure 6b). In particular, the average vehicle speed for $C = 1$ in the case of DWL2-ST-VSL is 85.8 [km/h], while the speed in the case of NO-VSL is 73.3 [km/h], an improvement of 17.1% (Figure 6c).

6.3.2. High Traffic Load

Similar results were obtained in the high traffic load experiment, where both DWL-ST-VSL configurations outperform the baseline controllers. The lowest TTS value in the cooperative agent case in DWL4-ST-VSL is 501.0 [veh·h] for $C = 0.25$. Compared to the NO-VSL case, ($TTS = 524.8$ [veh·h]); this is an improvement of 4.5% (Figure 7a). The density is 34.0 [veh/km/lane] for DWL2-ST-VSL ($C = 1$), while in the case of NO-VSL it is 38.5, a reduction of 11.7%. The density is reduced by 7.8% by using DWL4-ST-VSL (Figure 7b). In particular, in the case of DWL2-ST-VSL, the average vehicle speed for $C = 1$ is 73.8 [km/h], while in the case of NO-VSL the speed is 67.8 [km/h], an improvement of 8.8% (Figure 7c). In the case of DWL4-ST-VSL, the average speed is 10.9% higher (for $C = 1$).

6.4. Convergence of TTS during the Training Process

A comparison of the convergence of TTS measured per training episode (episode ≡ one simulation) during the learning process is shown in Figure 8. The graphs are created using the moving average over 10 episodes, while TTS was measured in the entire motorway network (including all on- and off-ramps). At the beginning of the learning process, all agent-based VSL approaches performed inferiorly compared to NO-VSL, since agents explore the environment by executing random actions with high probability. As simulations progress, the number of random actions taken reduces, and the exploitation of learned experiences increases. Consequently, TTS decreases, indicating progress in learning. Due to the different complexities of proposed RL-based multi-agent VSL controllers, the different decrease rate of TTS can be observed throughout the learning process. From Figure 8a, it can be seen that all approaches have stable decreasing learning curves; generally, DWL4-ST-VSL leads in TTS reduction over other strategies in the medium traffic scenario.

For the high traffic scenario (Figure 8b), the static VSL zones used in WL-VSL are prone to performing poorly compared to the dynamic VSL zones. Cases with dynamic VSL zone allocation via DWL2-ST-VSL and DWL4-ST-VSL need a higher number of training episodes to approach lower TTS values. As the learning process approaches 14,000 episodes, TTS in the case of DWL2-ST-VSL and DWL4-ST-VSL converges moderately towards and below the TTS value obtained in NO-VSL. Eventually, compared with the starting values, the overall TTS is gradually improved for all agent-based VSL strategies, favouring the learning rate of DWL4-ST-VSL in both traffic scenarios.

In the case of the high traffic scenario (Figure 8b), it can be seen that DWL4-ST-VSL needs a slightly longer time, i.e., higher number of training episodes (around 11,000) to reduce TTS below the value obtained by NO-VSL. Nevertheless, when converted in real-time, it takes roughly 90 [h] of training in a simulator (on an Intel(R) Core(TM) i7-10750H CPU processor). In case our simulated experiment represents actual recurrent traffic congestion observed online, DWL4-ST-VSL can be trained offline (on simulations) and deployed in a real application in a short period. Thus, DWL4-ST-VSL can be retrained offline to deal with traffic changes in the operating environment to ensure good performance in the newly observed traffic scenarios (similar to the continuous learning scheme for Q-Learning based VSL suggested in [17]).

The longer time needed for reaching the favorable level may be directly linked to the larger number of agents. They eventually need more training episodes to become aware of the interference they cause by their actions on their immediate neighbours and the controlled motorway system as a whole.

In the second half of the learning process, there are more pronounced oscillations in TTS. The possible contribution to this might be delaying W's convergence until Q is well known (see Section 5.4). Thus, W-values are more altered as Q-values are more learned. Consequently, this influences the policies' nomination (3) in the DWL process and eventually influences the cooperation strategies between agents. As a result, it might cause a change in a learned set of optimal policies, thus resulting in the different system responses during the second half of the learning process. The new policy can induce new rarely seen system states that have not been encountered before, thus affecting agents' poor decisions. Nevertheless, the function approximation techniques can address this problem by ensuring better generalization (reasonable outputs) for rarely seen states, thus stabilizing the training (learning) process.

7. Discussion

The outermost agents ($A3$ and $A4$) do not perceive congestion directly and, therefore, tend to exploit local stable traffic conditions by promoting higher speed limits and, in particular, favoring their local policy LP_{i1}. As a result, for small values of the cooperation coefficient C, they do not fully contribute to helping downstream agents to eliminate the congestion. This raises the question of whether C should be scaled differently depending on the spatial location of the agents rather than using uniformly distributed equal values for all agents. It might make sense to increase the coefficients of C the farther agents are from the location of the bottleneck so that they are more sensitive to the preferences of downstream agents and, therefore, give more priority to remote policies in the case of active congestion. The question then arises: to what extent?

The converse is also true, since the actions of the downstream agents affect the state variables (in particular, the measured average vehicle speed) of the upstream agents. The upstream agents always observe the average speed in their immediate downstream area (in the case of local policy LP_{i1}) and, possibly, the actions performed by the downstream agents (lower speed limits) reduce the chance of winning the LP_{i1}; therefore, a penalty by the measured TTS is more likely, even if the local environment is in free-flow conditions. This dependence is implicitly communicated to the downstream neighbouring agent Aj in the form of a higher W-value for remote policy RP_{ji1}, which complements the local policy LP_{i1} of the upstream agents Ai.

The above observation shows the possible trade-off in choosing optimal values for C. A feasible solution to make Cs adaptive is to use a scaling scheme used in (2) "learning Q (somewhat) before learning W" [43]. In this scheme, the updates of W-values are weighted differently. The weighting is higher when an agent is sure of what it is doing in a given state. Given that the underlying DWL process (WL algorithm) is considered as a "fair" resolution of competition, this leads to the question: can the W-values of local policies, together with the probability of nominating a particular action in a given state, be communicated between neighbors and used as input for computing C? This may trigger further research on adaptive cooperation coefficient C.

Furthermore, the overlapping states of the environment, including the downstream neighbourhood (see Figure 2), has positive and negative effects on the agent's learning behavior. The negative effect arises from the nonstationarity caused by the neighbours' actions, resulting in a moving learning target (particularly during the exploration phase in the training phase) since agents are learning simultaneously. Thus, each time, Ai's policy changes might cause other agents' policies to change, too [50]. The positive effect is the agent's ability to detect and respond to the early impulse of congestion in downstream traffic. All learning-based approaches were trained with the same number of simulations. However, due to nonstationarity, DWL4-ST-VSL may require more simulations to converge to better control policies for a given traffic scenario due to a higher number of agents. Therefore, DWL4-ST-VSL (and the final results) may be in a slightly unfavorable position compared to DWL2-ST-VSL.

In our experiments, we assumed that all measurements (traffic data) in our experiments are perfect. In reality, sensors are not ideal, and raw data needs to be analyzed and filtered before being used for traffic state estimation. Thus, accurate traffic states are important for real-time traffic control. Raw traffic flow data collected from sensors might be contaminated by different noises caused by the imperfection or damage of sensors. In [51], the authors introduced data denoising schemes to suppress the potential data outliers from raw traffic data for accurate traffic state estimation and prediction. This presents an open question for further research.

Additionally, the efficiency of DWL-ST-VSL is highly dependent on the learning process performed in traffic simulations. Since simulations themselves depend on the given initial parameters, not all possible relevant traffic conditions can be covered. A possible direction to improve the training process of DWL-ST-VSL by ensuring that all relevant traffic scenarios are covered is to use the idea of structured simulations. Originally proposed in [52], structured simulations are intended for testing the behavior of complex adaptive systems in general by changing the inputs into the simulations in a structured way. Such a framework might augment existing traffic scenarios (real or synthetic) with unprecedented scenarios that evoke or replicate important aspects of real traffic, such as rarely seen traffic states in which VSL agents performed poorly. Thus, a structured simulations approach can enrich the training data set and consequently minimize unexpected behavior of the RL-based VSL controller in practice.

Even under a medium load scenario, the resulting congestion on the motorway can be classified as a serious traffic problem. However, it has been shown that DWL2-ST-VSL and DWL4-ST-VSL can effectively resolve the congestion in this scenario due to their added ability to dynamically adjust the VSL zone configurations. Since the DWL agents could not fully handle the congestion in the high load scenario (even when using four agents), it might be useful to extend the DWL4-ST-VSL control, e.g., by integrating it with the merge control using the DWL multi-agent framework.

Experimental results confirmed the usefulness of using dynamic VSL zone allocation (the capability to adapt the VSL application area) while optimizing speed limits in traffic conditions with varying congestion. Similarly, in [42], a VSL strategy able to adjust each control cycle's length (duration) online, given the changes in traffic conditions, was shown to be superior compared to a fixed cycle length. Thus, integrating dynamic VSL zone allocation and dynamic control cycles can make VSL more adaptive, making VSL's performance more robust when operating in a nonstationary environment like a motorway. To accomplish the full benefits of adaptivity, the principal time constants of the system should be long enough for the system to ignore false disturbances and yet short enough to respond to indicative changes in the environment (the "stability-plasticity dilemma") [53]. Therefore, further research in this direction is desirable in DWL-ST-VSL.

The VSL control approaches with static VSL zone configuration performed poorer in high traffic scenarios than those with dynamic VSL zone allocation. Thus, results strongly indicate the need for the adaptive speed limit system in speed limit, length and position of VSL zones to efficiently cope with the unpredictable spatio-temporal varying congestion, which is more likely to be the case in a real traffic scenario.

8. Conclusions

This paper presented DWL-ST-VSL, a multi-agent RL-based VSL control approach for the dynamic adjustment of VSL zones and speed limits. In addition, an extended version, DWL4-ST-VSL, was analyzed for an urban motorway simulation scenario where four agents learn to jointly control four segments ahead of a congested area using the DWL algorithm on a longer motorway segment. The simulations show that DWL4-ST-VSL and the two-agent based DWL2-ST-VSL consistently perform better than our baseline solutions, WL-VSL and SPSC. The results do not differ significantly between DWL2-ST-VSL and DWL4-ST-VSL in terms of bottleneck parameters. In terms of system travel time, DWL4-ST-VSL gives better results. VSL control is improved by simultaneously adjusting speed limit values and VSL zone configuration in response to spatiotemporal changes in congestion intensity and the congestion's moving tail. In addition, performance is improved by DWL's ability to implement multiple different policies simultaneously and to use two sets of actions with different speed limit granularity, as well as to enable collaboration between agents implementing remote policies.

However, the efficiency of DWL-ST-VSL is highly dependent on the training process performed in simulations. To train DWL-ST-VSL in a structured way and ensure that all relevant traffic simulation scenarios are covered, we will use the structured simulations mentioned in the discussion. Using structured simulations and the nonlinear function approximation technique for better generalization together with sensitivity analysis of hyperparameters in DWL may reduce the poor performance of DWL in a nonstationary motorway environment, thus fostering DWL-ST-VSL to be closer to testing in reality. Eventually, this will enable the systematic evaluation of adaptive DWL-ST-VSL control.

Additionally, the results suggest that there may be multiple local optima for different coefficients of cooperation, which requires further analysis. How resilient the learning system would be to the loss of information exchange if one or more agents failed, which is often the case in a real scenario where sensors and equipment are imperfect and may break down, highlights the open research directions. We will consider implementing additional degrees of freedom to allow each agent of DWL-ST-VSL to adjust the length and position of the VSL zone in both directions, considering the constraints on the spatial difference of the speed limit between two adjacent VSL zones. Finally, we will consider integrating DWL-ST-VSL with dynamic control cycles and merge control, as this could further advance the VSL system toward instantaneous vehicle speed control in the presence of emerging vehicle-to-infrastructure technologies and traffic control on motorways in general.

Author Contributions: The conceptualization of this study was done by K.K. and E.I. Both also did the funding acquisition. The development of the control algorithm was done by K.K., E.I., and I.D. The writing of the original draft and preparation of the paper was done by K.K. and E.I. The supervision was done by E.I. and I.D. Visualizations were done by K.K. Preparation of the simulation models and simulation analysis was done by F.V. and M.G. All authors contributed to the writing review and final editing. All authors have read and agreed to the published version of the manuscript.

Funding: This work has been partly supported by the Science Foundation of the Faculty of Transport and Traffic Sciences under the project ZZFPZ-P1-2020 "Control system of the spatial-temporal variable speed limit in the environment of connected vehicles", the Croatian Science Foundation under the project IP-2020-02-5042, and the European Regional Development Fund under the grant KK.01.1.1.01.0009 (DATACROSS).

Institutional Review Board Statement: Not applicable.

Informed Consent Statement: Not applicable.

Data Availability Statement: Not applicable.

Acknowledgments: This research has also been carried out within the activities of the Centre of Research Excellence for Data Science and Cooperative Systems supported by the Ministry of Science and Education of the Republic of Croatia.

Conflicts of Interest: The authors declare no conflict of interest.

Abbreviations

The following abbreviations are used in this manuscript:

DWL	Distributed W-learning
DWL-ST-VSL	Distributed spatial-temporal multi-agent VSL
DWL2-ST-VSL	DWL-ST-VSL configuration with two agents
DWL4-ST-VSL	DWL-ST-VSL configuration with four agents
ISA	Intelligent speed assistance
MDPs	Markov decision processes
MTFC	Mainstream traffic flow control
NO-VSL	No control
RL	Reinforcement learning
RL-VSL	Reinforcement learning-based variable speed limit
TTS	Total time spent
SPSC	Simple proportional speed controller
SUMO	Simulation of urban mobility
VMS	Variable message sign
VSL	Variable speed limit
WL	W-learning
WL-VSL	W-learning VSL

References

1. Khondaker, B.; Kattan, L. Variable speed limit: An overview. *Transp. Lett.* **2015**, *7*, 264–278. [CrossRef]
2. Strömgren, P.; Lind, G. Harmonization with Variable Speed Limits on Motorways. *Transp. Res. Procedia* **2016**, *15*, 664–675. [CrossRef]
3. Carlson, R.C.; Papamichail, I.; Papageorgiou, M. Comparison of local feedback controllers for the mainstream traffic flow on freeways using variable speed limits. In Proceedings of the 2011 14th International IEEE Conference on Intelligent Transportation Systems (ITSC), Washington, DC, USA, 5–7 October 2011; pp. 2160–2167.
4. Shao-long, G.; Jun, M.; Jun-li, W.; Xiao-qing, S.; Yan, L. Methodology for Variable Speed Limit Activation in Active Traffic Management. *Procedia Soc. Behav. Sci.* **2013**, *96*, 2129–2137. [CrossRef]
5. Li, D.; Ranjitkar, P. A fuzzy logic-based variable speed limit controller. *J. Adv. Transp.* **2015**, *49*, 913–927. [CrossRef]
6. Li, D.; Ranjitkar, P.; Zhao, Y. Mitigating Recurrent Congestion via Particle Swarm Optimization Variable Speed Limit Controllers. *KSCE J. Civ. Eng.* **2019**, *23*, 3174–3179. [CrossRef]
7. Como, G.; Lovisari, E.; Savla, K. Convexity and robustness of dynamic traffic assignment and freeway network control. *Transp. Res. Part B Methodol.* **2016**, *91*, 446–465. [CrossRef]
8. Lu, X.Y.; Varaiya, P.; Horowitz, R.; Su, D.; Shladover, S.E. A new approach for combined freeway variable speed limits and coordinated ramp metering. In Proceedings of the IEEE Conference on Intelligent Transportation Systems, Proceedings, ITSC, Funchal, Portugal, 19–22 September 2010; pp. 491–498.
9. Zhang, Y.; Sirmatel, I.I.; Alasiri, F.; Ioannou, P.A.; Geroliminis, N. Comparison of Feedback Linearization and Model Predictive Techniques for Variable Speed Limit Control. In Proceedings of the 2018 21st International Conference on Intelligent Transportation Systems (ITSC), Maui, HI, USA, 4–7 November 2018.
10. Kušić, K.; Ivanjko, E.; Gregurić, M.; Miletić, M. An Overview of Reinforcement Learning Methods for Variable Speed Limit Control. *Appl. Sci.* **2020**, *10*, 4917. [CrossRef]
11. LA, P.; Bhatnagar, S. Reinforcement Learning With Function Approximation for Traffic Signal Control. *IEEE Trans. Intell. Transp. Syst.* **2011**, *12*, 412–421. [CrossRef]
12. Lu, C.; Huang, J.; Gong, J. Reinforcement Learning for Ramp Control: An Analysis of Learning Parameters. *PROMET Traffic Transp.* **2016**, *28*, 371–381. [CrossRef]
13. Gong, I.; Oh, S.; Min, Y. Train Scheduling with Deep Q-Network: A Feasibility Test. *Appl. Sci.* **2020**, *10*, 8367. [CrossRef]
14. Gueriau, M.; Cugurullo, F.; Acheampong, R.A.; Dusparic, I. Shared Autonomous Mobility on Demand: A Learning-Based Approach and Its Performance in the Presence of Traffic Congestion. *IEEE Intell. Transp. Syst. Mag.* **2020**, *12*, 208–218. [CrossRef]
15. Gosavi, A. *Parametric Optimization Techniques and Reinforcement Learning*, 2nd ed.; Springer: New York, NY, USA, 2015.
16. Zhu, F.; Ukkusuri, S.V. Accounting for dynamic speed limit control in a stochastic traffic environment: A reinforcement learning approach. *Transp. Res. Part C Emerg. Technol.* **2014**, *41*, 30–47. [CrossRef]
17. Li, Z.; Liu, P.; Xu, C.; Duan, H.; Wang, W. Reinforcement Learning-Based Variable Speed Limit Control Strategy to Reduce Traffic Congestion at Freeway Recurrent Bottlenecks. *IEEE Trans. Intell. Transp. Syst.* **2017**, *18*, 3204–3217. [CrossRef]
18. Walraven, E.; Spaan, M.T.; Bakker, B. Traffic flow optimization: A reinforcement learning approach. *Eng. Appl. Artif. Intell.* **2016**, *52*, 203–212. [CrossRef]
19. Zhou, W.; Yang, M.; Lee, M.; Zhang, L. Q-Learning-Based Coordinated Variable Speed Limit and Hard Shoulder Running Control Strategy to Reduce Travel Time at Freeway Corridor. *Transp. Res. Rec. J. Transp. Res. Board* **2020**, *2674*, 915–925. [CrossRef]
20. Gregurić, M.; Kušić, K.; Vrbanić, F.; Ivanjko, E. Variable Speed Limit Control Based on Deep Reinforcement Learning: A Possible Implementation. In Proceedings of the 2020 International Symposium ELMAR, Zadar, Croatia, 14–15 September 2020.

21. Schmidt-Dumont, T.; van Vuuren, J.H. A case for the adoption of decentralised reinforcement learning for the control of traffic flow on South African highways. *J. S. Afr. Inst. Civ. Eng.* **2019**, *61*, 7–19. [CrossRef]
22. Wang, C.; Zhang, J.; Xu, L.; Li, L.; Ran, B. A New Solution for Freeway Congestion: Cooperative Speed Limit Control Using Distributed Reinforcement Learning. *IEEE Access* **2019**, *7*, 41947–41957. [CrossRef]
23. Kušić, K.; Ivanjko, E.; Vrbanić, F.; Gregurić, M.; Dusparic, I. Dynamic Variable Speed Limit Zones Allocation Using Distributed Multi-Agent Reinforcement Learning. In Proceedings of the 2021 IEEE 24th International Conference on Intelligent Transportation Systems (ITSC), Indianapolis, IN, USA, 19–22 September 2021; pp. 1–8.
24. Kušić, K.; Dusparic, I.; Guériau, M.; Gregurić, M.; Ivanjko, E. Extended Variable Speed Limit control using Multi-agent Reinforcement Learning. In Proceedings of the 2020 IEEE 23rd International Conference on Intelligent Transportation Systems (ITSC), Rhodes, Greece, 20–23 September 2020; pp. 1–8.
25. Müller, E.R.; Carlson, R.C.; Kraus, W.; Papageorgiou, M. Microsimulation Analysis of Practical Aspects of Traffic Control With Variable Speed Limits. *IEEE Trans. Intell. Transp. Syst.* **2015**, *16*, 512–523. [CrossRef]
26. Martínez, I.; Jin, W.L. Optimal location problem for variable speed limit application areas. *Transp. Res. Part B Methodol.* **2020**, *138*, 221–246. [CrossRef]
27. Lai, F.; Carsten, O.; Tate, F. How much benefit does Intelligent Speed Adaptation deliver: An analysis of its potential contribution to safety and environment. *Accid. Anal. Prev.* **2012**, *48*, 63–72. [CrossRef] [PubMed]
28. Dusparic, I.; Cahill, V. Distributed W-Learning: Multi-Policy Optimization in Self-Organizing Systems. In Proceedings of the 2009 Third IEEE International Conference on Self-Adaptive and Self-Organizing Systems, San Francisco, CA, USA, 14–18 September 2009; pp. 20–29.
29. Lopez, P.A.; Behrisch, M.; Bieker-Walz, L.; Erdmann, J.; Flötteröd, Y.P.; Hilbrich, R.; Lücken, L.; Rummel, J.; Wagner, P.; Wießner, E. Microscopic Traffic Simulation using SUMO. In Proceedings of the 21st IEEE International Conference on Intelligent Transportation Systems, Maui, HI, USA, 4–7 November 2018.
30. Wang, Y. Dynamic Variable Speed Limit Control: Design, Analysis and Benefits. Ph.D. Thesis, University of Southern California, Los Angeles, CA, USA, 2011.
31. Chung, K.; Rudjanakanoknad, J.; Cassidy, M.J. Relation between traffic density and capacity drop at three freeway bottlenecks. *Transp. Res. Part B Methodol.* **2007**, *41*, 82–95. [CrossRef]
32. Papageorgiou, M.; Kosmatopoulos, E.; Papamichail, I. Effects of Variable Speed Limits on Motorway Traffic Flow. *Transp. Res. Rec. J. Transp. Res. Board* **2008**, *2047*, 37–48. [CrossRef]
33. Soriguera, F.; Martínez, I.; Sala, M.; Menéndez, M. Effects of low speed limits on freeway traffic flow. *Transp. Res. Part C Emerg. Technol.* **2017**, *77*, 257–274. [CrossRef]
34. Grumert, E.; Tapani, A.; Ma, X. Characteristics of variable speed limit systems. *Eur. Transp. Res. Rev.* **2018**, *10*, 21. [CrossRef]
35. Gao, C.; Xu, J.; Li, Q.; Yang, J. The Effect of Posted Speed Limit on the Dispersion of Traffic Flow Speed. *Sustainability* **2019**, *11*, 3594. [CrossRef]
36. van den Hoogen, E.; Smulders, S. Control by variable speed signs: Results of the Dutch experiment. In Proceedings of the Seventh International Conference on Road Traffic Monitoring and Control, London, UK, 26–28 April 1994; pp. 145–149.
37. Yang, Y.; Yuan, Z.Z.; Sun, D.Y.; Wen, X.L. Analysis of the factors influencing highway crash risk in different regional types based on improved Apriori algorithm. *Adv. Transp. Stud.* **2019**, *49*, 165–178.
38. Hegyi, A.; Hoogendoorn, S.P.; Schreuder, M.; Stoelhorst, H.; Viti, F. SPECIALIST: A dynamic speed limit control algorithm based on shock wave theory. In Proceedings of the 2008 11th International IEEE Conference on Intelligent Transportation Systems, Beijing, China, 12–15 October 2008; pp. 827–832.
39. Carlson, R.; Papamichail, I.; Papageorgiou, M. Local Feedback-Based Mainstream Traffic Flow Control on Motorways Using Variable Speed Limits. *Intell. Transp. Syst. IEEE Trans.* **2011**, *12*, 1261–1276. [CrossRef]
40. Kušić, K.; Ivanjko, E.; Gregurić, M. A Comparison of Different State Representations for Reinforcement Learning Based Variable Speed Limit Control. In Proceedings of the MED 2018—26th Mediterranean Conference on Control and Automation, Zadar, Croatia, 19–22 June 2018; pp. 266–271.
41. Vinitsky, E.; Parvate, K.; Kreidieh, A.; Wu, C.; Bayen, A. Lagrangian Control through Deep-RL: Applications to Bottleneck Decongestion. In Proceedings of the 2018 21st International Conference on Intelligent Transportation Systems (ITSC), Maui, HI, USA, 4–7 November 2018; pp. 759–765.
42. Zhang, Y.; Ma, M.; Liang, S. Dynamic Control Cycle Speed Limit Strategy for Improving Traffic Operation at Freeway Bottlenecks. *KSCE J. Civ. Eng.* **2021**, *25*, 692–704. [CrossRef]
43. Humphrys, M. Action Selection Methods Using Reinforcement Learning. Ph.D. Thesis, University of Cambridge, Cambridge, UK, 1996.
44. Tympakianaki, A.; Spiliopoulou, A.; Kouvelas, A.; Papamichail, I.; Papageorgiou, M.; Wang, Y. Real-time merging traffic control for throughput maximization at motorway work zones. *Transp. Res. Part C Emerg. Technol.* **2014**, *44*, 242–252. [CrossRef]
45. Lu, X.Y.; Varaiya, P.; Horowitz, R.; Su, D.; Shladover, S.E. Novel Freeway Traffic Control with Variable Speed Limit and Coordinated Ramp Metering. *Transp. Res. Rec.* **2011**, *2229*, 55–65. [CrossRef]
46. Jeon, S.; Park, C.; Seo, D. The Multi-Station Based Variable Speed Limit Model for Realization on Urban Highway. *Electronics* **2020**, *9*, 801. [CrossRef]

47. Wang, W.; Cheng, Z. Variable Speed Limit Signs: Control and Setting Locations in Freeway Work Zones. *J. Adv. Transp.* **2017**, *2017*, 1–13. [CrossRef]
48. Sutton, R.S.; Barto, A.G. *Reinforcement Learning: An Introduction*; The MIT Press: Cambridge, MA, USA, 1998.
49. Watkins, C.J.C.H.; Dayan, P. Technical Note: Q-Learning. In *Machine Learning*; Springer Nature: Berlin, Germany, 1992; pp. 279–292.
50. Busoniu, L.; Babuska, R.; De Schutter, B. A Comprehensive Survey of Multiagent Reinforcement Learning. *IEEE Trans. Syst. Man Cybern. Part C Appl. Rev.* **2008**, *38*, 156–172. [CrossRef]
51. Chen, X.; Chen, H.; Yang, Y.; Wu, H.; Zhang, W.; Zhao, J.; Xiong, Y. Traffic flow prediction by an ensemble framework with data denoising and deep learning model. *Phys. A: Stat. Mech. Its Appl.* **2021**, *565*, 125574. [CrossRef]
52. Schumann, R.; Taramarcaz, C. Towards systematic testing of complex interacting systems. In Proceedings of the First Workshop on Systemic Risks in Global Networks Co-Located with 14, Internationale Tagung WiRtschaftsinformatik (WI 2019), Siegen, Germany, 24 February 2019; pp. 55–63.
53. Haykin, S. *Neural Networks and Learning Machines*, 3rd ed.; Prentice Hall: Upper Saddle River, NJ, USA, 2008.

Article

ActressMAS, a .NET Multi-Agent Framework Inspired by the Actor Model

Florin Leon

Faculty of Automatic Control and Computer Engineering, "Gheorghe Asachi" Technical University of Iasi, Bd. Mangeron 27, 700050 Iasi, Romania; florin.leon@academic.tuiasi.ro

Abstract: Multi-agent systems show great promise in the actual state of increasing interconnectedness and autonomy of computer systems. This paper presents a .NET multi-agent framework for experimenting with agents and building multi-agent simulations. Its main advantages are conceptual simplicity and ease of use, which make it suitable for teaching agent-based notions. Several algorithms, protocols and simulations using this framework are also presented.

Keywords: multi-agent framework; .NET framework; simulations; agent-based systems; agent algorithms; software design

Citation: Leon, F. ActressMAS, a .NET Multi-Agent Framework Inspired by the Actor Model. *Mathematics* **2022**, *10*, 382. https://doi.org/10.3390/math10030382

Academic Editors: Ioannis G. Tsoulos and Jianquan Lu

Received: 5 November 2021
Accepted: 24 January 2022
Published: 26 January 2022

Publisher's Note: MDPI stays neutral with regard to jurisdictional claims in published maps and institutional affiliations.

Copyright: © 2022 by the author. Licensee MDPI, Basel, Switzerland. This article is an open access article distributed under the terms and conditions of the Creative Commons Attribution (CC BY) license (https:// creativecommons.org/licenses/by/ 4.0/).

1. Introduction

Currently computer systems are increasingly interconnected and the complexity of tasks that they solve requires less human intervention and an extended degree of autonomy. The promise of Internet of Things and autonomous cars and drones, including those aimed at delivering goods to customers, are prominent examples. There are also a large number of complex systems (e.g., social, economic, ecological) that can be studied using a bottom-up approach for modeling and simulation. These methods of analyzing the results of complex interactions are easier to apply than traditional, analytical models. Although multi-agent systems (MAS), arguably, still have to find some successful "show-off" applications, similar to the recent success of deep learning in artificial intelligence, they are an active area of research. Therefore, many MAS frameworks have been proposed with the goal of helping the user to focus on the high-level behavior rules and interaction protocols, rather than on the low-level details of concurrent and distributed programming.

The creation of ActressMAS [1,2] was fueled by two main reasons. First, it was based on the personal experience of the author in teaching multi-agent systems. Unfortunately, many popular frameworks require some effort of understanding a specific agent language, complex configurations of the platform itself, or some idiosyncrasies of the programming model. Therefore, the main goal of the proposed ActressMAS framework was simplicity. It requires minimal configuration, uses a mainstream programming language, and allows the user to focus on agent behavior, rather than learning the characteristics of the framework itself; therefore, it has proved successful in teaching MAS concepts. Secondly, it was found that while many multi-agent frameworks are based, e.g., on Java, not so many exist for the .NET ecosystem.

Thus, ActressMAS is a simple-to-use .NET multi-agent framework inspired by the actor model. This paper aims to present the philosophy of this framework, the main design decisions and the compromises that had to be made.

The rest of the article is organized as follows. In Section 2, some related work about other multi-agent frameworks available today is presented. Section 3 points out the similarities and differences between actors and agents. Section 4 describes the architecture of ActressMAS. Section 5 illustrates the performance of the framework on some benchmark problems and Section 6 contains a discussion about how ActressMAS relates to existing standards and to certain features found in other agent frameworks. The conclusions of this work are presented in Section 7, while Appendix A presents several applications (i.e., agent algorithms and protocols, as well as multi-agent simulations) together with specific details about the examples included in the publicly available GitHub repository and the ways of using the proposed framework in other projects.

2. Related Work

Many applications of multi-agent systems have been put forward. They can be grouped into several categories, of which one can mention the following [3]:

- Social simulations: the study of population dynamics, the evolution of social corruption, or models of civil violence;
- Mobility simulations: traffic situations with the goal of analyzing traffic jams, adaptive traffic lights, route choice, mobility planning systems, urban planning based on accessibility studies with dynamic populations, microscopic models of pedestrian crowds and evacuation of buildings, or air traffic control;
- Physical entities: robots and autonomous vehicles (cars, drones) seen as agents;
- Environment and ecosystems: simulations in ecology, biology, climate models, human and nature interaction (sometimes using geographic information systems), epidemiology (the spread of infections or diseases);
- Organizational simulations: planning and scheduling, enterprise and organizational behavior, workflow simulations;
- Economic studies: business, marketing, economics, e.g., price forecasting in real-world markets;
- Medical applications: personalized healthcare or hospital management;
- Industrial simulations: manufacturing and production, including the use of holons;
- Military applications: military combat simulations, air defense scenarios;
- Distributed computing, e.g., in cloud computing, virtualized data centers, large-scale parallel or distributed computing clusters and high performance supercomputers;
- Games or movie-making.

Consequently, many agent frameworks have been proposed, with largely different levels of scope, performance and adoption. In the following, some examples are referenced, which seem representative of their corresponding categories. It must be emphasized that no established consensus exists towards the degree of relevance of specific platforms, so this is not necessarily a ranking in terms of popularity. However, the frameworks included below have a consistent user base, as well as associated research papers and projects. The main categories are those identified in [3], which is a comprehensive review of both active and inactive multi-agent frameworks.

The first category includes general-purpose platforms. One representative is JADE [4], a FIPA-compliant middleware made in Java, where agents are programmed in terms of "behaviors", a specific way to handle concurrent execution. Another is MASON [5], also a Java framework focused on discrete-event multi-agent simulation, with 2D and 3D visualization capabilities. One can also mention Orleans [6], which is one of the few agent frameworks available to .NET programmers. It uses virtual actors whose activations are performed in a turn-based asynchronous manner. The fundamental entity is a "grain", which has user-defined identity, behavior, and state.

From the category of platforms for cognitive and social studies, one can point out two examples of cognitive architectures. ACT-R [7] is a hybrid architecture with a symbolic component–a production system–and a subsymbolic one in the form of a set of massively parallel processes modeled with mathematical equations, which control many symbolic processes and are generally responsible for learning. Procedures can be expressed similarly to the brain's action selection mechanisms. Soar [8] is another cognitive architecture that aims to identify the building blocks necessary for an agent with artificial general intelligence, i.e., an agent able to perform many tasks in various domains. Soar is the final point of an evolution that started with the Logical Theorist [9], often considered "the first artificial intelligence program", designed to perform symbolic automated reasoning (proofs of mathematical theorems) and shown during the Dartmouth workshop in 1956, the birthplace of the artificial intelligence field. One of the features of Soar is the use of "chunking" as a learning mechanism: once a (sub)goal is achieved, a rule or set of rules are added to the long-term memory expressed as a production system. Perhaps in the same category one can also include Jason [10], which implements a practical reasoning architecture (Belief-Desire-Intention, BDI [11]), using a special-purpose, logic-based programming language called AgentSpeak, the continuator of one of the first agent-oriented languages [12].

There are also platforms for artificial intelligence research, e.g., OpenAI Gym [13] that focuses on reinforcement learning environment simulations, control tasks, Atari games emulators that allow custom agents to play, and DeepMind Lab [14] that includes 3D navigation and puzzle-solving environments for intelligent agent experimentation, especially with deep reinforcement learning.

From the category of platforms for modeling and simulating natural and social phenomena, one can mention NetLogo [15]. It has a large library of "models", i.e., configurable multi-agent simulations with graphical user interface (GUI). Its main drawback is, arguably, its specific programming language based on Logo, which is quite different from other popular mainstream programming languages. Specific concepts are: "patches", i.e., cells in a grid similar to the cells in cellular automata, "turtles", i.e., agents that can move freely through space, and "links", which define connections between turtles and can be used to build network models.

Among the platforms for transport-related simulations, one can mention Carla [16], which focuses on autonomous driving systems and provides a physics engine for realistic 3D traffic scenarios simulations, and MATSim [17] for large-scale agent-based transport simulations.

Finally, a high-performance framework that uses the actor model is Akka [18], designed to build highly concurrent and distributed message-driven applications in Java and Scala. It can also be used with .NET languages by means of Akka.NET [19].

To summarize, all agent systems have certain commonalities, such as the need for support for parallel/distributed computations or communication, and the need to handle a reasonably large number of agents. The programmers should be able to focus on their specific tasks, not on these low-level details of the middleware, and this is the main goal of the various agent frameworks. However, they implement these requirements in very different ways, and integrate other concepts and ideas as well.

One can identify agent systems where the focus is on the autonomous behavior and on the interaction protocols, e.g., negotiation or coordination. These are supported by general purpose platforms, e.g., JADE. This somewhat contrasts with the multi-agent simulations, where many agents execute according to simple local rules, but the focus is on studying the complex interactions and the emergent behavior of the system. These are supported by platforms such as NetLogo. ActressMAS mainly belongs to the first type, but also allows the user to run simulations and build graphical user interfaces to observe the overall behavior of multi-agent systems.

3. Theoretical Aspects: Actors and Agents

Actors and agents are both entities that can be used for performing parallel and distributed computations. Both models rely on messages for exchanging information and do not recommend the use of shared memory structures. Still, there are conceptual differences between them that we point out as follows.

Actors can be described as computational processes that realize their functionality by sending and receiving messages to and from other actors in an asynchronous manner. These are the main operational axioms defined by the computational actor model [20]. In addition, an actor can create other actors. Actors are purely reactive entities because they can act only when they receive a message. In this computational model, the program flow is created by composing the individual behaviors of the actors in the system. They communicate only by sending messages and do not expose any part of their internal state to other actors. Each actor has an "inbox", i.e., a queue with the received messages, and processes them in order, one at a time. Many actors usually exist in a system and they run concurrently. However, the model avoids the need for synchronization because each actor processes its own messages sequentially. For example, one can use a proxy actor for a shared resource. If two other actors need to access the resource, they can only send request messages to the proxy actor, and the proxy handles these messages one at a time; therefore, the resource cannot be directly accessed by two or more actors simultaneously.

On the other hand, agents can be defined as autonomous entities situated in their execution environment (i.e., they are embedded in their virtual/software or physical/hardware environment). In addition, intelligent agents should be capable of reactive, proactive and social behavior [21,22]. From this point of view, agents focus more on the capabilities that may allow them to be human representatives in various interaction scenarios, including those that may involve reasoning and planning.

The actor model can be seen as "weaker" than the agent model. While the autonomy and reactiveness can be ensured by the actor model, the proactive behavior is not so straightforward to model using pure actors. For example, an agent may need to act even if it receives no message from the outside. In order to be able to demonstrate complex social behavior, there should also exist a richer set of message semantics that the agents could use. Also, reasoning capabilities are not a requirement for actors, although such algorithms can be part of their internal logic.

While agents are autonomous entities that perceive their environment and can act upon it, actors can be controlled by other actors and may lack sensors and effectors [23].

Actors and agents are separate concepts, but the basic characteristics of actors such as parallel execution and reactive behavior triggered by message-based communication can constitute a starting point for the development of an agent system. In many cases, the behavior of agents is also driven by the messages they receive from other agents. However, in other cases agents should be able to take the initiative and act even when no messages are received, e.g., when something in their internal state changes and requires a certain action to be performed. In addition, since they are situated entities, they need some mechanism to access their environment. The latter requirement can also be accomplished by the exchange of messages between the environment and the agent, but when large amounts of data need to be transferred, this option may be less efficient.

An alternative direction of applications for agents is related to multi-agent simulations, which are concerned with building systems of interacting entities in a bottom-up manner, whose aggregated behavior can provide useful insights about the dynamics of complex systems. Many such simulations contain agents whose behavior is based on simple rules; therefore, the time needed to decide an action at a certain moment is more or less equal for all agents. In this situation, especially when there is a large number of agents involved, a turn-based execution is preferred, which gives each agent a chance to act, without the need of complex scheduling. Even parallel behavior may be emulated with a sequential execution of agents in random order during each turn.

Although initially inspired by the actor model, ActressMAS includes some mechanisms to handle proactiveness and to offer support for multi-agent simulations, thus departing from the pure actor model. Thais is why a turn-based approach was considered for agent execution. Beside its role in simulations, it can also be used to handle most protocols and algorithms by designing general agent programs with fine-grained decision logic. Still, it must be stated that this is not suitable when time-specific actions are needed to be performed by agents or for real-time applications. The turn-based execution also helps to implement the proactiveness feature of agents. The framework identifies the situation when no messages are received at the end of a turn, and this can be optionally used to change the agent state or to perform an action. For scenarios in which agents need to access certain parts of their environment (e.g., for indirect communication by stigmergy), ActressMAS also provides a shared memory structure present in the environment.

4. ActressMAS Architecture

In this section, the architecture of ActressMAS is presented, both high-level and detailed. The representations are Unified Modeling Language (UML) class diagrams [24].

Figure 1 displays the general architecture. The main concepts of a multi-agent system can be observed: the agent and the environment. In order to handle parallel and sequential execution in an efficient, transparent way, the agents are stored either in a concurrent or "normal" C# dictionary. Agents communicate by sending messages. The upper part of the diagram addresses the distributed capabilities of the framework using the client-server model. On each machine there is a container that includes a "runnable" environment. Containers communicate through a server by means of a special type of messages. When agents migrate between containers, their state (or a part of their state) is serialized and sent to the destination container where the agent is restarted. The architecture also contains a class used by the agents to filter the agents with certain properties in their environment and to "observe" them automatically in a perception method that can be used to update their beliefs before acting.

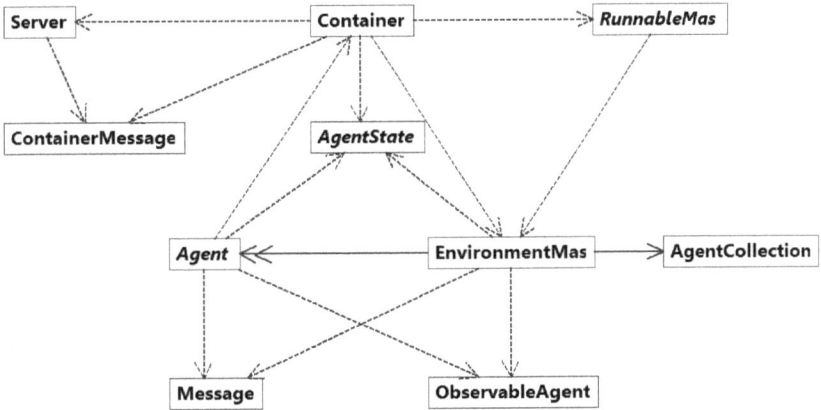

Figure 1. The general architecture of the ActressMAS framework.

All these classes will be detailed as follows.

4.1. Fundamental Features

Even if agents are the main way of expressing programming logic in an agent-oriented application, the description of ActressMAS will probably be clearer if we start with the description of the environment first. By definition, agents inhabit an environment of which they are a constituent part. They perceive it and act upon it. The entities involved in these operations may be other agents or properties of the environment.

In ActressMAS, the execution of the environment is based on turns. The maximum number of turns of a simulation is one of the parameters of the constructor of the environment class. The agents are not directly aware of this execution model, but the user can make this information available to the agents, as explained below. The turn-based execution was chosen in order to treat the two types of agent systems (the interaction of autonomous agents and multi-agent simulations) in the same way. The agents can be run sequentially or concurrently. In the latter case, the choice of turn-based execution (when the acting behavior of all agents is executed in parallel but all agents need to finish before starting the next turn) may be problematic if one agent takes much longer than the others to execute its behavior. However, there are at least two practical solutions for this situation. One solution is to place the long-running agent into a separate container which can also run on the same machine with the container that hosts the rest of the agents. The special agent can, e.g., perform intensive computations and report the results only at the end without blocking the others. Another solution is to design the acting behavior of all agents in a fine-grained manner so that only atomic computations should be done at one time, while responding to messages.

Thus, the agents execute in a turn-based manner in both sequential and concurrent settings, but if their actions are properly designed, the parallel execution benefits from the multiple processor cores, when available, and the overall performance is faster.

The parallel execution is performed by launching a *Task* for the current behavior of each agent. Tasks are a lightweight form of implementing asynchronous behavior in .NET. They use a thread pool which is managed transparently by the .NET framework and allows the execution of a large number of agents. Using this mechanism, the user can create, e.g., tens of thousands of agents. Attempting to create a similar number of threads would likely block the operating system.

In case of sequential execution, the user can also choose that the agents run in the order in which they have been added to the environment, or in a random order. While in most cases the random order is natural, there are protocols in which the user can ensure that, e.g., a manager agent receives messages from all the worker agents in a turn before starting another round of the protocol. Placing the manager as the last agent simplifies the implementation, because otherwise the messages of some agents may only be received in the next turn and the manager would have to include a mechanism to count the number of messages received so far, or to identify the actual agents that have responded.

The UML class diagram of the environment class is presented in Figure 2. The name of the class is *EnvironmentMas* to avoid a conflict with the .NET *System.Environment* class. If these classes from both *ActressMas* and *System* namespaces were used together, the user would have to use a namespace-qualified name such as *ActressMas.Environment* in his/her code. But since the other ActressMAS classes do not have this necessity, it was decided that *EnvironmentMas* was a more appropriate name.

The environment includes the typical methods for adding, removing or enumerating agents. Except for agent creation, these operations are usually done using agent names, not references to the agent objects.

The environment also acts as a bidirectional proxy for sending messages between agents and moving agents between containers.

It also has some special methods that allow the programmer to handle turns in an explicit manner. For this purpose, the user must create a subclass of *EnvironmentMas* and override the *TurnFinished* and/or *SimulationFinished* methods. This is especially useful for multi-agent simulations, where the user can compute, e.g., some statistics after each turn or introduce external conditions or events at special moments in the simulation.

The environment also has a shared memory in the form of a dictionary where agents can record any kind of object with a string key. While in multi-agent systems communication is normally done by messages, from the practical point of view there are cases when having a shared memory greatly reduces the communication overhead, e.g., when agents need to be aware of some changing properties of the environment encoded as large objects. The shared memory facility should not be abused; however, it corresponds to the situation where physical agents perceive and manipulate objects in their environment.

The internal methods marked with a tilde are accessible to the other classes of the *ActressMas* assembly, but invisible to the user programs.

The environment class uses a special structure to store the agents, named *AgentCollection*, displayed in Figure 3. Since the agents have unique names, a dictionary is used to handle the agent objects. However, the collection of agents may change dynamically during the execution of the user program, e.g., when agents are added or removed during the execution of a turn. When the agents run in parallel and perform this kind of behavior, the access to a simple C# *Dictionary* must be used in conjunction with a lock, which can degrade performance. That is why a *ConcurrentDictionary* (a thread-safe collection that can be accessed by multiple threads concurrently) is used to store the agents when the environment is set for a parallel execution.

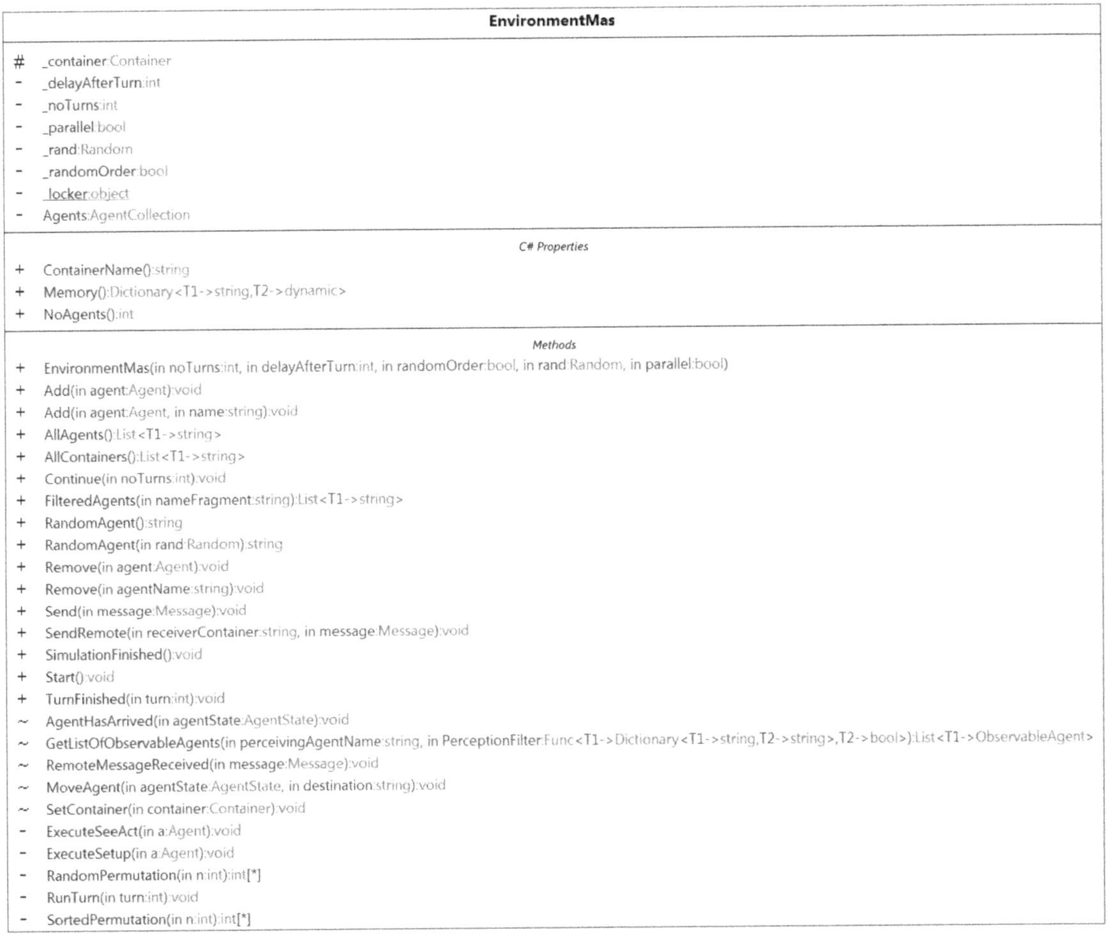

EnvironmentMas
_container:Container - _delayAfterTurn:int - _noTurns:int - _parallel:bool - _rand:Random - _randomOrder:bool - _locker:object - Agents:AgentCollection
C# Properties
+ ContainerName():string + Memory():Dictionary<T1->string,T2->dynamic> + NoAgents():int
Methods
+ EnvironmentMas(in noTurns:int, in delayAfterTurn:int, in randomOrder:bool, in rand:Random, in parallel:bool) + Add(in agent:Agent):void + Add(in agent:Agent, in name:string):void + AllAgents():List<T1->string> + AllContainers():List<T1->string> + Continue(in noTurns:int):void + FilteredAgents(in nameFragment:string):List<T1->string> + RandomAgent():string + RandomAgent(in rand:Random):string + Remove(in agent:Agent):void + Remove(in agentName:string):void + Send(in message:Message):void + SendRemote(in receiverContainer:string, in message:Message):void + SimulationFinished():void + Start():void + TurnFinished(in turn:int):void ~ AgentHasArrived(in agentState:AgentState):void ~ GetListOfObservableAgents(in perceivingAgentName:string, in PerceptionFilter:Func<T1->Dictionary<T1->string,T2->string>,T2->bool>):List<T1->ObservableAgent> ~ RemoteMessageReceived(in message:Message):void ~ MoveAgent(in agentState:AgentState, in destination:string):void ~ SetContainer(in container:Container):void - ExecuteSeeAct(in a:Agent):void - ExecuteSetup(in a:Agent):void - RandomPermutation(in n:int):int[*] - RunTurn(in turn:int):void - SortedPermutation(in n:int):int[*]

Figure 2. The environment class.

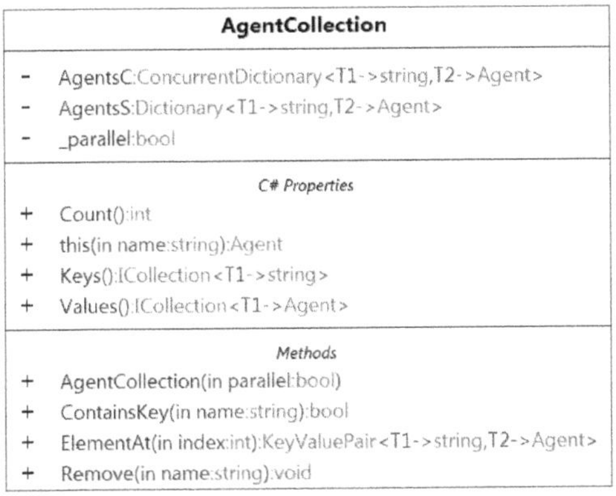

AgentCollection
- AgentsC:ConcurrentDictionary<T1->string,T2->Agent> - AgentsS:Dictionary<T1->string,T2->Agent> - _parallel:bool
C# Properties
+ Count():int + this(in name:string):Agent + Keys():ICollection<T1->string> + Values():ICollection<T1->Agent>
Methods
+ AgentCollection(in parallel:bool) + ContainsKey(in name:string):bool + ElementAt(in index:int):KeyValuePair<T1->string,T2->Agent> + Remove(in name:string):void

Figure 3. The agent collection class used by the environment to distinguish between concurrent and sequential behavior in a transparent way.

This class is designed as a kind of discriminated union. Currently (as of version 7), C# lacks support for this type of structure. The environment object uses either a *ConcurrentDictionary* or a "normal" *Dictionary*, depending on the *parallel* flag that shows whether the agents are run concurrently or sequentially. By using a dual collection, of which only one is allocated and actually used, an additional class hierarchy is avoided. All the public methods and properties are those common to dictionaries, therefore the *AgentCollection* object is used transparently as a generic dictionary, regardless of the underlying concurrent or sequential implementation.

Agents are the central entities of any program using the framework. The UML class diagram of the *Agent* class is presented in Figure 4. This is the base class for all the user-defined agents.

Figure 4. The abstract agent base class.

Each agent is registered in its environment by a unique name. This name is used in most operations, e.g., sending a message to another agent, which is the main form of communication in an agent-oriented program.

The typical methods that contain the logic of the agent are: *Setup*, *Act*, and *ActDefault*, which will be presented next. Although agents can execute in parallel, the code in these methods–for one agent–is always executed sequentially. This is one of the main strengths of the actor model, which avoids the need of synchronization for the access to critical resources, as explained in Section 1.

The *Setup* method is used for initialization. An agent class may also have a constructor, but as intended by the design of ActressMAS a constructor should be used to initialize the internal data structures such as lists, dictionaries, random number generators, etc. The *Setup* method should be used for agent-related logic, e.g., sending the initial messages that start the multi-agent protocol. Custom constructors may also be added to an agent class; a constructor with parameters can usually be employed when agents should be given some initial values for certain properties from the "outside" of the multi-agent system, i.e., when the agents are created and before the environment is started.

The *Act* method is activated when an agent receives a message. If there are more messages to be received, the *Act* method is activated once for each message. This is one of main tenets of actor-based programming, which is embedded in the ActressMAS framework. However, agents need not be purely reactive. They can maintain a state and they can update it after each message, and act based on their overall state, not only the current message. For example, a manager agent can know how many worker agents there are in the environment using the *FilteredAgents* method if the worker agents have a common part of their names, such as "worker*NN*-agent". Then the manager can decrease a counter for each message received from a worker. When the counter gets to zero, it knows that all workers have reported and may send them a new series of tasks.

The *ActDefault* method distinguishes ActressMAS from a purely actor-based model. In agent systems, there are situations when agents should act based on other conditions than responding to messages. For example, an English auction agent can designate the winner when no agent sends any more bids. This cannot be properly modeled within the pure actor paradigm because the acting condition is not the receipt of a message, but the lack of any message. Therefore, the *ActDefault* method was introduced to handle the cases when no messages have been received at the end of a turn. This increases the proactive capabilities of the agents, as their reactive capabilities are covered by the normal *Act* method. The agents also have the possibility to wait for a few turns–by counting the elapsed turns in *ActDefault*–and then act. In the initial version of ActressMAS, which did not contain the *ActDefault* method, a *Timer* object was used to send "wake up" messages that the agents could react to. However, a dedicated method within the framework seemed to be a much more elegant way to handle such issues.

The agents communicate directly by messages (detailed below). The main method used for this purpose is *Send*, where the receiver agent is designated by name. An agent can also send a message to all the other agents in the environment by using the *Broadcast* method.

An agent contains a *Stop* method, as well, which is called to deactivate the agent, which is thus removed from the multi-agent system. As it can be seen in Figure 2, the environment has a *Remove* method, which can be accessed by an agent through its *this.Environment* property. As intended by the design of ActressMAS, the *Stop* method should be used when the decision to be stopped belongs to the agent itself, while the *Environment.Remove* method should be used when the decision to stop an agent belongs to some other agent or to an external factor. The latter case is not common in the autonomous agent protocols, but it is encountered in multi-agent simulations, e.g., in a predator-prey simulation, a predator can "kill" a prey. As one can notice, there is no *Start* method in the *Agent* class. Agents start automatically when the environment starts, or when new agents are added later to a running environment. If the user wants the agent to suspend its execution for several turns, this can be easily accomplished, e.g., by using a Boolean flag in the agent class and conditioning any acting behavior on its value.

There are a few other methods implemented in an agent class, but they will be described in the following subsections, which present the observable properties and the distributed capabilities of ActressMAS.

Messages are the only direct method for inter-agent communication; therefore, the class corresponding to a message (Figure 5) deserves special attention. It has been designed with loose inspiration from the FIPA ACL specification (the Agent Communication Language proposed by the Foundation for Intelligent Physical Agents) [25] and the implementation of messages in the JADE framework [4]. However, the goal was to allow the use of ACL concepts while maintaining a very simple syntax. Therefore, a message has a compulsory sender and receiver, and an optional conversation identifier, which can be used in some protocols to identify an ongoing sequence of messages that form a unique conversation. The sender is automatically assigned when an agent sends a message. Usually, only the receiver and the content must be specified.

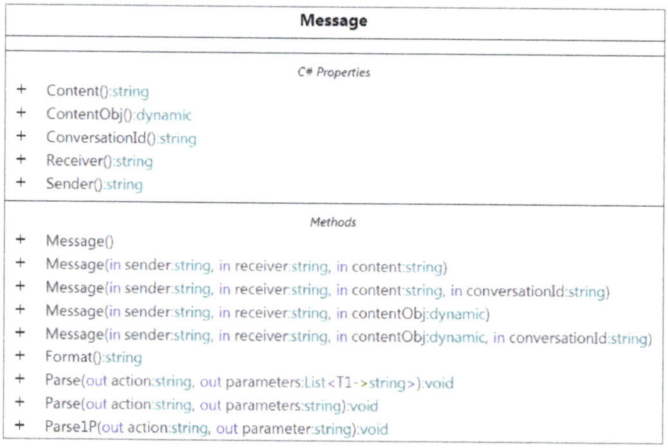

Message
C# Properties
+ Content():string + ContentObj():dynamic + ConversationId():string + Receiver():string + Sender():string
Methods
+ Message() + Message(in sender:string, in receiver:string, in content:string) + Message(in sender:string, in receiver:string, in content:string, in conversationId:string) + Message(in sender:string, in receiver:string, in contentObj:dynamic) + Message(in sender:string, in receiver:string, in contentObj:dynamic, in conversationId:string) + Format():string + Parse(out action:string, out parameters:List<T1->string>):void + Parse(out action:string, out parameters:string):void + Parse1P(out action:string, out parameter:string):void

Figure 5. The class corresponding to the messages passed between agents.

The content can be expressed in two ways. The first is in the form of a string. The space-delimited words express the meaning. Typically, the first word defines the main action or message type, somewhat similar to an ACL performative, which shows the type of the communicative act, as inspired from the speech acts theory [26]. However, in ActressMAS there are no constraints about the values of this message type, i.e., the user can choose any name. The rest of the words specify the parameters. For example, if an agent wishes to send another agent its position on a two-dimensional grid, the content of the message can be "position 2 5". Most of the time, an agent that receives a message needs to split the string to be able to interpret the content. For this goal, the *Message* class contains several *Parse* helper methods, which identify the first word, called the "action" and the rest, called the "parameters". Depending on the protocol defined by the user, the parameters can be identified as a list of strings, one for each parameter, or a single string with all the parameters concatenated. When the action has only one parameter, the *Parse1P* method can be used.

Using strings for the content of the messages is very flexible and in line with the philosophy of agent communication, but can be less efficient because of the need to split the whole string into parameters and convert them to their specific types, e.g., integers or double-precision real numbers. Therefore, the second way of encoding the content is directly as objects. The user should define one or more custom classes and assign their instances to the *ContentObj* property of the message. The receiving agent can directly cast this property to the corresponding object type.

Finally, the *Message* class also contains a *Format* method, which can be used to display the pretty-printed message, with its sender, receiver and content.

If the receiver agent is in another container (a situation that uses the distributed infrastructure described below in Section 4.3), the receiver name should be qualified with the name of the container, e.g., "agent1@container2".

The basic way to create a multi-agent system is illustrated in the code below. First, the environment is created, then the agents are created and added to the environment, and finally the environment is started. When an agent is added to the environment, it is customary to assign it a name, because several agents from the same class can exist in the environment.

```
public class Program
{
    private static void Main(string[] args)
    {
        var env = new EnvironmentMas();
        var a1 = new Agent1(); env.Add(a1, "a1");
        var a2 = new Agent2(); env.Add(a2, "a2");
        env.Start();
    }
}
```

In this example, the *Setup* method simply displays a message and introduces a short delay, which is needed to emphasize the concurrent behavior of the agents. A single message is sent during the setup by each agent, and the *Act* method responds to the message received from the peer. If several messages had been sent, the *Act* method would have been activated for each of them. The *ActDefault* method is used when, at the end of a turn, no messages have been received. When this happens, the agent stops.

When there are no more agents in the environment, the simulation stops, even before the maximum number of turns specified by the user is reached.

```
public class Agent1: Agent
{
    public override void Setup()
    {
        for (int i = 0; i < 10; i++)
        {
            Console.WriteLine($"Setup: {i + 1} from a1*");
            Thread.Sleep(100);
        }

        Send("a2", "msg");
    }

    public override void Act(Message message)
    {
        for (int i = 0; i < 3; i++)
        {
            Console.WriteLine($"Act: {i + 1} from a1*");
            Thread.Sleep(100);
        }
    }

            public override void ActDefault()
        {
    Console.WriteLine("ActDefault: no messages for a1*");
        Stop();
    }
}
```

A similar structure is used by the second agent, but the number of messages in *Setup* and *Act* are different (3 instead of 10 and vice versa), in order to break symmetry.

```
public class Agent2: Agent
{
    public override void Setup()
    {
        for (int i = 0; i < 3; i++)
        {
            Console.WriteLine($"Setup: {i + 1} from a2");
            Thread.Sleep(100);
        }

        Send("a1", "msg");
    }

    public override void Act(Message message)
    {
        for (int i = 0; i < 10; i++)
        {
            Console.WriteLine($"Act: {i + 1} from a2");
            Thread.Sleep(100);
        }
    }

    public override void ActDefault()
    {
        Console.WriteLine("ActDefault: no messages for a2");
        Stop();
    }
}
```

The output of this simple program can be seen in Figure 6. The asterisk is used to mark the reports of the first agent in order to help the reader to distinguish the behavior of the two agents more easily.

Setup: 1 from a2	Setup: 1 from a2
Setup: 1 from a1*	Setup: 2 from a2
Setup: 2 from a1*	Setup: 3 from a2
Setup: 2 from a2	Setup: 1 from a1*
Setup: 3 from a1*	Setup: 2 from a1*
Setup: 3 from a2	Setup: 3 from a1*
Setup: 4 from a1*	Setup: 4 from a1*
Setup: 5 from a1*	Setup: 5 from a1*
Setup: 6 from a1*	Setup: 6 from a1*
Setup: 7 from a1*	Setup: 7 from a1*
Setup: 8 from a1*	Setup: 8 from a1*
Setup: 9 from a1*	Setup: 9 from a1*
Setup: 10 from a1*	Setup: 10 from a1*
Act: 1 from a2	Act: 1 from a1*
Act: 1 from a1*	Act: 2 from a1*
Act: 2 from a1*	Act: 3 from a1*
Act: 2 from a2	Act: 1 from a2
Act: 3 from a1*	Act: 2 from a2
Act: 3 from a2	Act: 3 from a2
Act: 4 from a2	Act: 4 from a2
Act: 5 from a2	Act: 5 from a2
Act: 6 from a2	Act: 6 from a2
Act: 7 from a2	Act: 7 from a2
Act: 8 from a2	Act: 8 from a2
Act: 9 from a2	Act: 9 from a2
Act: 10 from a2	Act: 10 from a2
ActDefault: no messages for a2	ActDefault: no messages for a1*
ActDefault: no messages for a1*	ActDefault: no messages for a2

Figure 6. The output of the simple program used to exemplify the basic agent methods of ActressMAS. The results are displayed when agent behaviors are executed: left: in parallel; right: sequentially but in a random order.

Figure 6(right) shows the results when the environment is set to execute agents sequentially by using:

```
var env = new EnvironmentMas(parallel: false);
```

The agents can be executed sequentially and in the order in which they have been added to the environment by using:

```
var env = new EnvironmentMas(randomOrder:false, parallel: false);
```

In this case, agent *a1* would be the first to report in all three methods: *Setup*, *Act*, and *ActDefault*.

4.2. Observable Properties

Especially in multi-agent simulations, the next state of an agent may depend on the current state of its neighbors, e.g., simulations related to cellular automata. Perhaps agents may have a limited field of view and may perceive only a subset of the agents in their environment. In order to facilitate the handling of such situations, ActressMAS includes observable properties. They are implemented as a dictionary where the key is a string, i.e., the name of the property, and the value can be any kind of object. At initialization, e.g., in the *Setup* method, the agents define these properties. Then the agents override the *PerceptionFilter* predicate which defines the conditions that make other agents visible or observable. For example, if an agent can only see its neighbors within a certain radius, the predicate should express the condition that the Euclidian distance between the position of the "ego" agent (the *this* object) and the position of a neighbor agent (the *observed* parameter) be less than the specified radius. Initially, the agents should have defined an observable position property. Then the agent should implement the *See* method, called before *Act* or *ActDefault*, which provides the list of *ObservableAgent* objects (Figure 7) as a parameter. The agent can, for example, process or store this information in order to use it in the acting methods.

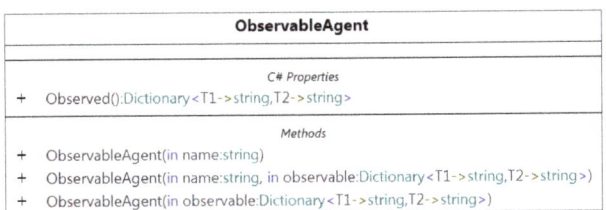

Figure 7. The class corresponding to observable agents.

The following code illustrates an example of using observables. The multi-agent system is defined in a similar fashion as in the previous example, but here the agents need to have the *UsingObservables* property explicitly set to be true.

```
public class Program
{
    public static void Main(string[] args)
    {
        var env = new EnvironmentMas(noTurns: 10, randomOrder: false, parallel: false);
        var a1 = new MyAgent(); a1.UsingObservables = true; env.Add(a1, "Agent1");
        var a2 = new MyAgent(); a2.UsingObservables = true; env.Add(a2, "Agent2");
        var a3 = new MyAgent(); a3.UsingObservables = true; env.Add(a3, "Agent3");
        env.Start();
    }
}
```

All three agents are instances of the same class, *MyAgent*. One can distinguish the *Perception Filter* method used to define the conditions that make a neighbor agent "visible" and the *See* method that provides the list of the agents that are observed in the current turn, before acting. Basically, each agent is assigned a random number between 0 and 30, and can only see the agents with similar numbers, i.e., when the difference between their corresponding numbers is less than 10. Since no messages are sent in this example, the main logic of the agents relies on the implementation of the *ActDefault* method.

```csharp
public class MyAgent: Agent
{
    private List<ObservableAgent> _observableAgents = null;

    public override void Setup()
    {
        Observables["Name"] = Name;
        Observables["Number"] = $"{Numbers.GenerateNumber():F2}";
    }

    public override bool PerceptionFilter(Dictionary<string, string> observed)
    {
        double myNumber = Convert.ToDouble(Observables["Number"]);
        double obsNumber = Convert.ToDouble(observed["Number"]);
        return (Math.Abs(myNumber − obsNumber) < 10);
    }

    public override void See(List<ObservableAgent> observableAgents)
    {
        _observableAgents = observableAgents;
    }

    public override void ActDefault()
    {
        Console.Write($"I am {Name}. ");

        if (_observableAgents == null || _observableAgents.Count == 0)
            Console.WriteLine("I didn't see anything interesting.");
        else
        {
            Console.WriteLine($"My number is {Observables["Number"]} and I saw:");
            foreach (var oa in _observableAgents)
                Console.WriteLine(
                    $"{oa.Observed["Name"]} with number = {oa.Observed["Number"]}");
        }

        Observables["Number"] = $"{Numbers.GenerateNumber():F2}";

        Console.WriteLine($"My number is now {Observables["Number"]}");
        Console.WriteLine("————————————————————————");
    }
}
```

The class that generates random numbers for the agents is also presented below.

```
public class Numbers
{
    private static Random _rand = new Random();
    public static double GenerateNumber() => _rand.NextDouble() * 30;
}
```

The output of this program is presented in Figure 8.

```
I am Agent1. My number is 5.85 and I saw:
Agent2 with number = 9.48
My number is now 22.72
-----------------------------------------------
I am Agent2. My number is 9.48 and I saw:
Agent3 with number = 17.06
My number is now 7.88
-----------------------------------------------
I am Agent3. My number is 17.06 and I saw:
Agent1 with number = 22.72
Agent2 with number = 7.88
My number is now 7.82
-----------------------------------------------
I am Agent1. I didn't see anything interesting.
My number is now 18.20
-----------------------------------------------
I am Agent2. My number is 7.88 and I saw:
Agent3 with number = 7.82
My number is now 22.65
-----------------------------------------------
I am Agent3. I didn't see anything interesting.
My number is now 8.71
```

Figure 8. The output of the program with observable agents.

4.3. Mobile Agents

Beside the ability to run agents concurrently, ActressMAS also supports mobile agents, which can stop their execution on one machine and resume their execution on a different one. In the following paragraphs, the distributed part of the architecture is presented, together with examples of using this capability.

The host of an environment on each machine is called a "container". This idea is inspired from the JADE framework [4]. However, in ActressMAS there is no distinction between a "main" container and "secondary" containers. Moreover, if the user does not intend to work with mobile agents, he/she does not need to define any container at all. Containers can be placed on different machines, but it is also possible to have multiple containers on the same machine.

Containers communicate by means of a server (Figure 9), which mainly keeps track of the active containers and passes messages between them.

The user must instantiate this class and may optionally define an event handler where the messages from the server can be accessed. In the example below, only the active containers are displayed.

A container (Figure 10) can be seen as a kind of proxy between an environment and the server. It manages the communication with the server (registers, deregisters and keeps a list of alive containers, received from the server). A container handles two main functions. First, when an agent wants to move, the container serializes the agent (actually, the desired part of its state, as explained below) and sends a corresponding message to the server, including the serialized state. The server routes the message to the destination container. There, the container deserializes the agent and informs the environment that an agent has arrived. Secondly, it routes remote messages between agents, i.e., from a container to another.

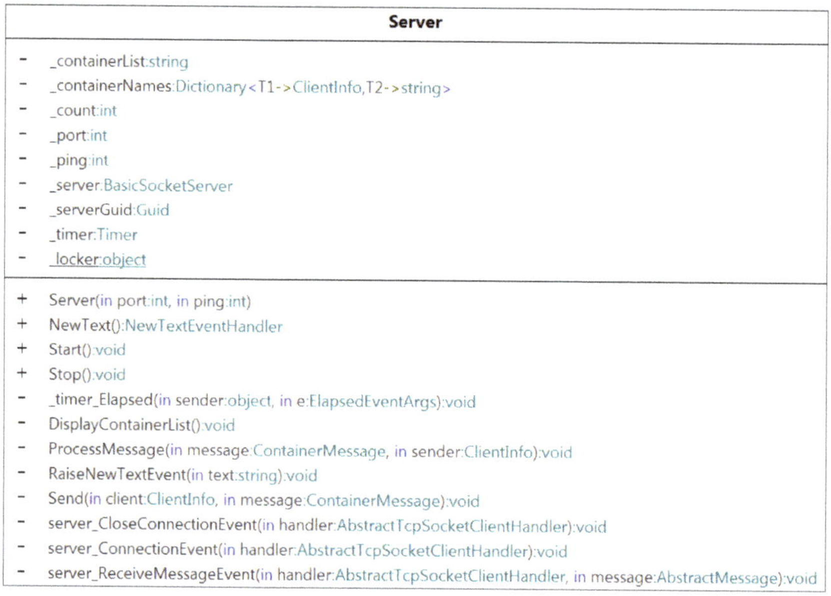

Server
− _containerList:string   − _containerNames:Dictionary<T1->ClientInfo,T2->string>   − _count:int   − _port:int   − _ping:int   − _server:BasicSocketServer   − _serverGuid:Guid   − _timer:Timer   − _locker:object
+ Server(in port:int, in ping:int)   + NewText():NewTextEventHandler   + Start():void   + Stop():void   − _timer_Elapsed(in sender:object, in e:ElapsedEventArgs):void   − DisplayContainerList():void   − ProcessMessage(in message:ContainerMessage, in sender:ClientInfo):void   − RaiseNewTextEvent(in text:string):void   − Send(in client:ClientInfo, in message:ContainerMessage):void   − server_CloseConnectionEvent(in handler:AbstractTcpSocketClientHandler):void   − server_ConnectionEvent(in handler:AbstractTcpSocketClientHandler):void   − server_ReceiveMessageEvent(in handler:AbstractTcpSocketClientHandler, in message:AbstractMessage):void

Figure 9. The server class.

```
public static class Program
{
    private static void Main()
    {
        var server = new Server(5000, 3000);
        Console.WriteLine("Server listening on port 5000.");
        server.NewText += server_NewText;
        server.Start();
        Console.WriteLine("Press ENTER to close the server.");
        Console.ReadLine();
        server.Stop();
    }

    private static void server_NewText(object source, NewTextEventArgs e)
    {
        if (e.Text.StartsWith("Containers:"))
        {
            Console.Clear();
            Console.WriteLine(e.Text);
        }
    }
}
```

The typical way of using the distributed capabilities of ActressMAS is first to initialize and connect the containers to the server. Then, the multi-agent environment in each container needs to be started. This is achieved by means of a simple class called *RunnableMas* (Figure 11), whose utilization is exemplified below.

The communication between a container and the server is done using a special type of message called *ContainerMessage* (Figure 12). Its structure is somewhat similar to an agent *Message*, but it includes the actual serialization and deserialization functionality, together with a special property, *Info*, which is used for handling the semantics of the message, e.g., "Request Register" (when a container wants to register to the server), "Inform Invalid

Name" (if the container name cannot be accepted by the server), "Inform Containers" (when the *Content* is a list with all available containers), "Request Move Agent" (when an agent has arrived), "Send Remote Message" (when an agent message is received in the *Content* property), etc.

The following paragraphs describe an example about creating a container with an environment that runs the agents. It includes the GUI in Figure 13, where one can see the relevant functionality: creating a container and starting it (connecting to the server), running the multi-agent system, and then stopping it (disconnecting from the server). For increased clarity and brevity, only the important parts of the methods are included, e.g., exception handling and the reading or writing of properties of GUI controls are omitted.

Container
- _allContainers:List<T1->string>
- _client:BasicSocketClient
- _clientGuid:Guid
- _environment:EnvironmentMas
- _name:string
- _serverIP:string
- _serverPort:int
- _locker:object
C# Properties
+ Name():string
Methods
+ Container(in serverIP:string, in serverPort:int, in name:string)
+ NewText():NewTextEventHandler
+ AllContainers():List<T1->string>
+ RunMas(in environment:EnvironmentMas, in mas:RunnableMas):void
+ Start():void
+ Stop():void
~ AgentHasArrived(in agentState:string):void
~ RemoteMessageReceived(in content:string):void
~ MoveAgent(in state:AgentState, in destination:string):void
~ SendRemoteAgentMessage(in receiverContainer:string, in message:Message):void
- client_CloseConnectionEvent(in handler:AbstractTcpSocketClientHandler):void
- client_ConnectionEvent(in handler:AbstractTcpSocketClientHandler):void
- client_ReceiveMessageEvent(in handler:AbstractTcpSocketClientHandler, in message:AbstractMessage):void
- ProcessMessage(in message:ContainerMessage):void
- RaiseNewTextEvent(in text:string):void
- Register():void
- Send(in message:ContainerMessage):void

Figure 10. The container class.

RunnableMas
+ RunMas(in env:EnvironmentMas):void

Figure 11. The class used to start running a multi-agent system in a container for distributed scenarios.

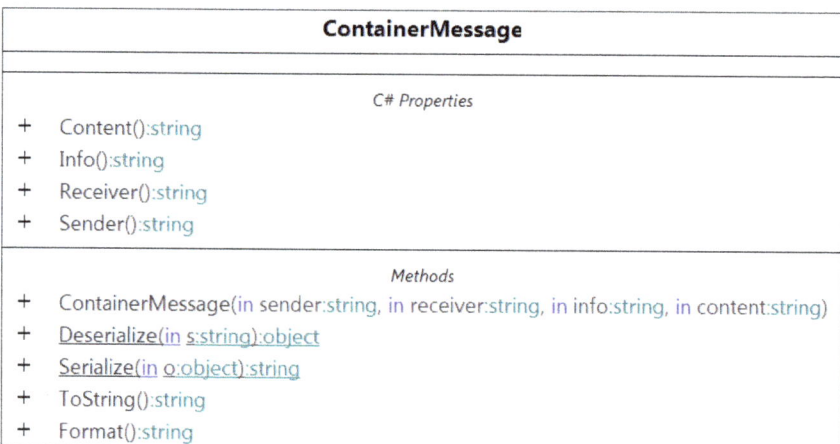

Figure 12. The class corresponding to the messages passed between containers and the server.

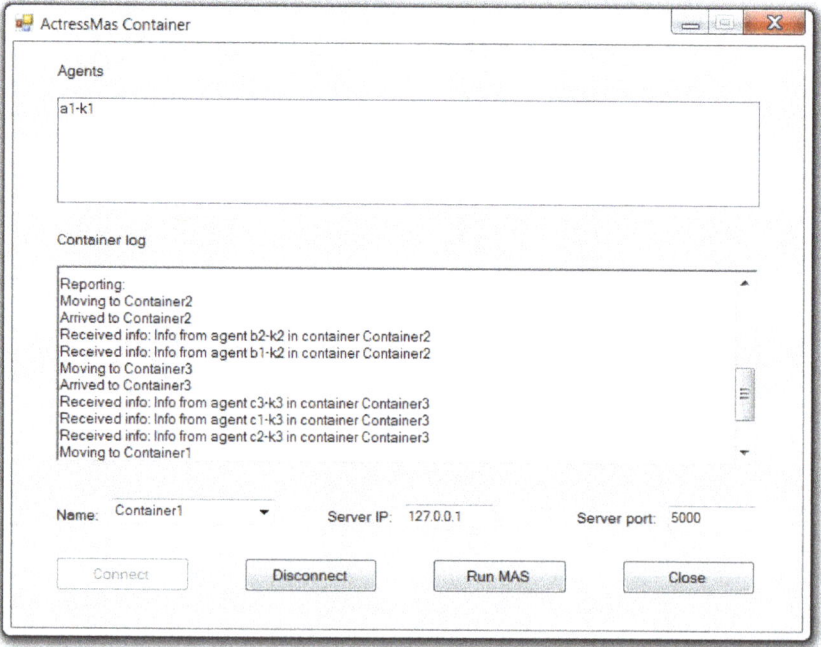

Figure 13. The graphical user interface of a program showing a multi-agent system with mobile agents.

```csharp
private void buttonStart_Click(object sender, EventArgs e)
{
    _container = new Container(_serverIP, Convert.ToInt32(_serverPort), _containerName);
    _container.Start();
}

private void buttonRunMas_Click(object sender, EventArgs e)
{
    _environment = new EnvironmentMas();
    _container.RunMas(_environment, new MasSetup());
}

private void buttonDisconnect_Click(object sender, EventArgs e)
{
    if (_container != null)
        _container.Stop();
}
```

The connection between a container and an environment is made by means of a class derived from the *RunnableMas* class. In this way, the user can make a single application which can be run on different machines or containers, and the setup of the multi-agent system will be different depending on the specific container.

The example in the code below considers three containers with static agents, which just provide some piece of information when being asked, and a mobile agent, starting in "Container1" and then moving to the other containers. It gathers information from the static agents along the way and then returns home to "Container1" and reports the aggregated information.

```csharp
public class MasSetup: RunnableMas
{
    public override void RunMas(EnvironmentMas env)
    {
        string home = env.ContainerName;

        switch (home)
        {
            case "Container1":
                // create a mobile agent and a static agent and add them to the environment env
                break;
            case "Container2":
                // create two static agents and add them to the environment
                break;
            case "Container3":
                // create three static agents and add them to the environment
                break;
        }

        env.Start();
    }
}
```

The next step is to create the agents, in the same way as described in Section 4.1. The following code specifies the mobile agent.

```csharp
public class MobileAgent: Agent
{
    private string _log; // stores the pieces of information received from the static agents
    private Queue<string> _moves; // the path to follow when moving between containers
    private bool _firstStart = true;
    private int _turnsToWaitForInfo;

    public override void Setup()
    {
        if (_firstStart) // Setup is also called when arriving to a new container
        {
            _firstStart = false;
            _moves = new Queue<string>();
            foreach (string cn in Environment.AllContainers())
                if (cn != Environment.ContainerName) // home
                    _moves.Enqueue(cn);
            // return home, get local info and report
            _moves.Enqueue(Environment.ContainerName);
        }
        else
        {
            // the agent has moved to Environment.ContainerName
            Broadcast("request-info");
            _turnsToWaitForInfo = 3;
        }
    }

    public override void Act(Message message)
    {
        _log += $"Received info: {message.Content}\r\n"; // info from static agents
    }

    public override void ActDefault()
    {
        if (_turnsToWaitForInfo-- > 0)
            return;

        if (_moves.Count > 0)
        {
            string nextDestination = _moves.Dequeue();
            // checks whether the destination container is still active
            if (CanMove(nextDestination))
            {
                _log += $"Moving to {nextDestination}\r\n";
                Move(nextDestination);
                return;
            }
        }

        _log += "Stopping\r\n";
        Stop();
    }
}
```

A static agent has a much simpler logic.

```
public class StaticAgent: Agent
{
    private string _info; // the piece of information it reports

    public override void Setup()
    {
        _info = $"Info from agent {Name} in container {Environment.ContainerName}";
    }

    public override void Act(Message message)
    {
        if (message.Content == "request-info")
            Send(message.Sender, _info);
    }
}
```

When an agent is supposed to move to a different container, its state is serialized and sent by means of a container message to the destination container. There, a new object is instantiated, its state is set and it is added to the new environment. Thus, ActressMAS employs the concept of weak mobility [27], i.e., the value of the internal fields are preserved during the move, but the execution flow is not preserved: the agent has to finish a method (e.g., *Act*) on the source container, and start from another method (e.g., *Setup*) on its arrival at the destination; it cannot move in the middle of the execution of a method and resume from the next instruction at the destination.

The framework does not impose that the user marks the entire agent class as serializable, in order not to add any constraints to the agent implementations, especially since the user may not want to use mobile agents at all. For example, if the user needs a *Timer* object in an agent to define a recurrent event, that class is no longer serializable. Therefore, the user can choose the specific state that he/she wishes to be transferred when an agent moves. This is achieved by subclassing the *AgentState* class (Figure 14). The derived class must be serializable. This process uses the Memento design pattern [28] to save and restore the internal state of an agent. In this way, the user is also able to send only the relevant parts of the agent state. However, if the agent class itself is serializable, the whole state of the agent can be sent as a specific agent object.

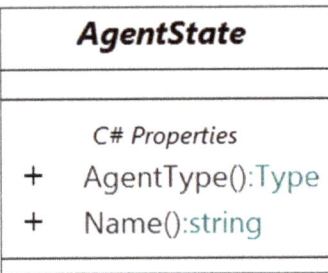

Figure 14. The class of the agent state used in conjunction with the Memento design pattern to ensure the movement of agents between containers in a transparent way.

For the mobile agent presented above, the following two methods can be added.

```
public class MobileAgent: Agent
{
    ...

    public override AgentState SaveState()
    {
        return new MobileAgentState
        {
            FirstStart = _firstStart,
            Log = _log,
            Moves = _moves
        };
    }

    public override void LoadState(AgentState state)
    {
        var st = (MobileAgentState)state;
        _firstStart = st.FirstStart;
        _log = st.Log;
        _moves = st.Moves;
    }
}


[Serializable]
public class MobileAgentState: AgentState
{
    public bool FirstStart;
    public string Log;
    public Queue<string> Moves;
}
```

5. Performance for Some Benchmark Problems

This section attempts to give the reader an idea about the capabilities of ActressMAS with the help of two benchmark problems. They are typically used for actor-based reactive systems; therefore, a direct comparison with other frameworks is not completely objective, because ActressMAS has not been optimized for message passing or agent creation and is not only actor-based. Still, these benchmarks can provide indicative information about the speed of the framework, and this can help potential users to decide whether ActressMAS is appropriate for their particular needs.

The first problem is "Ping Pong". Each agent sends an initial message to all the other agents in the multi-agent system. Then, when an agent receives a message, it replies to the sender. This continues until a maximum number of messages is reached. In our case, scenarios with 10 agents and 10 million messages were considered. A slightly different version of this problem also exists, where a pair of agents exchange only a specified number of messages (e.g., 100), but this was not addressed in our experiments.

The second problem is "Skynet", used to measure the performance of actor creation and basic calculations. Each agent creates 10 children, each child agent creates another 10 children and so on, until a maximum number of agents is eventually reached. Each agent is also assigned a number, incrementally. On the final level, the agents send their ordinal numbers to their parent. Each parent sums these numbers and transmits the sum to its own parent and so on, until the initial root agent sums all the partial sums and reports the final sum.

The results obtained for different scenarios are presented in Table 1. The last column shows the average values out of 10 runs for each configuration. They were obtained using a computer with a 4-core 2 GHz Intel processor and 8 GB of RAM.

Table 1. Performance of the ActressMAS framework for two benchmark problems with various settings.

Benchmark/ Common Settings	Scenario	Results
Ping Pong 10 agents, 10,000,000 messages	parallel execution	1.516 s 6,462,079 messages/s
	sequential execution	3.880 s 2,577,297 messages/s
Skynet 10 children	10,000 agents parallel execution	3.800 s
	20,000 agents parallel execution	12.221 s
	50,000 agents parallel execution	95.611 s
	100,000 agents parallel execution	443.956 s
	10,000 agents sequential execution	0.600 s
	20,000 agents sequential execution	1.523 s
	50,000 agents sequential execution	13.620 s
	100,000 agents sequential execution	92.017 s

The volume testing capacity of ActressMAS seems lower than other professional actor frameworks; e.g., for Akka [18] and Proto.Actor [29] there are reports [29] of approximately 40 million and 120 million messages per second for Ping Pong, and approximately 4 s and 1 s for Skynet with 1 million actors, respectively, although their benchmark results are obtained for different hardware configurations. Therefore, ActressMAS should be used for applications where high performance such as very fast execution speed or a very large number of actors are not critical requirements. It may not be recommended for problems based on very large numbers of simple reactive actors.

As the agent creation benchmark shows, the internal data structures used to store and access agents may be further optimized beyond what the .NET framework offers. Less crucially, the implementation of the queue used by the agents to receive messages may also be improved to increase the number of processed messages.

However, ActressMAS is not only an actor-based framework, but has an especially constructed infrastructure for multi-agent systems, where agents can communicate using messages with custom structure and are not restricted to purely reactive behaviors. Moreover, it is open-source software, and a deliberate implementation choice was to prefer clarity over optimization.

6. Discussion

6.1. Relationship with FIPA Standards

FIPA specifications represent a collection of standards which are intended to promote the interoperation of heterogeneous agents and the services that they can represent. They try to define multiple aspects of agent systems. In this section, we refer to the FIPA specification for an agent abstract architecture [30] that specifies the necessary components of a so-called "Agent Platform", presented in Figure 15. An agent must be registered on a platform in order to interact with other agents.

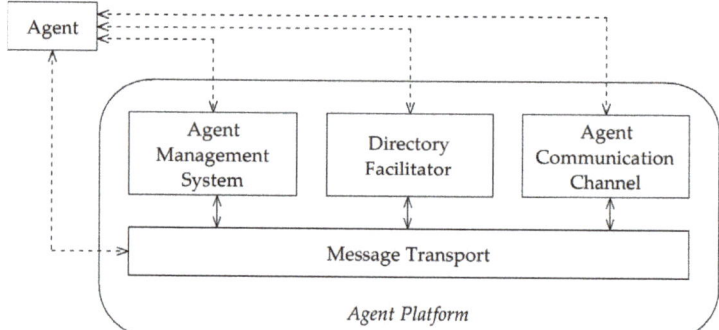

Figure 15. The components of the FIPA Agent Platform (adapted from [31]).

The FIPA specifications were initially intended to be used by various agent-based commercial systems that would need to address the issue of agent interoperability. However, over time such systems failed to materialize. Although more than one hundred agent frameworks have been created (some of them no longer under development) [3], not many had the compliance with the FIPA standards among their design objectives. JADE [4] is a prominent example of a FIPA-compliant agent framework, but a few others are FIPA-compliant as well.

Still, the abstract architecture proposed by FIPA deserves attention from a conceptual point of view. In the following, we discuss how ActressMAS relates to these four components, although it is not FIPA-compliant.

The Agent Management System (AMS) provides "white pages" services and life-cycle management, e.g., creation, deletion and migration of agents. Although this system is not explicitly defined as a separate entity, the environment in ActressMAS fulfils the role of the AMS because it stores the agents and is in charge of the operations mentioned above.

The Directory Facilitator (DF) provides "yellow pages" services for agents. In this case, the agents are seen as service providers and consumers. Some agents can register or deregister their services while others can look for specific services and attempt to use them. ActressMAS does not include a DF infrastructure, but the "yellow pages" functionality is implemented in one of the examples. A service broker agent maintains a collection of the service providers and the services they offer. The clients can interrogate it and receive the list of agents providing a certain service. Then, the clients (the service consumers) and the service providers can communicate directly.

The Agent Communication Channel is responsible for routing the messages to the agents located both on the current platform and on other platforms. The Message Transport service forwards the messages to the destination agents in order and is also responsible for the mapping between the logical names of the agents and their physical transport addresses. The messages are supposed to observe the Agent Communication Language (ACL) format, one of the major contributions of the FIPA specifications.

In ActressMAS, the environment is responsible for passing the messages between agents. If remote messages are needed, they are routed through the containers and the server, and arrive at the destination environment. A message is also a wrapper similar to an ACL message, containing fields such as sender, receiver, the actual content and a conversation identifier. The content can be either a string or an object, which ensures a higher efficiency in some applications. The first word in a string message may be used as a performative, but this is not enforced in any way. In the example of the contract net protocol, a custom class is used for messages, which represents exactly the structure of an ACL message. The protocol is implemented according to the corresponding FIPA specification [32].

6.2. Analysis of ActressMAS in Comparison with Other Multi-Agent Frameworks

In this section we discuss how some features of other agent frameworks are related to the present characteristics of ActressMAS. This can help potential users to better assess the advantages and disadvantages of the proposed framework.

In terms of communication architecture, JADE [4] uses a peer-to-peer approach, while ActressMAS uses a client-server approach for the communication between containers. This is completely transparent from the agent's point of view, because an agent can simply decide to migrate to another container or send a remote message to an agent in another container, while the environment, the containers and the server carry out these actions.

From the point of view of agent scheduling, ActressMAS assigns all agents the same priority. The acting methods of all agents are executed in a turn. Different priorities could be imposed if some agents were ignored during some turns. So far, there has been no intention of introducing such mechanism. ActressMAS agents are lightweight, e.g., one can create one million agents on a computer without any special memory capabilities.

Since the BDI architecture is closely related to agent research, Jason [10] and Jadex [33] have integrated support for it. ActressMAS does not, but provides an example with the BDI architecture. Moreover, it does not contain an internal planning engine; the plans need to be created by the user.

Akka [18] allows a hierarchical organization of actors. An actor can be created by another actor which is then considered its "parent". Each parent can then supervise the execution of the tasks assigned to its children. ActressMAS agents are not organized in a hierarchy; all agents are implicitly on the same level. However, as implemented in the Skynet example, an agent can store a reference to another agent considered to be its parent and report to it, and conversely, parents can store references to their children. These references hold agent names, not object references.

Frameworks such as JaCaMo [34] and MaDKit [35] have integrated support for organizations. For example, in the Agent-Group-Role (AGR) organizational model, agents play roles in groups and create organizations. The roles define some constraints on the agent actions, i.e., obligations, interdictions, and permissions. Although an ActressMAS example models workflows defined as RADs, there is currently no explicit support for roles; however, this is envisioned for future work.

In ActressMAS, agents can create new agents and can indirectly destroy them, through the environment. Therefore, it is not designed for adversarial scenarios with agents belonging to different owners that try to directly harm opposing agents.

Although one of the main tenets of both actors and agents is loose coupling, ActressMAS allows a form of shared memory, especially useful when agents in a multi-agent simulation need to access a large environment. This *Memory* property of the environment should not be used to store global variables for the main logic of the agents, but when large amounts of data need to be accessed frequently, this solution is much more efficient than sending the data in the form of messages. A typical scenario that may benefit from this is when agents communicate indirectly through the environment, by stigmergy. The perceptual function of the agent may also be helped by the custom automatic filtering of agents using the observable properties of ActressMAS.

7. Conclusions

The paper gave an in-depth description of the architecture of ActressMAS, originally intended as a simple-to-use .NET multi-agent framework. However, it proved to be adequate for the implementation of various algorithms, protocols, and simulations, as shown by the example applications.

Considering the categories presented in Section 2, perhaps many types of software agent systems may be implemented using ActressMAS, such as simulations in the mobility, organizational, social, economic, or environmental domains. So far, ActressMAS has not been considered to be used for games or for multi-agent learning scenarios, e.g., supporting multi-agent reinforcement learning algorithms.

Throughout the development of the framework, simplicity has been a main goal. Several methods have been eliminated in successive versions, following the idea that "if it is not important to be in the product, it is important not to be in the product". It was also intended to help the user achieve what other frameworks or specifications provide, but without imposing any constraints. Especially, some standards may be useful for large systems where efficiency and interoperability are necessary, but they may not be needed for simpler, e.g., academic, protocols.

However, recommendations and examples are provided, which can guide the developer to create functionality supported by other frameworks, e.g., using FIPA ACL messages with performatives, implementing a Directory Facilitator, or designing agent-based applications with a BDI or reactive architecture.

In the future versions of ActressMAS, it should be established whether some of these features should be abstracted and integrated into the platform itself. But in this case, the developer should not be forced to use intrusive or mandatory features that he/she does not need.

For example, the concept of agent roles and explicit support for workflow modeling can be added. The platform can be extended with additional capabilities related to various agent architectures. For example, a mechanism to register rules with specified priorities can be incorporated for reactive applications. The planning part of the BDI architecture can be integrated by means of forward reasoning based on pattern matching, already included in the FunCs functional programming library [36] created by the author. Deductive reasoning can also be based on this library and employed within a logical agent architecture.

Funding: This research received no external funding.

Institutional Review Board Statement: Not applicable.

Informed Consent Statement: Not applicable.

Data Availability Statement: The ActressMAS framework and the implementations of the algorithms are open-source and fully available at: https://github.com/florinleon/ActressMas (accessed on 1 November 2021).

Conflicts of Interest: The author declares no conflict of interest.

Appendix A

This appendix provides additional information about the use of the proposed framework and the examples available in its GitHub repository [1].

As stated in the Introduction, ActressMAS was designed to simplify the teaching of multi-agent protocols and algorithms, and it has been used since 2018 in the author's department. In the following paragraphs, several applications are briefly presented. Even if some of them are inspired by different sources, all the implementations are original.

The example in Figure A1 uses a reactive architecture. It implements an idea from [37], where a swarm of robot vehicles (the blue circles) explores an unknown region searching for rock samples (the cyan squares). When a sample is found, it needs to be delivered to a central base (the red circle), which also provides a radio signal that can be used to estimate the direction and distance to it from any other location. Each vehicle has a hierarchy of simple behavioral rules, such as: 1. If an obstacle is detected, then change direction; 2. If carrying samples and being at the base, then drop samples; 3. If carrying samples and not being at the base, then travel up signal gradient; 4. If a sample is detected, then pick it up; 5. Otherwise, move randomly. The ordering also defines the priority of the rules: the rules with smaller numbers have a higher priority than those with higher numbers. Although these rules are very simple, the aggregated behavior of the swarm can be very complex, as vehicles can be seen searching and actively delivering samples to the central base. This is a typical example of emergent behavior in a multi-agent system.

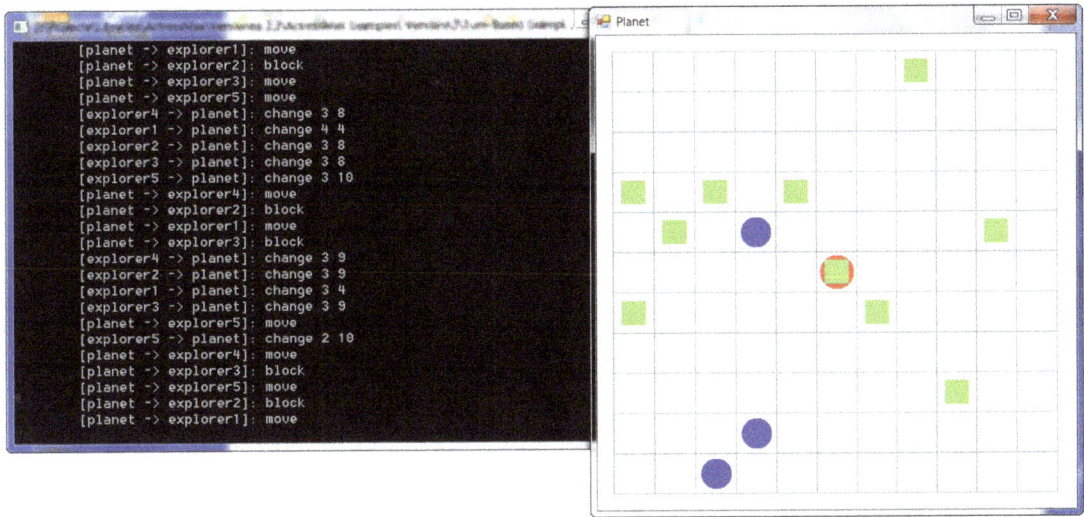

Figure A1. The implementation of a reactive architecture.

The next example implements a BDI architecture for agents [11] and is based on an idea from [38]. A helicopter, designated by a black ellipse in Figure A2, is patrolling for forest fires. Thus, its initial goals are to go from the first cell to the last cell and then reverse direction. When it detects a fire (the red rectangle) in its perceptual field, i.e., its current cell and two the adjacent ones, it adds a higher priority goal to get some water (the blue rectangle) from the first cell, move to the fire cell and drop the water. Therefore, the agent has beliefs such as: "position", "water", and "fire", and desires such as: "patrol right", "patrol left", and "extinguish fire". These goals are achieved by intentions, i.e., plans with a series of individual actions such as: "move left", "move right", "get water", and "drop water". The beliefs are updated based on the percepts received from the interaction with the terrain agent and its own actions.

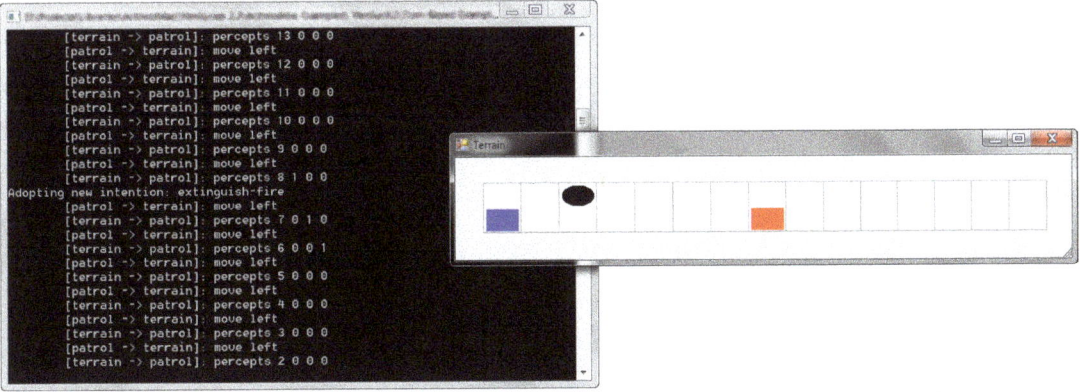

Figure A2. The implementation of a BDI architecture.

ActressMAS was also used to model business processes that define organizational activities. Such processes can be represented using role-activity diagrams [39]. The enactment of business agents for this purpose was previously achieved using the F# and Jason programming languages [40]. The F# implementation, which used actors with "mailbox processors", was easily converted to ActressMAS.

The application whose GUI is displayed in Figure A3 is a simulation of the famous "game of life" [41]. The simulation is driven by three local rules for each cell: 1. Any living cell with two or three living neighbors survives into the next generation; 2. Any dead cell with three living neighbors becomes a living cell; 3. All other living cells die and all other dead cells remain dead.

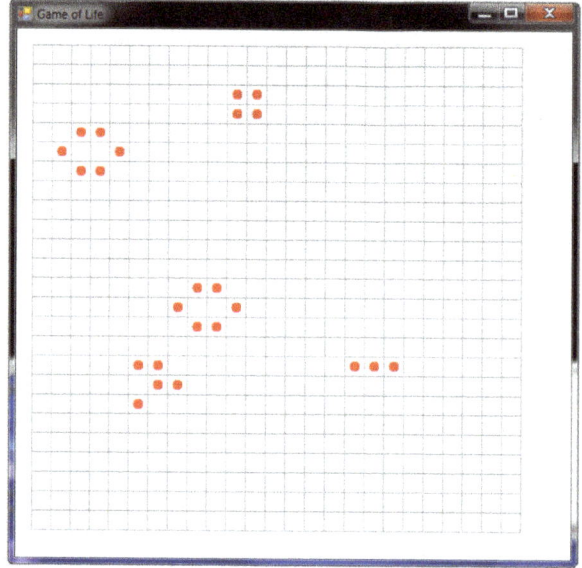

Figure A3. The implementation of the game of life using observables.

These rules can be easily implemented with observables, as explained in Section 4.2. Each cell is an agent that can register its position and its interest in perceiving only the neighboring cells in its Moore neighborhood, i.e., with eight surrounding cells. The state (living or dead) is also an observable property. Thus, each cell is aware of the state of its neighbors and can change its own state by applying the rules mentioned above.

ActressMAS was also used to simulate an environment with two species: predators (in this case, doodlebugs) and prey (in this case, ants). The rules of the simulation are as follows. An ant can: 1. Move randomly up, down, left, or right, if possible; 2. Breed: if an ant survives for three time steps, it creates an offspring in an adjacent cell, if it is free. Conversely, a doodlebug can: 1. Move: in each time step, the doodlebug moves to an adjacent cell containing an ant and eats it; otherwise, it moves randomly; 2. Breed: if a doodlebug survives for eight time steps, it creates an offspring; 3. Starve: if a doodlebug has not eaten an ant within three time steps, it dies.

For this simulation, the *Memory* property of the environment is used in order to store the ecological environment, with cells occupied by at most one type of insect. Again, each insect can perceive its neighboring cells, and when they need to reproduce, a new insect is created and added to the environment in an empty neighboring cell, if possible.

Figure A4 presents the result of the execution of a simulation, where one can see the oscillations of the two populations. This behavior has been theoretically modeled by the Lotka–Volterra equations.

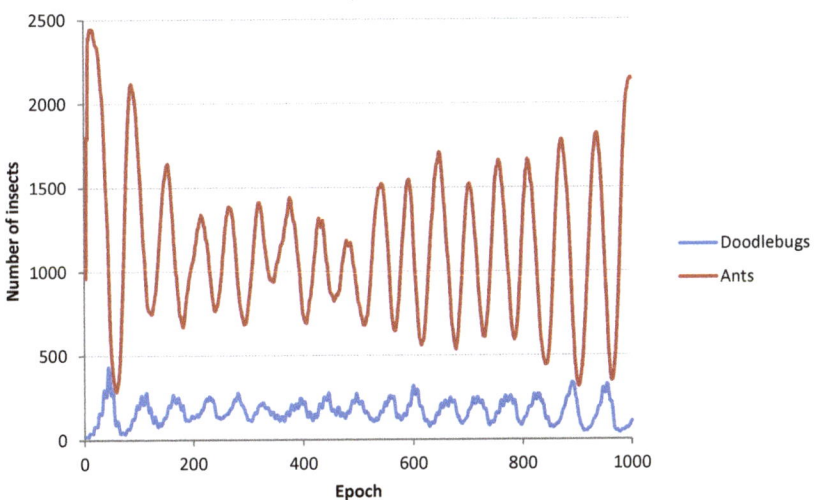

Figure A4. The results of a run of the predator-prey simulation.

Finally, a more complex simulation was made using the framework, i.e., a traffic simulator named "CarSim", which can be used to collect training data for a deep learning system intended for autonomous driving. The user can construct different types of road segments (a road segment is an agent) and place any number of cars (also agents) in different positions. As one can see in Figure A5, the white car with a black dot is the ego car (i.e., the autonomous vehicle), and the cars with other colors are the rest of traffic participants. The user can set several properties of the vehicles: the length, the initial speed, heading angle, acceleration, and the maximum speed, in order to simulate different driving behaviors (more cautious or more aggressive).

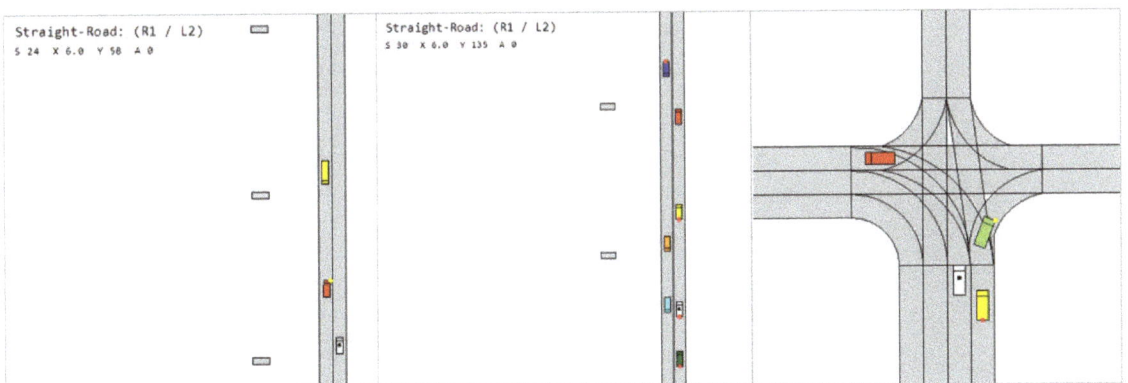

Figure A5. Different traffic simulations scenarios.

The simulator then computes the successive actions for each vehicle agent with a physics model combined with a symbolic model. An action is decided in each time step (e.g., each 0.2 s in the virtual time). The physics model is used to estimate the future trajectories of the traffic participants, based on their current positions, speeds and accelerations. The symbolic model contains rules to handle the interactions between agents. For example, if the physics model detects a possible collision, the vehicle slows down. If the maximum speeds of two or several vehicles require an overtaking to take place and if this is safe from the physics point of view, the current agent begins the overtaking. After it is completed, the vehicle returns to its normal lane.

The simulator can also include static obstacles, bicyclists and pedestrians. Although the main perspective in the simulator is that of the ego car, each agent decides independently, but taking into account its surroundings, as stated above. From a simulation, data can be exported in order to be used in other learning contexts for trajectory prediction.

ActressMAS is provided in the form of a .NET dynamic-link library (DLL) which can be downloaded and to which a developer should add a reference in his/her project and use it directly. No other external packages are needed.

The GitHub repository [1] also offers a single C# solution with all the implemented examples. It can be opened and explored, e.g., using Visual Studio 2017 or newer. The version of the .NET framework is 4.7.2. Table A1 presents the full list of available examples with the main concepts they address.

Table A1. A summary of the examples available in the GitHub repository of ActressMAS v3.0 (as of December 2021).

Project Name	Purpose and Learning Points
Simple Examples→Agents1, Agents2, Agents3, Agents4, Agents5	These examples show how agents should be created in ActressMAS and how to set up the environment. They show different possible execution types: parallel or sequential. The main focus is on sending messages and processing them in the *Act* method. The use of *ActDefault* method at the end of a turn with no received messages is also presented.
Simple Examples→ MultipleMessages	It shows a system where messages are exchanged between several worker agents and a manager agent in a way that would make it difficult for some messages to be delivered if the message passing infrastructure were not properly designed. This is also a test case for ActressMAS.
Simple Examples→ SendingObjects	It shows how user-defined objects can be directly sent in the content of messages.
Reactive Architecture	It presents an implementation of the reactive architecture where multiple behaviors can be activated based on the current state of the agent. The behaviors have priority levels, such that only the behavior with the highest priority defines the next action.
BDI Architecture	It presents an implementation of the Belief-Desire-Intention (BDI) architecture, where agents have explicit state information (beliefs), can have goals (desires) and make plans to achieve these goals (intentions).
LRTA Search	It presents an implementation of the Learning Real-Time A* (LRTA*) path finding algorithm. The search is designed as a continuous conversation between the search agent and the map agent, where the search agent is informed in each state about the neighboring states and the value of the heuristic function in that state. It reflects the behavior of an agent that discovers the map dynamically while performing the search.
Shapley Value	It presents a multi-agent system with worker agents with different skill values solving tasks with different difficulty levels. They divide their payoffs according to their marginal contributions to solving the tasks, using the game theoretic solution concept of the Shapley value.
Auctions→English with broadcast, English without broadcast	These projects implement the English auction protocol in two variants: when all the bids are broadcast to all the agents and when the bidders communicate only with the auctioneer, which in turn communicates the current best price after each round of bids.
Auctions→Vickrey	It implements the Vickrey auction, a sealed-bid protocol where the highest bidder is the winner but pays the second price. The dominant strategy of this auction protocol is bidding the true valuation, thus no agent has a motivation to bid a higher or a lower amount.

Table A1. Cont.

Project Name	Purpose and Learning Points
Yellow Pages	It presents a multi-agent system with service providers, clients (service consumers) and a service broker. Service providers can register or deregister at any time. This is the functionality envisioned by the FIPA Directory Facilitator.
Zeuthen Strategy	It implements the Zeuthen bargaining strategy based on the risks of breaking down the negotiation. The agent that has more to lose if the negotiation fails should be more willing to concede. At each step, the agent with a smaller risk needs to make a concession big enough to change the balance of risks, such that the other agent should concede in the next round.
Contract Net Protocol	It implements the FIPA specification for the contract net protocol. The agents communicate by messages that conform to the FIPA ACL structure. This example features some virtual postmen that should deliver letters and may exchange some of them in order to optimize their routes. Thus, it includes a heuristic travelling salesman problem solver as a subcomponent. The contract net protocol is used to find the (near-)optimal task allocation in terms of payoffs associated with letter delivery and the costs associated with tour length.
Mechanism Design	It implements the Clarke–Groves tax system which eliminates the incentive of an agent to lie about its true preference in a majority voting scenario.
Iterated Prisoner Dilemma	It implements the iterated prisoner dilemma game, which is a simple model that reveals deep questions related to human selfishness and cooperation. It includes multiple response strategies whose outcomes can be compared: acting randomly, always defecting, and "tit-for-tat" (where an agent first cooperates, then chooses the action chosen by the other agent in the previous round).
Predator-Prey→ PredatorPreyConsole, PredatorPreyGui	These projects present a simulation with two species, predators and prey, which emphasizes the oscillating evolution of the two populations. The grid world (i.e., the natural environment) is stored in the *Memory* property of the ActressMAS software environment and can be accessed by the individuals. Individuals can be created and removed from the environment.
Simple Observables→ ColorGame, NumberGame	These projects focus on the use of observable properties. Each agent perceives the other agents with a certain assigned color or with an assigned number in a specific range.
Game of Life	It is an implementation of Conway's "game of life" using observable properties of agents/cells to compute the number of alive neighbors.
Voting	It presents a voting protocol that first tries to identify the Condorcet winner using the Copeland's method, and if no Condorcet winner exists, it uses Borda count to produce the result.
Mobile Examples→ MyServerConsole, MasMobileConsole, MyServerGui, MasMobileGui	These projects present an example with mobile and static agents. A mobile agent visits the existing containers, collects information form the static agents running there and then returns to its original container and reports the information. There are two equivalent implementations, one using a console for displaying messages, i.e., a character user interface, and another using a Windows-based graphical user interface.
Mobile Examples→ RemoteMessages	It shows how two agents in different containers can communicate by sending remote messages.
Benchmarks→PingPong	This is a benchmark that measures the number of messages that can be exchanged between agents in a time unit in a communication-intensive scenario.
Benchmarks→Skynet, SkynetNumeric	This is a benchmark that measures the performance of agent creation and basic calculations. In Skynet, the numbers are encoded into string messages, while in SkynetNumeric, the numbers are encoded as 64-bit integer numbers. However, ActressMAS does not exhibit a significant difference in performance in the two cases.

References

1. Leon, F. ActressMAS Library. 2018–2021. Available online: https://github.com/florinleon/ActressMas (accessed on 1 November 2021).
2. Leon, F. ActressMAS. A .NET Multiagent Framework. 2021. Available online: http://florinleon.byethost24.com/actressmas (accessed on 1 November 2021).
3. Pal, C.V.; Leon, F.; Paprzycki, M.; Ganzha, M. A Review of Platforms for the Development of Agent Systems. *arXiv* **2020**, arXiv:2007.08961; 40 pages.
4. Bellifemine, F.L.; Caire, G.; Greenwood, D. *Developing Multi-Agent Systems with JADE*; Wiley Series in Agent Technology: Chichester, UK, 2007.
5. Luke, S.; Balan, G.C.; Panait, L.; Cioffi-Revilla, C.; Paus, S. MASON: A Java Multi-Agent Simulation Library. In Proceedings of the Agent 2003 Conference on Challenges in Social Simulation, Chicago, IL, USA, 2–4 October 2003.
6. Bernstein, P.; Bykov, S.; Geller, A.; Kliot, G.; Thelin, J. *Orleans: Distributed Virtual Actors for Programmability and Scalability*; Technical Report MSR-TR-2014-41; Microsoft: Redmond, WA, USA, 2014.
7. Ritter, F.E.; Tehranchi, F.; Oury, J.D. ACT-R: A cognitive architecture for modeling cognition. Wiley Interdisciplinary Reviews. *Cogn. Sci.* **2019**, *10*, e1488. [CrossRef]
8. Laird, J.E. *The Soar Cognitive Architecture*; MIT Press: Cambridge, MA, USA, 2012.
9. Newell, A.; Shaw, J.C.; Simon, H.A. Elements of a theory of human problem solving. *Psychol. Rev.* **1958**, *65*, 151–166. [CrossRef]
10. Bordini, R.H.; Hübner, J.F.; Wooldridge, M. *Programming Multi-Agent Systems in AgentSpeak Using Jason*; Wiley Series in Agent Technology: Chichester, UK, 2007.
11. Rao, A.S.; Georgeff, M.P. Modeling Rational Agents within a BDI-Architecture. In Proceedings of the 2nd International Conference on Principles of Knowledge, Representation and Reasoning, Cambridge, MA, USA, 22–25 April 1991; pp. 473–484.
12. Rao, A.S. AgentSpeak(L): BDI Agents Speak Out in a Logical Computable Language. In Proceedings of the Seventh European Workshop on Modelling Autonomous Agents in a Multi-Agent World (MAAMAW-96), Einhoven, The Netherlands, 22–25 January 1996.
13. Brockman, G.; Cheung, V.; Pettersson, L.; Schneider, J.; Schulman, J.; Tang, J.; Zaremba, W. OpenAI Gym. *arXiv* **2016**, arXiv:1606.01540; 4 pages.
14. Beattie, C.; Leibo, J.Z.; Teplyashin, D.; Ward, T.; Wainwright, M.; Küttler, H.; Lefrancq, A.; Green, S.; Valdés, V.; Sadik, A.; et al. DeepMind Lab. *arXiv* **2016**, arXiv:1612.03801; 11 pages.
15. Wilensky, U.; Rand, W. *An Introduction to Agent-Based Modeling: Modeling Natural, Social, and Engineered Complex Systems with NETLogo*; The MIT Press: Cambridge, MA, USA, 2015.
16. Dosovitskiy, A.; Ros, G.; Codevilla, F.; Lopez, A.; Koltun, V. CARLA: An Open Urban Driving Simulator. *arXiv* **2017**, arXiv:1711.03938; 16 pages.
17. Horni, A.; Nagel, K.; Axhausen, K.W. (Eds.) *The Multi-Agent Transport Simulation MATSim*; Ubiquity Press: London, UK, 2016. [CrossRef]
18. Bonér, J.; Klang, V.; Kuhn, R. Akka Library. 2021. Available online: https://akka.io (accessed on 1 November 2021).
19. Petabridge. Akka.NET Library. 2021. Available online: https://getakka.net (accessed on 1 November 2021).
20. Hewitt, C.; Bishop, P.; Steiger, R. A Universal Modular Actor Formalism for Artificial Intelligence. In Proceedings of the 3rd International Joint Conference on Artificial intelligence (IJCAI'73), Stanford, CA, USA, 20–23 August 1973; pp. 235–245.
21. Wooldridge, M. *Intelligent Agents. Multiagent Systems—A Modern Approach to Distributed Artificial Intelligence*; Weiss, G., Ed.; The MIT Press: Cambridge, MA, USA, 2000; pp. 27–77.
22. Wooldridge, M. *An Introduction to Multiagent Systems*, 2nd ed.; Wiley: Hoboken, NJ, USA, 2009.
23. Burgin, M. Systems, Actors and Agents: Operation in a Multicomponent Environment. *arXiv* **2017**, arXiv:1711.08319v1; 28 pages.
24. Rumbaugh, J.; Jacobson, I.; Booch, G. *Unified Modeling Language Reference Manual*, 2nd ed.; Pearson Education: Boston, MA, USA, 2005.
25. Foundation for Intelligent Physical Agents. FIPA Communicative Act Library Specification. 2002. Available online: http://www.fipa.org/specs/fipa00037/SC00037J.html (accessed on 1 November 2021).
26. Austin, J.L. *How to Do Things with Words*; Clarendon Press: Oxford, UK, 1975.
27. Cugola, G.; Ghezzi, C.; Picco, G.P.; Vigna, G. Analyzing Mobile Code Languages. In *Lecture Notes in Computer Science*; Springer: Berlin/Heidelberg, Germany, 1997; Volume 1222, pp. 94–109.
28. Gamma, E.; Helm, R.; Johnson, R.; Vlissides, J. *Design Patterns: Elements of Reusable Object-Oriented Software*; Addison-Wesley Professional: Boston, MA, USA, 1994.
29. Asynkron, A.B. Proto. Actor Library. 2021. Available online: https://proto.actor (accessed on 1 November 2021).
30. Foundation for Intelligent Physical Agents. FIPA Abstract Architecture Specification. 2002. Available online: http://www.fipa.org/specs/fipa00001/SC00001L.html (accessed on 1 November 2021).
31. Poslad, S.; Buckle, P.; Hadingham, R. The FIPA-OS agent platform: Open source for open standards. In Proceedings of the 5th International Conference and Exhibition on the Practical Application of Intelligent Agents and Multi-Agents, Manchester, UK, 10–12 April 2000; Volume 355.
32. Foundation for Intelligent Physical Agents. FIPA Contract Net Interaction Protocol Specification. 2002. Available online: www.fipa.org/specs/fipa00029/SC00029H.html (accessed on 1 November 2021).
33. Braubach, L.; Pokahr, A.; Lamersdorf, W. Jadex: A BDI-Agent System Combining Middleware and Reasoning. In *Whitestein Series in Software Agent Technologies*; Birkhäuser-Verlag: Basel, Switzerland, 2006; pp. 143–168. [CrossRef]

34. Boissier, O.; Bordini, R.H.; Hübner, J.F.; Ricci, A.; Santi, A. Multi-agent Oriented Programming with JaCaMo. *Sci. Comput. Program.* **2013**, *78*, 747–761. [CrossRef]
35. Gutknecht, O.; Ferber, J. The MadKit Agent Platform Architecture. In *Infrastructure for Agents, Multi-Agent Systems, and Scalable Multi-Agent Systems*; Springer: Berlin/Heidelberg, Germany, 2001; pp. 48–55. [CrossRef]
36. Leon, F. FunCs Library. 2018–2021. Available online: https://github.com/florinleon/FunCs (accessed on 1 November 2021).
37. Steels, L. Cooperation Between Distributed Agents Through Self-Organisation. In Proceedings of the IEEE International Workshop on Intelligent Robots and Systems, Towards a New Frontier of Applications, Ibaraki, Japan, 3–6 July 1990; pp. 8–14. [CrossRef]
38. Taillandier, P.; Gaudou, B.; Grignard, A.; Huynh, Q.; Marilleau, N.; Caillou, P.; Philippon, D.; Drogoul, A. Building, composing and experimenting complex spatial models with the GAMA platform. *Geoinformatica* **2019**, *23*, 299–322. [CrossRef]
39. Ould, M.A. *Business Process Management: A Rigorous Approach*; British Computer Society, Meghan Kiffer: Tampa, FL, USA, 2005.
40. Leon, F.; Bădică, C. A Comparison Between Jason and F# Programming Languages for the Enactment of Business Agents. In Proceedings of the International Symposium on INnovations in Intelligent SysTems and Applications (INISTA), Sinaia, Romania, 2–5 August 2016. [CrossRef]
41. Martin, G. The Fantastic Combinations of John Conway's New Solitaire Game "Life". *Math. Games. Sci. Am.* **1970**, *223*, 120–123. [CrossRef]

Article

Automatic Fingerprint Classification Using Deep Learning Technology (DeepFKTNet)

Fahman Saeed, Muhammad Hussain * and Hatim A. Aboalsamh

Department of Computer Science, King Saud University, Riyadh 11451, Saudi Arabia; fahmanali@gmail.com (F.S.); hatim@ksu.edu.sa (H.A.A.)
* Correspondence: mhussain@ksu.edu.sa

Abstract: Fingerprints are gaining in popularity, and fingerprint datasets are becoming increasingly large. They are often captured utilizing a variety of sensors embedded in smart devices such as mobile phones and personal computers. One of the primary issues with fingerprint recognition systems is their high processing complexity, which is exacerbated when they are gathered using several sensors. One way to address this issue is to categorize fingerprints in a database to condense the search space. Deep learning is effective in designing robust fingerprint classification methods. However, designing the architecture of a CNN model is a laborious and time-consuming task. We proposed a technique for automatically determining the architecture of a CNN model adaptive to fingerprint classification; it automatically determines the number of filters and the layers using Fukunaga–Koontz transform and the ratio of the between-class scatter to within-class scatter. It helps to design lightweight CNN models, which are efficient and speed up the fingerprint recognition process. The method was evaluated two public-domain benchmark datasets FingerPass and FVC2004 benchmark datasets, which contain noisy, low-quality fingerprints obtained using live scan devices and cross-sensor fingerprints. The designed models outperform the well-known pre-trained models and the state-of-the-art fingerprint classification techniques.

Keywords: multisensory fingerprint; interoperability; DeepFKTNet; deep learning; classification

MSC: 68T05

1. Introduction

A person can be recognized in security systems by a unique username and password, but they can be readily stolen [1]. The fingerprint is one of the first imaging modalities of biometric identification. It is more accurate and less expensive than other biometric modalities [2,3]. A fingerprint's surface has ridges and valleys, which do not change during a lifetime [4]. Fingerprint recognition can be used for authentication or identifying purposes. In verification, the fingerprint is compared to the templates of a particular subject in the database, but in identification, the unknown fingerprint is compared to the templates of all subjects in the database to ascertain the subject's identity [5]. Fingerprints are gaining in popularity and their datasets are becoming increasingly large. They are recorded utilizing a variety of low-cost embedded sensors in smart devices such as smartphones and computers. The high processing complexity of a fingerprint identification system is one of its primary drawbacks. One way to address this issue is to categorize fingerprints in a database to condense the search space. The existing classification methods are effective when fingerprints are recorded using the same sensor. However, when fingerprints are collected using various sensors (referred to as cross-sensor or sensor interoperability problem), classification performance is deteriorated; even verification of the same person's finger is degraded [6–8]. While considerable research has been conducted on cross-sensor fingerprint verification [8–12], there has been no study on cross-sensor fingerprint classification, which motivates us to work on this topic.

Numerous fingerprint categorization systems have been developed, some relying on non-conventional approaches and others on convolutional neural networks. The references provide an exhaustive overview of non-CNN methods [13,14]. The success rate of a fingerprint classification approach is highly dependent on the quality of the description of the discriminating information of a fingerprint. Directional ridge patterns and singularities are critical distinguishing characteristics of fingerprints, as demonstrated by the techniques proposed in [15–20], which utilize this information in a variety of ways to classify fingerprints. Gue et al. [15] employ the amount and kind of core points as fingerprint descriptors, as well as rule-based categorization, to classify fingerprints. Additionally, this approach classifies indistinguishable fingerprints using center-to-delta flow and balance arm flow. Its categorization accuracy is 92.7% on average. Jung and Lee [21] split a fingerprint into 16 × 16 pixel blocks, compute their representative directions, use Markov models to identify the core block, and then divide the fingerprint into four areas, each of which is represented using distributions of ridge directional values. This method has a classification accuracy of 97.4%. Dorasamy et al. [17] employed a simplified rule-based technique and two features: directional patterns and singular points for fingerprint description. The classification accuracy of this scheme is 92.2%. Saeed et al. [18] proposed a modified histogram of oriented gradients (HOG) fingerprint classification algorithm. The HOG descriptor's orientation field computation is not ridge pattern specific. In order to improve the HOG descriptor's ability to represent a fingerprint, we compute an orientation field that is suited to the ridge pattern. This technique achieved an average accuracy of 98.70% on the noisy fingerprint database FVC2004. Saeed et al. [19] suggested a new approach for classifying noisy fingerprints from live scan devices using statistical features (mean, standard deviation, kurtosis and skewness) from dense scale invariant feature transform (d-SIFT). This method achieved 97.6% accuracy using FVC2004, a noisy, low-quality live scanned fingerprint database. Sudhir et al. [22] employed GLCM, LBP, and SURF for feature extraction, while SVM and BoF classifiers were used for classification. Based on FVC2004, they got average accuracy of 74.50 using SVM and 84.75 using BoF.

Deep CNN has shown remarkable results in many applications [23–26]; it has been used to classify fingerprints [27–32] and has achieved encouraging results. Zia et al. [33] introduced the Bayesian DCNNs (B-DCNNs) by incorporating Bayesian model uncertainty to increase fingerprint categorization accuracy. They achieved 95.3% accuracy on FVC004 (5 class), showing a 0.8–1.0% improvement in model accuracy compared to the baseline DCNN. In Nguyen et al. [34], the CNN approach is suggested for the noise reduction stage of noisy fingerprint. Two main steps are involved in this procedure. Non-local information is used to construct a pre-processing phase for noisy image. Fingerprints are then separated into patches and utilized for CNN training, resulting in a model for CNN de-noising of future noisy images, which can subsequently be smoothed using Gaussian filtering to remove pixel artifacts. Fingerprints that have been pre-processed are separated into overlapping patches during the CNN training step. To train the convolutional neural network, they feed these patches into it. They've built a three-tiered network with distinct filters and operators at each level. Third layer convolutional layer predicts enhancing patches and reconstructs the output image. Using the Gaussian algorithm and a canny algorithm they strength the information edge, this approach is able to filter out noise. When all images have been processed by the morphological procedure, the result will be improved. They extracted features from pre-processed fingerprints (arch, loop and whorl) and classified them using for classifiers: random forest, SVM, CNN, and K-NN and obtained accuracies of 97.78%, 95.83%, 96.11%, and 92.05%, respectively.

Nahar et al. [35] designed CNN models based on the LeNet-5 design for fingerprint classification. They evaluated their method using the augmented subset (DB1) from the FVC2004 dataset. They got an accuracy of 99.1%. In deep models, layers and filters are defined by experiments, and no special rule is used to choose them; tuning the hyperparameters is tiring and time-consuming. Motivated by the difficulty in the design of CNN architectures, we propose a technique that determines automatically and adaptively the architecture of a CNN model using the fingerprints dataset. To begin, we use the LGDBP description Saeed, et al. [36] and K-medoid clustering algorithm [37] to choose representative fingerprints, and then we derive the layers filters using Fukunaga–Koontz Transform (FKT) [38]. To control the depth of a CNN model, we compute the ratio between traces of between-class scatter matrix S_b and within-class scatter matrix S_w.

The proposed fingerprint CNN classification system was evaluated against the state-of-the-art fingerprint classification schemes utilizing the benchmark multi-sensor datasets FingerPass and FVC2004. Specifically, the contributions of this work are as follows:

- We developed an efficient automatic method for classifying cross-sensor fingerprints based on a CNN model.
- We proposed a technique for the custom-designed building of a CNN model, which automatically determines the architecture of the model using the class discriminative information from fingerprints. The layers and their respective filters of an adaptive CNN model are customized using FKT, and the ratio of the traces of the between-class scatter matrix, and the within-class scatter matrix.
- We thoroughly evaluated the proposed method on two datasets. The proposed fingerprint classification scheme is quick, accurate, and performs well with noisy fingerprints obtained using live scan devices as well as cross-sensor fingerprints.

The rest of the paper is organized as follows. Section 2 presents the details of the proposed technique. The experimental results have been given in Section 3. Section 4 discusses the performance of the proposed method in detail. Section 5 concludes the article.

2. Proposed Method

The convolutional neural network (CNN) is one of the most widely used and popular deep learning networks [39]. Its general structure comprises different types of layers, including the CONV layer with different filters, pooling layer, activation function layer, fully connected layer, and loss function [40]. It has been used for a wide range of tasks, including image and video recognition [41], classification of images [42], medical image analysis [43], computer vision [44], and natural language processing [45].

Many advancements in CNN learning methods and architecture have a place, allowing the network to handle larger, diverse, more complicated, and multiclass issues [46]. Following AlexNet's outstanding performance on the ImageNet dataset in 2012, many applications used CNNs [47]. A layer-wise representation of CNN reversed the trend toward extraction of features at low spatial resolution in deep architecture, as achieved in VGG [48]. Most modern architectures follow VGG's simple and homogeneous topology idea. The Google deep learning group introduced the divide, transform, and merge concept with the inception block. The inception block introduced the concept of branching within a layer, allowing for feature abstraction at various spatial scales [49]. Skip connections, developed by ResNet [50] for deep CNN training, gained popularity in 2015. Others, like Wide ResNet, are exploring the influence of multilevel transformations on CNN's learning capacity by increasing cardinality or widening the network [51]. So, the research turned from parameter optimization to network architecture design. Thus, new architectural concepts like channel boosting, spatial and feature-map exploitation, and attention-based information processing emerged [52]. The main issue in the design of CNN models is to tune the architecture of CNN for a specific application.

2.1. Problem Formulation

The fingerprints are categorized into four types: arch, left loop, right loop, and whorl. Identifying the type of a fingerprint is a multiclass classification problem. Let there be N subjects, and K fingerprints are captured from each subject with M different sensors; these fingerprints are categorized into C classes. Let $\mathcal{F} = \left\{ F_{ij}^s \middle| 1 \leq i \leq K, 1 \leq j \leq N, 1 \leq s \leq M \right\}$, where F_{ij}^s represents the ith fingerprint of the jth subject captured with sth sensor, be the set of fingerprints, and $\mathbb{C} = \{1, 2, \ldots, C\}$, where C is the number of classes, be the set of fingerprint labels (classes). The problem of predicting the type of a fingerprint F_{ij}^s is to build a function $\psi : \mathcal{F} \to \mathbb{C}$ that takes a fingerprint $F_{ij}^s \in \mathcal{F}$ and assigns it a label $c \in \mathbb{C}$, i.e., $\psi\left(F_{ij}^s; \theta\right) = c$, where θ are the parameters. We design the function ψ using a CNN model, in this case θ represents the weights and biases of the model. The model is built adaptively. Its design process is shown in Figure 1, and the detail is given in the rest of the section.

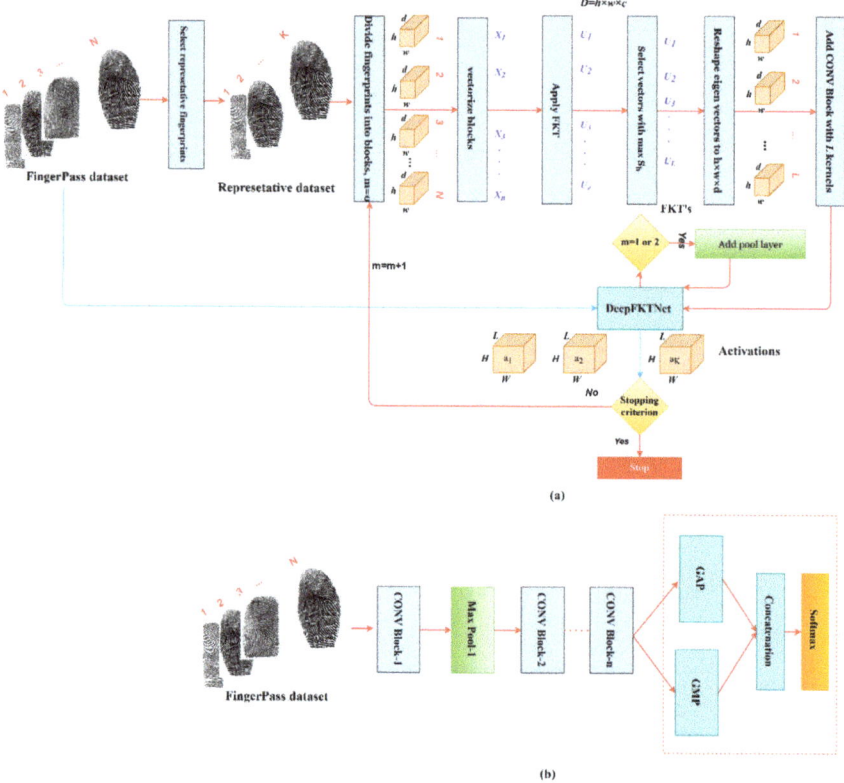

Figure 1. Design procedure of DeepFKTNet; (a) design of main DeepFKTNet architecture and (b) addition of global pooling and softmax layers and fine-tuning the model.

2.2. Adaptive CNN Model

The main constituent of a CNN model is a convolutional (CONV) layer. It extracts discriminative features from the input signal, applying convolution operation with filters of fixed size. CONV layers are stacked in a CNN model to extract a hierarchy of features. The number of filters in each CONV layer and the number of CONV layers in a CNN model are hyper-parameters, and finding the best configuration of a model for a specific application is a hard optimization problem; it entails the search of huge parameter space. In addition, the initialization of learnable parameters of a CNN model has a significant effect on the performance of the model when it is trained with an iterative optimization algorithm like Adam optimizer. Leveraging the discriminative content of fingerprints, we propose a simple method to find the best configuration of the model adaptively. Initially, we select the representative fingerprints from each type to guide the design process of a CNN model. The discriminative information in these fingerprints is used to determine the width (the number of filters) of each CONV layer and the depth (the number of CONV layers) of the model; it is also used for data-dependent initialization of the filters of CONV layers. An overview of the design process is shown in Figure 1. We employ clustering to select the representative fingerprints, the Fukunaga–Koontz Transform (FKT) [38], which exploits class-discriminative information, to determine the number of filters in a CONV layer, and the ratio of the between-class scatter matrix S_b to the within-class scatter matrix S_w to adjust the depth (i.e., the number of CONV layers) of the CNN model. Finally, to minimize the number of learnable parameters and avoid overfitting, global pooling layers are introduced. By decreasing the resolution of the feature maps, the pooling layer seeks to achieve shift-invariance, and the pooling layer's feature map is linked directly to SoftMax [53]. The design process is worked out in detail and discussed in the following subsections, and its overview is shown in Figure 1.

2.2.1. Selection of Representative Fingerprints

We extract discriminative information from fingerprints to specify the CONV layers and the depth of a CNN model adaptively. To do this, we cluster the training set to identify the most representative fingerprints of each class. For determining the representative fingerprints, discriminative features from fingerprints are extracted using the LGDBP descriptor [36] K-medoids [37] is used for clustering since it selects the instances as cluster centers and is suitable for finding the representative subset of the training set. The fingerprints corresponding to the cluster centers are chosen as the representative subset. The number of clusters for each class in the K-medoids algorithm is specified using the silhouette analysis [54]. Using this procedure, we select the set $X = \{X_1, X_2, \ldots, X_C\}$, where $X_i = \{RF_j, j = 1, 2, 3, \ldots, n_i\}$ is the set of representative fingerprints of ith class.

2.2.2. Design of the Main DeepFKTNet Architecture

The architectures of the state-of-the-art CNN models are usually not drawn from the data and are fixed and highly complex. On the contrary, we define a data-dependent architecture of DeepFKTNet. Its primary architecture is based on the answers to two questions: (i) how many CONV layers should be in the model and (ii) how many filters must be in each layer. These questions are addressed by an iterative algorithm that computes the number of filters in a CONV layer, adds it iteratively to the model, and terminates when a criterion is satisfied. We use the discriminative structural information embedded in fingerprints to determine the number of filters in a CONV layer and their initialization. The detail is given in Algorithm 1. We discuss the algorithm with motivation in the following paragraphs.

Initially, the set $X = \{X_1, X_2, \ldots, X_C\}$ is used to determine the number of filters of the first CONV layer and initialize them. Inspired by the filter size of the first CONV layer in the state-of-the-art CNN models like ResNet [50], DenseNet [55], and Inception [49], we fixed the size of filter size of the first layer to 7×7. We extract patches of size $w \times h$ from the representative fingerprints (steps 2–3 of Algorithm 1) and formulate the problem of determining the filters (f_i, $i = 1, 2, \ldots N$) as finding the optimal projection direction vectors u_i, $i = 1, 2, \ldots d$, which are determined by solving the following optimization problem:

$$U^* = \arg\max_U \frac{tr(U^T S_b U)}{tr(U^T S_w U)} \quad (1)$$

where S_b and S_w are the between-class and within-class scatter matrices (as computed in step 4 of the Algorithm 1). According to Fukunaga Koontz Discriminant Analysis (FKT) [38], the optimal projection direction vectors u_i are the eigenvectors of $\hat{S}_b$ i.e.,

$$\hat{S}_b u = \lambda u \quad (2)$$

where $\hat{S}_b = P^T S_b P$, $P = QD^{-1/2}$ and Q & D are obtained by the diagonalization of the sum $S_b + S_w$ i.e., $S_b + S_w = QDQ^T$ (steps 5–6 of Algorithm 1). The Equation (2) gives the optimal vectors, which simultaneously maximize $tr(U^T S_b U)$ and minimize $tr(U^T S_w U)$. Unlike Linear Discriminant Analysis (LDA) [56], the inversion of S_w is not needed in this approach, so it can tackle very high-dimensional data. Additionally, this approach seeks to find optimal vectors that are orthogonal. As the dimension of the patch vectors b_i related to the intermediate CONV layers is usually very high, and we need filters that are independent, so this approach is suitable for our design process. The problem of selecting the number of filters in the convolutional layer is to select the eigenvectors u_k, $k = 1, 2, \ldots L$ so that the ratio $\gamma_k = \frac{Trace(SF_b)}{Trace\ (SF_w)}$ attains maximum value. Here the between-class scatter matrix SF_b and within-class matrix SF_w are computed for each u_k by projecting all activations a_j^i in the space spanned by u_k (steps 7–8 of the Algorithm 1). It ensures to select the filters which extract discriminative features. After selecting u_k, $k = 1, 2, \ldots L$, the CONV block with L filters f_k, $k = 1, 2, \ldots, L$ initialized with u_k is introduced in DeepFKTNet. Then, a pooling layer is added if needed (step 8–10 of the Algorithm 1).

Using the current architecture of DeepFKTNet, the set of activations $Z = \{Z_1, Z_2, \ldots, Z_C\}$ of $X = \{X_1, X_2, \ldots, X_C\}$ is computed. These activations are used to determine whether to add more layers to the net. It is decided by calculating the trace ratio $TR = \frac{Trace(S_b')}{Trace\ (S_w')}$, where S_b' and S_w' are the between-class and within-class scatter matrices of the activations Z. If TR is greater than the previous TR (PTR), it means that the addition of the current block of layers introduced the discriminative potential to the network. This criterion ensures that the features generated by DeepFKTNet have large inter-class variation and small intra-class scatter. To add another CONV block, the steps 3–10 are repeated with Z. To reduce the size of feature maps for computational effectiveness, pooling layers are added after the first and second CONV blocks. As the kernels and their number are determined from the fingerprint images, each layer can have a different number of filters.

It is to be noted that the eigenvector u_k, which are used to specify the kernels of a CONV layer, have the maximum γ_k and capture most of the variability in input fingerprint images without redundancy in the form of independent features. The depth of a CNN model (number of layers) and the number of kernels for each layer are important factors that determine the model complexity. Step 7 of Algorithm 1 determines the best kernels that ensure the preservation of maximum energy of the input image, and step 8 initializes these kernels to be suitable for the fingerprint domain. The selected kernels extract the features from fingerprint images so that the variability of the structures in fingerprint images is maximality preserved. It is also important that the features must be discriminative (i.e., have large inter-class variance and small intra-class scatter as we go deeper in the network). It is ensured using the trace ration $TR = \frac{Trace(Sb)}{Trace\ (Sw)}$, the larger the value of the trace ratio, the larger the inter-class variance and the smaller the intra-class scatter [57]. Step 11 in Algorithm 1 allows adding CONV layers as long as TR is increasing and determines the data-dependent depth of DeepFKTNet, as shown in Figure 2.

Algorithm 1: Design of the main DeepFKTNet Architecture

Input: The set $X = \{X_1, X_2, \ldots, X_C\}$, where $X_i = \{RF_j, j = 1, 2, 3, \ldots, n_i\}$ is the set of representative fingerprints of ith class.
Output: The main DeepFKTNet Architecture.

Step 1: Initialize DeepFKTNet with input layer and set $w = 7, h = 7, d = 1$, and m (the number of filters) = 0 for the first layer; PTR (previous TR) = 0.

Step 2: For $i = 1, 2, 3, \ldots, C$
Compute $Z_i = \{a_j^i = RF_j, \text{ for each } RF_j \in X_i\}$

Step 3: For $i = 1, 2, 3, \ldots, C$
$A_i = \emptyset$
For each $a_j^i \in Z_i$
Extract patches $p_1^j, p_2^j, \ldots, p_m^j$ of size $w \times h$ with stride 1 from a_j^i, vectorize them into vectors of dimension $D = w \times h \times d$ and append to A_i.

Step 4: Using $A = [A_1, A_2, \ldots, A_C]$, compute
-between-class scatter matrix $S_b = \sum_{i=1}^{C} (\frac{1}{n_i} A_i J_i - \frac{1}{n} AJ)(\frac{1}{n_i} A_i J_i - \frac{1}{n} AJ)^T$, where J_i is an $n_i \times n_i$ matrix with all ones.
-within-class scatter matrices $S_w = \sum_{i=1}^{C} (A_i - \frac{1}{n_i} A_i J_i)(A_i - \frac{1}{n_i} A_i J_i)^T$

Step 5: Diagonalize the sum $\Sigma = S_b + S_w$ i.e., $\Sigma = QDQ^T$ and transform the scatter matrices using the transform matrix $P = QD^{-\frac{1}{2}}$. i.e., $\hat{S}_b = P^T S_b P, \hat{S}_w = P^T S_w P$.

Step 6: Compute eigenvectors $u_k, k = 1, 2, \ldots, D$ of $\hat{S}_b$ such that $\hat{S}_b u = \lambda u$

Step 7: For each eigenvector $u_k, k = 1, 2, \ldots, D$
-Reshape u_k to a filter f_k of size $w \times h \times d$
-Compute $Y = \{Y_1, Y_2, \ldots, Y_C\}$, where $Y_i = \{f_k * a_j^i, j = 1, 2, \ldots, n_i\}$
-Compute the between scatter matrix SF_b and within scatter matrix SF_w from Y.
-Compute the trace ratio $\gamma_k = \frac{Trace(SF_b)}{Trace(SF_w)}$

Step 8: Select L filters $f_k, k = 1, 2, \ldots, L$ corresponding to $\gamma_k > 0$ (as shown in Figure 2 for layer 1).

Step 9: Add the CONV block to DeepFKTNet with filters $f_k, k = 1, 2, \ldots, L$. Update $m = m + 1$.

Step 10: If $m = 1$ or 2, add a max pool layer with pooling operation of size 2×2 and stride 2 to Deep FKTNet.

Step 11: Compute $Z = \{Z_1, Z_2, \ldots, Z_C\}$, where $Z_i = \{a_j^i = \text{DeepFKTNet}(RF_j), \text{ for each } RF_j \in X_i\}$

Step 12: Using $Z = \{Z_1, Z_2, \ldots, Z_C\}$, compute the ratio $TR = \frac{Trace(S_b')}{Trace(S_w')}$
If $PTR \leq TR$, set $PTR = TR, w = 3, h = 3, d = L$ and go to Step 3, otherwise stop.

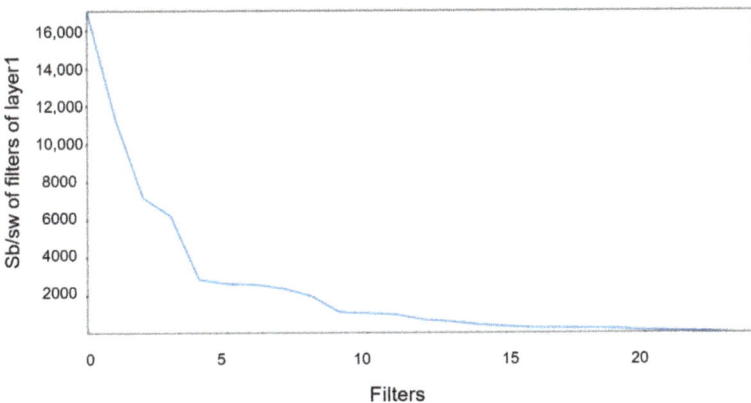

Figure 2. Selection of best filters for layer1 of DeepFKTNet model for FingerPass dataset.

2.3. Addition of Global Pool and Softmax Layers

Activation of the last CONV block is with dimension $h \times w \times L$, and after flattening, it is fed to FC layers; the number of parameters is huge and leads to overfitting. To reduce the number of parameters and spatial dimensions of the last CONV block activation, we feed it to global average pooling (GAP) and global max-pooling (GMP) layers [58]. The GAP average all the hw values, whereas the GMP takes into account the contributions of the neurons of maximum response; the number of neurons in the FC layer is $h \times w \times L$, and it is reduced to $1 \times 1 \times L$ when only GMP or GAP is introduced. We concatenate the output of GMP and GAP layers to overcome the shortcoming of each and then feed it to the FC layer, followed by the SoftMax layer.

2.4. Fine-Tuning the Model

The DeepFKTNet model is evaluated using the challenge multisensory FingerPass dataset [59], and it is compared to the well-known deep models: ResNet [50] and DenseNet [55] pre-trained on the ImageNet dataset and fine-tuned using the same dataset as DeepFKTNet. For further validation, we evaluated our method using the challenge FVC2004 dataset [60] and compared it to the state-of-the-art methods. For each dataset, we select the most representative fingerprint images from the training set using K-medoids and LGDBP descriptor and then built its adaptive DeepFKTNet architecture using Algorithm 1.

2.4.1. Datasets and the Adaptive Architectures

To verify the performance of the DeepFKTNet model on benchmark datasets, we used FingerPass and FVC2004 datasets. The FingerPass is a multi-sensor dataset; it was collected using nine different optical and capacitive sensors and two interaction types, i.e., press and sweep. The FingerPass contains a total of fingers separated into nine subsets based on sensors; each subset contains 12 impressions of 8 fingers from 90 persons.

FVC2004 dataset contains noisy images acquired by live scan devices. It has 4 sets: DB1 collected using optical V300 sensor, DB2 collected using optical U 4000, DB3 collected using thermal sweeping sensor, and DB4 is a synthetic fingerprint dataset. Each one contains 880 fingerprint images [60]. We categorized FVC2004 fingerprints into four categories: arch, left loop, right loop, and whorl. We merge the 4 sets of FVC2004 into one set of four classes; it is now a multi-sensor fingerprint dataset.

To setup best parameters for each DeepFKTNet model, the hyperparameter optimization software framework Optuna [61] is used to select the best hyperparameters for fine-tuning the DeepFKTNet model. Using Algorithms 1, the DeepFKTNet architecture obtained for the FVC2004 dataset consists of 5 CONV blocks, as shown in Figure 3a, whereas the architecture constructed for the FingerPass dataset has 11 blocks, as depicted in Figure 3b. The number of filters for each CONV block and the depth of each model for each fingerprint dataset are determined using Algorithm 1. Using the Optuna optimization algorithm, we fine-tuned the hyperparameters and tested three optimizers (Adam, SGD, and RMSprop), learning rate between 1×10^{-1}, and 1×10^{-5}, patch size (5, 10, 15, 20, 30, 50), activation functions (Relu, LRelu, and Sigmoid), and dropout between 0.25 and 0.50. After training for 10 epochs, the best hyper-parameters for each dataset are shown in Table 1.

Table 1. The optimized hyperparameters using Optuna algorithm.

Dataset	Activate Function	Learning Rate	Pach's Size	Optimizer	Dropout
FingerPass	Relu	0.0005	16	RMSprop	0.45
FVC2004	Relu	0.0008	10	RMSprop	0.38

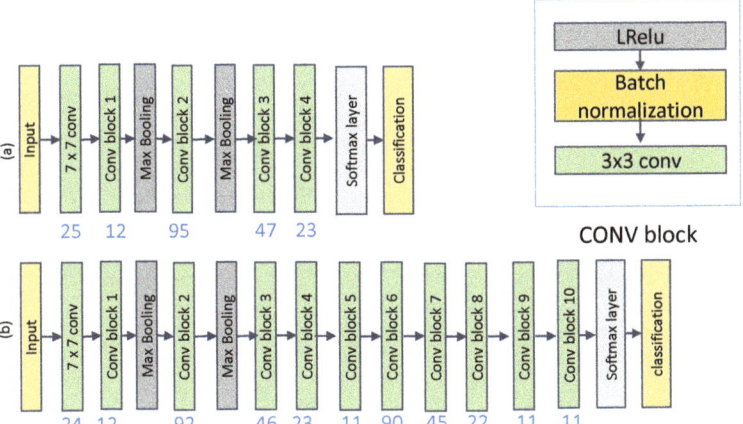

Figure 3. (**a**) FVC2004 FKTNET architecture. (**b**) Fingerprint FKTNET architecture.

2.4.2. Evaluation Procedure

For evaluation, we manually separated the FingerPass dataset into four classes (arch, left loop, right loop, and whorl). We divided the FingerPass dataset into three sets (80% training, 10% validation, and 10% testing) using two different scenarios. In scenario-1, the fingers from each sensor were divided into training, validation, and test sets. In scenario-2, fingers in the training, validation, and test sets are from different sensors.

For the FVC2004 dataset, we divided the dataset into training (80%), validation (10%), and testing (10%), keeping the balance. For performance evaluation, we used four commonly used metrics: accuracy (ACC), true positive rate (TPR), true negative rate (TNR), and Kappa [62–65]. The overall average of metrics has been computed. The used metrics [66,67] to evaluate the proposed system are:

$$ACC = \frac{TP + TN}{TP + FP + TN + FN} \quad (3)$$

$$TPR = \frac{TP}{TP + FN} \quad (4)$$

$$TNR = \frac{TN}{TN + FP} \quad (5)$$

$$Kappa = \frac{P_0 - P_e}{1 - P_e} \quad (6)$$

where TP, TN, FP, and FN are the numbers are true positives, true negatives, false positives, and false negatives; P_0 and P_e are calculated from the confusion matrix; the detail is given in [68]. To compute TP, TN, FP, and FN, one class, in turn, is taken as positive, the other classes are assumed to be negative, and the TPR and TNR are calculated. Finally, mean TPR and TNR are calculated by averaging TPR and TNR over all classes. In the results, the mean TPR and TNR are reported.

3. Experimental Results

This section presents the experimental results of the DeepFKTNet models designed for the two datasets.

We designed the DeepFKTNet model for each dataset and fine-tuned it using the training sets. We validated its performance on FingerPass and FVC2004 datasets and compared it with the widely used CNN models ResNet [50] and DenseNet [55], which were pre-trained on the ImageNet dataset and fine-tuned on the same training set that was used for the DeepFKTNet model. In the rest of the paper, we name the DeepFKTNet models as

DeepFKTNet-11 and DeepFKTNet-5, designed for the FingerPass and the FV2004 datasets, respectively.

The results of the three models DeepFKTNet-11, ResNet152, and DenseNet121 for scenario-1 are shown in Figure 4a and Table 2a. The DeepFKTNet-11 model generated adaptively on the FingerPass dataset outperforms the state-of-the-art ResNet152 and DenseNet121 models in terms of all metrics. Though DenseNet121 is not better than DeepFKTNet-11, it outperforms ResNet152 in terms of all metrics. Figure 4b and Table 2b show the results for scenario-2 on the FingerPass dataset. In this scenario, the results obtained with the DeepFKTNet-11 are almost similar to those obtained in scenario-1. The DeepFKTNet-11 outperforms ResNet152 and DenseNet121. Figure 5 illustrates the confusion matrices for both scenarios. These give insights into the system performance for different classes.

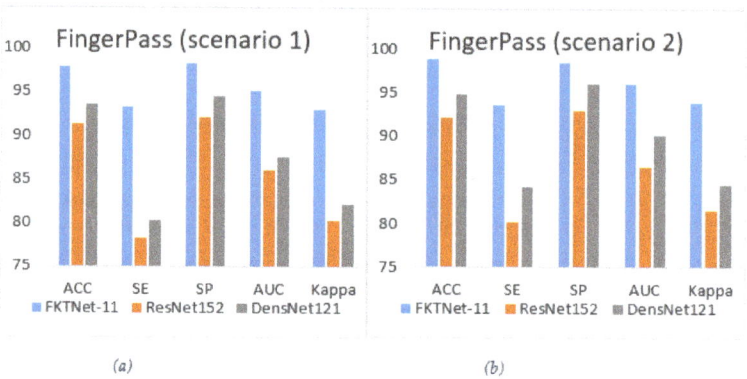

Figure 4. Comparison between FKTNET-11 and pre-trained ResNet-152 and DensNet-121 on Fingerprint dataset (4 classes) using scenario 1 (**a**) and scenario 2 (**b**).

Table 2. Comparison between FKTNET-11 and pre-trained ResNet-152 and DensNet-121 on Fingerprint dataset scenario 1 (**a**) and scenario 2 (**b**).

	(a)				
	ACC%	SE%	SP%	AUC%	Kappa%
FKTNet-11	97.84	93.25	98.28	95.21	93.05
ResNet152	91.22	78.22	92.05	86.11	80.32
DensNet121	93.55	80.22	94.44	87.55	82.11
	(b)				
	ACC%	SE%	SP%	AUC%	Kappa%
FKTNet-11	98.9	93.6	98.5	96.12	93.93
ResNet152	92.22	80.22	93.05	86.5	81.62
DensNet121	94.85	84.22	96.12	90.21	84.55

Scenario 1

Actual \ Predicted	0	1	2	3
0	73	6	3	2
1	2	314	2	16
2	1	8	268	7
3	0	8	11	809

Scenario 2

Actual \ Predicted	0	1	2	3
0	60	3	5	0
1	5	311	6	19
2	2	2	457	7
3	0	12	8	913

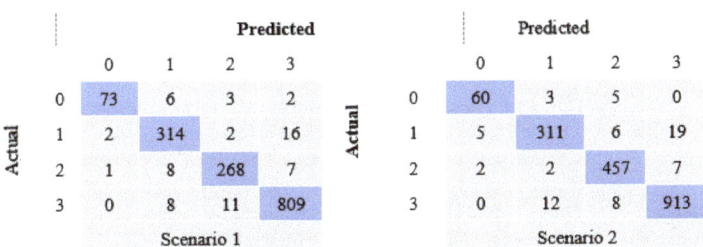

Figure 5. Confusion matrix based on FKTNET-11 model for scenario 1 and scenario 2.

The DeepFKTNet-5 model was adaptively designed for the challenge FVC2004 dataset; it was evaluated using the above evaluation procedure. We fine-tuned the developed DeepFKTNet-5 model and the pre-trained models ResNet152 and DenseNet121 using the same dataset. The results are shown in Figure 6; the DeepFKTNet-5 model outperforms the state-of-the-art ResNet152 and DenseNet121 models in terms of all metrics. Figure 7 illustrates the confusion matrices for the FVC2004 dataset. These give insights into the system performance for different classes.

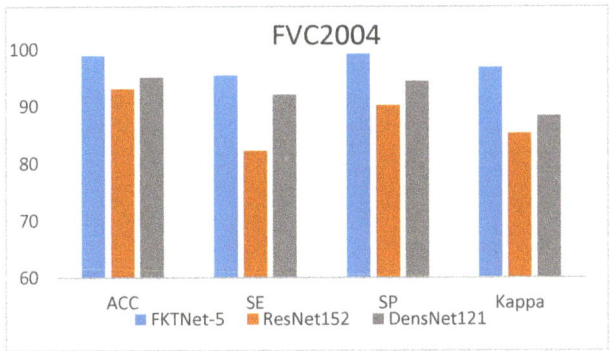

Figure 6. Comparison between FKTNET-5 and pre-trained ResNet-152 and DensNet-121 on FVC2004 dataset (four classes).

FVC 2004

Actual \ Predicted	0	1	2	3
0	6	0	1	0
1	0	34	0	0
2	0	0	23	0
3	0	1	0	25

Figure 7. Confusion matrix based on FKTNET-5 model for FVC2004 dataset.

4. Discussions

We addressed the multi-sensor fingerprint classification problem and proposed a novel method for automatically generating a custom-designed DeepFKTNet model from the target fingerprint dataset. The number of layers and filters for each layer are not specified randomly; they are determined from the best representative fingerprints selected using the K-medoids clustering algorithm and LDGBP descriptor from the fingerprint datasets.

The generated DeepFKTNet models are shallower than the state-of-the-art models, robust, involve a small number of learnable parameters, and suitable for fingerprint classification.

The results of the DeepFKTNet models on the FingerPass and FVC2004 datasets (Figures 4 and 6) indicate that they outperform the famous deep models ResNet152 and DenseNet121, which were pre-trained on the ImageNet dataset and fine-tuned using the same fingerprint datasets. The architecture of a DeepFKTNet model is drawn directly from the dataset; the internal structures of the data determine its design. For this reason, the DeepFKTNet model has a compact size and yields better classification results. Further, it does not suffer from the overfitting problem (see Table 3) since it involves a small number of learnable parameters (see Table 4), which is comparable with the number of training examples. If the number of learnable parameters is huge as compared to the training examples, the overfitting problem cannot be avoided. The training and testing accuracies shown in Table 3 indicate that the models do not suffer from overfitting. In addition, DeepFKTNet models are trained using the available training data, and the pre-training is not needed, unlike ResNet152 and DenseNet121.

Table 3. The train and test accuracy of DeepFKTNet-11 models for two scenarios.

Model	Train ACC	Test ACC
Scenarios 1	98.65	97.84
Scenarios 2	99.11	98.9

Table 4. The comparison between generated DeepFKTNet models from the two datasets and pre-trained ResNet152 and DenseNet121. K is for kilobyte and G is for Gigabyte.

Model	DeepFKTNet-5	DeepFKTNet-11	ResNet152	DenseNet121
number of params	58.456 k	119.599 k	60.19 M	7.98 M
FLOPs	0.5 G	0.9 G	5.6 G	1.44 G

The space complexity of a CNN model is measured in terms of the number of learnable parameters, whereas the number of FLOPS determines its time complexity. Table 4 gives the statistics of the space and time complexities of the models. Overall, the DeepFKTNet model got competitive performance with fewer layers and parameters. The DeepFKTNet models designed for the two datasets have a small number of parameters, in thousands against millions in ResNet152 and DensNet121 models. DeepFKTNet-5 and DeepFKTNet-11 have fewer FLOPs than ResNet152 and DensNet121 and better performance. The DeepFKTNet-11 is relatively more complex than DeepFKTNet-5; the reason is that the FingerPass dataset involves a large number of sensors as compared to the FVC2004 dataset, and there is more variety of patterns in the FingerPass dataset, and to encode the discriminative pattern, more rich structure is needed.

Further, for investigating which features the DeepFKTNet models focus on for decision making, we employed GradCam [69]. Figure 8 shows some heat maps generated with GradCam for DeepFKTNet-11. The fingerprint images from class arches and their GradCam visualizations are shown in Figure 8a,b, the fingerprint images from the class left loop and their GradCam visualizations are shown in Figure 8c,d. Figure 8e,f depicts fingerprint images from the class right loop and their GradCam visualization, whereas Figure 8g,h show fingerprint images from the class whorls and their GradCam visualizations. The visual analysis of the decision-making process of DeepFKTNet shows that it concentrates on the discriminative regions of fingerprints and extracts class discriminative features.

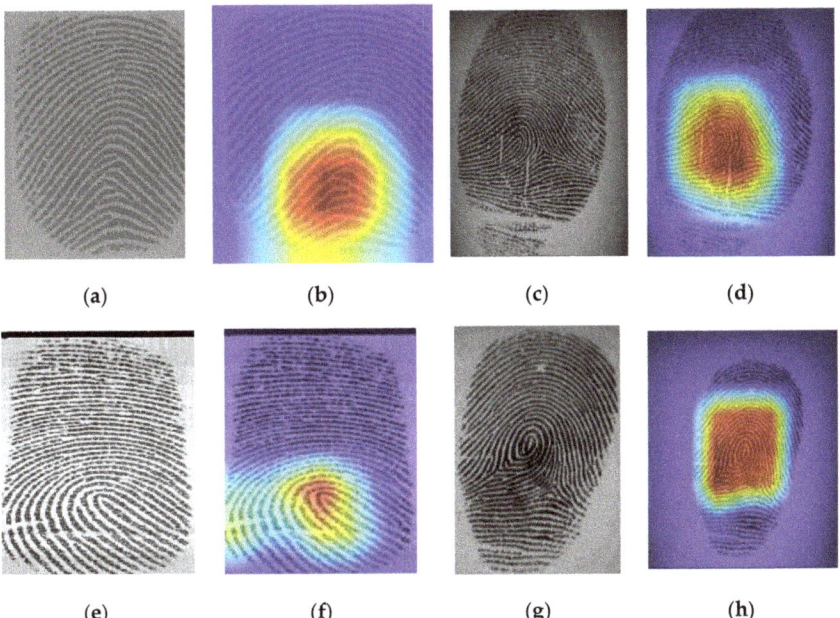

Figure 8. Visualizations of activation maps using the GradCam method for four samples from different classes of FingerPass dataset: (**a**) arches finger; (**b**) arches's gradcam; (**c**) left loop finger; (**d**) left loop's gradcam; (**e**) right loop finger; (**f**) right loop's gradcam; (**g**) whorls finger; and (**h**) whorls gradcam.

For a fair comparison, the DeepFKTNet-5 has been compared with the state-of-the-art fingerprint classification methods, which were validated on the benchmark public FVC2004 dataset; the comparison results are given in Table 5.

The DeepFKTNet-5 model outperforms the state-of-the-art methods (handcraft and CNN methods) on the same dataset in terms of accuracy. The method of Jeon et al. [70], despite being a complex ensemble of CNN models, got an accuracy of 97.2%, which is less than that of DeepFKTNet-5. Zia et al. [33] employed B-DCNNs with five convolution layers and two FC layers (with 1024 and 512 neurons) for fingerprint classification and validated on the FVC2004 dataset; it does not yield better accuracy than that of DeepFKTNet-5 (95.3% vs. 98.89%). Its complexity is high; it has more FLOPs (0.65 G vs. 0.5 G) and more learnable parameters (38.66 M vs. 58.456 k). Nguyen et al. [34] employed a two-stage CNN model for enhancing and then training and prediction. They used LBCNN [71] method in the first stage, which has 0.352 M learnable parameters, and then employed a three-ternary model for training and prediction. They got an accuracy of 96.1% based on FVC2004 (three classes), which is less than DeepFKTNet-5. Nahar et al. [35] used a modified LNet-5 model for fingerprint classification; they got 99.1% accuracy but with only a subset (DB1) from FVC2004, whereas the DeepFKTNet-5 model evaluated on the combined multi-sensor dataset of the four datasets (DB1, DB2, DB3, and DB4) from FVC2004. Also, the LNet-5 has a higher number of parameters, 19.25 M and 1.42 G FLOPs vs. 58.456 k and 0.5 G FLOPs of DeepFKTNet-5. The reason for the better performance and less complexity of DeepFKTNet-5 is that it is custom-designed, keeping in view the internal discriminative structures of fingerprints.

Table 5. Comparison between DeepFKTNet-5 and the state-of-the-art methods.

Paper	Method	Performance (%)			
		ACC	SE	SP	Kappa
Gupta et al. [72] 2015	Singular point	97.80	-	-	-
Darlow et al. [73] 2017	Minutiae and DL	94.55	-	-	-
Andono et al. [74] 2018	Bag-of-Visual-Words	90	-	-	-
Saeed et al. [19] 2018	statistics of D-SIFT descriptor	97.40	-	-	-
Saeed et al. [18] 2018	Modified HOG descriptor	98.70	-	-	-
Jeon et al. [70] 2017	Ensemble CNN model	97.2	-	-	-
Zia et al. [33] 2019	B-DCNNs	95.3			
Nguyen et al. [34] 2019	CNN (tested on 3 classes of FVC2004)	96.1			
Nahar et al. [35] 2022	Modified LeNet (tested on FVC2004-DB1)	99.1			
DeepFKTNet-5	DeepFKTNet model	98.89	95.46	99.18	96.82

5. Conclusions

We introduced a technique for automatically creating a custom-designed CNN model for multi-sensor fingerprint categorization. Since CNN models contain a large number of parameters and are designed randomly, we used the FKT approach to build a low-cost, high-speed CNN model tailored for the target fingerprint dataset. The developed DeepFKTNet model is data-dependent, with a distinctive architecture for each fingerprint dataset. DeepFKTNet-11 for the FigerPass dataset and DeepFKTNet-5 for FVC2004 outperform pre-trained deep ResNet152 and DenseNet121 models on identical datasets and assessment processes. The performance, complexity, and number of parameters of the DeepFKTNet models created are substantially fewer than those of ResNet152 and DenseNet. Compared to the state-of-the-art techniques on the FVC2004 dataset, the DeepFKTNet-5 model is simpler in terms of complexity and parameter count and achieves comparable performance. In future work, we will enhance DeepFKTNet to address the problem of cross-sensor fingerprint verification.

Author Contributions: Conceptualization, F.S. and M.H.; methodology, M.H. and F.S.; software, F.S.; validation, F.S., M.H. and H.A.A.; formal analysis, H.A.A. and M.H.; investigation F.S., M.H.; resources, F.S. and H.A.A.; data curation, F.S., M.H.; writing—original draft preparation, F.S.; writing—review and editing, M.H.; visualization, H.A.A.; supervision, M.H.; project administration, M.H.; funding acquisition, M.H and H.A.A. All authors have read and agreed to the published version of the manuscript.

Funding: This Project was funded by the National Plan for Science, Technology and Innovation (MAARIFAH), King Abdulaziz City for Science and Technology, Kingdom of Saudi Arabia, under Project no. 13-INF946-02.

Institutional Review Board Statement: Not applicable.

Informed Consent Statement: Not applicable.

Data Availability Statement: Used public domain datasets, FVC2004 dataset: available online: http://bias.csr.unibo.it/fvc2004/download.asp (accessed on 26 February 2022).

Conflicts of Interest: The authors declare no conflict of interest.

References

1. Grabatin, M.; Steinke, M.; Pöhn, D.; Hommel, W. A Matrix for Systematic Selection of Authentication Mechanisms in Challenging Healthcare related Environments. In Proceedings of the 2021 ACM Workshop on Secure and Trustworthy Cyber-Physical Systems, Virtually, TN, USA, 28 April 2021; pp. 88–97.
2. Maltoni, D.; Maio, D.; Jain, A.K.; Prabhakar, S. *Handbook of Fingerprint Recognition*; Springer Science & Business Media: Berlin/Heidelberg, Germany, 2009.
3. Pandey, F.; Dash, P.; Samanta, D.; Sarma, M. ASRA: Automatic singular value decomposition-based robust fingerprint image alignment. *Multimed. Tools Appl.* **2021**, *80*, 15647–15675. [CrossRef]

4. Khosroshahi, M.E.; Woll-Morison, V. Visualization and fluorescence spectroscopy of fingerprints on glass slide using combined 405 nm laser and phase contrast microscope. *J. Vis.* **2021**, *24*, 665–670. [CrossRef]
5. Banik, A.; Ghosh, K.; Patil, U.K.; Gayen, S. Identification of molecular fingerprints of natural products for the inhibition of breast cancer resistance protein (BCRP). *Phytomedicine* **2021**, *85*, 153523. [CrossRef] [PubMed]
6. Lugini, L.; Marasco, E.; Cukic, B.; Gashi, I. Interoperability in fingerprint recognition: A large-scale empirical study. In Proceedings of the 2013 43rd Annual IEEE/IFIP Conference on Dependable Systems and Networks Workshop (DSN-W), Budapest, Hungary, 24–27 June 2013; pp. 1–6.
7. Alrashidi, A.; Alotaibi, A.; Hussain, M.; AlShehri, H.; AboAlSamh, H.A.; Bebis, G. Cross-Sensor Fingerprint Matching Using Siamese Network and Adversarial Learning. *Sensors* **2021**, *21*, 3657. [CrossRef]
8. Priesnitz, J.; Rathgeb, C.; Buchmann, N.; Busch, C.; Margraf, M. An overview of touchless 2D fingerprint recognition. *EURASIP J. Image Video Process.* **2021**, *2021*, 1–28. [CrossRef]
9. AlShehri, H.; Hussain, M.; AboAlSamh, H.A.; AlZuair, M. A large-scale study of fingerprint matching systems for sensor interoperability problem. *Sensors* **2018**, *18*, 1008. [CrossRef]
10. Alshehri, H.; Hussain, M.; Aboalsamh, H.A.; Emad-Ul-Haq, Q.; Alzuair, M.; Azmi, A.M. Alignment-free cross-sensor fingerprint matching based on the co-occurrence of ridge orientations and Gabor-HoG descriptor. *IEEE Access* **2019**, *7*, 86436–86452. [CrossRef]
11. Marasco, E.; Feldman, A.; Romine, K.R. Enhancing Optical Cross-Sensor Fingerprint Matching Using Local Textural Features. In Proceedings of the 2018 IEEE Winter Applications of Computer Vision Workshops (WACVW), Lake Tahoe, NV, USA, 15 March 2018; pp. 37–43.
12. Lin, C.; Kumar, A. A CNN-based framework for comparison of contactless to contact-based fingerprints. *IEEE Trans. Inf. Forensics Secur.* **2018**, *14*, 662–676. [CrossRef]
13. Galar, M.; Derrac, J.; Peralta, D.; Triguero, I.; Paternain, D.; Lopez-Molina, C.; García, S.; Benítez, J.M.; Pagola, M.; Barrenechea, E. A survey of fingerprint classification Part I: Taxonomies on feature extraction methods and learning models. *Knowl.-Based Syst.* **2015**, *81*, 76–97. [CrossRef]
14. Galar, M.; Derrac, J.; Peralta, D.; Triguero, I.; Paternain, D.; Lopez-Molina, C.; García, S.; Benítez, J.M.; Pagola, M.; Barrenechea, E. A survey of fingerprint classification Part II: Experimental analysis and ensemble proposal. *Knowl.-Based Syst.* **2015**, *81*, 98–116. [CrossRef]
15. Guo, J.-M.; Liu, Y.-F.; Chang, J.-Y.; Lee, J.-D. Fingerprint classification based on decision tree from singular points and orientation field. *Expert Syst. Appl.* **2014**, *41*, 752–764. [CrossRef]
16. Bhalerao, B.V.; Manza, R.R. Development of Image Enhancement and the Feature Extraction Techniques on Rural Fingerprint Images to Improve the Recognition and the Authentication Rate. *IOSR J. Comput. Eng.* **2013**, *15*, 1–5.
17. Dorasamy, K.; Webb, L.; Tapamo, J.; Khanyile, N.P. Fingerprint classification using a simplified rule-set based on directional patterns and singularity features. In Proceedings of the 2015 International Conference on Biometrics (ICB), Phuket, Thailand, 19–22 May 2015; pp. 400–407.
18. Saeed, F.; Hussain, M.; Aboalsamh, H.A. Classification of live scanned fingerprints using histogram of gradient descriptor. In Proceedings of the 2018 21st Saudi Computer Society National Computer Conference (NCC), Riyadh, Saudi Arabia, 25–26 April 2018; pp. 1–5.
19. Saeed, F.; Hussain, M.; Aboalsamh, H.A. Classification of Live Scanned Fingerprints using Dense SIFT based Ridge Orientation Features. In Proceedings of the 2018 1st International Conference on Computer Applications & Information Security (ICCAIS), Riyadh, Saudi Arabia, 4–6 April 2018; pp. 1–4.
20. Dhaneshwar, R.; Kaur, M.; Kaur, M. An investigation of latent fingerprinting techniques. *Egypt. J. Forensic Sci.* **2021**, *11*, 1–15. [CrossRef]
21. Jung, H.-W.; Lee, J.-H. Noisy and incomplete fingerprint classification using local ridge distribution models. *Pattern Recognit.* **2015**, *48*, 473–484. [CrossRef]
22. Vegad, S.; Shah, Z. Fingerprint Image Classification. In *Data Science and Intelligent Applications*; Springer: Berlin/Heidelberg, Germany, 2021; pp. 545–552.
23. Gu, J.; Wang, Z.; Kuen, J.; Ma, L.; Shahroudy, A.; Shuai, B.; Liu, T.; Wang, X.; Wang, G.; Cai, J. Recent advances in convolutional neural networks. *Pattern Recognit.* **2018**, *77*, 354–377. [CrossRef]
24. Grigorescu, S.; Trasnea, B.; Cocias, T.; Macesanu, G. A survey of deep learning techniques for autonomous driving. *J. Field Robot.* **2020**, *37*, 362–386. [CrossRef]
25. Abou Arkoub, S.; El Hassani, A.H.; Lauri, F.; Hajjar, M.; Daya, B.; Hecquet, S.; Aubry, S. Survey on Deep Learning Techniques for Medical Imaging Application Area. In *Machine Learning Paradigms*; Springer: Berlin/Heidelberg, Germany, 2020; pp. 149–189.
26. Dong, S.; Wang, P.; Abbas, K. A survey on deep learning and its applications. *Comput. Sci. Rev.* **2021**, *40*, 100379. [CrossRef]
27. Mishra, A.; Dehuri, S. An experimental study of filter bank approach and biogeography-based optimized ANN in fingerprint classification. In *Nanoelectronics, Circuits and Communication Systems*; Springer: Berlin/Heidelberg, Germany, 2019; pp. 229–237.
28. Jian, W.; Zhou, Y.; Liu, H. Lightweight Convolutional Neural Network Based on Singularity ROI for Fingerprint Classification. *IEEE Access* **2020**, *8*, 54554–54563. [CrossRef]
29. Nahar, P.; Tanwani, S.; Chaudhari, N.S. Fingerprint classification using deep neural network model resnet50. *Int. J. Res. Anal. Rev.* **2018**, *5*, 1521–1537.

30. Rim, B.; Kim, J.; Hong, M. Fingerprint classification using deep learning approach. *Multimed. Tools Appl.* **2020**, 1–17. [CrossRef]
31. Ali, S.F.; Khan, M.A.; Aslam, A.S. Fingerprint matching, spoof and liveness detection: Classification and literature review. *Front. Comput. Sci.* **2021**, *15*, 1–18. [CrossRef]
32. Bolhasani, H.; Mohseni, M.; Rahmani, A.M. Deep learning applications for IoT in health care: A systematic review. *Inform. Med. Unlocked* **2021**, *23*, 100550. [CrossRef]
33. Zia, T.; Ghafoor, M.; Tariq, S.A.; Taj, I.A. Robust fingerprint classification with Bayesian convolutional networks. *IET Image Process.* **2019**, *13*, 1280–1288. [CrossRef]
34. Nguyen, H.T.; Nguyen, L.T. Fingerprints classification through image analysis and machine learning method. *Algorithms* **2019**, *12*, 241. [CrossRef]
35. Nahar, P.; Chaudhari, N.S.; Tanwani, S.K. Fingerprint classification system using CNN. *Multimed. Tools Appl.* **2022**, 1–13. [CrossRef]
36. Saeed, F.; Hussain, M.; Aboalsamh, H.A. Method for Fingerprint Classification. U.S. Patent 9,530,042, 13 June 2016.
37. Zhang, Q.; Couloigner, I. A new and efficient k-medoid algorithm for spatial clustering. In Proceedings of the International Conference on Computational Science and Its Applications, Singapore, 9–12 May 2005; pp. 181–189.
38. Huo, X. A statistical analysis of Fukunaga-Koontz transform. *IEEE Signal Process. Lett.* **2004**, *11*, 123–126. [CrossRef]
39. Dhillon, A.; Verma, G.K. Convolutional neural network: A review of models, methodologies and applications to object detection. *Prog. Artif. Intell.* **2020**, *9*, 85–112. [CrossRef]
40. Alzubaidi, L.; Zhang, J.; Humaidi, A.J.; Al-Dujaili, A.; Duan, Y.; Al-Shamma, O.; Santamaría, J.; Fadhel, M.A.; Al-Amidie, M.; Farhan, L. Review of deep learning: Concepts, CNN architectures, challenges, applications, future directions. *J. Big Data* **2021**, *8*, 1–74. [CrossRef]
41. Abdullah, S.M.S.A.; Ameen, S.Y.A.; Sadeeq, M.A.; Zeebaree, S. Multimodal emotion recognition using deep learning. *J. Appl. Sci. Technol. Trends* **2021**, *2*, 52–58. [CrossRef]
42. Jena, B.; Saxena, S.; Nayak, G.K.; Saba, L.; Sharma, N.; Suri, J.S. Artificial intelligence-based hybrid deep learning models for image classification: The first narrative review. *Comput. Biol. Med.* **2021**, *137*, 104803. [CrossRef]
43. Lu, J.; Tan, L.; Jiang, H. Review on convolutional neural network (CNN) applied to plant leaf disease classification. *Agriculture* **2021**, *11*, 707. [CrossRef]
44. Fang, W.; Love, P.E.; Luo, H.; Ding, L. Computer vision for behaviour-based safety in construction: A review and future directions. *Adv. Eng. Inform.* **2020**, *43*, 100980. [CrossRef]
45. Lavanya, P.; Sasikala, E. Deep learning techniques on text classification using Natural language processing (NLP) in social healthcare network: A comprehensive survey. In Proceedings of the 2021 3rd International Conference on Signal Processing and Communication (ICPSC), Coimbatore, India, 13–14 May 2021; pp. 603–609.
46. Khan, A.; Sohail, A.; Zahoora, U.; Qureshi, A.S. A survey of the recent architectures of deep convolutional neural networks. *Artif. Intell. Rev.* **2020**, *53*, 5455–5516. [CrossRef]
47. Krizhevsky, A.; Sutskever, I.; Hinton, G.E. Imagenet classification with deep convolutional neural networks. *Adv. Neural Inf. Process. Syst.* **2012**, *25*. [CrossRef]
48. Simonyan, K.; Zisserman, A. Very deep convolutional networks for large-scale image recognition. *arXiv* **2014**, arXiv:1409.1556.
49. Szegedy, C.; Liu, W.; Jia, Y.; Sermanet, P.; Reed, S.; Anguelov, D.; Erhan, D.; Vanhoucke, V.; Rabinovich, A. Going deeper with convolutions. In Proceedings of the IEEE Conference on Computer Vision and Pattern Recognition, Boston, MA, USA, 7–12 June 2015; pp. 1–9.
50. He, K.; Zhang, X.; Ren, S.; Sun, J. Deep residual learning for image recognition. In Proceedings of the IEEE Conference on Computer Vision and Pattern Recognition, Las Vegas, NV, USA, 27–30 June 2016; pp. 770–778.
51. Hamel, P.; Eck, D. Learning features from music audio with deep belief networks. In Proceedings of the ISMIR, Utrecht, The Netherlands, 9–13 August 2010; pp. 339–344.
52. Khan, A.; Sohail, A.; Ali, A. A new channel boosted convolutional neural network using transfer learning. *arXiv* **2018**, arXiv:1804.08528.
53. Lin, M.; Chen, Q.; Yan, S. Network in network. *arXiv* **2013**, arXiv:1312.4400.
54. Rousseeuw, P.J. Silhouettes: A graphical aid to the interpretation and validation of cluster analysis. *J. Comput. Appl. Math.* **1987**, *20*, 53–65. [CrossRef]
55. Huang, G.; Liu, Z.; Weinberger, K.; van der Maaten, L. Densely connected convolutional networks. CVPR 2017. *arXiv* **2016**, arXiv:1608.06993.
56. Izenman, A.J. Linear discriminant analysis. In *Modern Multivariate Statistical Techniques*; Springer: Berlin/Heidelberg, Germany, 2013; pp. 237–280.
57. Mika, S.; Ratsch, G.; Weston, J.; Scholkopf, B.; Mullers, K.-R. Fisher discriminant analysis with kernels. In Proceedings of the Neural Networks for Signal Processing IX: Proceedings of the 1999 IEEE Signal Processing Society Workshop (Cat. No. 98th8468), Madison, WI, USA, 25 August 1999; pp. 41–48.
58. Cook, A. Global Average Pooling Layers for Object Localization. 2017. Available online: https://alexisbcook.github.io/2017/globalaverage-poolinglayers-for-object-localization/ (accessed on 19 August 2019).
59. Jia, X.; Yang, X.; Zang, Y.; Zhang, N.; Tian, J. A cross-device matching fingerprint database from multi-type sensors. In Proceedings of the 21st International Conference on Pattern Recognition (ICPR2012), Tsukuba, Japan, 11–15 November 2012; pp. 3001–3004.

60. Maio, D.; Maltoni, D.; Cappelli, R.; Wayman, J.L.; Jain, A.K. FVC2004: Third fingerprint verification competition. In Proceedings of the International Conference on Biometric Authentication, Hong Kong, China, 15–17 July 2004; Springer: Berlin/Heidelberg, Germany, 2004; pp. 1–7.
61. Akiba, T.; Sano, S.; Yanase, T.; Ohta, T.; Koyama, M. Optuna: A next-generation hyperparameter optimization framework. In Proceedings of the 25th ACM SIGKDD International Conference on Knowledge Discovery & Data Mining, Anchorage, AK, USA, 4–8 August 2019; pp. 2623–2631.
62. Gao, Z.; Li, J.; Guo, J.; Chen, Y.; Yi, Z.; Zhong, J. Diagnosis of Diabetic Retinopathy Using Deep Neural Networks. *IEEE Access* **2019**, *7*, 3360–3370. [CrossRef]
63. Quellec, G.; Charrière, K.; Boudi, Y.; Cochener, B.; Lamard, M. Deep image mining for diabetic retinopathy screening. *Med. Image Anal.* **2017**, *39*, 178–193. [CrossRef] [PubMed]
64. Chowdhury, A.R.; Chatterjee, T.; Banerjee, S. A Random Forest classifier-based approach in the detection of abnormalities in the retina. *Med. Biol. Eng. Comput.* **2019**, *57*, 193–203. [CrossRef]
65. Zhang, W.; Zhong, J.; Yang, S.; Gao, Z.; Hu, J.; Chen, Y.; Yi, Z. Automated identification and grading system of diabetic retinopathy using deep neural networks. *Knowl. -Based Syst.* **2019**, *175*, 12–25. [CrossRef]
66. Haghighi, S.; Jasemi, M.; Hessabi, S.; Zolanvari, A. PyCM: Multiclass confusion matrix library in Python. *J. Open Source Softw.* **2018**, *3*, 729. [CrossRef]
67. Powers, D.M. Evaluation: From precision, recall and F-measure to ROC, informedness, markedness and correlation. *arXiv* **2011**, arXiv:2010.16061.
68. Fleiss, J.L.; Cohen, J.; Everitt, B.S. Large sample standard errors of kappa and weighted kappa. *Psychol. Bull.* **1969**, *72*, 323. [CrossRef]
69. Selvaraju, R.R.; Cogswell, M.; Das, A.; Vedantam, R.; Parikh, D.; Batra, D. Grad-cam: Visual explanations from deep networks via gradient-based localization. In Proceedings of the IEEE International Conference on Computer Vision, Venice, Italy, 22–29 October 2017; pp. 618–626.
70. Jeon, W.-S.; Rhee, S.-Y. Fingerprint pattern classification using convolution neural network. *Int. J. Fuzzy Log. Intell. Syst.* **2017**, *17*, 170–176. [CrossRef]
71. Juefei-Xu, F.; Naresh Boddeti, V.; Savvides, M. Local binary convolutional neural networks. In Proceedings of the IEEE Conference on Computer Vision and Pattern Recognition, Honolulu, HI, USA, 21–26 July 2017; pp. 19–28.
72. Gupta, P.; Gupta, P. A robust singular point detection algorithm. *Appl. Soft Comput.* **2015**, *29*, 411–423. [CrossRef]
73. Darlow, L.N.; Rosman, B. Fingerprint minutiae extraction using deep learning. In Proceedings of the 2017 IEEE International Joint Conference on Biometrics (IJCB), Denver, CO, USA, 1–4 October 2017; pp. 22–30.
74. Andono, P.; Supriyanto, C. Bag-of-visual-words model for fingerprint classification. *Int. Arab J. Inf. Technol.* **2018**, *15*, 37–43.

Article

Enhancement of Image Classification Using Transfer Learning and GAN-Based Synthetic Data Augmentation

Subhajit Chatterjee [1], Debapriya Hazra [1], Yung-Cheol Byun [1,*] and Yong-Woon Kim [2]

[1] Department of Computer Engineering, Jeju National University, Jeju 63243, Korea; subhajitchatterjee@stu.jejunu.ac.kr (S.C.); debapriyah@jejunu.ac.kr (D.H.)
[2] Centre for Digital Innovation, CHRIST University (Deemed to be University), Bengaluru 560029, Karnataka, India; jonathan.kim@christuniversity.in
* Correspondence: ycb@jejunu.ac.kr

Abstract: Plastic bottle recycling has a crucial role in environmental degradation and protection. Position and background should be the same to classify plastic bottles on a conveyor belt. The manual detection of plastic bottles is time consuming and leads to human error. Hence, the automatic classification of plastic bottles using deep learning techniques can assist with the more accurate results and reduce cost. To achieve a considerably good result using the DL model, we need a large volume of data to train. We propose a GAN-based model to generate synthetic images similar to the original. To improve the image synthesis quality with less training time and decrease the chances of mode collapse, we propose a modified lightweight-GAN model, which consists of a generator and a discriminator with an auto-encoding feature to capture essential parts of the input image and to encourage the generator to produce a wide range of real data. Then a newly designed weighted average ensemble model based on two pre-trained models, inceptionV3 and xception, to classify transparent plastic bottles obtains an improved classification accuracy of 99.06%.

Keywords: deep learning; generative adversarial networks; image classification; transfer learning; plastic bottle

MSC: 68U10

1. Introduction

Due to flexibility in terms of cost, light weight, processing, and ease of carrying, plastic bottles are the most widely used material in daily life and industrial fields. Every day, tons of plastic bottles are dumped as waste, and in addition, toxic, hazardous materials in the trash are polluting the environment day by day [1]. An essential strategy for dealing with this issue is the recycling process. Recycling plastic bottles can be used further in new products, automobiles, textile, etc. Plastic bottle recycling has recently emerged as a significant part of the plastic bottle industry, potentially saving fossil fuels while simultaneously lowering greenhouse gas emissions [2].

The recycling task involves a lot of labor cost, and the DL approach helps in the way to automatically classify waste plastic bottles for recycling tasks [3]. Much research has been conducted to find a category of cost-effective PET bottle classifiers. PET bottles can be divided into several categories based on chemical resins, transparency, and color [4]. PET plastic bottles have the highest recycling values compared to other plastic bottles. The Ministry of Environment announced on 5 February that it would start a pilot project for the separation and disposal of transparent plastic bottles from this month. At the beginning of this month, five regions were phased out individually, including Busan, Cheonan (Chungnam), Gimhae (Gyeongnam), Jeju, and Seogwipo. One of the changes will require companies to label plastic bottles that are easy to remove. Legislative changes will also bring system reforms to make recycling more convenient. Plastic bottles with easy-to-tear labels are produced in Japan. Designed to protect the environment from plastic pollution, it promotes the growth and innovation of industry and human life through comprehensive transformation: the production, use, and recycling of plastic bottles. PET bottles must be colorless and unlabeled to be completely recycled. It is only possible to crush transparent plastic bottles without labels into thin plastic flakes. These materials can be utilized to create new plastic items.

Plastics are an inextricable aspect of human life, particularly in countries experiencing rapid economic growth. Drinking water bottles and beverage bottles are two of the most common plastic applications in everyday life. Plastic bottles must be separated according to recyclable and non-recyclable to improve plastic bottle waste management. Recycling is the process of rebirth; plastic bottles that have been discarded are recycled into high-quality consumer goods. Recycled clear plastic bottles have been resurrected as garments, eco-friendly purses, and cosmetic bottles, among other high-quality items. Previously, all discarded plastic bottles used to make garments and other products in South Korea were imported from abroad. Only 10% of the old plastic bottles collected in the community were recycled into high-quality consumer goods. Another point to consider is that the production of plastic emits a substantial quantity of greenhouse gases, which contributes to global warming. Because recycling reduces crude oil and energy consumption, greenhouse gas emissions, such as carbon dioxide, also decrease significantly. Transparent plastic bottles are mainly used to make fiber materials for clothing, with polar fleece, a polyester material that has lately gained popularity, being a notable example. However, the foreign matter in waste bottles collected in South Korea throughout the disposal and composing procedure raises concerns about their suitability for recycling. According to the application requirements, the sorting equipment only needs to pick transparent plastic bottles in a sorting process. So correct bottle classification is crucial in the sorting system based on machine vision.

This paper proposes a GAN-based model, modified lightweight-GAN, to generate synthetic images using a small dataset containing real plastic bottle images. The main contribution is as follows:

- A new technique that enhances the imbalanced data problem using image data augmentation is proposed based on a GAN-based framework, named modified lightweight-GAN, that can generate high-quality images using a few original images.
- We propose a weighted average ensemble transfer learning-based method, IncepX-Ensemble, to classify six types of plastic bottle images.
- We construct a computationally efficient model and demonstrate its resilience based on the two presented strategies.

2. Related Works

Deep learning with a small training dataset leads to overfitting issues. The capacity to generalize data expansion was examined using deep neural network training data extensions. Instead of using traditional data augmentation techniques, GAN can generate more stable and realistic images.

A computer-aided machine learning-based plastic bottle classification technique was proposed by [5]. Specifically, the authors performed feature extraction for classification tasks by achieving 80% accuracy. The authors also proposed classification with the region of interest segmentation technique with PET and non-PET plastic bottle dataset with two classes and achieved 80% of accuracy [6]. Ref. [7] proposed an automated classification of plastic bottles based on SVM for recycling purposes and achieved 97.3% of accuracy based on the best computation time. A real-time application was designed for plastic bottle identification, and the proposed system achieved an accuracy of 97% [8]. Generative adversarial networks are an advanced technique for data augmentation and use semi-supervised cycleGAN to augment the training data. Hazra et al. proposed generating synthetic images for bone marrow cells using GAN and the classification approach using the transfer learning model [9]. The proposed model achieved 95% precision and 96% recall.

The authors of [10] proposed an inception-cycleGAN model that will classify COVID-19 X-ray images and achieved 94.2% of accuracy. An artificial intelligence-based plastic bottle color classification system was proposed by [11] and achieved 94.14% of accuracy. Wang et al. [12] proposed the recycling of used plastic bottles based on a support vector machine algorithm, and accuracy reached 94.7%. In [13], medical image classification is a famous approach; the researcher applied data augmentation using GAN and using three transfer learning models to overcome the training-time constraints. They achieved 86.45% of accuracy using the inceptionV3 model. Srivastav et al. [14] proposed an approach of generating a synthetic image using GAN to improve the diagnosis of pneumonia disease using chest X-ray image classification and achieved 94.5% accuracy. Waste management and waste classification are essential issues for the environment and human health. Recycling is one most basic forms of waste management; we need to classify the particular waste that can be recycled. There are few publicly available datasets for waste classification; for this reason, Alsabei et al. [15] proposed a model that can classify waste using pre-trained models, and for generating data, they applied the GAN approach. In [16], an intelligent system for waste sorting using a skip-connection-based model was proposed, and the novel model achieved 95% of accuracy. Pio et al. [17] hypothesized that combining a transfer learning approach with the metabolic features developed will deliver a considerable improvement in reconstruction accuracy. A new combined methodology was proposed for a higher recognition rate and robustness to enhance a low-resolution video [18]. GAN and transfer learning are used to deal with license plate image recognition in various challenging situations. Mohammed et al. [19] suggested an ensemble classifier that decreases both the space and temporal complexity of the generated ensemble members while classifying an instance by improving prediction time while maintaining significant accuracy.

3. Dataset

In our experiment, we collected plastic bottle images from the industry in South Korea. However, it is not a publicly available dataset. We intend to build models that correctly classify plastic bottle images before deploying them into a plastic bottle recycling machine. The precise detection and identification of plastic bottles is the most significant challenge when designing a recycling machine in preventing fraud. It depends on precision and cost.

There are few publicly plastic bottle datasets available. Trashnet [20] is a dataset used for trash classification that has plastic bottle images in it. Each image in the PET bottle dataset contains only one object, a plastic bottle, and a plain background. The human eye more easily perceives this but not by a recycling machine. There are no other objects in the image that could provide additional information.

Our dataset, named the PET-bottle dataset, has six classes, having a total number of 1667 plastic bottle images. We divided the plastic bottle images according to the design and bottle specification; we uniquely named three classes, Bottle_ShapeA, Bottle_ShapeB, and Bottle_ShapeC, and the other three classes are called Masinda, Pepsi, and Samdasoo, respectively. Plastic bottles which do not have a label on them but have black caps are named Bottle_ShapeA. Plastic bottles with a design on the body and a white cap but without a label are named Bottle_ShapeB. Plastic bottles that do not have any design or label on them but have a red cap are designated as Bottle_ShapeC. Masinda is a drinking water bottle company whose class depicts a company label and sky-colored cap. Pepsi is a well-known soft drink manufacturing company whose class represents a label with a company logo and black cap. Jeju Samdasoo is a mineral water brand developed by the Jeju Province Development Corporation; this plastic bottle image depicts a label with a company logo and white cap. Details of the original dataset are given in Table 1. The Sl number represents the numerical value for six classes, from 0 to 5; the class name depicts all the six classes we have used for our experiment. The images per class section describes the images containing each class.

It is noticeable that the dataset is small, and classes are primarily imbalanced in the original dataset, with most data labeled as the Samdasoo class. Training a deep neural network to categorize the data into six categories will over-fit the data with this unbalanced dataset.

Table 1. Detailed specification of original dataset.

Sl Number	Class Name	Images per Class
0	Bottle_ShapeA	169
1	Bottle_ShapeB	238
2	Bottle_ShapeC	41
3	Masinda	249
4	Pepsi	339
5	Samdasoo	631
	Total	1667

4. Methodology

The proposed method is discussed in this section. Figure 1 depicts the proposed method's block diagram. Our proposed method can be divided into five parts. The first block (a) shows the overview of the original dataset with the class label. In the second block, (b) synthetic images are generated using a modified lightweight-GAN model for data augmentation. The third block (c) is traditional data augmentation based on basic image manipulation techniques. In the fourth block, the (d) pre-trained ImageNet model is fine-tuned on our dataset for plastic bottle classification. In the last part, (e) is the evaluation metrics for classification. A detailed explanation is given in the following subsections.

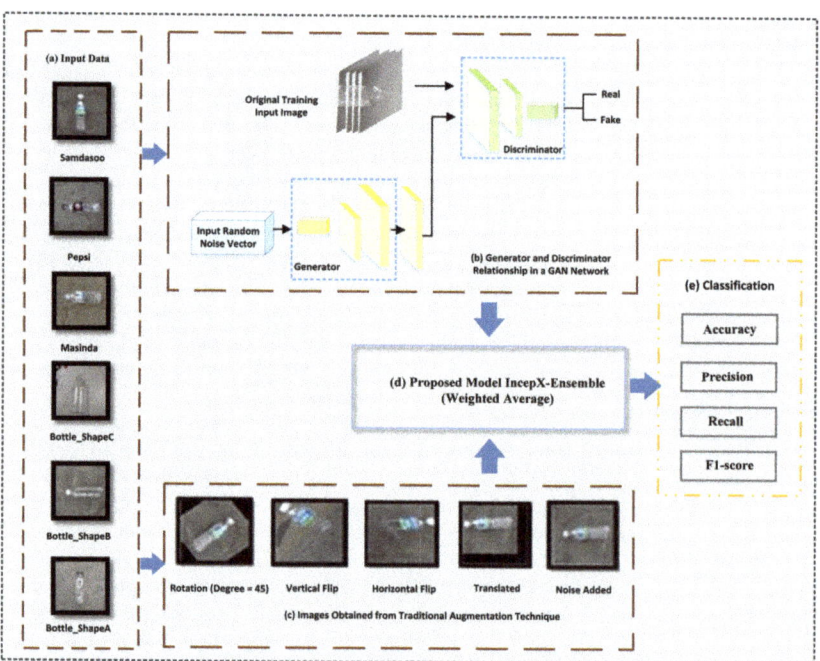

Figure 1. Workflow of the proposed framework. (**a**) shows the overview of the original dataset with the class label; (**b**) synthetic images are generated using a modified lightweight-GAN model for data augmentation; (**c**) is traditional data augmentation based on basic image manipulation techniques; (**d**) pre-trained ImageNet model is fine-tuned on our dataset for plastic bottle classification; (**e**) is the evaluation metrics for classification.

4.1. Original Dataset Description

Our dataset contains 1667 images of plastic bottles, which are segmented into six classes. The PET bottle dataset is divided into six types according to the bottle specification details.

4.2. Synthetic Image Generation Using Modified Lightweight-GAN Model

Recently, researchers have focused on combining GANs with other models or techniques that allow for superior data reconstruction. We improvised a new approach to our model. We used convolution layers compatible with high-resolution images for both G and D. The basic GAN architecture for the generator and discriminator are graphically depicted in Figure 2. The model structure of G and D and a description of the component layers are shown in Figures 3 and 4.

4.2.1. Generative Adversarial Networks

The generative adversarial network (GAN) was developed by Goodfellow et al. in 2014 [21]. This intriguing invention has been gaining interest in various machine learning fields. GAN consists of two interacting neural networks. It is a generator (G) and a discriminator (D). The generator network is trained to map points in the latent space to generate new data instances. The discriminator network is trained to distinguish between the actual and plausible images produced by the generator network. Eventually, the generator generates images that resemble actual training samples. The generator is updated based on the discriminator's predictions to have better images at the training time. The discriminator increases its ability to distinguish between actual and fake images. The difference between real and counterfeit labels determines the discriminator loss. The label specifies whether the image is artificial or natural. The general diagram of GAN is shown in Figure 2.

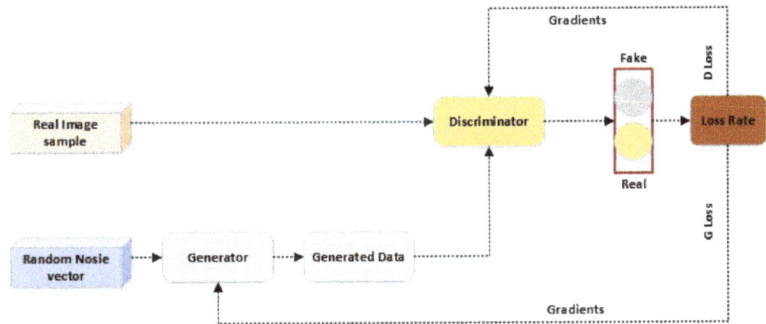

Figure 2. Generative adversarial networks architecture.

The main objective of GAN theory can be painted as a two-player min–max game which can be defined by,

$$\min_G \max_D V(D,G) = \mathbb{E}_{x \sim P_d(x)}[log D(x)] + \mathbb{E}_{rnv \sim P_{rnv}(rnv)}[log(1 - D(G(rnv)))] \quad (1)$$

The discriminator and the generator are involved in a min–max game with the value function $V(D,G)$. The discriminator is trying to minimize its reward $V(D,G)$, and the generator is attempting to reduce the discriminator's reward or, in other words, maximize its loss.

The generator always tries to minimize the following loss function; on the other hand, the discriminator always maximizes it. In GAN, the generator receives the original input data x, adds random noise variable $P_{rnv}(rnv)$ and generates samples $G(rnv)$. $D(x)$ is the discriminator's estimate of the probability that real instance x is real over the data distribution P_d. $D(G(rnv))$ is the discriminator's estimate of the probability that a fake instance is real. The generator tries to create almost perfect images to fool the discriminator. In contrast, the discriminator tries to improve the performance by distinguishing real and fake samples until the time that the samples generated from the generator cannot be distinguished from real data samples.

4.2.2. Generator Network

The generator needs to be impending with the deeper network to generate good synthesized images to orchestrate with high images. A deeper network means more of a convolution layer and more training time for up-sampling. Considering GPU for the training, we first fed the original image data and resized it into $128 \times 128 \times 3$. The image was scaled to $[-1, 1]$ pixel values to match the generator. It was issued because it uses the tanh activation function. The generator network inputs a 100×1 noise vector and generates fake samples. We used four convolution layers with ReLU activation to create high-quality synthesized images. Figure 3 illustrates the generator model architecture.

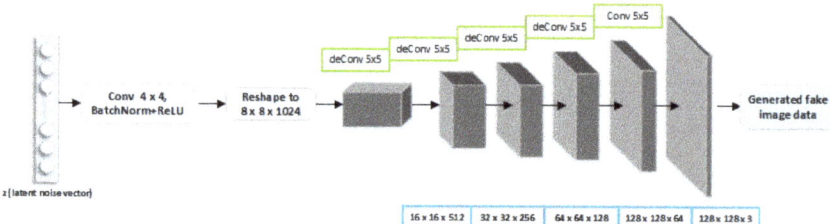

Figure 3. The architecture of the generator.

4.2.3. Discriminator Network

Following the assumption that the encoder and discriminator network information overlaps, we partially amalgamated the encoder into the discriminator [22]. The main objective of the encoder is to learn the representation feature, whereas the discriminator aims to discover the discriminating feature.

$$\mathcal{L}_{\text{recons}}^{pixel} = \mathbb{E}_{q \sim D_{encoder}(x), x \sim I_{real}}[||\kappa(q) - \tau(x)||] \tag{2}$$

where the discriminator's feature map is q, the κ function processes q, and the decoder's function τ reflects processing on sample x from real images I_{real}.

Figure 4 illustrates the discriminator model architecture. Firstly, we resize the original image to produce the I part. Then, the main part of our discriminator acts as an encoder to extract a good image feature map, and the decoder can produce a good reconstruction I'. The decoder consists of four convolutional layers to create the image 128×128. Finally, the discriminator and decoder are trained together to minimize the reconstruction loss by matching the part I' to I. The auto-encoding technique used is a common strategy for self-supervised learning that has been shown to improve model robustness and generalization capabilities [23–25].

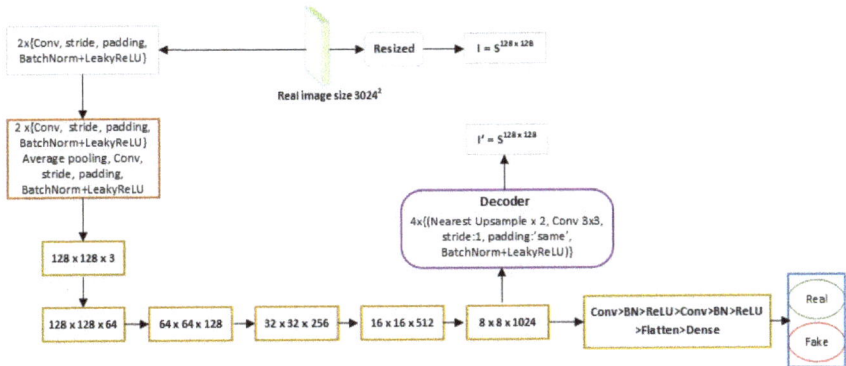

Figure 4. The architecture of the discriminator.

Recently, generative models have focused on combining new strategies with the GAN model. In many approaches, the authors combined GAN and VAE to generate a good image [22]. On the other hand, our proposed model is a pure GAN with a significantly more simple generator and discriminator and an auto-encoding function. The auto-encoding training is exclusively used for discriminator regularization and does not include the generator [26].

Here, a hinge adversarial loss for GAN is suggested, incorporating SVM margins and considering actual and fake samples falling within the margins while calculating the loss. Artificial samples outside of the boundaries that partially incorporate false local patterns are ignored in the generator training stage [27,28].

$$\mathcal{L}_D = -\mathbb{E}_{x \sim I_{real}}[min(0, -1 + D(x))] - \mathbb{E}_{z \sim P(z)}[min(0, -1 - D(G(z)))] + \mathcal{L}_{recons}^{pixel} \quad (3)$$

$$\mathcal{L}_G = -\mathbb{E}_{z \sim P(z)}[D(G(z))] \quad (4)$$

4.3. Traditional Data Augmentation Techniques

In this section, we describe traditional data augmentation based on basic image manipulation techniques [29]. Additionally, consider issues with limited datasets and how imbalances and data expansion can be helpful for oversampling solutions [30]. Class imbalance describes the dataset as a biased ratio of the majority to a sample of the minority.

- Flipping :
  There are two types of flipping used for image transformation; horizontal flipping is more common than vertical flipping. This augmentation is one of the simplest to employ and has shown to be effective on various datasets.

$$\begin{bmatrix} p' \\ q' \\ 1 \end{bmatrix} = \begin{bmatrix} -1 & 0 & 0 \\ 0 & 1 & 0 \\ 0 & 0 & 1 \end{bmatrix} \times \begin{bmatrix} p \\ q \\ 1 \end{bmatrix} \quad (5)$$

$$p' = -p, q' = q \quad (6)$$

Horizontal flipping formulas are depicted in Equations (5) and (6).

$$\begin{bmatrix} p' \\ q' \\ 1 \end{bmatrix} = \begin{bmatrix} 1 & 0 & 0 \\ 0 & -1 & 0 \\ 0 & 0 & 1 \end{bmatrix} \times \begin{bmatrix} p \\ q \\ 1 \end{bmatrix} \quad (7)$$

$$p' = p, q' = -q \quad (8)$$

Vertical flipping formulas are depicted in Equations (7) and (8).

- Rotation :
  The image is rotated right or left on an axis between [0–360] degree for rotation augmentations. The rotation degree parameter significantly impacts the safety of rotation augmentations. Outside of the rotating area, pixels are be filled with 0, and the formula of rotation is given in Equation (9).

$$R = \begin{bmatrix} cos(q) & sin(q) & 0 \\ -sin(q) & cos(q) & 0 \\ 0 & 0 & 1 \end{bmatrix} \quad (9)$$

  where q specifies the angle of rotation.

- Translation :
  To avoid data-position bias, shifting the image left, right, up, or down is a valuable adjustment, so the neural network looks everywhere in the image to capture it. The original image is translated into the [0–255] value range.

$$t = \begin{bmatrix} 1 & 0 & 0 \\ 0 & 1 & 0 \\ t_x & t_y & 1 \end{bmatrix} \quad (10)$$

  where in Equation (10), t_x specifies the displacement along the x axis, and t_y specifies the displacement along the y axis.

- Noise added :
  Noise is an exciting augmentation technique; noise injection injects a matrix of random values usually drawn from a Gaussian distribution. Stochastic data expansion is applied when the neural network sees the same image, which is slightly different. This difference can be seen as adding noise to the data sample and letting the neural network learn generalized features rather than overfitting the dataset.

4.4. Transfer Learning

Transfer learning techniques are used to improve the performance of machine learning algorithms using labeled data. TL efforts learn and apply one or more source tasks to enhance learning in related fields. It has been studied as a machine learning process to solve problems. TL includes pre-training models that have already been trained on large datasets and models that have been retrained at several levels of the model on a small training set. The initial layer of the pre-training network will be changed if necessary. You can use the final layer of the model's fine-tuning parameters to learn the capabilities of the new dataset [31]. According to the new task, models that have already been trained will be retrained with a smaller new dataset, and the model weights will be modified. Newly developed neural networks parameters are not built from scratch. The DL algorithm can achieve higher functionality or performance for many problems, but they need a lot of data for training time.

As a result, it can be helpful to reuse pre-trained models for similar tasks. We used two pre-trained models named inceptionV3 and Xception. The PET bottle dataset is used to fine-tune the models once they have been pre-trained with the ImageNet dataset [32]. The most common method for fine-tuning is to delete the last completely connected layer of pre-trained CNN models and replace it with a new fully connected layer (the same size as the number of classes in our dataset). Our PET bottle dataset contains six categories. Finally, the suggested method meets the goal of providing excellent classification results with a small dataset.

4.4.1. InceptionV3 and Xception

The pre-trained network models InceptionV3 and Xception were trained on millions of images from the ImageNet dataset. The InceptionV3 [33] and Xception [34] networks include 48 and 71 layers, respectively, and require a 299 × 299 × 3-pixel input image. The structure of the InceptionV3 and Xception are shown in Figures 5 and 6. While Inception considers typical congestion and yield issues, efficient results can be obtained by using asymmetric filters and bottlenecks and replacing large filters with smaller ones. Xception is simpler and more efficient. Using cross-channel and spatial correlations independently, Xception provides more specific and efficient outcomes. For the Xception model, depth-wise separable convolution is also proposed, as well as the use of cardinality to develop better abstractions.

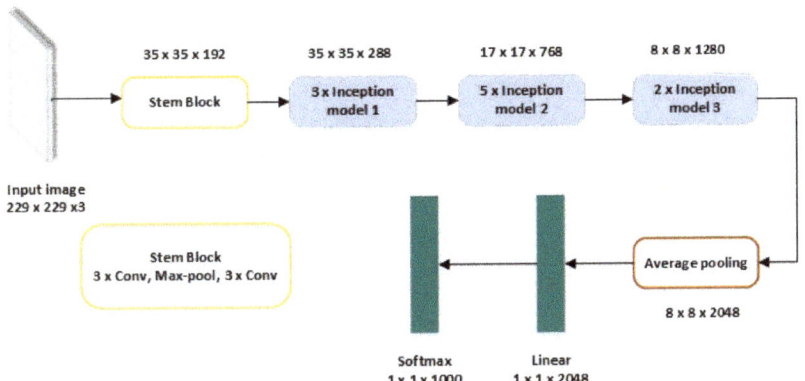

Figure 5. InceptionV3 model architecture.

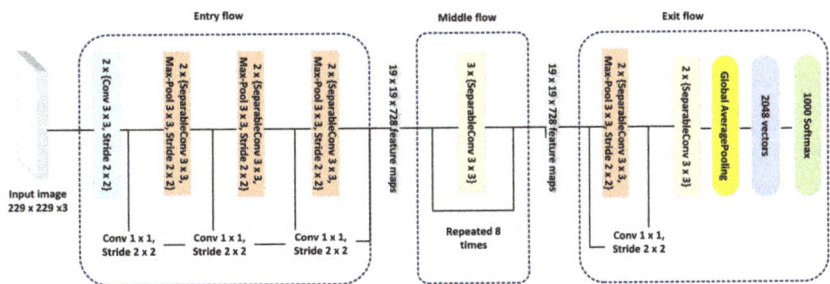

Figure 6. Xception model architecture.

4.4.2. Ensemble Learning

Ensemble learning is a way of combining multiple models to benefit in terms of computation and performance. The results of an ensemble of deep neural networks are always superior to those of a single model. The average ensemble learning was used in this study, with the same weights allocated to each model.

$$P = \frac{\sum M_i}{N} \tag{11}$$

where, in Equation (11), M_i is the probability of model i, and N is the total number of models.

DL models have different architectures and complexity; they do not provide the same result. Therefore, assigning more weights to the model performing better is convenient. By this, the maximum output can be extracted from any model. The challenge is to find the correct combination of model weights. We used the grid search technique to solve this challenge, as shown in Figure 7. A total of 1000 weight combinations were used. The search procedure continues until all varieties have been checked. The approach finally provided us with the ideal weight combination for the maximum of our given evaluation metric.

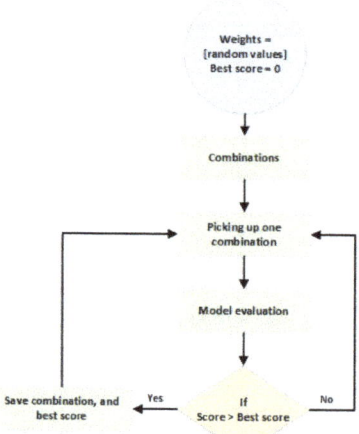

Figure 7. Grid search method for finding the weights.

4.5. Evaluation Metrics

The performance of our model was evaluated, using accuracy, precision, and recall, and the F1-score based on the confusion matrix; it includes four indicators, true positive (TP), false positive (FP), false negative (FN), and true negative (TN).

Accuracy is calculated by dividing the number of true positives and true negatives by the total number of instances. Precision is calculated with actual positive classes from the total predicted classes. The recall is derived by dividing the real positive values by the actual positive values. The F1-score is simply the average of precision and recall. Equations (12)–(15) show the accuracy, precision, recall, and F1-score calculations.

$$Accuracy = \frac{(TP+TN)}{(TP+TN+FP+FN)} \quad (12)$$

$$Precision = \frac{TP}{(TP+FP)} \quad (13)$$

$$Recall = \frac{TP}{(TP+FN)} \quad (14)$$

$$F1\text{-}score = 2 \times \frac{Precision \times Recall}{(Precision + Recall)} \quad (15)$$

5. Results

5.1. Experimental Setup

In this study, the first part of the experiments, the modified lightweight-GAN model was trained in 500 epochs and generated synthetic images of PET bottles for each of the six categories. The weights of the generator and discriminator models were updated after each epoch to produce a composite image as close as possible to the actual image. After network training, the PET bottle dataset has 4200 images, including original and synthetic images generated from the modified lightweight-GAN model and traditional augmentation methods. In the second series of experiments, the pre-trained Inception V3 and Xception models were trained using the original training set and a combination of the training set and the image of the generated plastic bottle. Later, we employed a weighted average ensemble to enhance the classification performance using the IncepX-Ensemble model. The samples of real plastic bottle images and synthetic images generated by the modified lightweight-GAN model are shown in Figure 8. For training hyperparameter settings, we used binary cross-entropy as the cost function, a learning rate of 0.0001, and Adam as the optimizer. We used 100 epochs and a batch size of 32 for every model.

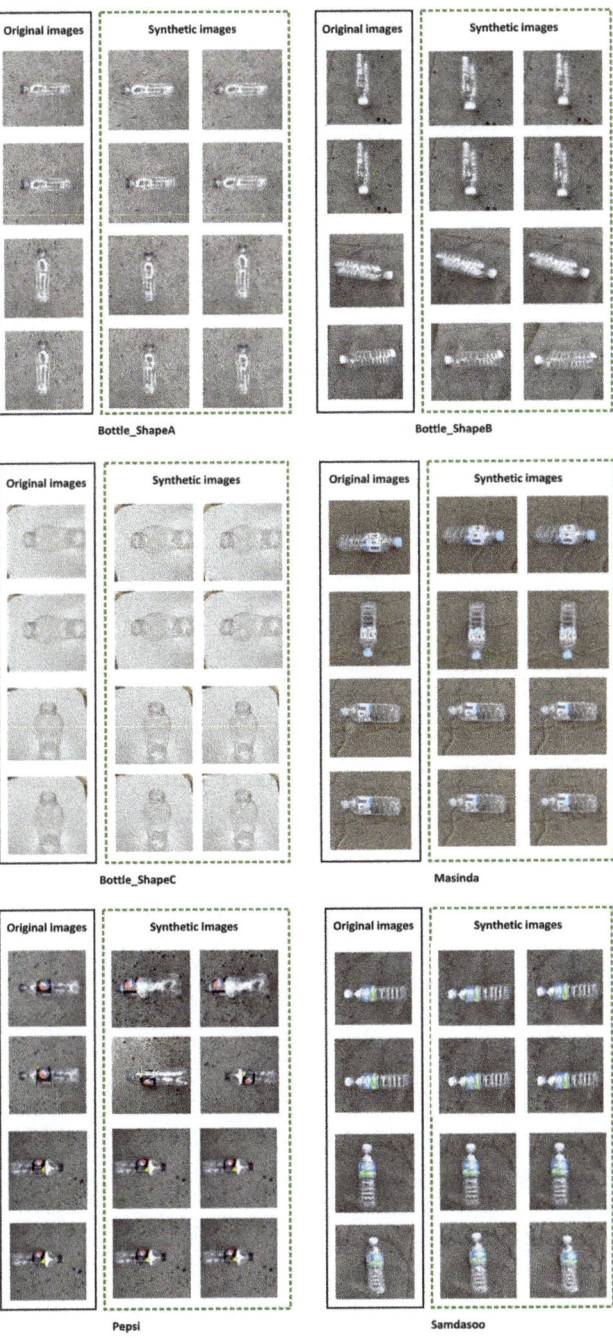

Figure 8. Original plastic bottle images and synthetic plastic bottle images generated by modified lightweight-GAN.

We divided our dataset that has 4200 images, which includes original plastic bottle images and generated images by the GAN model. Further, we split our dataset into training, validation, and testing sets for training. The training set is given to the machine learning model to analyze and learn the feature; the validation dataset is a sample of the data retained from the model training and is used to estimate the model's method while optimizing the model's hyperparameters. The test set is not used for training, and it is used to determine whether the model's hypothesis is correct. In the experiment, we first divided the dataset into 60% for training and 40% for test data. In addition, the holdout test data were split into 10% for validation (0.25% of total holdout test data) and 30% for testing (0.75% of entire holdout test data). Details of the experimental dataset are given in the Table 2.

Table 2. Details of the dataset after data augmentation using both augmentation techniques.

Sl No.	Class Name	Images per Class	Training (60%)	Validation (10%)	Testing (30%)
0	Bottl_ShapeA	700	420	70	210
1	Bottl_ShapeB	700	420	70	210
2	Bottle_ShapeC	700	420	70	210
3	Masinda	700	420	70	210
4	Pepsi	700	420	70	210
5	Samdasoo	700	420	70	210
	Total	4200	2520	420	210

5.2. Performance Metrics of GAN

We used two metrics to measure the model performance, as shown in Table 3.

- The IS is an objective metric for assessing the quality of synthetic images generated by the generative adversarial networks model. The IS was proposed by [35], and it captures the two properties of generated images: image quality, and image diversity.
- The FID is a metric that measures the overall semantic realism that compares the distance between feature vectors calculated for real and generated images. FID score was proposed by [36] to improve the performance over inception score.

Table 3. Quantitative comparison on our dataset—inception score (IS), Frechet inception distance (FID).

Sl No.	Accuracy	IS	FID
1	DCGAN	12.36	73.4
2	LSGAN	10.06	67.6
3	WGAN-GP	9.67	72.3
4	TrGAN	9.82	65.4
5	ACGAN	9.47	76.3
6	CGAN	9.89	70.0
7	Modified lightweight-GAN	9.42	64.7

5.3. Implementation Details

Specification details for performing the experiments are given below in Table 4. We used the Windows operating system with a single GPU and 32 GB of RAM. We trained our model on Tensorflow 2.6.0 version, CUDA Toolkit version 11.2 and cuDNN version 8.1.

Table 4. System components and specification.

Component	Description
Operating system	Windows 10 64 bit
Browser	Google Chrome
CPU	Intel(R) Core(TM) i5-8500K CPU @ 3.70 GHz
RAM	32 GB
Programming language	Python 3.8.5
GPU	NVIDIA GeForce RTX 2070
CUDA	CUDA Toolkit version 11.2
cuDNN	cuDNN version 8.1
Tensorflow	Tensorflow version 2.6.0
IDE	jupyter
Machine learning algorithm	Modified lightweight-GAN
Machine learning algorithm	Xception
Machine learning algorithm	InceptionV3

5.4. Classification Performance Details

In Table 5, we show how the performance of pre-trained models, such as InceptionV3 [37], Xception [38] and our ensemble model IncepX-Ensemble, may be used to determine how well classifiers can classify plastic bottle types after being trained with both original and synthetic data. The results show that the accuracy of the models is enhanced when synthetic data generated by GAN models are used to train the model. Among all the GAN models, our proposed IncepX-Ensemble model produced the best accuracy value of 99.06%.

Table 5. Comparison of IncepX-Ensemble with other existing models.

Model/Classifier	InceptionV3				Xception				IncepX-Ensemble			
	Acc	Pre	Rec	F1	Acc	Pre	Rec	F1	Acc	Pre	Rec	F1
Original Data	86.6	89.2	88.6	90.1	92.8	87.2	93.2	90.1	93.5	93.7	92.8	93.8
DCGAN	81.2	82.4	79.6	80.4	90.8	92.1	92.6	91.5	92.4	94.7	95.2	94.6
LSGAN	83.2	81.9	85.4	83.6	85.4	86.3	90.6	86.4	84.4	85.3	84.0	85.4
WGAN-GP	93.1	92.6	94.2	93.9	93.6	93.2	94.2	94.4	97.2	97.4	96.4	97.6
ACGAN	89.9	89.1	90.1	90.5	91.4	91.2	92.0	91.6	95.5	95.7	94.5	96.2
CGAN	97.1	98.3	96.5	97.9	98.4	97.2	98.3	97.9	97.1	98.6	98.7	98.7
Modified Lightweight-GAN	98.8	98.2	99.0	98.6	98.9	97.4	98.7	98.5	99.0	99.1	99.3	99.2

Acc, Pre, Rec, and F1 refer to accuracy, precision, recall, and f1-score, respectively.

We also assessed the performance of classification models that use original data and actual and synthetic data. We employed two different combinations of augmentation procedures for the augmentation of plastic bottle images. To produce synthetic data, Augmentation-1 employs a modified lightweight-GAN. Flipping, rotation, translation, and noise addition are all used in Augmentation-2. We kept the total number of images for each example to ensure a fair comparison.

In Table 6, we show the performance of the traditional augmentation technique with transfer learning models. We also examined classification model performance utilizing original, augmented data and a synthetic image generated by our model, which produces better quality images and performs better. We can notice that in the case of noise addition, accuracy is fairly low because of overfitting.

Table 6. Accuracy, precision, recall, and F1-score of different classification using traditional augmentation methods and a combination of original with synthetic data.

Tradition Augmentation/Classifier	InceptionV3				Xception				IncepX-Ensemble			
	Acc	Pre	Rec	F1	Acc	Pre	Rec	F1	Acc	Pre	Rec	F1
Original Data	86.2	75.0	86.1	86.0	86.2	75.2	89.0	86.8	88.2	87.1	94.2	89.0
Flipping	87.1	83.2	91.0	86.0	88.0	91.1	79.8	84.5	87.1	88.1	93.0	89.1
Rotation	88.5	79.7	86.5	82.2	86.1	82.0	83.5	75.8	87.0	87.1	84.1	73.0
Translation	85.1	76.5	88.1	80.2	86.2	82.2	85.1	87.5	88.1	81.1	88.0	82.2
Noise Addition	75.2	72.0	77.1	75.6	75.6	76.0	77.0	77.1	75.8	75.2	77.2	76.1
Modified Lightweight-GAN	89.8	87.4	83.7	83.3	91.3	89.3	88.5	88.7	93.1	89.6	92.9	92.1

Acc, Pre, Rec, and F1 refer to accuracy, precision, recall, and f1-score, respectively.

We evaluated our IncepX-Ensemble model with the ImgaeNet dataset in Table 7. We first trained the models with the original imageNet data and tested the model with actual data. The model can be easily adapted to support fine-tuning for classification tasks. We used the dataset for 60% for training and 40% for testing, and further testing data were split into 0.75% of the total holdout test data and 0.25% validation. The performance of the classification models using synthetic data, augmented data and a mix of original and synthetic data was then determined using the same procedure. The images created by our suggested improved lightweight-GAN model are of higher quality. It performs quantitatively better than existing GAN models, as can be seen from all of the findings.

Table 7. Evaluation of our proposed model on the ImageNet dataset.

Original + Synthetic Image/Classifier	InceptionV3				Xception				IncepX-Ensemble			
	Acc	Pre	Rec	F1	Acc	Pre	Rec	F1	Acc	Pre	Rec	F1
Original Data	93.9	92.5	95.8	94.3	94.4	94.6	92.9	92.9	96.2	95.8	96.1	95.6
Rotation	95.6	94.7	97.9	95.6	95.9	91.1	94.9	96.2	96.9	95.3	95.6	97.1
Translation	94.6	94.9	93.0	95.4	94.5	93.9	92.6	94.9	95.2	93.8	93.2	95.7
ACGAN	95.3	87.3	91.3	92.2	95.2	87.0	91.0	94.1	95.6	94.2	93.6	94.0
WGAN-GP	95.6	95.4	96.1	96.0	96.2	95.9	89.6	95.5	96.8	95.4	96.2	96.1
CGAN	94.6	95.0	96.1	95.3	75.6	76.0	77.0	77.1	95.8	92.5	95.4	96.0
Modified Lightweight-GAN	96.2	95.2	93.7	96.3	97.6	96.3	97.5	98.2	98.9	96.6	95.9	99.1

Acc, Pre, Rec, and F1 refer to accuracy, precision, recall, and f1-score, respectively.

6. Conclusions

The aim is to develop an application-based system that automatically detects plastic bottle images. Our proposed approach is simple: to overcome the small and imbalanced dataset, we first applied a modified lightweight-GAN method to generate synthetic images of plastic bottles. Next, we developed a transfer learning-based model, IncepX-Ensemble, classifying different plastic bottle images. Therefore, we developed a new system using the transfer learning technique, and a new framework was developed by integrating with modified lightweight-GAN. Modified lightweight-GAN was used for data augmentation enhancement of the dataset, and the proposed transfer learning-based model was trained and evaluated using original and generated images. Finally, we designed a weighted average ensemble model named IncepX-Ensemble, tuning the influence of the base models using the grid search technique. However, the two transfer learning models show excellent performance, though in some cases, the two models fail to classify plastic bottles correctly. To obtain an improved performance, we used a combination of transfer learning and the weighted average technique to boost the application performance. The obtained results indicate the algorithm's efficacy with 99.06% accuracy. Future work may validate the proposed model to evaluate recycling performance using more diverse big data. We plan to use the model we developed to explore other datasets and waste management applications in the future. We hope that this will play a positive role in plastic bottle waste management and environmental growth.

Author Contributions: Conceptualization, S.C. and D.H.; Formal analysis, S.C.; Funding acquisition, Y.-C.B.; Methodology, S.C. and D.H.; Writing—review and editing, S.C.; Investigation, Y.-C.B.; Resources, Y.-C.B.; Project administration, Y.-C.B. and Y.-W.K.; Supervision, Y.-C.B. and Y.-W.K. All authors have read and agreed to the published version of the manuscript.

Funding: This research was financially supported by the Ministry of SMEs and Startups (MSS), Korea, under the "Startup growth technology development program (R&D, S3125114)".

Institutional Review Board Statement: Not applicable.

Informed Consent Statement: Not applicable.

Data Availability Statement: Not applicable.

Conflicts of Interest: The authors declare no conflict of interest.

Abbreviations

The following abbreviations are used in this manuscript:

DL	Deep Learning
GAN	Generative Adversarial Networks
CNN	Convolutional Neural Network
TL	Transfer Learning
VAE	Variational Autoencoders
PET	Polyethylene Terephthalate
IS	Inception Score
FID	Frechet Inception Distance
DCGAN	Deep Convolutional GAN
LSGAN	Least Squares GAN
WGAN-GP	Wasserstein GAN-Gradient Penalty
ACGAN	Auxiliary Classifier GAN
CGAN	Conditional GAN

References

1. Huth-Fehre, T.; Feldhoff, R.; Kowol, F.; Freitag, H.; Kuttler, S.; Lohwasser, B.; Oleimeulen, M. Remote sensor systems for the automated identification of plastics. *J. Near Infrared Spectrosc.* **1998**, *6*, A7–A11. [CrossRef]
2. Zhang, H.; Wen, Z.G. The consumption and recycling collection system of PET bottles: A case study of Beijing, China. *Waste Manag.* **2014**, *34*, 987–998. [CrossRef] [PubMed]
3. Vo, A.H.; Vo, M.T.; Le, T. A novel framework for trash classification using deep transfer learning. *IEEE Access* **2019**, *7*, 178631–178639. [CrossRef]
4. Hammaad, S. 7.25 Million AED is the Cost of Waste Recycling. *Al-Bayan Newspaper*, 11 March 2005.
5. Ramli, S.; Mustafa, M.M.; Hussain, A.; Wahab, D.A. Histogram of intensity feature extraction for automatic plastic bottle recycling system using machine vision. *Am. J. Environ. Sci.* **2008**, *4*, 583. [CrossRef]
6. Ramli, S.; Mustafa, M.M.; Hussain, A.; Wahab, D.A. Automatic detection of 'rois' for plastic bottle classification. In Proceedings of the 2007 5th Student Conference on Research and Development, Selangor, Malaysia, 11–12 December 2007; pp. 1–5.
7. Shahbudin, S.; Hussain, A.; Wahab, D.A.; Marzuki, M.; Ramli, S. Support vector machines for automated classification of plastic bottles. In Proceedings of the 6th International Colloquium on Signal Processing and Its Applications (CSPA), Melaka, Malaysia, 21–23 May 2010; pp. 1–5.
8. Scavino, E.; Wahab, D.A.; Hussain, A.; Basri, H.; Mustafa, M.M. Application of automated image analysis to the identification and extraction of recyclable plastic bottles. *J. Zhejiang Univ.-Sci. A* **2009**, *10*, 794–799. [CrossRef]
9. Hazra, D.; Byun, Y.C.; Kim, W.J.; Kang, C.U. Synthesis of Microscopic Cell Images Obtained from Bone Marrow Aspirate Smears through Generative Adversarial Networks. *Biology* **2022**, *11*, 276. [CrossRef] [PubMed]
10. Bargshady, G.; Zhou, X.; Barua, P.D.; Gururajan, R.; Li, Y.; Acharya, U.R. Application of CycleGAN and transfer learning techniques for automated detection of COVID-19 using X-ray images. *Pattern Recognit. Lett.* **2022**, *153*, 67–74. [CrossRef] [PubMed]
11. Tachwali, Y.; Al-Assaf, Y.; Al-Ali, A. Automatic multistage classification system for plastic bottles recycling. *Resour. Conserv. Recycl.* **2007**, *52*, 266–285. [CrossRef]

12. Wang, Z.; Peng, B.; Huang, Y.; Sun, G. Classification for plastic bottles recycling based on image recognition. *Waste Manag.* **2019**, *88*, 170–181. [CrossRef] [PubMed]
13. Zulkifley, M.A.; Mustafa, M.M.; Hussain, A. Probabilistic white strip approach to plastic bottle sorting system. In Proceedings of the 2013 IEEE International Conference on Image Processing, Melbourne, Australia, 15–18 September 2013; pp. 3162–3166.
14. Srivastav, D.; Bajpai, A.; Srivastava, P. Improved classification for pneumonia detection using transfer learning with gan based synthetic image augmentation. In Proceedings of the 2021 11th International Conference on Cloud Computing, Data Science & Engineering (Confluence), Noida, India, 28–29 January 2021; pp. 433–437.
15. Alsabei, A.; Alsayed, A.; Alzahrani, M.; Al-Shareef, S. Waste Classification by Fine-Tuning Pre-trained CNN and GAN. *Int. J. Comput. Sci. Netw. Secur.* **2021**, *21*, 65–70.
16. Bircanoğlu, C.; Atay, M.; Beşer, F.; Genç, Ö.; Kızrak, M.A. RecycleNet: Intelligent waste sorting using deep neural networks. In Proceedings of the 2018 Innovations in Intelligent Systems and Applications (INISTA), Thessaloniki, Greece, 3–5 July 2018; pp. 1–7.
17. Pio, G.; Mignone, P.; Magazzù, G.; Zampieri, G.; Ceci, M.; Angione, C. Integrating genome-scale metabolic modelling and transfer learning for human gene regulatory network reconstruction. *Bioinformatics* **2022**, *38*, 487–493. [CrossRef] [PubMed]
18. Du, X. Complex environment image recognition algorithm based on GANs and transfer learning. *Neural Comput. Appl.* **2020**, *32*, 16401–16412. [CrossRef]
19. Mohammed, A.M.; Onieva, E.; Woźniak, M. Selective ensemble of classifiers trained on selective samples. *Neurocomputing* **2022**, *482*, 197–211. [CrossRef]
20. Yang, M.; Thung, G. Classification of trash for recyclability status. *CS229 Proj. Rep.* **2016**, *2016*, 3.
21. Goodfellow, I.; Pouget-Abadie, J.; Mirza, M.; Xu, B.; Warde-Farley, D.; Ozair, S.; Courville, A.; Bengio, Y. Generative adversarial nets. In Proceedings of the Advances in Neural Information Processing Systems, Montreal, QC, Canada, 8–13 December 2014; Volume 27.
22. Munjal, P.; Paul, A.; Krishnan, N.C. Implicit discriminator in variational autoencoder. In Proceedings of the 2020 International Joint Conference on Neural Networks (IJCNN), Glasgow, UK, 19–24 July 2020; pp. 1–8.
23. Hendrycks, D.; Mazeika, M.; Kadavath, S.; Song, D. Using self-supervised learning can improve model robustness and uncertainty. In Proceedings of the Advances in Neural Information Processing Systems, Vancouver, BC, Canada, 8–14 December 2019; Volume 32.
24. Jing, L.; Tian, Y. Self-supervised visual feature learning with deep neural networks: A survey. *IEEE Trans. Pattern Anal. Mach. Intell.* **2020**, *43*, 4037–4058. [CrossRef] [PubMed]
25. Goyal, P.; Mahajan, D.; Gupta, A.; Misra, I. Scaling and benchmarking self-supervised visual representation learning. In Proceedings of the IEEE/CVF International Conference on Computer Vision, Seoul, Korea, 27–28 October 2019; pp. 6391–6400.
26. Liu, B.; Zhu, Y.; Song, K.; Elgammal, A. Towards faster and stabilized gan training for high-fidelity few-shot image synthesis. In Proceedings of the International Conference on Learning Representations, Addis Ababa, Ethiopia, 26–30 April 2020.
27. Lim, J.H.; Ye, J.C. Geometric gan. *arXiv* **2017**, arXiv:1705.02894.
28. Kim, S.; Lee, S. Spatially Decomposed Hinge Adversarial Loss by Local Gradient Amplifier. In Proceedings of the ICLR 2021 Conference, Vienna, Austria, 4 May 2020.
29. Shorten, C.; Khoshgoftaar, T.M. A survey on image data augmentation for deep learning. *J. Big Data* **2019**, *6*, 60. [CrossRef]
30. Hao, R.; Namdar, K.; Liu, L.; Haider, M.A.; Khalvati, F. A comprehensive study of data augmentation strategies for prostate cancer detection in diffusion-weighted MRI using convolutional neural networks. *J. Digit. Imaging* **2021**, *34*, 862–876. [CrossRef] [PubMed]
31. Kamishima, T.; Hamasaki, M.; Akaho, S. TrBagg: A simple transfer learning method and its application to personalization in collaborative tagging. In Proceedings of the 2009 Ninth IEEE International Conference on Data Mining, Miami Beach, FL, USA, 6–9 December 2009; pp. 219–228.
32. ImageNet Dataset. 2016. Available online: https://image-net.org/ (accessed on 12 July 2021).
33. Szegedy, C.; Vanhoucke, V.; Ioffe, S.; Shlens, J.; Wojna, Z. Rethinking the inception architecture for computer vision. In Proceedings of the IEEE Conference on Computer Vision and Pattern Recognition, Las Vegas, NV, USA, 27–30 June 2016; pp. 2818–2826.
34. Chollet, F. Xception: Deep learning with depthwise separable convolutions. In Proceedings of the IEEE Conference on Computer Vision and Pattern Recognition, Honolulu, HI, USA, 21–26 July 2017; pp. 1251–1258.
35. Salimans, T.; Goodfellow, I.; Zaremba, W.; Cheung, V.; Radford, A.; Chen, X. Improved techniques for training gans. In Proceedings of the Advances in Neural Information Processing Systems, Barcelona, Spain, 5–10 December 2016; Volume 29.
36. Heusel, M.; Ramsauer, H.; Unterthiner, T.; Nessler, B.; Hochreiter, S. Gans trained by a two time-scale update rule converge to a local nash equilibrium. In Proceedings of the Advances in Neural Information Processing Systems, Long Beach, CA, USA, 4–9 December 2017; Volume 30.
37. Xia, X.; Xu, C.; Nan, B. Inception-v3 for flower classification. In Proceedings of the 2017 2nd International Conference on Image, Vision and Computing (ICIVC), Chengdu, China, 2–4 June 2017; pp. 783–787.
38. Wu, X.; Liu, R.; Yang, H.; Chen, Z. An xception based convolutional neural network for scene image classification with transfer learning. In Proceedings of the 2020 2nd International Conference on Information Technology and Computer Application (ITCA), Guangzhou, China, 18–20 December 2020; pp. 262–267.

MDPI
St. Alban-Anlage 66
4052 Basel
Switzerland
Tel. +41 61 683 77 34
Fax +41 61 302 89 18
www.mdpi.com

Mathematics Editorial Office
E-mail: mathematics@mdpi.com
www.mdpi.com/journal/mathematics

www.ingramcontent.com/pod-product-compliance
Lightning Source LLC
LaVergne TN
LVHW070233100526
838202LV00015B/2125